1990 LECTURES IN COMPLEX SYSTEMS

1990 LECTURES IN COMPLEX SYSTEMS

THE PROCEEDINGS OF THE 1990 COMPLEX SYSTEMS SUMMER SCHOOL SANTA FE, NEW MEXICO JUNE, 1990

Editors

Lynn Nadel
Department of Psychology
University of Arizona

Daniel L. Stein
Department of Physics
University of Arizona

Lectures Volume III

SANTA FE INSTITUTE
STUDIES IN THE SCIENCES OF COMPLEXITY

Addison-Wesley Publishing Company
The Advanced Book Program
Redwood City, California • Menlo Park, California • Reading, Massachusetts
New York • Don Mills, Ontario • Wokingham, United Kingdom • Amsterdam
Bonn • Sydney • Singapore • Tokyo • Madrid • San Juan

Publisher: *Allan M. Wylde*
Production Manager: *Pam Suwinsky*
Marketing Manager: *Laura Likely*

Director of Publications, Santa Fe Institute: *Ronda K. Butler-Villa*
Technical Assistant, Santa Fe Institute: *Della L. Ulibarri*

Library of Congress Cataloging-in-Publication Data

Complex Systems Summer School (1990: Santa Fe Institute)
   1990 lectures in complex systems: the proceedings of the 1990
   Complex Systems Summer School, Santa Fe, New Mexico, June 1990/
   edited by Lynn Nadel and Daniel Stein
     p. cm.– (Santa Fe Institute studies in the sciences of complexity.
   Lectures: v. 3.)
   Includes bibliographical references and index.
   1. Computational complexity–Congresses. 2. Chaotic behavior
   in systems–Congresses. 3. Self-organizing systems–Congresses.
   I. Nadel, Lynn. II. Stein, Daniel L.   III. Title. IV. Series
   QA267.7.C66 1990   003–dc20   91-15013
   ISBN 0-201-52575-5

This volume was typeset using TEXtures on a Macintosh II computer. Camera-ready output from an Apple LaserWriter Plus Printer.

1 2 3 4 5 6 7 8 9 10-MA-95 94 93 92 91

About the Santa Fe Institute

The *Santa Fe Institute* (SFI) is a multidisciplinary graduate research and teaching institution formed to nurture research on complex systems and their simpler elements. A private, independent institution, SFI was founded in 1984. Its primary concern is to focus the tools of traditional scientific disciplines and emerging new computer resources on the problems and opportunities that are involved in the multidisciplinary study of complex systems—those fundamental processes that shape almost every aspect of human life. Understanding complex systems is critical to realizing the full potential of science, and may be expected to yield enormous intellectual and practical benefits.

All titles from the *Santa Fe Institute Studies in the Sciences of Complexity* series will carry this imprint which is based on a Mimbres pottery design (circa A.D. 950–1150), drawn by Betsy Jones.

Santa Fe Institute Studies in the Sciences of Complexity

PROCEEDINGS VOLUMES

Volume	Editor	Title
I	David Pines	Emerging Syntheses in Science, 1987
II	Alan S. Perelson	Theoretical Immunology, Part One, 1988
III	Alan S. Perelson	Theoretical Immunology, Part Two, 1988
IV	Gary D. Doolen et al.	Lattice Gas Methods of Partial Differential Equations, 1989
V	Philip W. Anderson et al.	The Economy as an Evolving Complex System, 1988
VI	Christopher G. Langton	Artificial Life: Proceedings of an Interdisciplinary Workshop on the Synthesis and Simulation of Living Systems, 1988
VII	George I. Bell & Thomas G. Marr	Computers and DNA, 1989
VIII	Wojciech H. Zurek	Complexity, Entropy, and the Physics of Information, 1990
IX	Alan S. Perelson & Stuart A. Kauffman	Molecular Evolution on Rugged Landscapes: Proteins, RNA and the Immune System, 1990
X	John A. Hawkins & Murray Gell-Mann	Evolution of Human Landscapes, 1990
XI	C. Langton et al.	Artificial Life II

LECTURES VOLUMES

Volume	Editor	Title
I	Daniel L. Stein	Lectures in the Sciences of Complexity, 1988
II	Erica Jen	1989 Lectures in Complex Systems
III	Lynn Nadel & Daniel L. Stein	1990 Lectures in Complex Systems

LECTURE NOTES VOLUMES

Volume	Author	Title
I	J. Hertz, R. Palmer, & A. Krogh	Introduction to the Theory of Neural Computation, 1990
II	Gérard Weisbuch	Complex Systems Dynamics

Contributors to This Volume

Charles R. Doering, Department of Physics and Institute for Nonlinear Studies, Clarkson University, Potsdam, New York 13699-5820

Bruce Bayly, Department of Mathematics, University of Arizona, Tucson, Arizona 85721

Wing Yim Tam, Physics Department, University of Arizona, Tucson, Arizona 85721

Sidney R. Nagel, The James Franck Institute and the Department of Physics, The University of Chicago, Chicago, IL 60637

J. F. Traub, Department of Computer Science, Columbia University

H. Woźniakowski, Institute of Informatics, University of Warsaw

Andrew G. Barto, Department of Computer and Information Science, University of Massachusetts, Amherst, MA 01003

Z. Hasan, Department of Physiology and of Electrical and Computer Engineering, University of Arizona, Tucson, AZ 85724

Timothy R. Thomas, Computing Division, Mail Stop B258, Los Alamos National Laboratory, Los Alamos, NM 87545

Mychael Trong Vo, Department of Electrical Engineering, California Polytechnical Institute, Pomona, CA 91768

Leif H. Finkel, Department of Bioengineering and Institute of Neurological Sciences, University of Pennsylvania, Philadelphia, Pa 19104-6392

Gerald M. Edelman, The Neurosciences Institute and the Rockefeller University, 1230 York Avenue, New York, NY 10021

Richard E. Michod, Department of Ecology and Evolutionary Biology, University of Arizona, Tucson, AZ 85721

Mark E. Nelson, Computation and Neural Systems Program, Division of Biology 216-76, California Institute of Technology, Pasadena, CA 91125

James M. Bower, Computation and Neural Systems Program, Division of Biology 216-76, California Institute of Technology, Pasadena, CA 91125

M. A. V. Gremillion, Center for Nonlinear Studies, Los Alamos National Laboratory, Los Alamos, NM 87545 and Department of Neurosciences, University of California at San Diego, La Jolla, CA 92093

Hans B. Sieburg, HIV-Neurobehavioral Research Center, Department of Psychiatry, M-003, University of California, San Diego, CA 92093

André Longtin, Complex Systems Group and Center for Nonlinear Studies, Theoretical Division, Mail Stop B213, Los Alamos National Laboratory, Los Alamos, NM 87545

Walter Fontana, Theoretical Division and Center for Nonlinear Studies, Los Alamos National Laboratory, Los Alamos NM 87545 and Santa Fe Institute, 1120 Canyon Road, Santa Fe, NM 87501

Andreas S. Weigend, Physics Department, Stanford University, Stanford, CA 94305

Bernardo A. Huberman, Dynamics of Computation Group, Xerox PARC, Palo Alto, CA 94304

David E. Rumelhart, Psychology Department, Stanford University, Stanford, CA 94305

Russel E. Caflisch, Department of Mathematics, University of California, Los Angeles, CA 90024-1555

Kevin Atteson, GRASP Lab, University of Pennsylvania, 3401 Walnut Street, Suite 300C, Philadelphia, PA 19104

Joseph L. Breeden, Center for Complex Systems Research–Beckman Institute and the Physics Department, University of Illinois, 405 North Mathews Avenue, Urbana, IL 61801

Alfred Hübler, Center for Complex Systems Research–Beckman Institute and the Physics Department, University of Illinois, 405 North Mathews Avenue, Urbana, IL 61801

John C. Crepeau, Department of Mechanical Engineering, University of Utah, Salt Lake City, UT 84112

Daniel R. Greening, IBM T. J. Watson Research Center and the University of California, Los Angeles, CA 90089-0272

J. A. Hoffnagle, IBM Research Division, Almaden Research Center, Almaden, CA CA 10598

G. Reyna, IBM Research Division, T. J. Watson Research Center, Yorktown Heights, NY 10598

J. R. Sobĕhart, Instituto Balseiro, 8400 San Carlos de Bariloche, Argentina

Jan E. Hutson, Department of Mathematics, University of Illinois, Urbana-Champaign, Il 61801

Jérôme Losson, Department of Physics and Center for Nonlinear Dynamics, McGill University, McIntyre Building, Room 1123a, 3655 Drummond, H3G-IY6, Montréal P.Q. CANADA

Preface

When the preface to the first volume of this proceedings series was written, it was far from certain—indeed, it appeared unlikely—that it would ever be necessary to repeat the task. The notion of complexity was (and remains) ill-defined at best, and extremely wishful thinking was required to convince oneself that the tremendously high level of enthusiasm and interest which accompanied the first year of the school could be maintained indefinitely by a limitless pool of potential high-caliber participants, already overburdened faculty, and beleaguered funding agencies. Nevertheless, as the King in *Amadeus* was fond of saying—there it is. In three short years, the Complex Systems Summer School has become something of an institution, and its contributions to both education and new research in an emerging field have become increasingly substantial.

Having acquired a momentum of its own, the summer school has begun to display those qualities of self-organization and nonlinear feedback symptomatic of the systems studied under its auspices. The idea of a "complex system" remains nebulous, despite attempts at characterization and classification. As long as a hard, clear-cut set of defining characteristics is lacking, the field retains a good deal of fluidity, contributing in no small part to the excitement many researchers and students bring to the school. Much discussion was devoted to just such a description of complexity in the prefaces to the previous two volumes of proceedings in this series (see *Lectures in the Sciences of Complexity: Proceedings of the 1988 Summer School*, edited by Daniel Stein, and *1989 Lectures in Complex Systems*, edited by Erica Jen). However, the evolving set of topics in the summer school, as recorded in this series, has contributed in, perhaps, a more lasting way to what it means to be a "complex system". The topics, in a real sense, speak for themselves.

We consider this to be one of the more important roles the summer school has to play; the cumulative effect of the proceedings series has been not only to record what has already been done in the field, but also contributes to defining its future evolution. Needless to say, one cannot cover the whole field in a single summer or a single book; the value of the enterprise can only be regarded in taking the series to date as a whole, and in the anticipation of its future course. (It's amusing to note that one feature of complexity that all commentators agree on is the impossibility of understanding the behavior of a system by breaking it down into its component parts!).

The Summer School itself has evolved over the three years of its existence. In 1989 a "research week" replaced the fourth week of lectures. In 1990 the students played an even stronger role in the school, organizing their own seminar series in which they discussed either their own research or some topic of broad interest to the participants. For the first time, this volume includes contributions from the students themselves, much of which is original research of high caliber. We would very much like to see this become a tradition in future years. The strongest aspects of the school continue: first-rate lecture series and seminars, numerous private and group discussions between participants and lecturers, an extensive computational laboratory, an increasingly well-stocked library, and organized trips to Los Alamos. Unfortunately, there has been little improvement in the food at St. John's.

Like any good complex system, the Summer School always contains some unpredictability. Research projects are started, new ideas generated, and collaborations are formed which last well beyond the brief duration of the school. One of the unanticipated outgrowths of this summer's session was an initiative of two of our participants, Neil Gershenfeld and Andreas Weigend. They proposed a competition on time series analysis and prediction, to be sponsored by the Santa Fe Institute, and a follow-up workshop. The results of the competition may serve as a useful measure of progress in the field, and perhaps lead to improved techniques.

As in previous volumes, the contents of this book reflect the topics discussed in the 1990 school, although a few of the lectures are not included (those given by R. Linsker, and J. Ford). All the chapters have been written to reflect both the nature and substance of the lectures and seminars themselves. The first chapter, by Charles Doering, discusses stochastic processes, an understanding of which is crucial for making sense of the behavior of many complex systems. The nature of stochasticity is discussed, several relatively simple processes are introduced and analyzed, and mathematical tools and techniques are given extensive treatment. In a similar vein, Bruce Bayly introduces the reader to some of the intriguing phenomena and deep problems arising from the study of fluid flow. The many intricacies of the problems discussed, and the complicated interplay between physical intuition and careful mathematical treatment, convey a sense of both the liveliness and mystery of the field.

In studying any system, it is crucial to get a strong feeling for the physical characteristics and behaviors exhibited therein, and in the following two chapters Wing Tam and Sidney Nagel provide valuable perspectives on their systems of interest gained from their own first-hand experience. Nagel discusses two different systems. In his first set of lectures, the rich phenomenology of glassy systems is presented, with special emphasis on the long-known but poorly understood phenomena of slow relaxation, rapidly increasing viscosities, and multiplicity of time scales. The section ends with an intriguing new result on universality of dielectric response (and presumably other kinds of response as well). The second part of his lectures deals with both experimental results and numerical simulations on sand piles, avalanches, and self-organized criticality.

Wing Tam discusses his, and others', work on the subject of pattern formation in reaction-diffusion processes in open reactors. Questions dealing with pattern stability, transitions among patterns, and the types of patterns themselves, both spatial and temporal structure, are analyzed in detail. The physics is striking, and the subject a central one in anyone's list of problems in complexity.

The connection between complexity in physical, chemical, and biological systems with the better-defined—and understood—notions of complexity in computer science is dealt with in the elegant set of lectures of Traub and Woźniakowski. The idea of information-based complexity is introduced, and its applications to computational, mathematical, and physical problems presented. The lectures conclude with a discussion of the limits and potentialities of what is "knowable" in science.

The nature of adaptive change, or learning, is central to many of the complex systems studied by biologists, and the principles underlying learning systems are applicable to problems in a wide range of domains. In his lectures Andrew Barto describes a collection of learning tasks within a *control* framework, distinguishing between classes of tasks in terms of the kind of information available during learning. He argues quite convincingly that learning performance can be improved dramatically when the right kind of information is available to the learning system. Barto also makes the critical point that an understanding of learning systems will most likely follow from an analysis of realistic models of actual tasks organism face in the world, rather than the study of learning as an abstraction.

This emphasis on specific biological systems is represented in the next three chapters. The first two of these concentrate on problems in the motor system. Ziaul Hasan discusses what appears at first glance to be a relatively simple problem: moving a human or robot arm from one spot to another. He breaks the problem down into a series of black boxes that together could solve this problem, providing a beautiful example of a productive interaction between mathematical considerations and biological constraints. A somewhat more specific problem, that of the generation of speech, is discussed by Thomas and Vo. They apply a tongue-movement model based on a second-order dynamical system. The purpose of this model is to illuminate how a listener perceives speech by discerning the speaker's intended gestures from those actually generated. The model succeeds in accounting for some, but not all, aspects of speech, and provides a promising start to this problem.

Finkel and Edelman shift the emphasis to the sensory side, discussing the organization of the somatosensory system. In particular, they focus on the way in which a *topographic* map is created within this system, and how such maps remain dynamic throughout the life of the organism. Their model depends on a theory of neuronal groups, organized during development through the action of selective principles quite similar to those familiar to molecular biologists. In addition to presenting their own model, the authors review four other models, and discuss the kinds of data that might help us decide between the various schemes. This chapter provides an excellent example of how computational and biological approaches can interact in the study of a complex system.

What is more complex than sex? Michod considers the purpose of sex, and attempts to resolve two of Darwin's deepest dilemmas: missing links in habitat space, and missing links in time. He applies dynamical systems techniques to the study of evolution and the role of sex, concluding that sex played two roles early in evolutionary history, recombination and out-crossing, both of which are necessary for the kind of damage repair critical for simple replicating molecules.

The next section of the book consists of a series of seminars. Russel Caflisch presents a discussion of singularities in complex (here referring to numbers which include the square root of -1 !) hyperbolic partial differential equations. The development of singularities (related to shock waves) in Burger's equation is analyzed. Walter Fontana presents a new approach to self-organization which can be represented via manipulation of logocal strings but whose principles may be much more far-reaching. Finally, we conclude with a selection of student contributions.

ACKNOWLEDGMENTS

Many people contributed to the success of the summer school. The planning for the school, its day-by-day functioning, and the follow-up after the school finishes, are all a reflection of the efforts of a number of people at the Santa Fe Institute. George Cowan and Mike Simmons helped with the organization; Ginger Richardson and Andi Sutherland were indispensable from start to finish as usual; Ronda Butler-Villa, and Della Ulibarri played a major role in getting this volume together; Marcella Austin handled the rather complex financial side, and Robin Justice got the computational laboratory up and running, and kept it that way. David Campbell helped organize the school's interaction with the laboratory at Los Alamos. Stuart Kauffman, of the Santa Fe Institute, committed much of his time to the students, and could often be seen in heated discussion with groups of them in the courtyard of the Institute. Most critical of all, Valerie Gremillion, also of Los Alamos, served as social organizer, mother confessor, seminar presenter, and several other roles, in providing the glue necessary to keep the school running smoothly.

We thank also our advisory board, and several institutions that provided computers and associated peripherals. Finally, we must thank those agencies that contributed the funds needed to make the school a reality—The Department of Energy, The National Science Foundation, The Research Corporation, and the Office of Naval Research.

Lynn Nadel   Daniel Stein
University of Arizona   University of Arizona
Tucson, AZ   Tucson, AZ

March 31, 1991

Contents

I Courses

Charles R. Doering
Department of Physics and Institute for Nonlinear Studies, Clarkson University, Potsdam, New York 13699-5820

Modeling Complex Systems: Stochastic Processes, Stochastic Differential Equations, and Fokker-Planck Equations

This course consists of a brief mathematical introduction to the theory and applications of Markov diffusion processes. We discuss Brownian motion, white noise, stochastic differential equations, and Fokker-Planck equations. We approach the subject with the aim of developing techniques for analyzing models of complex interactions as stochastic dynamic systems and obtaining reduced descriptions of those models in certain limits. Toward the latter end, we discuss adiabatic elimination of fast noise as a singular perturbation analysis of higher-dimensional stochastic processes. The distinction between equilibrium and nonequilibrium stationary states is stressed as we illustrate the technology by application to a simple example from the natural sciences.

1990 Lectures in Complex Systems, SFI Studies in the Sciences of Complexity, Lect. Vol. III, Eds. L. Nadel and D. Stein, Addison-Wesley, 1991

3

---

1. INTRODUCTION

Complex systems, by definition, consist of many interacting components. In some situations we are concerned with the behavior of the system as a whole, and sometimes we are concerned with the behavior of a small part of it, focusing on just a few specific aspects of the phenomena at hand. The canonical example is perhaps equilibrium thermodynamics, where the behavior of large numbers of atoms or molecules are, for many purposes, adequately described by just a few macroscopic variables—energy, entropy, pressure, etc. Moreover, those macroscopic variables obey a compact set of fundamental laws independent of the precise constituents of the system or the precise nature of their interactions.

The task of reducing the number of variables in a complex dynamical system to some manageable level, and the search for some general rules controlling the behavior of complex systems, cuts across all disciplinary boundaries—thermodynamics, for example, is fundamental to all branches of physics, as well as chemistry, biology, and engineering. More specifically, even in the best of cases, it is impossible to completely divorce a real sub-system from its environment, whether that subsystem is a solid-state physics experiment, a chemical reaction, or a living entity. The influence of many factors are brought to bear on any "isolated" system, and we really have no hope of completely accounting for them theoretically. It is often useful then, in the modeling process, to lump all of these unknown, uncontrollable, and essentially stochastic factors into some noise acting on the system. In this way at least some of the overwhelming complexity of the whole is accounted for in the description of the part. The cost of our parody of these effects as a random noise is the sacrifice of complete predictability. Once we start talking about random variables, we necessarily restrict our predictions to probabilities and averages, rather than to the definite outcome of any specific experiment or observation.

The more general goal of obtaining faithful reduced descriptions of complicated networks of interacting components doesn't end with the introduction of a stochastic model. For example, the approximately 10^{23} coupled differential equations describing the motion of particles that make up a fluid can reduced to the Boltzmann equation by making some probabilistic assumptions about the dynamics. The Boltzmann equation description is then further reduced to one of mesoscopic average motions, or fluctuating hydrodynamics. On larger and slower scales, i.e., at the macroscopic level, fluctuations are accounted for by simple dissipation and pressure terms in a Navier-Stokes equation. Even the dynamics in the Euler and Navier-Stokes equations can, in certain situations, be accounted for in simpler amplitude equations like the Korteweg-deVries, nonlinear Schrödinger, or Ginzburg-Landau equations. Although few of these simplifying steps are taken with full mathematical rigor, there are some tested methods which are used over and over at different levels. Chief among them is the "averaging over" of fast variables, leaving a smaller set of the remaining slowly changing variables to describe the system.

Several aspects of stochastic modeling will be discussed here. Besides representing complex influences by random noise and exploring some elementary implications of noise in a simple nonlinear model, we will also take the next step and see how one can, in certain situations, reduce the level of description even further by the identification, separation, and elimination of rapidly evolving dynamical variables.

In these lectures we will present some basic ideas and methods in the theory of stochastic processes and stochastic differential equations. The aim is to study the behavior of noise-driven dynamical systems to gain some insight into the variety of phenomena possible in (generally nonlinear) systems coupled to a complex environment that can be modeled as random fluctuations, or noise. It is *not* the mission of these lectures either to present a rigorous mathematical formulation or to exhaust the subject of mathematical modeling. Rather, some fundamental concepts and derivations in the theory of continuous-time Markov processes will be introduced and developed to the point that these techniques may become part of the students' technical arsenal for the modeling and analysis of stochastic dynamic systems. Calculations are presented in detail here, with missing steps or natural extensions stated explicitly as exercises. The assumed mathematical background is a post-calculus-level undergraduate course in probability theory. The reader who can complete Exercise 1 (below) should be adequately prepared. I have very much tried to present the theory with as little mystery and in as much detail as the space and time allow.

Most of the material discussed herein may be found in greater mathematical detail in monographs by Arnold,[1] van Kampen,[6] Horsthemke and Lefever,[3] and Risken.[5] These books also develop myriad applications of stochastic processes and stochastic differential equations in engineering and the sciences—from signal processing to superconductivity to the electrical activity of nerve membranes. Recent reviews of many aspects of the interplay of noise and nonlinear dynamics are given in Moss and McClintock[4] and Doering, Brand, and Ecke.[2] Readers who wish to continue their studies in this area are referred to these works as a starting point into the current literature.

Beginning with the next section, we develop the mathematical framework for the analysis of continuous-time stochastic processes, focussing on the Wiener process, a.k.a. Brownian motion. We then discuss the concept of white noise as a prelude to the study of stochastic differential equations driven by gaussian white noise. We derive the relationship between the coefficients of a stochastic differential equation with white noise and those of the associated Fokker-Planck equation for the transition density, interpreting the stochastic differential equation (in the sense of Itô) as the continuous time limit of a discrete time problem. The Fokker-Planck equation is a linear partial differential equation, amenable to comprehensive analysis and, at times, exact solution.

Armed with these preliminaries, we introduce the notion of detailed balance in the stationary state of a stochastic dynamic system, and discuss the concepts of equilibrium vs. non-equilibrium stationary states. We then go on to present the general mathematical ideas of the reduction of the number of variables via adiabatic elimination and the overdamped approximation. This procedure is developed as a singular perturbation analysis of a multidimensional Fokker-Planck equation. The difference between the Itô and Stratonovich interpretations of stochastic differential equations with white noise is presented in this way by comparing the white-noise limit of a continuous-time real noise problem with the continuous time limit of the discrete time problem. Along the way these concepts are illustrated with a specific example, the Verhulst model of population biology.

EXERCISE 1. Let X_1 and X_2 be independent identically distributed (i.i.d) random variables, uniformly distributed over the interval (0,1). Let

$$G_1 = \sqrt{-2\ln[X_1]} \cos(2\pi X_2) \text{ and}$$
$$G_2 = \sqrt{-2\ln[X_1]} \sin(2\pi X_2).$$

Show that G_1 and G_2 are i.i.d. random variables, with a mean zero, unit variance gaussian distribution.

2. STOCHASTIC PROCESSES

Stochastic processes, also known as random processes, are to be thought of as random functions of a variable which we will call *time*. That is, if $X(\bullet)$ is a random process, then for each value t of its argument, $X(t)$ is a random variable characterized by a probability density function, $\rho(x,t)$. The argument, or *index*, of the

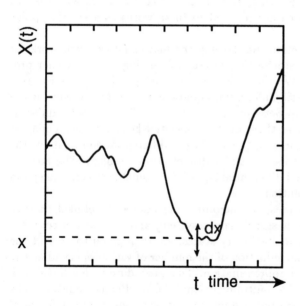

FIGURE 1 A "random" function. The single time probability density $\rho(x,t)$, times dx, gives the probability that the random function passes through a window of width dx about x.

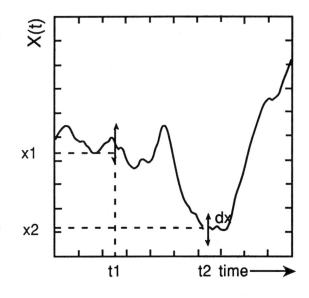

FIGURE 2 The joint density $\rho(x_2, t_2; x_1, t_1)$, times dx^2, gives the probability that the random function falls through both windows of width dx about x_1 at time t_1, and about x_2 at time t_2.

process t may be a discrete or a continuous variable; we concentrate on the case where t is a continuous variable. An actual realization of the process is called a sample path. The probability density function times the width of some small window dx around the value x at each instant, i.e., $\rho(x, t)dx$, gives the relative frequency that a random function falls within that window, as illustrated in Figure 1. The probability density is positive and for each time t is related to the probability of $X(t)$ taking some values by

$$\int_a^b \rho(x, t)dx = \text{Prob}[a < X(t) < b], \qquad (2.1)$$

for $a < b$, and where the $\text{Prob}[\bullet]$ means the probability of the specified event.

It is not enough to know the characteristics of the random variables $X(t)$ for each t alone. For example, in order to contemplate the probability that one of the random functions passes through two windows, one of width dx_1 around x_1 at time t_1 *and* one of width dx_2 around x_2 at time t_2, as illustrated in Figure 2, we must consider the *joint* probability distribution $\rho(x_2, t_2; x_1, t_1)$. Our notation is that the semicolon (;) is read "and" in the Boolean sense. Only if the values of the random functions are *independent* at different times, i.e., if

$$\rho(x_2, t_2; x_1, t_1) = \rho(x_2, t_2)\rho(x_1, t_1), \qquad (2.2)$$

does the single-time density function specify the statistics of the process as a whole. Processes where the values at different times are statistically independent are called *white-noise* processes.

In order to answer more complicated questions about a stochastic process we must know the joint probability distributions for the values of the functions at an arbitrary number of times. That is, we must consider the complete set of joint densities, or the finite-dimensional distributions, $\rho(x_1, t_1; \ldots; x_n, t_n)$, for all n. The joint distribution functions cannot be specified arbitrarily, but must satisfy the compatibility conditions

$$\int \rho(x_n, t_n; \ldots; x_i, t_i; \ldots; x_1, t_1) dx_i = \rho(x_n, t_n; \ldots; x_{i+1}, t_{i+1}; x_{i-1}, t_{i-1}; \ldots; x_1, t_1)$$
$$(2.3)$$

where the integral is over the allowed values of X, and the ith window has been "removed" on the right-hand side. This condition merely states that the random functions fall somewhere—anywhere—between $t = t_{i-1}$ and $t = t_{i+1}$.

A fundamental theorem of Kolmogorov states that this hierarchy of joint density functions is just what is needed to completely specify the stochastic process. This is not a trivial statement because it asserts that we need only know the probabilities that the random functions fall through any *discrete* number of windows, rather than the uncountable number of possibilities allowed by the continuous time variable. More precisely, Kolmogorov's theorem states that, for each consistent hierarchy of finite-dimensional distributions, there exists a probability space equipped with a sigma algebra, a measure, and a family of random variables $X(t)$, defined for each t, whose joint probability density functions are those originally given.

The *average*, or *expectation*, of a random variable X with probability density $\rho(x)$ is denoted $E\{X\}$,[1] and is defined as

$$E\{X\} = \int x \rho(x) dx \,. \qquad (2.4)$$

The average of the nth power of X, $E\{X^n\}$, is called the nth *moment* of X, and the expectation of a function $f\{X\}$ is

$$E\{f(X)\} = \int f(x) \rho(x) dx \,. \qquad (2.5)$$

For a stochastic process $X(t)$, the average of n products of the process at various times is called the *n-point correlation function*:

$$E\{X(t_n) \ldots X(t_1)\} = \int x_n \ldots x_1 \rho(x_n, t_n; \ldots; x_1, t_1) dx_n \ldots dx_1. \qquad (2.6)$$

The expectation of the product of the process at two times, i.e., the two-point correlation function

$$E\{X(t)X(s)\} = \int xy \rho(x, t; y, s) dx dy \,, \qquad (2.7)$$

---

[1]A common notation for the expectation of a random variable X is $\langle X \rangle$, with the exact same meaning as $E\{X\}$.

will be referred to simply as the *correlation function* or the *covariance*. Note that while the density functions determine the moments and the correlation functions, the converse is not true; i.e., the process is *not* uniquely specified by the moments and the correlation functions.

The probability density that a random function passes through x_n at time t_n *given* that it passed through x_{n-1} at time t_{n-1}, x_{n-2} at time t_{n-2}, etc., is a *conditional* density. In terms of the finite-dimensional distributions, this conditional density is defined by

$$\rho(x_n, t_n | x_{n-1}, t_{n-1}; \ldots; x_1, t_1) = \frac{\rho(x_n, t_n; x_{n-1}, t_{n-1}; \ldots; x_1, t_1)}{\rho(x_{n-1}, t_{n-1}; \ldots; x_1, t_1)}. \tag{2.8}$$

for an ordered set of times $t_1 < \ldots < t_n$. Our notation is that the vertical bar ($|$) is read "given." In particular, the *transition density* for the process to go from x_{n-1} at t_{n-1} to x_n at t_n is

$$\rho(x_n, t_n | x_{n-1}, t_{n-1}) = \frac{\rho(x_n, t_n; x_{n-1}, t_{n-1})}{\rho(x_{n-1}, t_{n-1})}. \tag{2.9}$$

For the rest of these lectures we will be concerned with a restricted class of stochastic processes known as *Markov* processes. A Markov process is a stochastic process with the property that, in plain language, "the future is independent of the past given the present." In terms of the joint density functions, this means that

$$\rho(x_n, t_n | x_{n-1}, t_{n-1}; x_{n-2}, t_{n-2}; \ldots; x_1, t_1) = \rho(x_n, t_n | x_{n-1}, t_{n-1}), \tag{2.10}$$

and this property will serve as our definition of a Markov process. Once the value of the process $X(t)$ is specified at time t_{n-1}, its future evolution for times $t > t_{n-1}$ is statistically independent of its history before time t_{n-1}. The joint distribution functions for a Markov process may be written in terms of the single-time density and the transition density:

$$\rho(x_n, t_n; x_{n-1}, t_{n-1}; x_{n-2}, t_{n-2}; \ldots; x_2, t_2; x_1, t_1)$$
$$= \rho(x_n, t_n | x_{n-1}, t_{n-1}) \rho(x_{n-1}, t_{n-1} | x_{n-2}, t_{n-2}) \ldots \rho(x_2, t_2 | x_1, t_1) \rho(x_1, t_1). \tag{2.11}$$

Equation (2.11) says that, for Markov processes, the probability of passing through the n windows is the product of the probabilities to go from one to the next, times the probability of passing through the first window.

Markov processes are, in a practical sense, just one step more general than white-noise processes. That is, white-noise processes are defined by the single-time density function alone, from which all higher joint density functions may be built. Markov processes are defined by the single-time density function and the transition density function (or the two-time joint density alone), from which all higher joint density functions may be built. This extra degree of generality attributed to Markov processes makes a world of difference in applications. To a certain extent, white-noise processes are trivial; they lack any structure to justify their consideration as much more than just a collection of independent random variables. White noise is not useless, however, as some natural processes can be modeled as being nearly statistically independent at successive instants. Often the first step in modeling random phenomena is to determine the level on which some variables, or some aspect of the problem, can be modeled as white noise. On the other hand, Markov processes, with their one more degree of complexity, are a reasonable model of many stochastic systems where the variables are correlated at different times, but they "lose memory" of their history in the sense that knowledge of the current state is sufficient to predict the future (statistically) no matter how that current state was achieved. As we will discuss later in these lectures, Markov processes are the solutions of differential equations with white-noise coefficients and thus arise naturally in applications. More generally, as we will also discuss below, they can be good approximations to non-Markovian processes on long time scales.

EXERCISE 2. Using the definitions of the conditional density, Eq. (2.8), and Markov processes, Eq. (2.10), verify Eq. (2.11).

3. BROWNIAN MOTION AND WHITE NOISE

A fundamental example of a Markov process is *Brownian motion*, also referred to as the *Wiener process*, which we denote $W(t)$. It is the Markov process with a δ-function single-time density at $t = 0$,

$$\rho(w, t = 0) = \delta(w),$$ (3.1)

and a gaussian transition density, of variance Δt, between times t and $t + \Delta t$,

$$\rho(w, t + \Delta t | w', t) = \frac{1}{\sqrt{2\pi \Delta t}} e^{-\frac{1}{2}\frac{(w - w')^2}{\Delta t}}.$$ (3.2)

The single-time density is explicitly computed from these definitions according to

$$
\begin{aligned}
\rho(w, t) &= \int \rho(w, t; w', 0) dw' \\
&= \int \rho(w, t | w', 0) \rho(w', 0) dw' \\
&= \int \frac{1}{\sqrt{2\pi t}} e^{-\frac{1}{2}\frac{(w - w')^2}{t}} \delta(w') dw' \\
&= \frac{1}{\sqrt{2\pi t}} e^{-\frac{1}{2}\frac{w^2}{t}}.
\end{aligned}
$$ (3.3)

Hence $W(t)$ is a mean zero, variance t, normally distributed random variable. The finite-dimensional distributions for Brownian motion are, according to Eq.(7),

$$\rho(w_n, t_n; w_{n-1}, t_{n-1}; w_{n-2}, t_{n-2}; \ldots; w_2, t_2; w_1, t_1)$$

$$= \frac{1}{\sqrt{2\pi(t_n - t_{n-1})}} e^{-\frac{1}{2}\frac{(w_n - w_{n-1})^2}{(t_n - t_{n-1})}} \frac{1}{\sqrt{2\pi(t_{n-1} - t_{n-2})}} e^{-\frac{1}{2}\frac{(w_{n-1} - w_{n-2})^2}{(t_{n-1} - t_{n-2})}} \tag{3.4}$$

$$\cdots \frac{1}{\sqrt{2\pi(t_2 - t_1)}} e^{-\frac{1}{2}\frac{(w_2 - w_1)^2}{(t_2 - t_1)}} \frac{1}{\sqrt{2\pi t_1}} e^{-\frac{1}{2}\frac{w_1^2}{t_1}} .$$

Brownian motion is a *gaussian* process because the random variables $W(t_1) \ldots$ $W(t_n)$ are jointly gaussian random variables. A typical realization of the Wiener process is shown in Figure 3. The time-dependent probability density, Eq. (3.3), is plotted in Figure 4 for several time values.

The moments of the Wiener process are easy to compute because of its gaussian distribution:

$$E\{W(t)^{2n+1}\} = \int_{-\infty}^{\infty} w^{2n+1} \frac{1}{\sqrt{2\pi t}} e^{-\frac{1}{2}\frac{w^2}{t}} dw = 0 \tag{3.5}$$

and

$$E\{W(t)^{2n}\} = \int_{-\infty}^{\infty} w^{2n} \frac{1}{\sqrt{2\pi t}} e^{-\frac{1}{2}\frac{w^2}{t}} dw = \frac{(2n)!}{2^n n!} t^n . \tag{3.6}$$

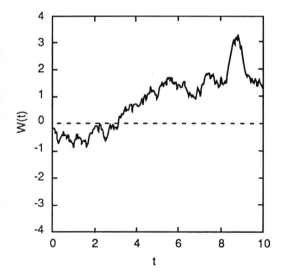

FIGURE 3 A typical realization of the Wiener process.

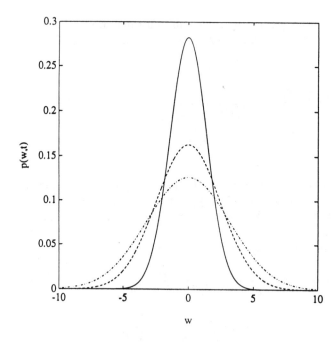

FIGURE 4 Several single-time probability densities for the Wiener process. These are the densities at times $t = 2$ (solid), $t = 6$ (dashed), and $t = 10$ (dash-dot).

The two-point correlation function—the covariance—is, for $t \geq s$,

$$
\begin{aligned}
E\{W(t)W(s)\} &= \int_{-\infty}^{\infty} dw \int_{-\infty}^{\infty} dw'\, ww' \frac{1}{\sqrt{2\pi(t-s)}} e^{-\frac{1}{2}\frac{(w-w')^2}{(t-s)}} \frac{1}{\sqrt{2\pi s}} e^{-\frac{1}{2}\frac{w'^2}{s}} \\
&= \int_{-\infty}^{\infty} \frac{dw'}{\sqrt{2\pi s}} w' e^{-\frac{1}{2}\frac{w'^2}{s}} \int_{-\infty}^{\infty} \frac{dz}{\sqrt{2\pi(t-s)}} (z+w') e^{-\frac{1}{2}\frac{z^2}{(t-s)}} \\
&= \int_{-\infty}^{\infty} \frac{dw'}{\sqrt{2\pi s}} w'^2 e^{-\frac{1}{2}\frac{w'^2}{s}} \\
&= s .
\end{aligned}
$$

(3.7)

For an arbitrary ordering of t and s, the covariance is thus

$$
E\{W(t)W(s)\} = \min\{t, s\} .
\tag{3.8}
$$

A quantity of particular interest is the change

$$
\Delta W(t) = W(t + \Delta t) - W(t)
\tag{3.9}
$$

of the Wiener process over a time interval Δt. The *increment* $\Delta W(t)$ is a gaussian random variable of mean zero, and variance

$$
\begin{aligned}
E\{\Delta W(t)^2\} &= E\{(W(t + \Delta t) - W(t))^2\} \\
&= E\{W(t + \Delta t)^2\} - 2E\{W(t + \Delta t)W(t)\} + E\{W(t)^2\} \\
&= t + \Delta t - 2t + t \\
&= \Delta t .
\end{aligned}
\tag{3.10}
$$

The change in $W(t)$ over a short time interval Δt is, on average, of the order $\sqrt{\Delta t}$. This suggests that the sample paths of Brownian motion are continuous functions—which is in fact true—but that they are highly irregular. These properties of Brownian motion are seen in Figure 3. The irregularity over short time scales results from the fact that the increments of the Wiener process over disjoint intervals are independent random variables. Jointly gaussian (with mean zero) random variables are independent if their covariance vanishes, and for $t > s + \Delta t$ the jointly gaussian random variables $\Delta W(t)$ and $\Delta W(s)$ satisfy

$$
\begin{aligned}
E\{\Delta W(t)\Delta W(s)\} &= E\{[W(t + \Delta t) - W(t)][W(s + \Delta t) - W(s)]\} \\
&= E\{W(t + \Delta t)W(s + \Delta t) - W(t)W(s + \Delta t) \\
&\qquad - W(t + \Delta t)W(s) + W(t)W(s)\} \\
&= (s + \Delta t) - (s + \Delta t) - s + s \\
&= 0 .
\end{aligned}
\tag{3.11}
$$

If we consider the approximate derivative $\Delta W(t)/\Delta t$ of the Wiener process as a stochastic process itself, with discrete index set (time), then it is a white-noise process as defined earlier. This white-noise process consists of independent, identically distributed, gaussian random variables with mean zero and variance

$$
E\left\{\left(\frac{\Delta W}{\Delta t}\right)^2\right\} = \frac{\Delta t}{\Delta t^2} = \frac{1}{\Delta t} .
\tag{3.12}
$$

If we try to think of the $\Delta t \to 0$ limit of the approximate derivatives of the Wiener process as a stochastic process itself, then we run into trouble because the variance in Eq. (3.12) diverges as $\Delta t \to 0$. In Figure 5(a), (b), and (c) we plot the approximate derivatives—of the particular realization in Figure 3—for several values of the time interval Δt, showing their divergence. The derivative of Brownian motion does not make sense as an ordinary stochastic process as we discussed in the last section, because it is not a well-defined random variable for each time t; the calculation above shows that it is a "gaussian random variable with infinite variance." The sample paths of the Wiener process are continuous but nondifferentiable functions.

(a)

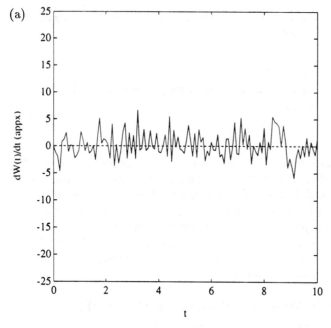

(b)

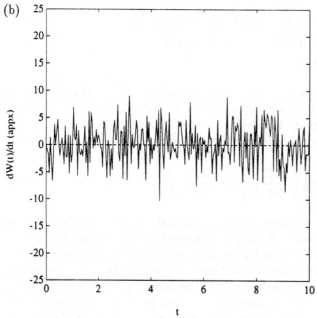

FIGURE 5
Approximate
derivatives—finite
differences—of the
Wiener process.
These are for the
realization shown
in Figure 3, and
are for time steps
(a) $\Delta t = 0.08$,
(b) $\Delta t = 0.04$,
(c) $\Delta t = 0.02$, and
(d) $\Delta t = 0.01$.

(c)

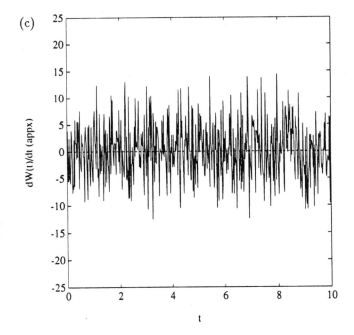

(d)

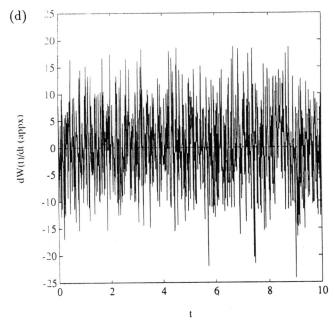

FIGURE 5
(continued)

We can make sense of the derivative of the Wiener process as a *generalized stochastic process*. That is, rather than considering the stochastic process as a collection of random variables indexed by time, we enlarge the index set to a space of functions. The idea here is that of a distribution-valued random process $\xi(\cdot)$, where in order to get a well-defined random variable we need to specify not just a number (like t) but a whole function $f(t)$. Think of the *test function* $f(t)$ as the linear response function of a probe which is sensitive to the value of ξ, and $\xi(f)$ as a measurement of ξ according to

$$\xi(f) = \int \xi(t) f(t) dt. \tag{3.13}$$

For smooth, bounded, and rapidly vanishing functions f, this integral makes perfect sense if the sample paths of $\xi(\cdot)$ are (almost surely) locally integrable functions of t; it even makes sense if the realizations of $\xi(\cdot)$ are more general objects like δ-functions. The more restricted the space of test functions, the more pathological the objects ξ may be. On the other hand, if the space of test functions includes distributions like the δ-functions, then ξ must be defined pointwise and we recover our original definition of the stochastic process.

The derivative of the Wiener process can be interpreted in this "smeared" way by noting that if $\xi = dW/dt$, then ξ should be gaussian and mean zero (because every approximation to it is) with variance

$$
\begin{aligned}
E\{\xi(f)^2\} &= E\left\{ \left[\int_0^\infty \frac{dW}{dt} f(t) dt \right]^2 \right\} \\
&= E\left\{ \left[-\int_0^\infty \frac{df}{dt} W(t) dt \right]^2 \right\} \quad \text{(integrating by parts)} \\
&= \int_0^\infty dt \int_0^\infty ds\, E\{W(t)W(s)\} f'(t) f'(s) \\
&\qquad \text{(interchanging integrals and the expectation)} \\
&= \int_0^\infty dt \int_0^\infty ds\, \min(t,s) f'(t) f'(s) \\
&= -\int_0^\infty f'(t) dt \int_0^t f(s) ds \quad \text{(integrating by parts in the s integral)} \\
&= \int_0^\infty f(t)^2 dt. \quad \text{(integrating by parts once more)}
\end{aligned}
\tag{3.14}
$$

In the above we assume that the test function $f(t)$ vanishes fast enough as $t \to \infty$ to allow the integrations by parts without introducing any boundary terms. Thus, the variance of the derivative of Brownian motion *smeared* with a square integrable function $f(t)$ is finite and $\xi(f)$ may be interpreted as a perfectly well-behaved, gaussian random variable.

The white-noise process $\xi(\cdot)$, defined as the derivative of the Wiener process in the generalized sense described above, is known as *gaussian white noise*. Often we (and many others, to be sure) will write gaussian white noise as if it was an ordinary gaussian stochastic process, $\xi(t)$, with the moments

$$E\{\xi(t)\} = 0\,,$$
$$E\{\xi(t)\xi(s)\} = \delta(t - s)\,. \tag{3.15}$$

The δ-function covariance indicates that $\xi(t)$ is not really a random variable for each t, but this is what we must put in to reproduce the calculation in Eq. (3.14) in a shorthand notation:

$$\begin{aligned} E\{\xi(f)^2\} &= E\left\{ \left[\int_0^\infty \xi(t)f(t)dt \right]^2 \right\} \\ &= \int_0^\infty dt \int_0^\infty ds\, E\{\xi(t)\xi(s)\}f(t)f(s) \\ &= \int_0^\infty dt \int_0^\infty ds\, \delta(t - s)f(t)f(s) \\ &= \int_0^\infty f(t)^2 dt\,. \end{aligned} \tag{3.16}$$

Gaussian white noise is not just an abstract mathematical construction. We may think of it as the continuous time limit of a discrete time model of some stochastic process which is very rapidly varying in time, so much so that its values at successive instants are essentially independent. The gaussian statistics of the process makes it very useful as a model of processes which result from the sum of many nearly independent random effects: the central limit theorem then ensures, under very general conditions, that the resulting random variables will be normally distributed.

EXERCISE 3. Show that the transition density for the Wiener process, Eq. (3.2), for $t > s$, satisfies the diffusion equation

$$\partial_t \rho(x, t|y, s) = \frac{1}{2}\partial_x^2 \rho(x, t|y, s) \tag{3.17}$$

and that

$$\lim_{t \downarrow s} \rho(x, t|y, s) = \delta(x - y)\,. \tag{3.18}$$

4. STOCHASTIC DIFFERENTIAL EQUATIONS AND FOKKER-PLANCK EQUATIONS

We have discussed one specific example of a continuous time Markov process at this point, Brownian motion $W(t)$. Brownian motion is in the class of continuous time Markov processes whose sample paths are (almost surely) continuous but not differentiable functions, called *diffusion processes*. This is a natural name for these processes because, as we will see below, their transition densities satisfy partial differential equations related to the diffusion equation. For example, by direct calculation in Exercise 3, we found that the transition density for the Wiener process satisfies the diffusion equation

$$\partial_t \rho(x,t|y,s) = \frac{1}{2}\partial_x{}^2 \rho(x,t|y,s),\tag{4.1}$$

with the initial condition

$$\rho(x,s|y,s) = \delta(x-y)\tag{4.2}$$

at time $t = s$. In this section we will show how Markov diffusion processes arise naturally as the solutions of differential equations with white-noise coefficients, i.e., *stochastic differential equations*. Moreover, we will show how to write down the partial differential equation for the transition density, generally called the *Fokker-Planck equation*, starting from the stochastic differential equation.

Brownian motion may be described as the (one-dimensional) position as a function of time of a particle whose velocity is gaussian white noise. That is, by definition of gaussian white noise as the derivative of the Wiener process, $\xi(t) = dW(t)/dt$, we may consider the reverse logical order of this equation, writing

$$\frac{dW(t)}{dt} = \xi(t),\tag{4.3}$$

and think of $W(t)$ as the solution of this stochastic differential equation. The associated Fokker-Planck equation for transition density of the solution is exactly Eq. (4.1) and, given some initial conditions, its solution contains all possible information about the stochastic process (remember, from Eq. (2.11), that all the joint density functions are built up from the transition density and the initial one-point density for Markov processes).

A more general stochastic differential equation is one where the derivative of the solution depends on the solution itself and the white-noise process appears as a coefficient. Because gaussian white noise is really a generalized stochastic process, it makes no sense to perform nonlinear operations on it, so the most general such stochastic differential equation for a process $X(t)$ is

$$\frac{dX(t)}{dt} = f\big(X(t),t\big) + g\big(X(t),t\big)\xi(t)\,.\tag{4.4}$$

This is a model of a system subject to a deterministic driving force, $f(X, t)$, as well as a rapidly fluctuating random force, $g(X, t)\xi(t)$, whose influence depends both on the state of the system and on the time. For locations in the state space and intervals of time where g is rather large in absolute value, the fluctuations ξ have more influence on the dynamical variable's evolution than those locations and times where it is small in absolute value. Such white-noise-driven differential equations are often called *Langevin equations* in the physics and chemistry literature. In the specific case $f = 0$ and $g = 1$, Eq. (4.4) reduces to Eq. (4.3) and we know that its corresponding Fokker-Planck equation is Eq. (4.1). Our task now is to start from Eq. (4.4), stating somewhat more carefully how we will interpret it, and then to derive the Fokker-Planck equation for the transition density of the process $X(t)$.

We will consider the stochastic differential equation Eq. (4.4) to be the continuous time limit of the discrete time problem

$$\Delta X(t) = X(t + \Delta t) - X(t) = f\big(X(t), t\big)\Delta t + g\big(X(t), t\big)\Delta W(t), \qquad (4.5)$$

where $\Delta W(t) = W(t + \Delta t) - W(t)$ is the increment of the Wiener process in the time interval between t and $t + \Delta t$. This is the kind of discretization that we could contemplate solving numerically on a computer. If we divide through by Δt and formally take the limit $\Delta t \to 0$, we recover Eq. (4.4). This discrete formulation also explicitly displays the Markov character of the solution process $X(t)$: if the process is given (i.e., if its value is specified) at time t_0, then we may evolve the process some integral number of time steps into the future using only the statistically independent random variables $\Delta W(t)$ which we generate along the way, completely independently of the past history of the process. Thus the future of $X(t_0)$, i.e., $X(t_0 + n\Delta t)$ is independent of the past when we are given the present value of the process. As the Markov property holds for each discretization of $X(t)$ with ever decreasing intervals Δt, it is not surprising that this characteristic survives the limit to continuous time.

We use the time-sliced formulation in Eq. (4.5) to view the information contained in the stochastic differential equation from a slightly different angle. Here we need to introduce the concept of the *conditional expectation* of a function of a stochastic process. For a Markov process $X(t)$ with transition density $\rho(x, t|y, s)$, we define the "expectation of $F(X(t))$ given $X(s) = y$" for $t > s$, denoted $E\{F(X(t))|X(s) = y\}$, as the expectation computed using the conditional probability density:

$$E\{F(X(t))|X(s) = y\} = \int F(x)\rho(x, t|y, s)dx, \qquad (4.6)$$

where, as usual, the integral covers the process' state space. The concept is simple; we ask for the average value of $F(X(t))$ over those realizations of the process that started at position y at time s. In particular, for $s = t$, the process is taking on the given value, $X(s) = y$, so that

$$
\begin{aligned}
E\{F(X(t))|X(t) = y\} &= \int F(x)\rho(x,t|y,t)dx \\
&= \int F(x)\delta(x - y)dx = F(y).
\end{aligned}
\tag{4.7}
$$

Now we may use the discrete formula Eq. (4.5) to find the conditional expectation of the jump in $X(t)$ given its starting point:

$$
\begin{aligned}
E\{\Delta X(t)|X(t) = x\} &\\
&= E\{X(t + \Delta t)|X(t) = x\} - E\{X(t)|X(t) = x\} \\
&= \int (x' - x)\rho(x',t + \Delta t|x,t)dx' \\
&= E\{f(X(t),t)\Delta t|X(t) = x\} + E\{g(X(t),t)\Delta W(t)|X(t) = x\}.
\end{aligned}
\tag{4.8}
$$

Clearly, $E\{f(X(t),t)\Delta t|X(t) = x\} = f(x,t)\Delta t$. Also, because the increment of the Wiener process $\Delta W(t)$ is *independent* of $X(t)$ and all the previous history of the process,

$$
\begin{aligned}
E\{g(X(t),t)\Delta W(t)|X(t) = x\} &= E\{g(X(t),t)|X(t) = x\}E\{\Delta W(t)|X(t) = x\} \\
&= g(x,t)E\{\Delta W(t)\} \\
&= 0.
\end{aligned}
\tag{4.9}
$$

Hence, the expectation of the increment in the process $X(t)$ is

$$
E\{\Delta X(t)|X(t) = x\} = f(x,t)\Delta t.
\tag{4.10}
$$

We also need the second moment of the increments of the process, and these can all be computed from the difference equation. Indeed, recalling that the variance of ΔW is Δt,

$$
\begin{aligned}
E\{\Delta X(t)^2|X(t) = x\} &= \int (x' - x)^2\rho(x',t + \Delta t|x,t)dx' \\
&= E\left\{\left[f(X(t),t)\Delta t + g(X(t),t)\Delta W(t)\right]^2 |X(t) = x\right\} \\
&= f(x,t)^2\Delta t^2 + 2f(x,t)\Delta t g(x,t)E\{\Delta W(t)\} \\
&\quad + g(x,t)^2 E\{\Delta W(t)^2\} \\
&= g(x,t)^2\Delta t + O(\Delta t^2).
\end{aligned}
\tag{4.11}
$$

All the higher jump moments are of higher than linear order in the time increment Δt. For our purposes, the stochastic differential equation $dX/dt = f + g\xi$ will be interpreted in terms of the jump moments derived above; $X(t)$ is a Markov process whose first and second jump moments are of order Δt and specified as in Eq. (4.10) and Eq. (4.11), with all higher moments of higher order in Δt.

To derive the Fokker-Planck equation, we begin with the assumption that the process $X(t)$ possesses a differentiable transition density, and the Chapman-Kolmogorov relation

$$\rho(x,t|y,s) = \int \rho(x,t|z,u)\rho(z,u|y,s)dz \qquad \text{for } s < u < t, \qquad (4.12)$$

which is simply a rewriting of Eqs.(2.3) and (2.11). This says that the probability to go from y at time s, to x at time t, is the sum of the probabilities of the mutually exclusive events of making the transition via different points z at the intermediate time u. The probability of each of these events is the product of the probabilities of two independent events for Markov processes.

Introduce an arbitrary smearing function $R(x)$ on the state space which is smooth (so we have as many derivatives as we wish) and rapidly vanishing at the boundaries of the state space (so that no boundary terms arise from integrations by parts). Then, for $t > s$ and $\Delta t > 0$, the Chapman-Kolmogorov relation says that the conditional expectation of $R(X(t + \Delta t))$ given $X(s) = y$ is

$$\int R(x)\rho(x,t+\Delta t|y,s)dx = \int dx \int dz \rho(x,t+\Delta t|z,t)\rho(z,t|y,s)R(x)$$
$$= \int dz\rho(z,t|y,s) \int dx R(x)\rho(x,t+\Delta t|z,t). \qquad (4.13)$$

Now expand $R(x)$ in a Taylor series around z. The motivation here is that for small Δt, $\rho(x,t+\Delta t|z,t)$ is "almost" $\delta(x-z)$, so that only the values of $R(x)$ near $x = z$ are significant in the second integral above. Using the normalization of the transition density,

$$\int \rho(x,t+\Delta t|z,t)dx = 1,$$

we have

$$\int R(x)\rho(x,t+\Delta t|y,s)dx$$
$$= \int dz\rho(z,t|y,s)R(z)$$
$$+ \int dz\rho(z,t|y,s)R'(z) \int dx(x-z)\rho(x,t+\Delta t|z,t)$$
$$+ \int dz\rho(z,t|y,s)\frac{1}{2}R''(z) \int dx(x-z)^2\rho(x,t+\Delta t|z,t) + \dots \qquad (4.14)$$

We have now cast the conditional expectation of $R(X(t + \Delta t))$ in terms of conditional expectations of $R(X(t))$ and its derivatives, and the jump moments for the process. The jump moments were computed from the stochastic differential equation above, and inserting Eqs.(4.10) and (4.11) into Eq. (4.14) above,

$$\int R(x)\rho(x,t+\Delta t|y,s)dx = \int dz\rho(z,t|y,s)\Big\{R(z) + R'(z)f(z,t)\Delta t$$
$$+ \frac{1}{2}R''(z)g(z,t)^2\Delta t + O(\Delta t^2)\Big\}. \tag{4.15}$$

Change the z to an x on the right-hand side above, place all the terms on one side of the equality sign, and divide through by Δt:

$$0 = \int dx\Big\{ R(x)\frac{\rho(x,t+\Delta t|y,s) - \rho(z,t|y,s)}{\Delta t}$$
$$- \Big[R'(x)f(x,t) + \frac{1}{2}R''(x)g(x,t)^2 + O(\Delta t)\Big]\rho(x,t|y,s)\Big\}. \tag{4.16}$$

In the limit $\Delta t \to 0$, we find that for any function $R(\cdot)$,

$$0 = \int dx\Big\{ R(x)\frac{\partial\rho(x,t|y,s)}{\partial t} - \Big[R'(x)f(x,t) + \frac{1}{2}R''(x)g(x,t)^2\Big]\rho(x,t|y,s)\Big\}. \tag{4.17}$$

Integrate by parts in the last two terms above to find

$$0 = \int dx\,R(x)\Big\{ \partial_t\rho(x,t|y,s) + \partial_x[f(x,t)\rho(x,t|y,s)] - \frac{1}{2}\partial_x^2[g(x,t)^2\rho(x,t|y,s)]\Big\}. \tag{4.18}$$

Because R is an arbitrary function, the term in brackets must vanish so the transition density $\rho(x,t|y,s)$ satisfies the Fokker-Planck equation

$$\partial_t\rho(x,t|y,s) = \Big[-\partial_x f(x,t) + \frac{1}{2}\partial_x^2 g(x,t)^2\Big]\rho(x,t|y,s). \tag{4.19}$$

(The reader is reminded at this point that the differential operator ∂_x acts on everything to their right.) Along with this evolution equation for $\rho(x,t|y,s)$ goes the initial condition at time $t = s$

$$\rho(x,t = s|y,s) = \delta(x - y). \tag{4.20}$$

This is the central goal of this section. The solution to the stochastic evolution in Eq. (4.4) is a Markov process defined by its transition density which satisfies the Fokker-Planck equation in Eq. (4.19). This is really how we are interpreting the random dynamical evolution law in the original stochastic differential equation. The solution of the associated Fokker-Planck equation provides us with the transition density defining the Markov process which we take to be the solution $X(t)$.

The natural generalization of these considerations allows us to formulate the same ideas for stochastic processes which are not just scalar valued, but also multi-component objects. For example, let ξ_i, $i = 1,\ldots,n$, be n independent gaussian white-noise processes, and $\mathbf{X}(t) \in R^n$ be an n-dimensional vector-valued process with components $X_i(t)$ satisfying the stochastic differential equations

$$\frac{dX_i}{dt} = f_i(\mathbf{X}) + g_{ij}(\mathbf{X})\xi_j \, . \tag{4.21}$$

Here we use the summation convention on repeated indices; $\mathbf{f}(\mathbf{X})$ is a vector-valued function of $\mathbf{X}$ with components f_i and $g_{ij}(\mathbf{X})$ are the components of an $n \times n$ matrix-valued function of $\mathbf{X}$. The transition density $\rho(\mathbf{x},t|\mathbf{y},s)$ for the vector-valued Markov process $\mathbf{X}(t)$ then satisfies the Fokker-Planck equation

$$\frac{\partial \rho(\mathbf{x},t|\mathbf{y},s)}{\partial t} = \left\{ -\frac{\partial}{\partial x_i} f_i(\mathbf{x}) + \frac{1}{2}\frac{\partial}{\partial x_i}\frac{\partial}{\partial x_j} D_{ij}(\mathbf{x}) \right\} \rho(\mathbf{x},t|\mathbf{y},s) \tag{4.22}$$

where the positive semi-definite diffusion matrix $D_{ij}(\mathbf{x})$ is

$$D_{ij} = g_{ik}g_{jk} \, , \tag{4.23}$$

and the transition density satisfies the initial condition

$$\rho(\mathbf{x},t|\mathbf{y},s) = \delta(\mathbf{x} - \mathbf{y}) \, . \tag{4.24}$$

Along with the Fokker-Planck equation and its initial condition go some boundary conditions. The issue of boundary conditions for these processes is rather involved, and we will for the most part neglect the subtle issues here. Our practical criteria will be the condition that a proposed solution to the Fokker-Planck equation is positive and integrable, and thus interpretable as a probability density. We refer the reader to the discussion of boundary conditions in Horsthemke and Lefever[3] for details.

EXERCISE 4. The Ornstein-Uhlenbeck process, $U(t)$, is defined by the stochastic differential equation

$$\frac{dU}{dt} = -\gamma U + \sigma\xi \tag{4.25}$$

with γ and σ constant parameters. It is Newton's law for the acceleration (dU/dt) of a particle in some medium subject to frictional retarding force proportional to its velocity $(-\gamma U)$ and a rapidly fluctuating random force due to collisions with the "fluid" particles in the medium $(\sigma\xi)$.

a.   Show that the associated Fokker-Planck equation is

$$\frac{\partial \rho(x,t|y,s)}{\partial t} = \left[\gamma\frac{\partial}{\partial x}x + \frac{\sigma^2}{2}\frac{\partial^2}{\partial x^2} \right] \rho(x,t|y,s) \, . \tag{4.26}$$

b. Verify that the transition density is

$$p(x,t|y,s) = \frac{1}{\sqrt{2\pi\Sigma(t-s)}} \exp\left\{-\frac{(x - ye^{-\gamma(t-s)})^2}{2\Sigma(t-s)}\right\} \qquad (4.27)$$

where

$$\Sigma(t) = \frac{\sigma^2}{2\gamma}(1 - e^{-2\gamma t}). \qquad (4.28)$$

c. The *stationary state* of the Ornstein-Uhlenbeck process is achieved as $t \to \infty$. It is a gaussian Markov process with the time-independent one-time density (the *stationary* density ρ_{stat})

$$\rho_{\text{stat}}(x) = \lim_{t\to\infty} \rho(x,t) = \sqrt{\frac{\gamma}{\pi\sigma^2}} \exp\left\{-\frac{\gamma x^2}{\sigma 2}\right\}. \qquad (4.29)$$

Show that the covariance in the stationary state is

$$E\{U(t)U(s)\} = \int_{-\infty}^{\infty} dx \int_{-\infty}^{\infty} dy\, xy\rho(x,t|y,s)\rho_{\text{stat}}(y)$$
$$= \frac{\sigma^2}{2\gamma}e^{-\gamma|t-s|}. \qquad (4.30)$$

The exponential decay time $\tau = \gamma^{-1}$ of the correlation function is called the *correlation time* of the process.

5. APPLICATION: THE VERHULST MODEL OF POPULATION DYNAMICS

The example we will develop here is a simple model of population dynamics. Let $X(t) > 0$ be the population of the species, obeying the *Verhulst equation*

$$\frac{dX}{dt} = \mu X - X^2. \qquad (5.1)$$

The linear term on the right-hand side is the net rate of change of population due to birth and death; the growth rate μ is the difference between the birth and death rates. The nonlinear saturation term roughly models the effect of overcrowding which limits the total population. We have chosen the units so that the coefficient of the nonlinear term is 1.

The deterministic dynamics of this equation are very simple. If $\mu < 0$ (i.e., death rate > birth rate), then the population always dies out because the right-hand side of Eq. (5.1) is always negative. Thus $X = 0$ is a stable steady state of the system for $\mu < 0$, and it is in fact the attractor. All solutions approach this attracting state at a uniform asymptotic rate:

$$X(t) \sim e^{\mu t} \rightarrow 0, \qquad \text{as } t \rightarrow \infty \text{ for } \mu < 0. \tag{5.2}$$

If $\mu > 0$, then birth outpaces death and a small population will grow—at first exponentially—until the nonlinear saturation takes over. On the other hand, if the population starts out very large, then the nonlinear drives it down until it balances with the positive growth rate. Both of these cases are illustrated in Figure 6. The solution $X = 0$ is linearly unstable when $\mu > 0$, and the steady new solution $X = \mu$ is a stable fixed point. All solutions are attracted to this state at a steady rate,

$$|X(t) - \mu| \sim e^{-\mu t} \rightarrow 0, \qquad \text{as } t \rightarrow \infty \text{ for } \mu > 0. \tag{5.3}$$

The long time dynamics are simply summarized in the bifurcation diagram in Figure 7, where we show the stable steady population as a function of μ.

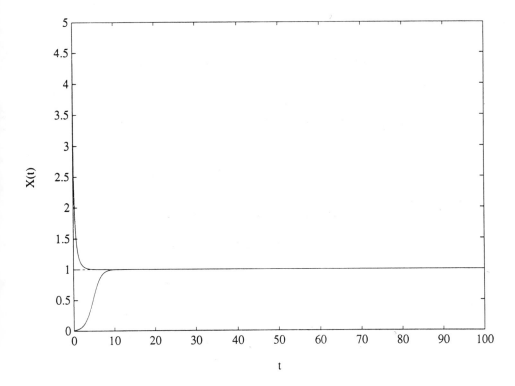

FIGURE 6   Two solutions of the deterministic Verhulst equation, starting from $X(0) = 5$ (top) and $X(0) = .01$ (bottom). The growth rate is $\mu = 1$. At long times the solutions approach the steady state $X = \mu$.

X steady state

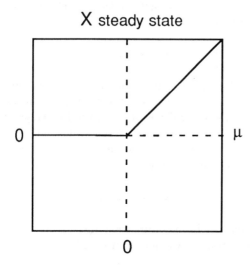

FIGURE 7 Bifurcation diagram of the stable steady state for the Verhulst equation.

Now the question is the following: what is the effect of environmental fluctuations on this population? That is, imagine that the birth and death rates are not constant (more specifically, their difference μ is not constant), but rather they fluctuate from generation to generation due to environmental effects like the weather, predator populations, disease, etc. We will model these fluctuations as essentially random from the point of view of the species under study, so the "noisy" growth rate $\mu(t)$ is a stochastic process. Specifically, we write

$$\mu(t) = \langle\mu\rangle + \sigma\xi(t),\tag{5.4}$$

where $\langle\mu\rangle$ is the time averaged rate, σ is a parameter that we will refer to as the noise amplitude, and ξ is gaussian white noise (normalized as usual by $E\{\xi(t)\xi(s)\} = \delta(t-s)$). Physically, in modeling the fluctuations as white noise we are assuming that the variations are very fast on the system's deterministic time scale, given by the relaxation time $\langle\mu\rangle^{-1}$. Our stochastic differential equation for the population is then

$$\frac{dX}{dt} = \langle\mu\rangle X - X^2 + \sigma X\xi(t),\tag{5.5}$$

which we interpret, as in the last section, as the continuous time limit of the discrete time process defined by

$$\Delta X(t) = (\langle\mu\rangle X - X^2)\Delta t + \sigma X\Delta W(t).\tag{5.6}$$

This discrete time process is already a sensible model if we consider nonoverlapping generations with environmental fluctuations affecting only the birth rate. The continuous time limit then describes the population dynamics on time scales much longer than the life of any one generation.

Let us begin by making some phenomenological observations based on numerical simulations of the discrete time process in Eq. (5.6). Not unexpectedly, a very small amount of noise simply causes the population to vary about the deterministic solution, as illustrated in Figure 8. We are interested in the long-term steady-state behavior of the population, characterized by the statistics of the stochastic dynamic system after any initial transients have died away. In Figures 9 and 10 we show the steady-state population for two higher noise values. The qualitative behavior is similar to the low noise dynamics, although it is clear in Figure 10 that the average population is, at this noise level, less than the deterministic steady population set by the average growth rate. The time series starts to look significantly different for higher noise as shown in Figures 11 and 12. Not only is the average population lower than that expected from the average growth rate, but the species spends a substantial amount of time near zero population, occasionally making large deviations to high population states. These deviations appear to be increasingly sporadic as the environmental noise amplitude is increased. In Figure 13 we observe that, when the noise is large enough, the population completely vanishes at some finite time. In this case the steady-state behavior is extinction!

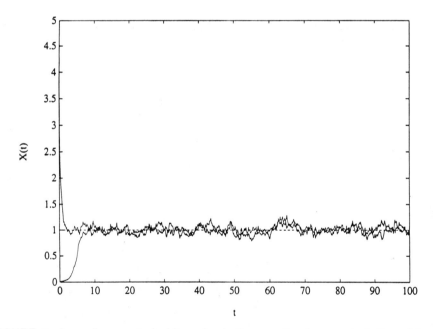

FIGURE 8 A small amount of white noise in the growth rate, a perturbation of the system in Figure 6. The noise amplitude is $\sigma^2 = .01$, and the average growth rate is $\langle \mu \rangle = 1$.

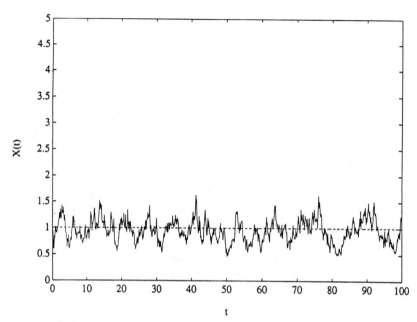

FIGURE 9   Steady-state time series for $\sigma^2 = .1$, average growth rate $\langle \mu \rangle = 1$.

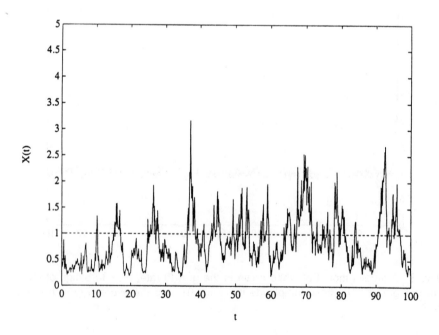

FIGURE 10   Steady-state time series for $\sigma^2 = .5$, average growth rate $\langle \mu \rangle = 1$.

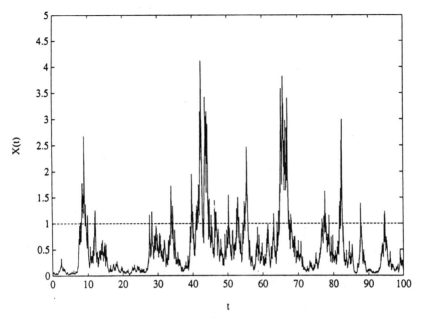

FIGURE 11   Steady-state time series for $\sigma^2 = 1$, average growth rate $\langle \mu \rangle = 1$.

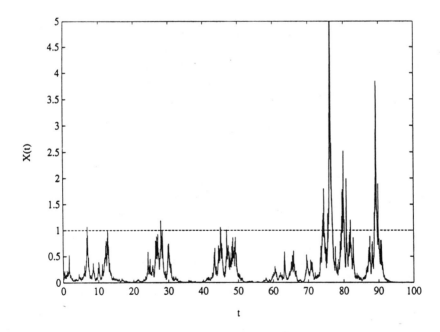

FIGURE 12   Steady-state time series for $\sigma^2 = 1.5$, average growth rate $\langle \mu \rangle = 1$.

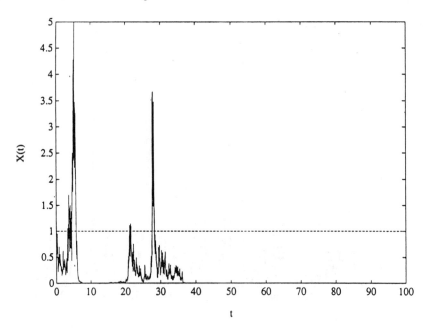

FIGURE 13 Time series for $\sigma^2 = 2$, average growth rate $\langle\mu\rangle = 1$. The population becomes extinct (in this simulation) at some time between $t = 30$ and $t = 40$.

Let us turn now to a more quantitative analysis of the process. The stochastic differential equation, Eq. (5.5), fits into the analytical framework of the last section: the drift function is $f(x) = (\langle\mu\rangle x - x^2)$, and the diffusion function is $g(x)^2 = \sigma^2 x^2$. The solution, $X(t)$, is a Markov diffusion process whose transition density, $\rho(x, t|y, s)$, obeys the Fokker-Planck equation

$$\frac{\partial\rho}{\partial t} = \frac{\partial}{\partial x}\left\{x^2 - \langle\mu\rangle x + \frac{\sigma^2}{2}\frac{\partial}{\partial x}x^2\right\}\rho. \tag{5.7}$$

The steady state of a stochastic process is characterized by (among other things) the stationary probability density, $\rho_{\text{stat}}(x)$, defined by

$$\rho_{\text{stat}}(x) = \lim_{t\to\infty}\rho(x, t|y, s), \tag{5.8}$$

assuming that this limit makes sense as a probability distribution independent of the starting point $X(s) = y$. This steady-state probability density is an invariant density and so satisfies the time-independent Fokker-Planck equation

$$0 = \frac{\partial}{\partial x}\left\{x^2 - \langle\mu\rangle x + \frac{\sigma^2}{2}\frac{\partial}{\partial x}x^2\right\}\rho_{\text{stat}}. \tag{5.9}$$

This equation has the formal solution

$$\rho_{\text{stat}}(x) = N x^{2((\langle\mu\rangle/\sigma^2-1)}e^{-2x/\sigma^2}, \tag{5.10}$$

which is an acceptable probability density on the positive real line as long as it's integrable, i.e., so long as the normalization condition

$$1 = \int_0^\infty \rho_{\text{stat}}(x)dx = N\left(\frac{\sigma^2}{2}\right)^{(2\langle\mu\rangle/\sigma^2-1)}\Gamma\left(\frac{2\langle\mu\rangle}{\sigma^2}-1\right) \tag{5.11}$$

can be satisfied for some finite, nonvanishing normalization constant N. Hence, the function in Eq. (5.10) is a stationary probability distribution for the process as long as the average growth rate and the noise amplitude satisfy

$$\frac{2\langle\mu\rangle}{\sigma^2} > 1. \tag{5.12}$$

This condition says that, if the environmental fluctuations are not too strong, on the scale of the average growth rate, the process may achieve a stochastic stationary state where the population $X(t)$ is described by the time-independent probability density in Eq. (5.10). This probability density is plotted for several values of the noise amplitude in Figures 14–17. The qualitative behavior of the process is clearly indicated by the features of the stationary density: at small noise the process fluctuates in a relatively tight region around its most probable value (compare Figures 9 and 10), and under the influence of larger variations, the most probable value is actually the zero population state (compare Figures 11 and 12). There is an interesting change in the population dynamics precisely when $\sigma^2 = \langle\mu\rangle$, the point at which the zero population state becomes the most probable value. This kind of qualitative change in a system's behavior, brought on by the influence of fluctuations alone—the mean growth rate is constant here—is called a *noise-induced transition*, akin to bifurcations in deterministic systems and phase transitions in equilibrium statistical mechanics, and is studied in detail in the monograph of Horsthemke and Lefever.[3]

What happens if the noise is even stronger, i.e., if the inequality in Eq. (5.12) is violated? We must keep in mind that the trivial stochastic process $X(t) = 0$ is an exact solution of the stochastic differential equation, corresponding to the stationary density $\rho_{\text{stat}}(x) = \delta(x)$. This singular distribution function will certainly describe the system if the initial population was zero, but it also describes the steady state if the population eventually dies out for any reason. In fact, this is just how this model behaves. If the amplitude of the growth rate fluctuations are too

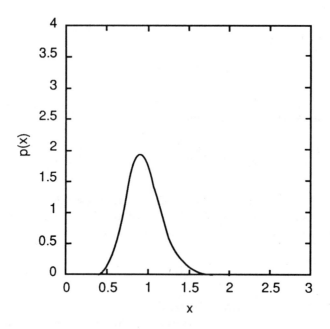

FIGURE 14 Stationary probability distribution for $\sigma^2 = .1$, average growth rate $\langle \mu \rangle = 1$.

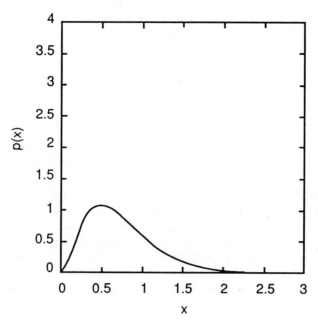

FIGURE 15 Stationary probability distribution for $\sigma^2 = .5$, average growth rate $\langle \mu \rangle = 1$.

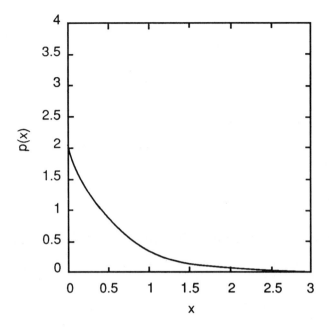

FIGURE 16 Stationary probability distribution for $\sigma^2 = 1$, average growth rate $\langle\mu\rangle = 1$.

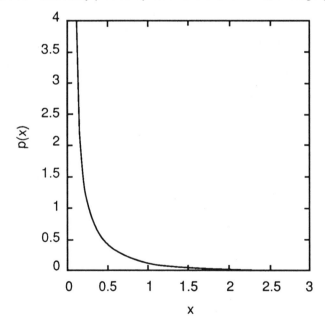

FIGURE 17 Stationary probability distribution for $\sigma^2 = 1.5$, average growth rate $\langle\mu\rangle = 1$.

strong, specifically if $\sigma^2 \geq 2\langle\mu\rangle$, then the population eventually becomes extinct and the steady state density is

$$\rho_{\text{stat}}(x) = \delta(x) \qquad (\text{for } \sigma^2 \geq 2\langle\mu\rangle) . \tag{5.13}$$

EXERCISE 5. Using the stationary probability density Eq. (5.10) or Eq. (5.13), compute the average steady-state population as a function of the mean growth rate and the noise amplitude.

6. EQUILIBRIUM VS. NONEQUILIBRIUM STATIONARY STATES

The Fokker-Planck equation is a continuity equation for the flow of the probability density of the variables of the stochastic process. Indeed, for the system

$$\frac{dX_i}{dt} = f_i(\mathbf{X}) + g_{ij}(\mathbf{X})\xi_j , \tag{6.1}$$

the Fokker-Planck equation

$$\frac{\partial\rho}{\partial t} = \left\{ -\frac{\partial}{\partial x_i} f_i + \frac{1}{2}\frac{\partial}{\partial x_i}\frac{\partial}{\partial x_j} D_{ij} \right\} \rho , \tag{6.2}$$

where $D_{ij}(\mathbf{x}) = g_{ik}(\mathbf{x})g_{jk}(\mathbf{x})$, can be written in the form of the continuity equation

$$\frac{\partial\rho}{\partial t} + \frac{\partial J_i}{\partial x_i} = 0 , \tag{6.3}$$

with the current vector field $\mathbf{J}$ whose components are

$$J_i(\mathbf{x}, t) = \left\{ f_i - \frac{1}{2}\frac{\partial}{\partial x_j} D_{ij} \right\} \rho(\mathbf{x}, t) . \tag{6.4}$$

The current vector gives the flow of probability in the system's state space.

A stationary state of the stochastic process is achieved when the probability density of the variable $\mathbf{X}(t)$ becomes time independent. In terms of the Fokker-Planck equation, the stationary state is described by a time-independent probability distribution, $\rho_{\text{stat}}(\mathbf{x})$, that solves the equation

$$0 = \frac{\partial}{\partial x_i} \left\{ -f_i + \frac{1}{2}\frac{\partial}{\partial x_j} D_{ij} \right\} \rho_{\text{stat}}(\mathbf{x}) . \tag{6.5}$$

The approach to stationary behavior may also be of interest in some situations, but in this section we will restrict our considerations to the nature of time-independent solutions of the Fokker-Planck equation.

The stationary Fokker-Planck equation, Eq. (6.5), implies that the stationary current vector field, $\mathbf{J}_{\text{stat}}(\mathbf{x})$, defined by

$$J_{\text{stat},i}(\mathbf{x}) = \left\{ f_i(\mathbf{x}) - \frac{1}{2} \frac{\partial}{\partial x_j} D_{ij}(\mathbf{x}) \right\} \rho_{\text{stat}}(\mathbf{x}), \qquad (6.6)$$

is a divergence-free vector field:

$$0 = \nabla \cdot \mathbf{J}_{\text{stat}} . \qquad (6.7)$$

In general, the current need not vanish in a stationary state—only its divergence necessarily vanishes.

We define an *equilibrium* stationary state to be a solution of the stationary Fokker-Planck equation which satisfies not only the stationary Fokker-Planck equation, but also

$$\mathbf{J}_{\text{stat}} = 0 . \qquad (6.8)$$

Physically, the vanishing of the probability current in the stationary state implies that the rate at which probability flows from any one point in the state space to another is equal to the rate at which probability flows in the opposite direction between the two points. This is the condition of *detailed balance* in the stationary state, and this is a general feature of systems in true thermodynamic equilibrium. In fact, if the stochastic dynamics described by the Fokker-Planck equation is the reduced description of an underlying system with time reversal invariant dynamics, such as Hamiltonian dynamics of classical mechanics, then the condition of detailed balance follows for the stationary state. Stochastic dynamic systems that satisfy the detailed balance condition can for many purposes be considered as thermal equilibrium systems. Mathematically, the property of detailed balance in equilibrium stationary states facilitates solution of the Fokker-Planck equation. Indeed, rather than solving the second-order differential equation in Eq. (6.5), we need only solve the first-order equation in Eq. (6.8).

As an example, consider a diffusion process defined by the stochastic differential equations

$$\frac{dX_i}{dt} = f_i(\mathbf{X}) + \sigma \xi_i(t), \qquad i = 1, \ldots, n , \qquad (6.9)$$

where the ξ_i are independent gaussian white-noise processes. These dynamics describe the effect of additive noise, of equal magnitude in each component, on an n-dimensional dynamical system. The stationary Fokker-Planck equation is

$$0 = \frac{\partial}{\partial x_i} \left\{ -f_i + \frac{\sigma^2}{2} \frac{\partial}{\partial x_i} \right\} \rho_{\text{stat}}(\mathbf{x}) , \qquad (6.10)$$

and if the stationary state is to be an equilibrium stationary state, then we must also have

$$0 = \left\{ -f_i + \frac{\sigma^2}{2} \frac{\partial}{\partial x_i} \right\} \rho_{\text{stat}}(\mathbf{x}) . \qquad (6.11)$$

What are the conditions on the dynamics so that this is possible? Because the stationary density is a positive function, we may write

$$\rho_{\text{stat}}(\mathbf{x}) = \exp\left\{-\frac{2\Phi(\mathbf{x})}{\sigma^2}\right\}, \tag{6.12}$$

introducing the "potential" $\Phi(\mathbf{x})$. Inserting Eq. (6.12) into Eq. (6.11), we find

$$f_i(\mathbf{x}) = -\frac{\partial \Phi(\mathbf{x})}{\partial x_i}. \tag{6.13}$$

This says that the drift vector field $\mathbf{f}(\mathbf{x})$ must be a gradient vector field—in particular it must be curl free—in order for the detailed balance to hold in the stationary state. This is a strong restriction on the unperturbed deterministic dynamics so that the stochastic version possesses an equilibrium stationary state; the deterministic dynamics is certainly not chaotic in this case, for example. Such finite-dimensional "gradient flow" systems generically display few interesting dynamics.

More generally, the condition of detailed balance in the stationary state places severe constraints on the functional forms of the drift vector and diffusion matrix, but allows for the solution of the stationary distribution up to quadratures. For systems with a positive definite diffusion matrix $D_{ij}(\mathbf{x})$, if there is a zero current solution for Eq. (6.6), then we may write $\rho_{\text{stat}}(\mathbf{x}) = \exp\{-\Phi(\mathbf{x})\}$ so that

$$\frac{\partial \Phi(\mathbf{x})}{\partial x_j} = D_{ij}^{-1}(\mathbf{x})\left\{2f_j(\mathbf{x}) - \frac{\partial}{\partial x_k}D_{jk}(\mathbf{x})\right\}. \tag{6.14}$$

Here the condition for detailed balance is that the vector field defined by the diffusion matrix and the drift vector above is a gradient, which is not true for arbitrary dynamics. Should this condition hold, then the generalized potential $\Phi(\mathbf{x})$ can be computed by just one integration (in each variable).

We refer to stationary states with a nonvanishing probability current as *non-equilibrium* stationary states. Physical examples are steady current flow in a wire, steady heat flow across a slab of material, or even steady convection rolls in a fluid heated from below. Mathematical examples are any time-independent solutions to our Fokker-Planck equation which do not satisfy the detailed balance condition. We do not generally know the functional form of the stationary probability distribution in terms of the drift and diffusion for these systems, and thus we are generally ignorant of their stationary properties. One of the fundamental open questions in statistical physics is to develop techniques for the analysis of nonequilibrium steady states with the goal of identifying some common—and if possible universal—rules by which they are organized. Further research in this area is left as a challenge to the reader.

7. ADIABATIC ELIMINATION: COLORED NOISE, ITÔ VS. STRATONOVICH, AND KRAMERS VS. SMOLUCHOWSKI

We have defined the solution of a stochastic differential equation with gaussian white noise as a Markov process defined by the transition density obtained as the solution of the Fokker-Planck equation. The Fokker-Planck equation associated with a given stochastic differential equation was derived based on a specific interpretation of the stochastic differential equation as the continuous time limit of a discrete time problem. The analysis in section 4 corresponds to what is called the *Itô interpretation* of the stochastic differential equation. Our interpretation of the stochastic differential equation is not unique, however, and other interpretations (which may be appropriate for some problems) may lead to different results—both qualitative as well as quantitative! In this section we will develop the *Stratonovich interpretation* of the stochastic differential equation as the white-noise limit of a "real" noise problem.

The approach we will take here is to start with the Itô interpretation of a multi-dimensional system, one of whose components corresponds to a rapidly fluctuating noise, and the other of which is the system variable. In the limit that the ratio of the time scale associated with the fast noise to the system time scale vanishes, we may consider the noise to be white. Using a singular perturbation analysis of the Fokker-Planck equation for the full system, we will perform an *adiabatic elimination* of the noise variable and derive a reduced Fokker-Planck equation for the system variable alone. We will see that this interpretation of the white-noise problem, as the "white-noise limit" of a continuous-time fast noise problem, may lead to a different Fokker-Planck equation than before. The adiabatic elimination procedure that we will develop may also be applied to other problems where there is a clear separation of time scales between several processes. The same mathematical approach will be used to reduce Kramers' equation, the Fokker-Planck equation associated with Newton's law for the velocity and position of a particle subject to both a random and a deterministic force, to Smoluchowski's equation—a Fokker-Planck equation for the position of the particle alone.

We begin with a somewhat general formulation. Consider a system variable $X(t)$ which obeys the stochastic differential equation

$$\frac{dX}{dt} = f(X) + g(X)\frac{1}{\sqrt{\tau}}\zeta(t) \qquad (7.1)$$

where $f(X)$ is the deterministic component of the evolution, and $\tau^{-1/2}\zeta(t)$ is an approximate white noise, i.e., a stochastic process which we assume varies on a time scale much faster than $X(t)$. The state-dependent diffusion factor $g(X)$ describes the sensitivity of the system variable to the noise at different locations in the state space. The noise process $\zeta(t)$ evolves separately from the system variable, and we have to make some specific assumptions regarding it in order to proceed. It is not unreasonable for many applications to consider the noise to be gaussian because this would be the case (under very general conditions) if it was the result of the sum of many separate, nearly independent effects. Moreover, it is not unreasonable to model the noise as a Markov process, that is, we will assume that the future state of the noise process cannot be predicted any better by supplying more of its history than simply its current state. Finally, we assume that the noise is stationary. Then its one-time probability density is time independent and its transition density between times t and s is only a function of the time difference $|t - s|$. We may thus model $\zeta(t)$ as a stationary Ornstein-Uhlenbeck process as introduced in Exercise 4.

To establish notations and normalizations, we take $\zeta(t)$ to be a Markov process with the time-independent one-time probability density that is normal with mean zero and variance one,

$$\rho(z) = \frac{1}{\sqrt{\pi}} e^{-z^2}, \tag{7.2}$$

and transition density

$$\rho(z, t | z_0, s) = \frac{1}{\sqrt{2\pi \Sigma(t - s)}} \exp \left\{ -\frac{(z - z_0 e^{-(t-s)/\tau})^2}{2\Sigma(t - s)} \right\} \tag{7.3}$$

with

$$\Sigma(t) = \frac{1}{2}(1 - e^{-2t/\tau}). \tag{7.4}$$

The stationary covariance function of $\zeta(t)$ is thus

$$E\{\zeta(t)\zeta(s)\} = \frac{1}{2} e^{-|t-s|/\tau}. \tag{7.5}$$

In Figure 18 we show a typical realization of the Ornstein-Uhlenbeck process on a time scale much shorter than the relaxation time τ, $\zeta(t)$ then looks like Brownian motion (compare Figures 18 and 3). The difference between the $\zeta(t)$ and the Wiener process starts to appear on time scales of the order of the relaxation time. The Ornstein-Uhlenbeck process does not tend to wander so far from its starting point, and this is the property that allows for the existence of a nontrivial stationary state for the process. On a time scale much longer than τ, the process looks very much like a white noise because it appears to almost instantly decorrelate from itself. Compare Figures 19 and 5. The ζ–ζ correlation function in Eq. (7.5) vanishes almost immediately for short τ so the correlation function of $\tau^{-1/2}\zeta$ is an approximate δ-function with the correct normalization,

$$\int_{-\infty}^{\infty} \tau^{-1} E\{\zeta(t)\zeta(s)\} dt = 1, \tag{7.6}$$

like the white-noise correlation function. The Ornstein-Uhlenbeck process $\zeta(t)$ is the solution of the stochastic differential equation

$$\frac{d\zeta}{dt} = -\frac{1}{\tau}\zeta + \frac{1}{\sqrt{\tau}}\xi \tag{7.7}$$

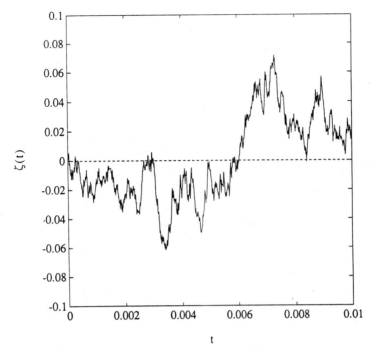

FIGURE 18 A realization of the Ornstein-Uhlenbeck process on s short time scale. The correlation time here is $\tau = 1$.

with $\xi(t)$ the usual gaussian white-noise process. In fact, $\zeta(t)$ is just a "filtered" white noise as can be seen by writing the solution to Eq. (7.7) in integral form:

$$\zeta(t) = e^{-(t-t_0)/\tau}\zeta(t_0) + \frac{1}{\sqrt{\tau}}\int_{t_0}^{t} e^{-(t-s)/\tau}\xi(s)ds . \qquad (7.8)$$

In the stationary state (as the initial time $t_0 \to -\infty$), $\zeta(t)$ becomes the steady-state response of a linear low-pass filter of bandwidth τ^{-1} to a white-noise signal. Thus, for ever decreasing relaxation time τ, the Ornstein-Uhlenbeck process approaches the unfiltered white-noise process itself.

The two-dimensional Markov process $(X(t), \zeta(t))$, describing the state variable and the noise simultaneously, satisfies the system of stochastic differential equations

$$\begin{aligned} \frac{dX}{dt} &= f(X) + g(X)\frac{1}{\sqrt{\tau}}\zeta \\ \frac{d\zeta}{dt} &= -\frac{1}{\tau}\zeta + \frac{1}{\sqrt{\tau}}\xi \end{aligned} \qquad (7.9)$$

(a)

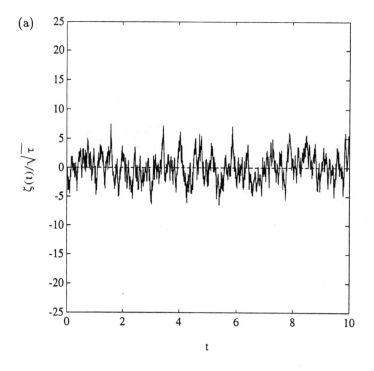

(b)

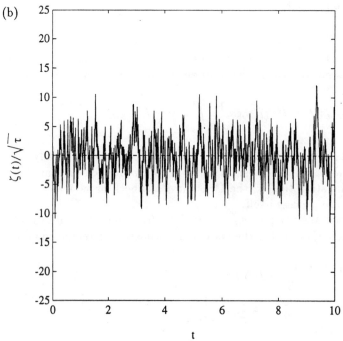

FIGURE 19 The approximate white-noise process $\zeta(t)/\sqrt{\tau}$, for several decreasing correlation times: (a) $\tau = .08$, (b) $\tau = .04$, (c) $\tau = .02$, and (d) $\tau = .01$. Compare Figure 5.

(c)

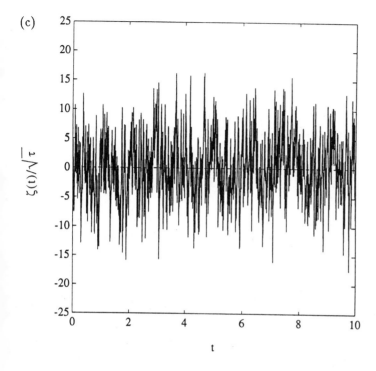

(d)

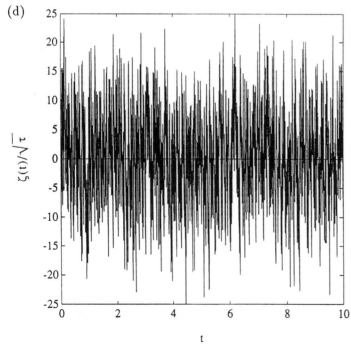

FIGURE 19
(continued)

According to Eqs.(4.21)–(4.23), the transition density $\rho(x, z, t | x_0, z_0, t_0)$ satisfies the Fokker-Planck equation

$$\partial_t \rho = \left\{ -\partial_x f(x) - \frac{1}{\sqrt{\tau}} z \partial_x g(x) + \frac{1}{\tau} \partial_z \left(z + \frac{1}{2} \partial_z \right) \right\} \rho. \qquad (7.10)$$

Our goal now is to derive a Fokker-Planck equation for the "reduced," or "marginal" transition density $r(x, t | x_0, t_0)$ of the process $X(t)$ *alone* in the limit $\tau \to 0$. For each $\tau > 0$, the reduced density is defined as

$$r(x, t | x_0, t_0) = \int_{-\infty}^{\infty} dz \int_{-\infty}^{\infty} dz_0 \rho(x, z, t | x_0, z_0, t_0) \frac{1}{\sqrt{\pi}} e^{-z_0^2}. \qquad (7.11)$$

The marginal transition density does not satisfy a (closed) Fokker-Planck equation for $\tau > 0$, but it does in the limit $\tau \to 0$. We cannot generally solve the full Fokker-Planck problem in Eq. (7.10) even in a stationary state, because the process does not generically satisfy the condition for detailed balance. Hence, we are forced to resort to some kind of perturbation theory.

Let $\varepsilon = \sqrt{\tau}$, so that the Fokker-Planck equation in Eq. (7.10) is written

$$0 = \left\{ (-\partial_t + F_2) + \frac{1}{\varepsilon} F_1 + \frac{1}{\varepsilon^2} F_0 \right\} \rho, \qquad (7.12)$$

where

$$\begin{aligned} F_2 &= -\partial_x f(x), \\ F_1 &= -z \partial_x g(x), \\ F_0 &= \partial_z \left(z + \frac{1}{2} \partial_z \right). \end{aligned} \qquad (7.13)$$

We make the ansatz that the full transition density can be expanded in a power series in ε:

$$\rho = \rho_0 + \varepsilon \rho_1 + \varepsilon^2 \rho_2 + \dots \qquad (7.14)$$

The marginal density is then also assumed to be a power series in ε as we have

$$\begin{aligned} r &= \int_{-\infty}^{\infty} dz \int_{-\infty}^{\infty} dz_0 \left\{ \rho_0 + \varepsilon \rho_1 + \varepsilon^2 \rho_2 + \dots \right\} \frac{1}{\sqrt{\pi}} e^{-z_0^2}, \\ &= r_0 + \varepsilon r_1 + \varepsilon^2 r_2 + \dots \end{aligned} \qquad (7.15)$$

This is a *singular perturbation* analysis, because of the singular appearance of the expansion parameter in the partial differential equation.

Inserting this expansion for the transition density into the Fokker-Planck equation and collecting coefficients of like powers of ε, we find

$$\begin{aligned} O(\varepsilon^{-2}): \quad & 0 = F_0 \rho_0 \\ O(\varepsilon^{-1}): \quad & 0 = F_1 \rho_0 + F_0 \rho_1 \\ O(\varepsilon^0): \quad & 0 = (-\partial_t + F_2) \rho_0 + F_1 \rho_1 + F_0 \rho_2 \end{aligned} \qquad (7.16)$$

and so on. These terms above are all that we will need for the analysis.

Of primary importance in this calculation is the properties of the "noise evolution operator," $F_0 = \partial_z(z + (1/2)\partial_z)$. This operates only on the noise variable z, and has the following spectrum of eigenvalues and eigenfunctions $p_n(z)$:

$$F_0 p_n = -n p_n, \tag{7.17}$$

where

$$p_n = H_n(z)\frac{1}{\sqrt{\pi}}e^{-z^2}, \tag{7.18}$$

and the $H_n(z)$ are the Hermite polynomials defined by

$$H_n(z) = (-1)^n e^{z^2}\frac{d^n}{dz^n}e^{-z^2}. \tag{7.19}$$

The stationary distribution of $\zeta(t)$ is the $n = 0$ eigenfunction of the operator, p_0. The first few Hermite polynomials are

$$\begin{aligned} H_0(z) &= 1 \\ H_1(z) &= 2z \\ H_2(z) &= 2(2z^2 - 1). \end{aligned} \tag{7.20}$$

and they, as well as the eigenfunctions p_n, satisfy the recursion relation

$$z H_n = \frac{1}{2}H_{n+1} + n H_{n-1}. \tag{7.21}$$

The $O(\varepsilon^{-2})$ equation implies that ρ_0 must be decomposed as (suppressing the initial data)

$$\rho_0(x, z, t) = r_0(x, t)p_0(z), \tag{7.22}$$

where $r_0(x, t)$, whose evolution equation is yet to be determined, is the leading term in the reduced density as $\varepsilon \to 0$. Then, the second equation in the perturbation expansion, of order ε^{-1}, is rewritten

$$\begin{aligned} F_0 \rho_1 &= -F_1 \rho_0 \\ &= z p_0(z)\partial_x g(x)r_0(x, t) \\ &= \frac{1}{2}p_1(z)\partial_x g(x)r_0(x, t), \end{aligned} \tag{7.23}$$

where the recursion relation in Eq. (7.21) was used to write $zp_0(z) = p_1(z)/2$. Using Eq. (7.17) for $n = 0$ and 1, we see that the solution for ρ_1 is

$$\rho_1(x, z, t) = -\frac{1}{2}\partial_x g(x)r_0(x, t)p_1(z) + r_1(x, t)p_0(z), \tag{7.24}$$

where $r_1(x,t)$ is also undetermined at this point. Now, the $O(\varepsilon^0)$ equation gives

$$
\begin{aligned}
F_0\rho_2 &= -F_1\rho_1 + (\partial_t - F_2)\rho_0 \\
&= z\partial_x g(x)\left\{-\frac{1}{2}\partial_x g(x)r_0(x,t)p_1(z) + r_1(x,t)p_0(z)\right\} \\
&\quad + \{\partial_t + \partial_x f(x)\}r_0(x,t)p_0(z) \\
&= -\frac{1}{2}\partial_x g\partial_x g r_0 z p_1(z) + \partial_x g r_1 z p_0(z) + \{\partial_t + \partial_x f\}r_0 p_0(z).
\end{aligned}
\tag{7.25}
$$

Using the recursion relation for the terms zp_1 and zp_0 above, we thus have have

$$
F_0\rho_2 = \left\{\partial_t + \partial_x f - \frac{1}{2}\partial_x g\partial_x g\right\}r_0 p_0 + \frac{1}{2}\partial_x g r_1 p_1 - \frac{1}{4}\partial_x g\partial_x g r_0 p_2.
\tag{7.26}
$$

To solve Eq. (7.26) for ρ_2, the coefficient of $p_0(z)$ on the right-hand side above must vanish. That is, because $F_0 p_0 = 0$, the operator F_0 is not invertible on the functional subspace spanned by p_0. This is the central logical step in this singular perturbation theory, and this integrability condition finally determines the evolution equation for the leading term in the reduced density:

$$
\partial_t r_0(x,t|x_0,t_0) = \left\{-\partial_x f + \frac{1}{2}\partial_x g\partial_x g\right\}r_0.
\tag{7.27}
$$

This is the main result of this section. In the white-noise limit, $\tau \to 0$, the state variable $X(t)$ has a transition probability $r(x,t|x_0,t_0) = r_0(x,t|x_0,t_0)$ that satisfies the Fokker-Planck equation above. This procedure of *adiabatic elimination* of the noise variable is a useful way of simplifying a problem by "averaging over" the fast variables, leaving an effective dynamics for the "slow" components. It should be clear that the procedure can be carried on and we may derive dynamical equations for succeeding corrections to the leading white-noise behavior of the marginal probability density.

Note that this Fokker-Planck equation is *not* the same that we would have written down for the white-noise problem

$$
\frac{dX}{dt} = f(X) + g(X)\xi,
\tag{7.28}
$$

which is

$$
\partial_t r(x,t|x_0,t_0) = \left\{-\partial_x f + \frac{1}{2}\partial_x^2 g^2\right\}r.
\tag{7.29}
$$

The Fokker-Planck equation in Eq. (7.27), regarded as the evolution equation for the transition density of the white-noise limit of a "colored," or "real," noise problem, corresponds to the *Stratonovich* interpretation of the stochastic differential equation. The Itô interpretation, and its Fokker-Planck equation in Eq. (7.29), that we have been using up to now, is valid as the continuous time limit of a discrete time problem. The two interpretations are not necessarily the same—if the solutions of the Fokker-Planck equations in Eqs.(7.27) and (7.29) are different, then we are talking about two different Markov processes. This means that it may be crucial in the modeling process to be aware of the exact sense in which fluctuations are to be regarded as white noise. The only case in which the evolution operators in Eqs.(7.27) and (7.29) are the same is if the diffusion coefficient $g(x)$ is constant as a function of x. In this case we say that the stochastic differential equation is driven by *additive* noise, as opposed to *multiplicative*, or *state-dependent*, noise.

The same philosophy of adiabatic elimination, or the elimination of fast variables, can be used to simplify the description of the effect of additive noise of systems with inertia. By this we mean systems where the *acceleration* of the state variable is proportional to a white noise. Consider a model of a particle of mass m moving (without loss of generality in one dimension) in a dissipative medium, and subject to a conservative force field with potential $U(x)$. The dissipative medium is a model of a gas or fluid whose effect on the particle is to provide a force due to the rapid bombardment of the particle by the particles that make up the medium. This force is made up of two components: (1) a viscous drag force proportional in magnitude to the velocity, and (2) a fast, mean zero, stochastic force due to fluctuations caused by the discrete nature of the medium. Writing the particle's position-velocity coordinates as $X(t)$ and $V(t)$, we have from Newton's third law

$$
\begin{aligned}
\frac{dX}{dt} &= V, \\
m\frac{dV}{dt} &= -U'(X) - \gamma V + \sigma \xi,
\end{aligned}
\tag{7.30}
$$

where γ is the linear friction coefficient, and σ is the strength of the δ-correlated white-noise force ξ. We will analyze this model now without saying any more about its validity in specific situations. In fact, a great part of the field of statistical mechanics is concerned with the derivation of such a system of equations from first principles.

The Fokker-Planck equation for the phase-space probability density of the joint process is called the *Kramers' equation* or *Klein-Kramers equation*. Explicitly, it is

$$
\partial_t \rho(x,v,t) = \left\{ -\partial_x v + \partial_v \frac{U'(x)}{m} + \frac{\gamma}{m}\partial_v \left(v + \frac{\sigma^2}{2m\gamma}\partial_v \right) \right\} \rho.
\tag{7.31}
$$

We require that the equilibrium (stationary) probability density of the process is the Gibbs distribution of equilibrium statistical mechanics,

$$
\rho_{eq}(x,v) = Z^{-1} \exp\left\{ \frac{-H}{kT} \right\},
\tag{7.32}
$$

where the Hamiltonian function H is

$$
H = \frac{mv^2}{2} + U(x),
\tag{7.33}
$$

the partition function Z is the normalization constant for the probability density, T is the temperature of the medium in which the particle lies, and k is Boltzmann's constant. Then, the friction and noise amplitude coefficients are not independent, but are connected by setting

$$0 = \left\{ -\partial_x v + \partial_v \frac{U'(x)}{m} + \frac{\gamma}{m} \partial_v \left(v + \frac{\sigma^2}{2m\gamma} \partial_v \right) \right\} \rho_{eq}(x, v). \qquad (7.34)$$

This is true when

$$\frac{\sigma^2}{2\gamma} = kT. \qquad (7.35)$$

This relationship between the effective noise strength σ^2, the friction coefficient γ, and the temperature T, is known as a *fluctuation-dissipation* relation. In line with our intuition, the temperature T is proportional to the strength of the stochastic force from the medium (at fixed friction coefficient).

Redefining the noise amplitude in terms of the temperature and γ, we have the Fokker-Planck equation

$$\partial_t \rho(x, v, t) = \left\{ -\partial_x v + \partial_v \frac{U'(x)}{m} + \frac{\gamma}{m} \partial_v \left(v + \frac{kT}{m} \partial_v \right) \right\} \rho. \qquad (7.36)$$

At a given temperature then, the frictional rate γ/m plays a similar mathematical role as the inverse time scale of the noise, τ^{-1}, in Eq. (7.10). For large friction coefficients, the noise—the velocity variable $V(t)$—is "fast" on the time scale of the evolution of the position variable. What we would like to do is to find a Fokker-Planck equation for the position variable alone in the high friction limit. To systematize the procedure, let us change variables to a dimensionless velocity measured in units of the thermal velocity $\sqrt{kT/m}$:

$$w = \sqrt{\frac{m}{kT}} v, \qquad (7.37)$$

and then let us change to a long time scale

$$s = \frac{m}{\gamma} t. \qquad (7.38)$$

Kramers' equation becomes

$$\partial_s \rho(x, w, s) = \varepsilon^{-1} \left\{ -\sqrt{\frac{kT}{m}} w \partial_x + \frac{U'(x)}{\sqrt{mkT}} \partial_w \right\} \rho + \varepsilon^{-2} \partial_w (w + \partial_w) \rho, \qquad (7.39)$$

where $\varepsilon = (m/\gamma)$. In the limit $\varepsilon \to 0$, we would like to find an effective Fokker-Planck equation for the reduced density

$$r(x, t | x_0, t_0) = \int_{-\infty}^{\infty} dw \int_{-\infty}^{\infty} dw_0 \rho(x, w, s | x_0, w_0, s_0) \frac{1}{\sqrt{2\pi}} e^{-\frac{1}{2} w_0^2}, \qquad (7.40)$$

where we take the marginal distribution of the (fast) velocity process to be the equilibrium Boltzmann distribution, $(2\pi)^{-1/2} \exp\{-w^2/2\}$.

As before, we expand the joint density in a power series

$$\rho = \rho_0 + \varepsilon\rho_1 + \varepsilon^2\rho_2 + \ldots \tag{7.41}$$

and collect terms order by order in powers of ε:

$$
\begin{aligned}
O(\varepsilon^{-2}): \quad & 0 = \partial_w(w + \partial_w)\rho_0 \\
O(\varepsilon^{-1}): \quad & 0 = \left\{-\sqrt{\tfrac{kT}{m}}w\partial_x + \tfrac{U'(x)}{\sqrt{mkT}}\partial_w\right\}\rho_0 + \partial_w(w + \partial_w)\rho_1 \\
O(\varepsilon^0): \quad & \partial_s\rho_0 = \left\{-\sqrt{\tfrac{kT}{m}}w\partial_x + \tfrac{U'(x)}{\sqrt{mkT}}\partial_w\right\}\rho_1 + \partial_w(w + \partial_w)\rho_2,
\end{aligned}
\tag{7.42}
$$

and so on. The role of F_0 above is played by the closely related operator $\partial_w(w+\partial_w)$. Its spectrum is the same as F_0's, with slightly different eigenfunctions:

$$\partial_w(w + \partial_w)R_n(w) = -nR_n(w) \tag{7.43}$$

where the eigenfunctions are

$$R_n = He_n(w)\frac{1}{\sqrt{2\pi}}e^{-\frac{1}{2}w^2}, \tag{7.44}$$

where the Hermite polynomials $He_n(w)$ are defined by

$$He_n(w) = (-1)^n e^{\frac{1}{2}w^2}\frac{d^n}{dw^n}e^{-\frac{1}{2}w^2}. \tag{7.45}$$

The first few of these Hermite polynomials are

$$
\begin{aligned}
He_0(w) &= 1 \\
He_1(w) &= w \\
He_2(w) &= w^2 - 1.
\end{aligned}
\tag{7.46}
$$

and they, as well as the eigenfunctions R_n, satisfy the recursion relation

$$wHe_n = He_{n+1} + nHe_{n-1}. \tag{7.47}$$

The $O(\varepsilon^{-2})$ equation tells us that

$$\rho_0(x, w, s) = r_0(x, s)R_0(w), \tag{7.48}$$

where, as before, the function $r_0(x, s)$ is the leading order reduced density whose evolution equation we seek. Using the recursion relation in Eq. (7.47), along with the fact that $\partial_w R_0 = -R_1$, the $O(\varepsilon^{-1})$ equation becomes

$$
\begin{aligned}
\partial_w(w + \partial_w)\rho_1 &= \sqrt{\frac{kT}{m}}\partial_x r_0(x, s)w R_0(w) - \frac{U'(x)}{\sqrt{mkT}}r_0(x, s)\partial_w R_0(w) \\
&= \left[\sqrt{\frac{kT}{m}}\partial_x r_0(x, s) + \frac{U'(x)}{\sqrt{mkT}}r_0(x, s)\right]R_1(w).
\end{aligned}
\tag{7.49}
$$

In light of Eq. (7.43), we find the expression for ρ_1

$$
\rho_1 = -\left[\sqrt{\frac{kT}{m}}\partial_x r_0(x, s) + \frac{U'(x)}{\sqrt{mkT}}r_0(x, s)\right]R_1(w) + r_1(x, s)R_0(w),
\tag{7.50}
$$

where, again as before, $r_1(x, s)$ is undetermined at this stage. Now the $O(\varepsilon^0)$ equation is

$$
\begin{aligned}
\partial_w(w + \partial_w)\rho_2 &= \partial_s\rho_0 + \left\{\sqrt{\frac{kT}{m}}w\partial_x - \frac{U'(x)}{\sqrt{mkT}}\partial_w\right\}\rho_1 \\
&= \partial_s r_0(x, s)R_0(w) - \left\{\sqrt{\frac{kT}{m}}w\partial_x - \frac{U'(x)}{\sqrt{mkT}}\partial_w\right\} \\
&\quad \times \left[\sqrt{\frac{kT}{m}}\partial_x r_0(x, s) + \frac{U'(x)}{\sqrt{mkT}}r_0(x, s)\right]R_1(w) \\
&\quad + \left\{\sqrt{\frac{kT}{m}}w\partial_x - \frac{U'(x)}{\sqrt{mkT}}\partial_w\right\}r_1(x, s)R_0(w).
\end{aligned}
\tag{7.51}
$$

The idea is to isolate the coefficient of $R_0(w)$ on the right-hand side of Eq. (7.51); it must vanish as the integrability condition for us to solve for ρ_2. The recursion relation and the fact that $\partial_w R_1 = -R_2$ lead us to

$$
\partial_w(w + \partial_w)\rho_2 = R_0(w)\left\{\partial_s - \partial_x\frac{U'(x)}{m} - \frac{kT}{m}\partial_x^2\right\}r_0 \\
+ [\text{terms proportional to } R_1(w) \text{ and } R_2(w)],
\tag{7.52}
$$

so, in the limit $\varepsilon \to 0$, we have the reduced dynamics for $r(x, s)$,

$$
\partial_s r(x, s) = \partial_x\left\{\frac{U'(x)}{m} + \frac{kT}{m}\partial_x\right\}r.
\tag{7.53}
$$

This Fokker-Planck equation for the marginal density of the position process $X(t)$ alone, in the high friction limit, is called the *Smoluchowski equation*. It is the Fokker-Planck equation corresponding to the stochastic differential equation

$$\frac{dX(s)}{ds} = -\frac{U'(X)}{m} + \sqrt{\frac{2kT}{m}}\xi(s), \qquad (7.54)$$

or in terms of the original time variable $t = \gamma s/m$,

$$\frac{dX(t)}{dt} = -\frac{U'(X)}{\gamma} + \sqrt{\frac{2kT}{\gamma}}\xi(t). \qquad (7.55)$$

EXERCISE 6. Show that white noise must rescale as $\xi(t) \rightarrow \sqrt{\alpha}\xi(\alpha t)$ under the change of variables $t \rightarrow \alpha t$.

8. APPLICATION: THE STOCHASTIC VERHULST EQUATION REVISITED

To illustrate the differences between the Itô and Stratonovich interpretations of white-noise stochastic differential equations, we will now reconsider the model of section 6 interpreting the stochastic dynamics as the white-noise limit of a colored-noise process. In the Stratonovich interpretation, the solution of the stochastic differential equation

$$\frac{dX}{dt} = \langle\mu\rangle X - X^2 + \sigma\xi \qquad (8.1)$$

has a transition density satisfying the Fokker-Planck equation

$$\frac{\partial\rho}{\partial t} = \frac{\partial}{\partial x}\left\{x^2 - \langle\mu\rangle x + \frac{\sigma^2}{2}x\frac{\partial}{\partial x}x\right\}\rho$$
$$= \frac{\partial}{\partial x}\left\{x^2 - \langle\mu\rangle x - \frac{\sigma^2}{2}x + \frac{\sigma^2}{2}\frac{\partial}{\partial x}x^2\right\}\rho. \qquad (8.2)$$

This Fokker-Planck equation is different from that considered before, Eq. (5.7), but for this example the difference can be absorbed into a renormalization of the parameters as shown by the second line above. That is, our previous results for the stationary probability density all carry over from section 5 with the replacement $\langle\mu\rangle \rightarrow \langle\mu\rangle + \sigma^2/2$ in those formulæ.

The stationary probability density is thus

$$\rho_{\text{stat}}(x) = Nx^{(2\langle\mu\rangle/\sigma^2 - 1)}e^{-2x/\sigma^2}, \qquad (8.3)$$

so long as

$$\frac{\langle \mu \rangle}{\sigma^2} > 0. \tag{8.4}$$

This condition holds whenever the average birth rate exceeds the average death rate, for any finite value of the noise amplitude. The other solution,

$$\rho_{\text{stat}}(x) = \delta(x), \tag{8.4}$$

holds true for the initial condition $X(0) = 0$, or for average death rate exceeding average birth rate ($\langle \mu \rangle \leq 0$). The nontrivial density in Eq. (8.4) displays the noise-induced transition from the state where the most probable value is nonzero, to the state where it is zero, at the critical noise amplitude $\sigma^2 = 2\langle \mu \rangle$.

Contrary to Itô interpretation, in the Stratonovich interpretation there is never a "noise-induced extinction" phenomenon where the average growth rate is positive, but very strong fluctuations may drive the population to zero. These distinctions represent very different qualitative behaviors coming from the two interpretations of the model, highlighting the crucial role that the details of a white-noise model play in its predictions.

9. SUMMARY

We have introduced some of the fundamentals of stochastic processes, focusing on Markov diffusion processes as the solutions of stochastic differential equations driven by gaussian white noise. The three central achievements of these lectures has been (1) to show how one may consistently and practically interpret differential equations with white-noise coefficients, (2) to illustrate by example that some interesting problems can be completely and exactly solved within this formalism, and (3) to show that the question of modeling is a crucial one for these systems. In particular, the example illustrated explicitly that the interplay of noise and nonlinear dynamics is not a trivial one. Even using the idealization of white noise with its infinitely fast fluctuation time scale, the effects of the variations may not just "average out," but can deeply modify the system's qualitative behavior.

ACKNOWLEDGMENTS

I thank Benjamin Luce and Paul Laub for careful and thoughtful readings of the manuscript. I also thank all the students of the 1990 Complex Systems Summer school for their interest, their questions, and for the warm yet challenging reception that these lectures incited. Preparation of the lectures and these lecture notes was supported by NSF grants PHY-8958506 (Presidential Young Investigator Award) and PHY-8907755. The numerical simulations and computations were performed on the MATLAB software system, supplied by The Math Works, Inc.

REFERENCES

1. Arnold, L. *Stochastic Differential Equations.* New York: Wiley, 1974.
2. Doering, C., H. Brand, and R. Ecke, eds. "External Noise and Its Interaction with Spatial Degrees of Freedom." *J. Stat. Phys.* **54(5/6)** (1989).
3. Horsthemke, W., and R. Lefever. *Noise Induced Transitions.* New York: Springer, 1984.
4. Moss, F., and P. McClintock. *Noise in Nonlinear Dynamical Systems,* vols. 1–3. New York: Cambridge, 1989.
5. Risken, H. *The Fokker-Planck Equation,* 2nd ed. New York: Springer, 1989.
6. van Kampen, N. *Stochastic Processes in Physics and Chemistry.* Amsterdam: North Holland, 1981.

Bruce Bayly
Department of Mathematics, University of Arizona, Tucson, Arizona 85721

Complexity in Fluids

PHYSICS OF FLUIDS AND FLOWS

Fluids are basically very simple things. The fluids we encounter all the time—air, water, milk, coffee, etc.—are undramatic. One blob of a given fluid looks much like any other, except for such gross properties as volume or mass. Of course, blobs of fluid come in different shapes. However, it's easy to change the shape of a blob of fluid, with the result that we rarely think of the shape of a fluid blob as a defining property. In fact, a blob that starts with one shape can be deformed into almost any other shape, with arbitrarily small input of energy.

When we talk about lumps of a solid, in contrast, shape is important. This is because it takes work, i.e., energy, to change the shape of a solid. Making a small deformation from some rest configuration takes a small amount of energy, and a large deformation takes a lot of energy. Sometimes, as in idealized elastic systems, the required energy goes to infinity as the deformation becomes unbounded. Real solids usually break if you deform them enough; all subsequent deformations cost no energy. Basically, a finite deformation requires finite energy.

Complexity arises in fluid systems *because* the shape of a blob of fluid is indeterminate. Nothing prevents an initially simple fluid blob from deforming into the weirdest shape imaginable. It is the absence of any kind of blob-shape constraint that allows complexity to enter fluid science.

In view of the absence of shape constraints on blobs, it's surprising that we can say anything at all about fluid motion. But all blobs are subject to Newton's laws of motion and the laws of thermodynamics. The fact that *all* blobs obey these laws actually makes it possible to derive equations that do not make explicit reference to blobs. These partial differential equations are the Navier-Stokes equations, some of the nastiest equations ever to appear in the physical sciences. However, they do provide a starting point for quantitative understanding of fluid motion.

A major redeeming feature of the Navier-Stokes equations is that they apply to virtually all fluids. The same conservation laws apply to flows in coffee cups, oceans, interstellar molecular clouds, and blood in our bodies. Fluid equations have even been used to model the dynamics of large atomic nuclei or to investigate unusual modes of radioactive decay. Insight into any one fluid problem often yields better understanding of problems in quite different fields.

During these lectures I shall describe briefly a few areas in which complexity arises and has to be dealt with. These lectures will be roughly divided as follows:

1. Physical and mathematical description of fluids and flows.
2. Flow transport and ergodic theory.
3. Magnetic dynamos and related problems.
4. Flow instabilities.
5. Turbulence.

I am delighted to learn that I am lecturing to an audience whose backgrounds cover almost the entire range of academic disciplines. I will therefore try, and I know I will be unsuccessful, to avoid presenting complicated formulas. This is one area of complexity we can do without! When I present a mysterious array of symbols, it will be nothing more than a precise statement of something I have already said in words.

All the topics I will discuss are areas of vigorous current research. I am unfortunately far from able to give a satisfactory bibliography. Instead, I shall refer to those books which give good introductions to the topics I talk about. You will then have a springboard from which to vault into serious literature searches on particular things that interest you. Apart from these books, journals whose titles contain the word "Fluid" always contain fascinating papers. For example, *Annual Reviews of Fluid Mechanics*, *The Journal of Fluid Mechanics*, and *The Physics of Fluids* are full of good stuff.

STATICS

Fluid mechanics simplifies when the fluid is not moving. But this is not as trivial as it sounds.

Since a fluid deforms under any arbitrarily small force, all forces must be in perfect balance for a motionless equilibrium. Such a balance can only occur when the whole body of fluid takes a very special shape. A typical hydrostatics problem usually involves determining the shape of a fluid body for which all forces balance. This requires solving a set of simultaneous nonlinear equations in an uncountably infinite set of variables! Even hydrostatics involves complexity.

A hydrostatic equilibrium is often a configuration that minimizes some potential energy function. In such a case, the solution can be found by techniques from the calculus of variations. For example, consider a dish of water rotating about a vertical axis (Figure 1).

If the dish rotates with constant angular velocity Ω for a long time, the water will gradually pick up the rotation of the dish until there is negligible relative motion between the water and its container. From the point of view of an observer on the dish, the potential energy of the configuration shown in Figure 1 is

$$\Phi[h] = \int_D dx dy \int_0^{h(x,y)} dz \left[gz - \frac{1}{2}\Omega^2(x^2 + y^2) \right]. \tag{1.1}$$

It would be a simple matter to minimize $\Phi[h]$ over all functions $h(x, y)$, but that would be *wrong*.

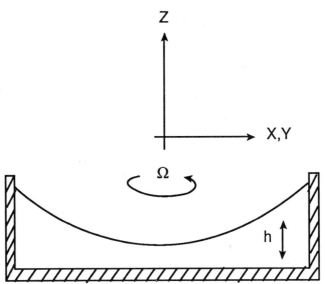

FIGURE 1 Rotating dish of liquid.

We have left out the amount of water in the dish, which is an important part of the problem. For the configuration above, the volume of water is

$$V[h] = \int_D dx\,dy \int_0^{h(x,y)} dz\,.$$

We must minimize $\Phi[h]$ over only those functions $h(x,y)$ for which $V[h]$ equals the given volume V_{given}. The details of the solution are left for the student, but I'll reveal here that the minimizing configuration is a paraboloid of revolution.

Extremely accurate paraboloids may be obtained by spinning dishes of fluid; all that is required is a steady motor and some care taken to ensure that the rotation axis really is vertical. A rigid paraboloid may be made by spinning a dish of hot water with a layer of melted wax on top. As the water cools, the wax solidifies, and the paraboloid can then be lifted out and used in demonstrations on solar energy or acoustic focusing.

Surface tension affects the shape of any fluid system with a free boundary. Surface tension comes from the fact that the potential energy includes interaction energy from every interface between two materials in the system. The energy contained in each interface is proportional to the area of the interface. The constant of proportionality, which is a property of the materials meeting at the interface, is the surface tension.

In Figure 1, there is surface energy from the air-water interface, the air-dish interface, and the water-dish interface. If the bulk energy of the rotating fluid is much greater than the interface energies, the effects of surface tension will be confined to the neighborhood of the points where the air, water, and dish meet. Very near these points, surface forces dominate all others and determine things like the angle between the air-water interface and the solid dish boundary.

DYNAMICS

The way to tell if you're dealing with hydrostatics or hydrodynamics is to look at the fluid and see if it's moving. But that could be hard—you can't tell if it's windy outside unless you see trees bending or dust blowing. You have to make something in the fluid visible, maybe put some smoke in the air, or dye in the water. You note its position when you introduce it, look away for some time interval, and see whether its position has changed. If a particle of added substance is dynamically neutral, it moves just as if it were a particle of working fluid. The additive marks the fluid so its motion can be observed.

To describe this mathematically, we introduce a vector $\mathbf{a}$ denoting the initial position (at time t_0) of a marked fluid particle. We denote its later position at time t by the vector $\mathbf{X}(\mathbf{a}, t)$. The vectors $\mathbf{a}$ and $\mathbf{X}$ can be specified by three numbers in any reasonable coordinate system. Unless otherwise specified, we shall use the Cartesian coordinates (a_1, a_2, a_3) for $\mathbf{a}$, (X_1, X_2, X_3) for $\mathbf{X}$, and so on.

The vector **a** is sometimes referred to as the Lagrangian label of the particle. If we knew $\mathbf{X}(\mathbf{a}, t)$ for all **a** and all $t > t_0$, few would argue with our claim that we understand the flow completely.

The transformation from **a** to $\mathbf{X}(\mathbf{a}, t)$ takes particle locations to particle locations, and is called the *flow map*. The flow map makes sense for t values less than t_0 also; in this case $\mathbf{X}(\mathbf{a}, t)$ means the position at time t of the particle that will subsequently be at **a** at time t_0. The flow map is invertible: the inverse is the flow map obtained by switching t with t_0, and **X** with **a**.

The so-called Lagrangian velocity vector of the marked particle is

$$\mathbf{U}(\mathbf{a}, t) = (\partial \mathbf{X} / \partial t)_{\mathbf{a}} \, . \tag{1.2}$$

The subscript indicates that the label **a** stays fixed while $\partial / \partial t$ is taken. If we define the Eulerian velocity

$$\mathbf{u}(\mathbf{X}(\mathbf{a}, t), t) \equiv \mathbf{U}(\mathbf{a}, t) \, , \tag{1.3}$$

then $\mathbf{u}(\mathbf{x}, t)$ can be interpreted as the velocity vector measured at time t by an observer fixed in space at the point **x** (Figure 2.). The Eulerian velocity as a function of position is often just called the velocity field.

The Lagrangian velocity is what you see when you watch a smoke particle moving in an airstream; the Eulerian velocity is proportional to the rotation rate

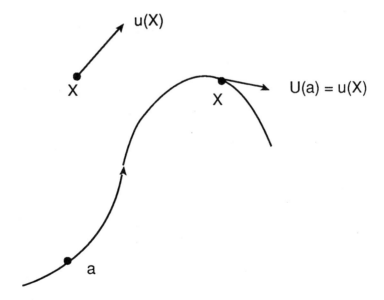

FIGURE 2 Flow, flow map, and velocity vectors.

of a pinwheel fixed at position $\mathbf{x}$ in the laboratory frame. Equation (1.2) shows how the velocity field can be obtained if we know the flow map. Conversely, if the velocity field $\mathbf{u}(\mathbf{x},t)$ is completely known, the flow map can be recovered by solving

$$\left(\frac{\partial \mathbf{X}}{\partial t}\right)_{\mathbf{a}} = \mathbf{u}(\mathbf{X}(\mathbf{a},t),t) \quad , \quad \mathbf{X}(\mathbf{a},t=t_0) = \mathbf{a} \tag{1.4}$$

as an ordinary differential equation initial value problem. Thus, full knowledge of the flow map is conceptually equivalent to full knowledge of the velocity field.

The equations of motion for the fluid, although tough to solve, are simple in appearance. The most basic equation is Newton's law $F = ma$ applied to arbitrarily small blobs of fluid. In order to obtain a closed set of equations, however, we also need a mass-conservation equation and two thermodynamic equations. One of the thermodynamic equations is always the equation of state for the material, but there are various choices for the other. We shall give the Second Law of Thermodynamics applied to arbitrarily small fluid blobs. One could also use an equation reflecting the conservation of total energy (the First Law) or an equation simply describing the temperature field (the Zeroth Law).

There is always a force associated with pressure differences from point to point in the fluid, and there is always some kind of viscous force. Both the pressure and viscosity are ultimately due to the microscopic interactions among the molecules making up the fluid. Pressure comes from the strong dislike the molecules have of being squeezed together tighter than they want to be, while viscosity arises from their comparatively mild desire to be at rest with respect to each other. It frequently turns out that the viscous forces can be neglected in comparison with other forces acting on the fluid, but the pressure forces are always important.

In order to present the equations, we need some notation. Let $\rho(\mathbf{x},t)$ be the fluid density, $p(\mathbf{x},t)$ the pressure, $T(\mathbf{x},t)$ the temperature, and $S(\mathbf{x},t)$ the entropy per unit mass, all measured at the point $\mathbf{x}$ at time t. We need to be able to differentiate the flow map, so we introduce the matrix $\mathbf{J} = \partial \mathbf{X}/\partial \mathbf{a}$ whose components are

$$J_{ij} = \left(\frac{\partial X_i}{\partial a_j}\right)_t \quad , \quad i,j = 1,2,3 . \tag{1.5}$$

The matrix $\mathbf{J}$ is a direct measure of the deformation of the fluid surrounding the moving point $\mathbf{X}(\mathbf{a},t)$.

Mass conservation simply says that if the volume of a fluid blob decreases, the density increases. In symbols,

$$\rho(\mathbf{X}(\mathbf{a},t),t) = \frac{\rho(\mathbf{a},t_0)}{det(\mathbf{J})} . \tag{1.6}$$

Momentum conservation, i.e., $F = ma$, becomes

$$\rho(\mathbf{X}(\mathbf{a},t))\left(\frac{\partial^2 \mathbf{X}(\mathbf{a},t)}{\partial t^2}\right)_{\mathbf{a}} = -\nabla p + \nabla \cdot [\mu \mathbf{\Sigma}(\mathbf{X}(\mathbf{a},t),t)] + \mathbf{f}(\mathbf{X}(\mathbf{a},t),t) , \tag{1.7}$$

and the Second Law of Thermodynamics is

$$\rho T \left(\frac{\partial S(\mathbf{X}(\mathbf{a},t),t))}{\partial t} \right)_{\mathbf{a}} = \nabla \cdot (k \nabla T) + \rho Q(\mathbf{X}(\mathbf{a},t),t) \,. \tag{1.8}$$

The quantity Σ in (1.7) is the trace-free part of the fluid strain-rate tensor, and μ is the coefficient of viscosity. The term $\mathbf{f}$ represents any external forces acting on the fluid, such as the forces that make the fluid move in the first place. These can be mechanical, gravitational, or electromagnetic, for example.

The parameter k is the coefficient of heat conduction in the fluid, and Q represents irreversible heat generation in the fluid. Viscous dissipation of mechanical energy always generates heat irreversibly. Chemical reactions and electric currents also create entropy.

The momentum equation (1.7) is hard to work with, because the stress tensor and pressure gradient involve derivatives taken with respect to the current positions of the particles. Factors of $\mathbf{J}$ enter when the $\mathbf{X}$ derivatives are transformed into a derivatives. Not only does this complicate the appearance of the equations, but the equations are likely to develop problems if $\mathbf{J}$ grows unboundedly. And $\mathbf{J}$ is almost sure to grow unboundedly, since it measures the total fluid deformation since time t_0.

The equations simplify if we rewrite them from the point of view of observers fixed in space rather than moving with the fluid. The only complication is that the time-derivative $(\partial/\partial t)_{\mathbf{a}}$ turns into $(\partial/\partial t)_{\mathbf{x}} + (\mathbf{u}(\mathbf{x},t) \cdot \nabla)$. With this modification, and using the shorthand $\partial_t \equiv (\partial/\partial t)_{\mathbf{x}}$, we get mass conservation:

$$(\partial_t + \mathbf{u} \cdot \nabla)\rho = -(\nabla \cdot \mathbf{u})\rho \tag{1.9}$$

momentum conservation:

$$\rho(\partial_t + \mathbf{u} \cdot \nabla)\mathbf{u} = -\nabla p + \nabla \cdot [\mu \Sigma] + \mathbf{f} \tag{1.10}$$

and entropy generation:

$$\rho T(\partial_t + \mathbf{u} \cdot \nabla)S = \nabla \cdot (k \nabla T) + \rho Q \,. \tag{1.11}$$

Coupled with state equations that determine p, T, μ, and k as functions of ρ and S, these equations form a closed set. These are the *Navier-Stokes equations*.

Equations (1.9)–(1.11) are incomplete without a description of the stirring forces in the momentum equation and the dissipative processes in the entropy equation. Sometimes additional equations are required to describe the extra physics.

For example, if there is salt dissolved in the fluid with a space-time dependent concentration $C(\mathbf{x},t)$, then we may expect an added gravitational force term $C\mathbf{g}$ in the momentum equation. The salt will most likely just be carried around with the fluid, perhaps diffusing a little, thus satisfying

$$(\partial_t + \mathbf{u} \cdot \nabla)C = -C(\nabla \cdot \mathbf{u}) + \nabla \cdot (\kappa \nabla C) \,. \tag{1.12}$$

The Navier-Stokes equations augmented by the buoyancy force and the equation for C are used in studying the dynamics of the ocean.

An electrically conducting fluid can have currents $\mathbf{j}(\mathbf{x}, t)$ and magnetic fields $\mathbf{B}(\mathbf{x}, t)$ taking part in the dynamics. The new equations here are Maxwell's equations applied to a moving conductor:

$$\partial_t \mathbf{B} = \nabla \times (\mathbf{u} \times \mathbf{B} + \mathbf{j}/\sigma) \tag{1.13}$$

where $\chi \mathbf{j} = \nabla \times \mathbf{B}$, χ is the magnetic permeability of the fluid, and σ is the resistivity. The added force on the fluid is the electromagnetic Lorentz force $\mathbf{j} \times \mathbf{B}$.

The texts of Landau and Lifshitz,[7] Batchelor,[3] and Yih[17] discuss the fundamental physics of fluids from different points of view. Some aspects of the derivation of the Navier-Stokes equations are quite subtle, and experienced fluid researchers as well as students will benefit from going over the introductory chapters in these books. Pedlosky[13] and Parker[12] show how whole disciplines are built on fluid dynamics with special added physics.

FLUID MIXING AND RELATED TOPICS

Let's start with a non-fluid example. Consider a roving researcher (R) for a consumer group. R's job is to go to town x, write down the price $E(x)$ of washing machines, and then move on. R has no say about where s/he goes next; s/he just gets the name from the local office of the group and goes wherever s/he is told.

In this example, the staff of the local offices are fantastically unimaginative; they always give the same instructions. Therefore, we can denote by $T(x)$ the town that the office in x sends people to. T may be viewed as a function from the set of towns into itself.

Despite the lack of imagination of the officials, complicated trips are possible. A travel agent might like to know what functions T give the most scenic routes. The consumer group is likely to be interested in the simpler questions of whether R visits all the available towns or just some of them, and whether R can come up with a sensible average price for washing machines.

The answer to the first question depends on what "scenic" means to the travel agent. In this Summer School, scenic means complex. The appropriate definition of complexity is currently being figured out by the participants of the School. I don't know what it will be, but it should be available by July.

The questions that the consumer group is interested in are easier to answer, especially if the total number of towns is finite. The basic observation is that, if there is a finite number of towns, R must eventually hit a town s/he has been to before. Since the instructions in the local offices are assumed never to change, R has no choice but to retrace his or her steps forever afterward! Hence, whether or not R hits every town can be determined in finite time. As for the last question, if R just keeps a running average of washing machine prices, the running average *must* converge to $E(x)$ averaged over the finite number of towns in R's cycle.

RECURRENCE AND AVERAGES

Ergodic theory is devoted to answering questions like those above when the set of "towns" is infinite. Since an infinite town-set is not such a useful concept, we shall just let x denote a point in some infinite "domain" set D, and T will be some map from D into D. Obviously, we can't make useful statements about completely arbitrary maps on arbitrary domains. However, if we assume there is a measure defined on D, such that D has finite total measure, that T is invertible, and that T preserves the measure of subsets of D, then we can prove essentially the same kind of results for the infinite system as for the finite model system.

As far as these lectures are concerned, we shall be concerned with maps that take the location of a fluid particle at one moment in time, and give you its location some time interval τ later (Figure 3). If the flow is incompressible, this means that the volume of any blob of fluid does not change as it moves around. Thus, if we let the measure of a fluid region simply be its volume, then the flow map is a perfectly good measure-preserving transformation to which all the results of ergodic theory can be directly applied.

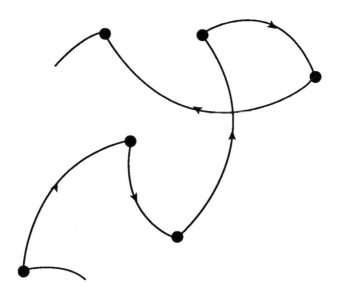

FIGURE 3 Particle positions after time intervals τ, 2τ, 3τ, etc.

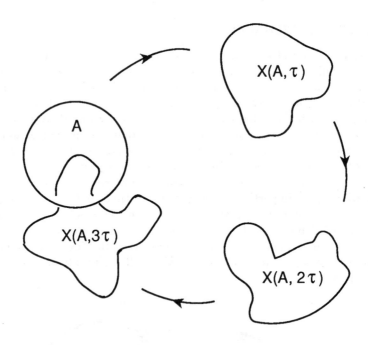

FIGURE 4   Initial blob A and subsequent images.

The flow should be periodic in time with period τ, so that a particle starting at x always ends up at the same point $T(x)$ after time τ. An example of such a flow is blood flow in your circulatory system, with τ being the interval between heartbeats. It is all too well known that the circulation is an overwhelmingly complex system. However, we can still observe empirically that blood flow appears to be both ergodic and mixing (see below). The evidence in favor of this view is that we are all alive.

Let us now take a blob A of fluid, of positive volume, and see where it goes (Figure 4). As drawn here, $\mathbf{X}(A, \tau)$ and $\mathbf{X}(A, 2\tau)$ just happen to be disjoint from A and each other, but $\mathbf{X}(A, 3\tau)$ has some points in common with A. A little thought convinces you that the sets

$$A, \mathbf{X}(A, \tau), \mathbf{X}(A, 2\tau), \cdots, \mathbf{X}(A, N\tau) \qquad (2.1)$$

have to intersect each other if you take N large enough. For if they did not, the total volume occupied by these sets would be $N \times [Volume\ of\ A]$, which would eventually exceed the finite volume of the flow domain D.

In fact, as $N \to \infty$, these sets have to keep intersecting each other, over and over, infinitely often. This, roughly, is the *Poincaré Recurrence Theorem* (PRT). As described here, it's obvious. The PRT is less obvious, but true for exactly the same reasons, when applied to the phase flow of a Hamiltonian dynamical system. When applied to a room of interacting gas molecules, the PRT says that if you start with all the molecules in one corner of the room, there will be an infinite sequence of times when all the molecules simultaneously return to their starting corner. Think about it, but don't worry about it.

The PRT says that almost all fluid particle trajectories mimic the roving researcher R in that they keep visiting the same regions of fluid over and over. Hence the long-time average

$$\bar{E}(\mathbf{a}) = \lim_{t\to\infty} \frac{1}{t} \int_0^t E(\mathbf{X}(\mathbf{a},t'))dt' \qquad (2.2)$$

ends up just averaging the function $E(\mathbf{x})$ over the "region" filled by the trajectory $\mathbf{X}(\mathbf{a},t)$. This "region" can be a single point, if $\mathbf{X}(\mathbf{a},t) \equiv \mathbf{a}$ for all t, or a closed curve, or a two-dimensional surface, a three-dimensional region of finite volume, or something even more weird.

The *Birkhoff Ergodic Theorem* (BET) says that the limit (2.2) actually exists for almost all points $\mathbf{a}$ in the flow domain D, provided only that the function E is integrable. The geometry of the region filled by $\mathbf{X}(\mathbf{a},t)$ does not enter (explicitly) into the result. If the region filled by $\mathbf{X}(\mathbf{a},t)$ is too bizarre, the long-time average might not exist. But the BET assures us that such badly behaved trajectories constitute a set of measure zero. This usually means we can forget about them.

ERGODICITY

It's obvious what we mean when we ask whether the roving researcher R eventually visits every member of a finite set of towns. Given a flow on a domain containing an uncountably infinite number of points, it's nontrivial to formulate the same question. The following does so precisely.

We say that a set K of points in the fluid is *invariant* under the flow if $\mathbf{X}(K,n\tau) \subset K$ for all positive and negative integers n. Negative n means the locations at earlier times of the particles that are in K at $t = 0$. Plenty of sets are invariant. The domain D had better be invariant, and the null set is always vacuously invariant.

If a particle returns to its starting point after time $p\tau$ (and not before), for some integer p, then the p discrete points on its orbit constitute an invariant set. Invariant sets may also take the form of curves, surfaces, three-dimensional regions, or even stranger sets. Since the union of invariant sets is also an invariant set, invariant sets can get pretty complicated.

A flow is said to be *ergodic* if each invariant set either has the same volume as D, or zero volume. This is equivalent to saying that almost all flow trajectories $\mathbf{X}(\mathbf{a},t)$ pass arbitrarily close to all points of D. Points, curves, and other zero-volume objects are still allowed as invariant sets in an ergodic flow. As long as they do not interfere with other trajectories, such invariant sets do not prevent a flow from being ergodic.

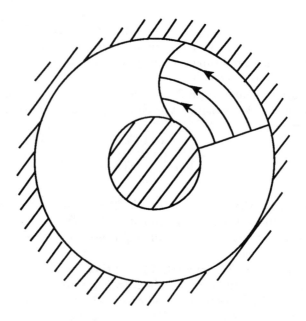

FIGURE 5   The circular vortex.

If we suspect that a given is ergodic, we would like to be able to demonstrate this mathematically. Unfortunately, there is no simple algorithm into which we can plug the flow and get the answer. There are a few tricks which can settle the question for flows with particular properties, but in general one has to examine each flow individually. Unless the flow is special in some way, it is usually impossible to determine whether a given flow is ergodic. In the following examples, the ergodicity or non-ergodicity of the flow can be determined by simple arguments.

Circular Vortex Flow: Consider the two-dimensional flow with circular symmetry, in which particles at distance r from the origin are rotating around the origin with angular velocity $\Omega(r)$ (Figure 5). D is the annular region between the circles $r = R_1$ and $r = R_2$. This is a trivial example of a non-ergodic flow, since the sets $R_1 \leq r < (R_1 + R_2)/2$ and $(R_1 + R_2)/2 < r \leq R_2$ are disjoint invariant sets, both with nonzero area.

Unidirectional Flow on a 2-Torus: The flow domain D in Figure 6 is the square $\{0 \leq x \leq 2\pi, 0 \leq y \leq 2\pi\}$ in the $x - y$ plane. The fluid particles all move with the same (constant) velocity (α, β). When a particle leaves through an edge of the square, we immediately reintroduce it at the corresponding point on the opposite edge of the square. This reintroduction rule essentially turns the square into a doughnut (or torus).

To describe the flow map associated with this velocity field, let (a, b) be the Cartesian components of the vector **a** and (X, Y) be the components of **X**. Then the flow map is given by

$$X(\mathbf{a}, t) = a + \alpha t \; (mod \, 2\pi), \qquad Y(\mathbf{a}, t) = b + \beta t \; (mod \, 2\pi) \qquad (2.3)$$

Figure 7 shows a typical orbit.

A little thought reveals that if α and β are rationally related, the orbit eventually closes on itself and spends the rest of eternity retracing the same curve (Figure 8a). The spacing between branches of this curve is $2\pi\sqrt{1/m^2 + 1/n^2}$, where m, n are the smallest positive integers for which $m\alpha = n\beta$. Clearly, for any trajectory there are points in the square which are not approached by the trajectory. Hence, the flow is non-ergodic.

If α/β is irrational, no particle trajectory closes, and every trajectory fills the square densely. The flow is ergodic in this case (Figure 8b).

Ergodicity or non-ergodicity has been easy to show in the foregoing examples. It is also straightforward to show that for each of these examples, the long-time average of any function $E(\mathbf{x})$ along every trajectory converges. An interesting problem is to estimate how fast the averages converge. Subtle aspects of rational approximation of irrationals emerge!

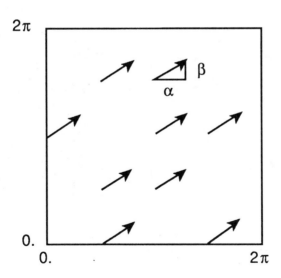

FIGURE 6   Unidirectional flow on the torus.

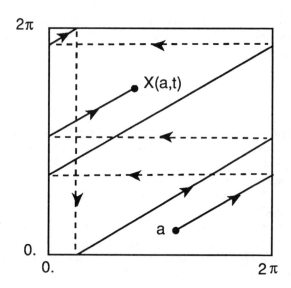

FIGURE 7   Trajectory in unidirectional flow on torus.

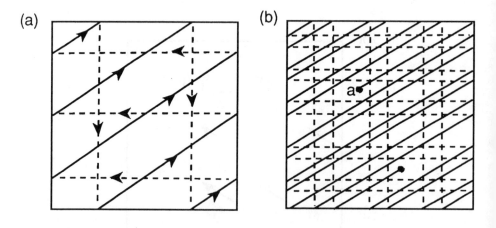

FIGURE 8   (a) Periodic nonergodic trajectory when α and β are rationally related, and (b) Nonperiodic ergodic trajectory when α and β are in irrational ratio.

Pulsed Sinusoid Flows: These examples are tough to analyze. As far as I know, no one has been able to prove that the ergodic-looking ones are ergodic, nor that the non-ergodic-looking ones are non-ergodic. Nonetheless, these flows are easy to investigate numerically and they illustrate all the kinematic effects described in these lectures.

These flows are defined on the same domain as the unidirectional torus flows, with the same particle reintroduction rule. Again we define a velocity field, but now there's a difference. For a time-interval of length $\tau/2$, we let the fluid velocity at (x, y) be $(0, 2\cos(x + \phi))$. Then, for a time-interval of length $\tau/2$ we let the fluid velocity at (x, y) be $(-2\sin(y + \phi), 0)$. This cycle, shown in Figure 9, repeats over and over.

The phase shift ϕ is chosen to be $(1 + \sqrt{5})/2$. Otherwise, the exact center of the square would be an exact fixed point of the flow map. This coincidence causes interesting but irrelevant numerical problems in computations without the phase shift.

The flow map corresponding to one cycle of flow pulses is easy to calculate. Once we have the flow map, it is straightforward to take a point in D and plot its location at the end of every cycle of flow pulses, for a large number of pulses.

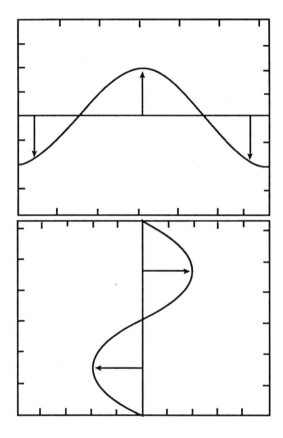

FIGURE 9   Pulsed sinusoid flow.

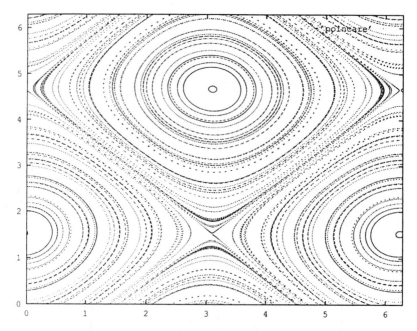

FIGURE 10   Streamlines of exactly solvable pulsed sinusoid flow in limit $\tau \to 0$.

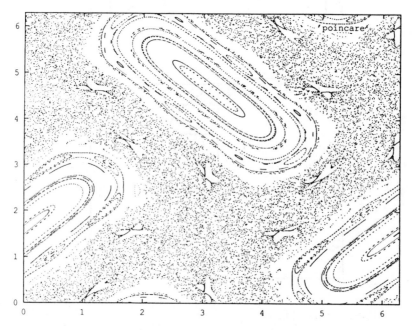

FIGURE 11   Poincaré section of pulsed sinusoid flow when $\tau = 2$.

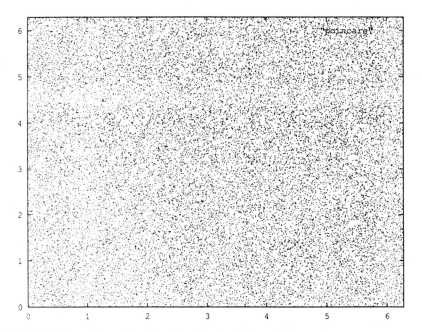

FIGURE 12   Poincaré section of pulsed sinusoid flow with $\tau = 5$.

In other words, instead of plotting the continuous trajectory, we plot a discrete sequence of points on the trajectory. We get the same information, but plotting points rather than curves gives a much less cluttered picture. We can actually learn things about the system by looking at it. This type of picture is called a *Poincaré section* of the flow.

The pictures that follow are each superpositions of 144 Poincaré sections. That is, 144 separate trajectories were calculated, each beginning on a different corner of a 12×12 lattice in the square. The resulting superposition of Poincaré sections gives a good global representation of the flow and its properties.

The pictures that follow show what happens with various values of τ. Unfortunately, the scale has been altered by the graphics device, so that lengths have been squashed in the y direction. The topology of the orbits is unchanged.

In the limit $\tau \to 0$ (Figure 10), the flow approaches the steady time-independent velocity field $\mathbf{u}(\mathbf{x}) = (-\sin(x + \phi), \cos(x + \phi))$. This flow is "exactly integrable" in the sense that the stream function $\psi = \cos(X + \phi) + sin(Y + \phi)$ is constant on every particle trajectory, and every trajectory can be solved explicitly in terms of elliptic functions.

$\tau = 2$. (Figure 11) most of the Poincaré sections still fill curves, but a substantial fraction fill a large two-dimensional area. Note the appearance of "necklaces" of small closed curves amongst the larger curves. These necklaces were present in the previous figure, but were too small to be detected by the relatively coarse survey of only 144 trajectories.

When $\tau = 5$. (Figure 12), all 144 trajectories fill a large two-dimensional region. The region filled appears to be the whole square, though of course we cannot be certain on the basis of a mere 144 trajectories of finite length. The available evidence is that, for this values of τ, the flow is truly ergodic.

MIXING

Ergodicity if a powerful concept, even though proving that a particular flow is ergodic can be difficult. Our gut feeling is that plenty of flows in the real world are ergodic. After all, if there is nothing *preventing* a fluid particle from exploring the whole domain D, we expect it to do so. If we put milk in our coffee, it takes just a few stirs with a spoon to generate a flow in which every particle explores essentially the whole cup.

Mixing is an even more powerful concept. When you stir your coffee-and-milk, the milk region does not just travel all around the inside of the cup; it also contorts and deforms into a much more complex shape than it had originally. The end result of all the distortions is a milk region that interpenetrates the coffee region, which is also highly distorted, on such a fine scale that we only see a uniform mixture. The flow generated by the spoon is said to be *mixing*.

The mathematical definition of a mixing flow is as follows. Let A and B be subregions of the flow domain D with positive volume. That is, A and B cannot just be lines or planes or some other zero-volume weirdness; they have to be real blobs. $\mathbf{X}(B,t)$ is the set of locations at time t of the points that started at B at time zero, and $\mathbf{X}(B,t) \cap A$ is the set of points that start at B at time zero and are in A at time t. We say that the flow is *mixing* if

$$Volume[\mathbf{X}(B,t) \cap A] \sim Volume[A]Volume[B]/Volume[D] \qquad (2.4)$$

as $t \to \infty$. This means that as t increases, the flow smears B uniformly over D, and the amount that ends up in A should be determined solely by the volume of A and the average density $Volume[B]/Volume[D]$ of points starting in B.

It is simple, although thought-provoking, to show that all mixing flows have to be ergodic. However, an ergodic flow need not be mixing. An example of such a flow is the unidirectional torus flow described earlier. The pulsed-sinusoid flows with $\tau \leq 3$ are almost certainly not mixing, because they are non-ergodic. The pulsed sinusoid flow with $\tau = 5$. seems to be mixing as well as ergodic, but this has not been proven.

The circular vortex flow is of course not mixing, since it is not ergodic. It's instructive to think about mixing here anyway, and we shall use a physical model. Suppose $T^{initial}(r, \theta)$ is the initial temperature field in the fluid, and suppose that its angular dependence is smooth enough that we can write a Fourier series

$$T^{initial}(r, \theta) = \sum_{n=-\infty}^{\infty} T_n(r)e^{in\theta} \qquad (2.5)$$

where

$$T_n(r) = \frac{1}{2\pi} \int_0^{2\pi} T^{initial}(r,\theta) e^{-in\theta} d\theta . \tag{2.6}$$

Assume that while the fluid is moving, we can ignore the effects of heat diffusion and possible sources of heat. The temperature thus remains constant on each moving fluid particle, and the temperature field at later time t is simply

$$T(r,\theta,t) = T^{initial}(r,\theta - t\Omega(r)) \tag{2.7}$$

Now, consider introducing the small region $A = \{r_1 < r < r_2, \theta_1 < \theta < \theta_2\}$. What is the average temperature over A? It is just

$$\bar{T}_A(t) = [\theta_2 - \theta_1]^{-1}[r_2^2 - r_1^2]^{-1} \int_A ddrd\theta \sum_{-\infty}^{\infty} T_n(r) e^{in(\theta - t\Omega(r))} . \tag{2.8}$$

Interchanging the summation and r-integral (which is not strictly justified, but you can check it afterwards) we get

$$\bar{T}_A(t) = [\theta_2 - \theta_1]^{-1}[r_2^2 - r_1^2]^{-1} \int_{\theta_1}^{\theta_2} d\theta \sum_{-\infty}^{\infty} e^{in\theta} \int_{r_1}^{r_2} e^{-int\Omega(r)} T_n(r) ddr . \tag{2.9}$$

Now, let's assume that $\Omega(r)$ is a smooth function with no intervals of constancy. This ensures that the fluid is nontrivially deformed at every point. The integrand of the r integral is then a very wiggly function for all $n \neq 0$. Indeed, it gets more and more wiggly as $t \to \infty$ for any nonzero n. By the Riemann-Lebesgue lemma, therefore, all the r-integrals vanish as $t \to \infty$ except the one with $n = 0$.

The conclusion of the above calculation is that the average temperature over the region A converges to a limit as $t \to \infty$, and the limit is

$$\bar{T}_A(\infty) = [r_2^2 - r_1^2]^{-1} \int_{r_1}^{r_2} (2\pi)^{-1} \int_0^{2\pi} T^{initial}(r,\theta) d\theta \, rdr . \tag{2.10}$$

This flow is semi-mixing in the sense that any initial distribution of temperature eventually gets smeared out and homogenized in the angular direction, while different temperatures on different radii never mix. This kind of mixing also appears to occur on the invariant curves in the pulsed sinusoid flows with $\tau = 0$ and $\tau = 2$.

The topics considered in this lecture are discussed in most textbooks on ergodic theory. Halmos[5] is exceptionally concise and well written. Arnold and Avez[2] is also well written, and describes ergodicity and mixing in precise but non-technical language. Ottino[11] describes applications of these concepts to technological applications and real fluids. His book is full of beautiful pictures.

ALGAE, MAGNETIC DYNAMOS, DYNAMICAL THERMODYNAMICS

Ergodic theory is an appealing branch of mathematics. The systems studied are often derived from real applications and the results are easy to interpret in terms of the applications that we started with.

An interesting example is a tank of algae that is stirred in such a way that we know the flow exactly. Perhaps we should imagine algae suspended in a very viscous fluid which only moves when it is forced to do so by the stirring agent. Suppose one side of the tank gets more light than the other, nutrients are injected somewhere, and romantic music piped into a certain part of the tank to encourage reproduction there.

These combined effects can be modeled by postulating a *reproductive speed function* $R(\mathbf{x})$ that roughly says how favorable the conditions at $\mathbf{x}$ are for population growth. In the absence of flow, the algae population density $C(\mathbf{x}, t)$ at the point $\mathbf{x}$ satisfies

$$\frac{\partial C}{\partial t} = R(\mathbf{x})C. \tag{3.1}$$

The reproductive speed R may be negative if there are regions in which the romantic music is replaced by government messages encouraging responsible family planning. The population of algae at the point $\mathbf{x}$ will grow or decay exponentially depending on the value of $R(\mathbf{x})$.

If there is a flow, a good way to see what happens is to follow a clump of algae as it moves through the flow, and track its population density. The population will increase when the clump is in a region of positive R, and decay when the clump is in a negative-R region.

Mathematically, if $C^{initial}(\mathbf{a})$ is the initial density at the point $\mathbf{a}$, then the subsequent evolution satisfies the Eulerian equation

$$(\partial_t + \mathbf{u} \cdot \nabla)C(\mathbf{x}, t) = R(\mathbf{x})C(\mathbf{x}, t) \;\;,\;\; C(\mathbf{x}, t_0) = C^{initial}(\mathbf{x}), \tag{3.2}$$

which is equivalent to the Lagrangian equation

$$\left(\frac{\partial C(\mathbf{X}(\mathbf{a}, t), t)}{\partial t} \right)_{\mathbf{a}} = R(\mathbf{X}(\mathbf{a}, t))C(\mathbf{X}(\mathbf{a}, t), t) \;\;,\;\; C(\mathbf{a}, t_0) = C^{initial}(\mathbf{a}) \tag{3.3}$$

where $\mathbf{X}(\mathbf{a}, t)$ is the flow map. Given the speed function R, different flows can result in wildly different algae distributions.

If we follow a particular clump, we can write down the solution to (3.3) as

$$C(\mathbf{X}(\mathbf{a}, t), t) = \exp\left(\int_{t_0}^{t} R(\mathbf{X}(\mathbf{a}, t'))dt' \right) C^{initial}(\mathbf{a}). \tag{3.4}$$

Note that this is not a complete solution unless we know the flow map *and its inverse* for all t. In general, the flow map gets complex as t gets large, with the result that Eq. (3.4) can represent extremely complicated algae distributions. Indeed, as $t \to \infty$, the algae distribution cannot even be described using ordinary functions.

However, not everything is complicated in (3.4). The integral in the exponential is just $(t - t_0)$ times the average value of R over the trajectory $\mathbf{X}(\mathbf{a}, t)$. The Birkhoff Ergodic Theorem (BET) tells us that the average of R over a trajectory converges as the length of the trajectory goes to infinity. Hence, very roughly,

$$C(\mathbf{X}(\mathbf{a}, t), t) = s(t - t_0) \cdot \exp\left(\bar{R}(\mathbf{a})(t - t_0)\right), \tag{3.5}$$

where $s(t)$ is a function with the property that $t^{-1} \log s(t) \to 0$ as $t \to \infty$. In words, the concentration C following the trajectory $\mathbf{X}(\mathbf{a}, t)$ decays or grows exponentially, with a decay- or growth-rate given by the trajectory average of R.

LYAPUNOV EXPONENTS

Using slightly different notation, we sometimes write

$$\lambda(\mathbf{a}) \equiv \bar{R}(\mathbf{a}) = \lim_{t \to \infty} \frac{1}{t} \log C(\mathbf{X}(\mathbf{a}, t)). \tag{3.6}$$

Here λ stands for Lyapunov; the quantity $\lambda(\mathbf{a})$ is called the Lyapunov exponent. Note that it depends on both the function $R(\mathbf{x})$ and the flow. There are many ways to define Lyapunov exponents, but all reduce to measuring the grow or decay rates of solutions to linear equations whose coefficients are functions of the position of a moving particle.

Lyapunov exponents can also be defined for systems of equations. Suppose $\mathbf{M}(\mathbf{x})$ is an $n \times n$ function of position in the flow domain (i.e., all elements are functions of position). Consider the system of coupled linear equations

$$\frac{d\mathbf{v}}{dt} = M(\mathbf{X}(\mathbf{a}, t))\mathbf{v}(t), \quad \mathbf{v}(t_0) \; given. \tag{3.7}$$

Because matrices and vectors do not behave like scalars, we cannot use the BET directly to conclude that $\mathbf{v}(t) \to \infty$ exponentially as $t \to \infty$. It is nonetheless true, and we can define the analogous Lyapunov exponent by

$$\lambda(\mathbf{a}) = \lim_{t \to \infty} \frac{1}{t} \log |\mathbf{v}(t)|. \tag{3.8}$$

There typically exist n distinct Lyapunov exponents when the matrix function $\mathbf{M}(\mathbf{x})$ is $n \times n$. These Lyapunov exponents play the role of generalized eigenvalues. Indeed, in the absence of flow, the Lyapunov exponents $\lambda(\mathbf{a})$ are just the eigenvalues of the matrix $\mathbf{M}(\mathbf{a})$.

A special system giving interesting Lyapunov exponents arises when the matrix $\mathbf{M}$ is the velocity gradient matrix $[\nabla \mathbf{u}]$. Recall that the ij entry of this matrix is $\partial u_i / \partial x_j$. Equation (3.7) then describes how the separation vector between two very close fluid particles evolves. Equation (3.8) then states that in a very coarse sense, the separation will increase or decrease exponentially. The Lyapunov exponents are the rate constants in the exponentials. These Lyapunov exponents are so widely used that they are frequently called just "the" Lyapunov exponents, without explicitly saying that they are associated with the matrix [$\nabla \mathbf{u}$].

If one of the Lyapunov exponents is positive, this indicates that nearby particles separate exponentially quickly. If all exponents are zero, this indicates a comparatively slow increase, if any, in the particle separation. Since rapid particle separation implies fast loss of information about a flow, one might expect chaos to accompany the existence of positive Lyapunov exponents. It turns out that chaos and positive Lyapunov exponents are almost perfectly correlated. Since Lyapunov exponents are easy to calculate from Eq. (3.8), researchers often use the positivity of Lyapunov exponents as evidence that a flow is chaotic.

MAGNETIC DYNAMOS

Starting from the non-relativistic Maxwell equations, all quantities can be expressed in terms of the magnetic field and its derivatives, and an equation for the magnetic field alone can be condensed out. This is the so-called induction equation:

$$(\partial_t + \mathbf{u} \cdot \nabla)\mathbf{B}(\mathbf{x}, t) = [\nabla \mathbf{u}]\mathbf{B} + \eta \nabla^2 \mathbf{B}. \tag{3.9}$$

This equation can be derived from Eq. (1.13) and the property $\nabla \cdot \mathbf{u} \equiv 0$ of incompressible flows.

The magnetic diffusivity η is proportional to the square root of the fluid resistivity. If the fluid is a very good conductor, or if the domain D is very large and the flow very fast, then the diffusion term can be neglected. The resulting equation can be written in Lagrangian terms as

$$\left(\frac{\partial \mathbf{B}(\mathbf{X}(\mathbf{a}, t), t)}{\partial t}\right)_{\mathbf{a}} = [\nabla \mathbf{u}(\mathbf{X}(\mathbf{a}, t), t)]\mathbf{B}(\mathbf{X}(\mathbf{a}, t), t). \tag{3.10}$$

Already, a connection with Lyapunov exponent theory begins to appear.

Equation (3.10) says that the magnetic field vectors evolve exactly like very small line segments embedded in the fluid. If the flow stretches line segments, the flow also intensifies magnetic fields, and vice versa. If the flow rotates vectors locally, the magnetic field vectors will also rotate. And if the flow has a tendency to deform simple-looking blobs of fluid into funny shapes, it will also take simple configurations of magnetic field and turn them into highly weird patterns.

This geometric interpretation of magnetic fields is the miracle that makes theoretical work in MHD (magnetohydrodynamics) possible. The result goes under the name of Alfvén's Theorem. On the strength of this theorem and related work in the early 1940's, Hannes Alfvén was awarded the 1970 Nobel Prize in physics.

The geometric picture of MHD is also useful in the so-called *dynamo problem*. We know that the Earth and other planets, and the Sun and other stars, have magnetic fields. These fields cannot be due to permanent magnetization in the interior (it's too hot), nor are they the passive remnants of stronger fields that were present a long time ago. Indeed, observations show that the magnetic fields of these objects are active, dynamic entities that interact strongly with the fluid motion and other physical processes happening in the interior.

Now we apply Sherlock Holmes's principle, that when you have eliminated the impossible whatever remains must be the truth. The only remaining sensible possibility is that the field-fluid interaction is responsible for maintaining the field, and the required energy comes from transforming mechanical energy into electromagnetic energy. This is exactly what happens in electric power generating stations, where the generators are sometimes called dynamos. The task of finding an explicit example of the same process in a homogeneous body of fluid is called the *dynamo problem.*

At first sight, the dynamo problem seems trivial. Since magnetic fields behave like line-segments, if we find a flow in which line segments are stretched rapidly, ths same flow should intensify magnetic fields. If the flow is chaotic, with at least one positive Lyapunov exponent, we might even expect exponential growth of the total field. We might even expect the growth rate of the field to be somehow related to the biggest Lyapunov exponent. Unfortunately, none of these expectations is realized.

The reason chaotic flow with positive Lyapunov exponent is insufficient to guarantee dynamo action is *cancellation.* In a chaotic flow, not only does magnetic field get stretched rapidly, it gets mixed around. There is nothing that necessarily prevents blobs of fluid in which the field is pointing one way from mixing with blobs of fluid in which the field is pointing the opposite way (Figure 13). In fact, this *always* happens if the flow is either two-dimensional or axisymmetric.

If the fluid has any electrical resistance at all, the field-mixing process leads to enormous energy dissipation by currents in the fluid, and the magnetic field energy is rapidly depleted.

Field-mixing can diminish the apparent field seen by a distant observer. We know from observations that the Sun has a mean magnetic field that influences things like the solar wind. The mean field is no more than a few gauss after averaging over the whole Sun. But at the surface (and probably below) the fluctuations can be enormous. Regions of field in the thousands of gauss can be seen with current high-resolution equipment.

The same difficulty appears, even more strongly, when observing the Earth's magnetic field. Since we cannot even see the Earth's core, we are completely unable to pick out any small-scale but high-amplitude fluctuations. If exponential field-growth with cancellation resulting from field mixing were happening in the Earth's core, we would never see it.

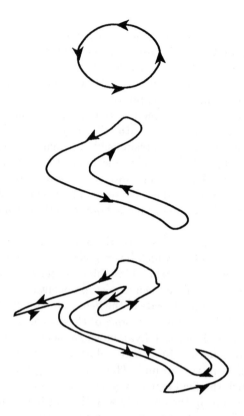

FIGURE 13   Magnetic field mixing and cancellation in two dimensions.

Since we see fairly strong magnetic fields up here at ground level, cancellation must not be occurring—at least, not total cancellation. There is also a theorem due originally to Cowling, that says if you have either a two-dimensional or axisymmetric flow, then field growth *must* be accompanied by total cancellation. Hence, the flow in the Earth's core cannot be purely axisymmetric. While it's almost obvious that the Earth's core flow would not be purely axisymmetric, we cannot even model the flow in the core as axisymmetric, as a simplifying assumption. Thus, to study magnetic field generation in large bodies, we have to consider flows with nontrivial three-dimensional structure.

DYNAMICAL THERMODYNAMICS

The dynamo problem and the algae-population problem have very similar mathematical structures. First, the quantity of interest is carried around by the fluid. Second, as it is being carried, the rate of change of the quantity of interest is determined as the product of some known function of position and the quantity itself.

A big question in both problems is: does the total amount of stuff, integrated over the whole domain, grow exponentially in time?

When formulated mathematically, these questions become almost identical in appearance to statistical-mechanical formulas for thermodynamics. Since the resemblance is purely formal, the use of tricks and facts from real thermodynamics in the context of flows treated as dynamical systems is called *thermodynamic formalism*. Since Ruelle has used this name extensively for a particular application which is somewhat different from what I will present here, I will call this material *dynamical thermodynamics*. The reader should be aware that there are several different thermodynamic formalisms in the literature, and they are not always carefully distinguished.

Let us make some elementary changes to Eq. (3.2), so that

$$(\partial_t + \mathbf{u} \cdot \nabla)C(\mathbf{x},t) = \beta R(\mathbf{x})C(\mathbf{x},t), \quad C(\mathbf{x},t_0) = 1. \tag{3.11}$$

β can be thought of as a global parameter saying how strongly we are encouraging or discouraging the algae. If $\beta = 0$, this means the algae distribution will remain steady in a state of uniform concentration. If $\beta > 0$, the algae population will grow or decay exponentially. The total number of algae at time t will be

$$Z_R(\beta,t) = \int_D \exp\left(\int_{t_0}^t R(\mathbf{X}(\mathbf{a},t'))dt'\right) d\mathbf{a}. \tag{3.12}$$

The resemblance of Eq. (3.12) to the thermodynamic partition function is too strong to be ignored. So let's define, *purely by analogy and with no deeper significance,*

$$F_R(\beta) = \beta^{-1} \lim_{t\to\infty} t^{-1} \left[\log Z_R(\beta,t)\right]. \tag{3.13}$$

This will be called the dynamical Helmholtz free energy. The analogous dynamical internal energy is

$$U_R(\beta) = \frac{d}{d\beta} \lim_{t\to\infty} t^{-1} \left[log Z_R(\beta,t)\right]. \tag{3.14}$$

The internal energy can be written explicitly as

$$U_R(\beta) = \lim_{t\to\infty} \frac{\int_D \left[t^{-1}\int_0^t R(\mathbf{X}(\mathbf{a},t'))dt'\right] \exp\left(\int_{t_0}^t R(\mathbf{X}(\mathbf{a},t'))dt'\right) d\mathbf{a}}{\int_D \exp\left(\int_{t_0}^t R(\mathbf{X}(\mathbf{a},t'))dt'\right) d\mathbf{a}}. \tag{3.15}$$

Thus, $U_R(\beta)$ is the weighted average of all the time averages of $R(\mathbf{X}(\mathbf{a},t))$ over different particle trajectories. When $\beta = 0$, the weight becomes the uniform measure, and $U_R(\beta = 0)$ is just the average of $R(\mathbf{x})$ over D. However, when $\beta \neq 0$, the weighting shifts attention to zero-measure subsets of D on which the long-time averages of R are unusually large or unusually small.

The specific heat is the derivative with respect to temperature of the internal energy:

$$C_R(\beta) = \frac{d^2}{d\beta^2} \lim_{t \to \infty} t^{-1} log Z_R(\beta, t).$$ (3.16)

I won't give the explicit formula here, but the specific heat is simply the weighted *variance* of the time-averages of $R(\mathbf{X}(\mathbf{a}, t))$ over different trajectories, computed using the same weight as in Eq. (3.15).

It's a general property of variances that they always be positive, which implies that $U_R(\beta)$ is a monotonically increasing function of β. Also, $Z_R(\beta, t)$ must be convex in β. However, nothing says that $C_R(\beta)$ has to be finite! Points of infinite specific heat correspond to *phase transitions* in the thermodynamics in real physical systems. In dynamical thermodynamics, a phase transition indicates that something interesting and not perfectly understood is happening.

If we want to extend thermodynamic formalism to the dynamo problem, a new problem raises its head: non-commuting variables. The exponential analogous to that appearing in Eqs. (3.12)–(3.16) now has to be considered in the time-ordered sense because the 3×3 matrices appearing in it do not commute. We can no longer use ordinary thermodynamics as our guide; we have to use some horrible kind of non-abelian field theory. Perhaps it is not so surprising that the dynamo is challenging!

Thermodynamic formulas for flows and other dynamical systems can probably tell us all kinds of interesting things about the flows. But the results are still being figured out. I expect and encourage the members of the audience to make important contributions to this new and exciting subject.

Magnetic dynamo theory is a highly developed subject. Moffatt[9] devotes a whole book to the subject. Parker[12] and Priest[14] have chapters on dynamo theory. Priest's book is a comprehensive account of magnetohydrodynamics and its specific application to the Sun. Ruelle[15] is a *tour de force* of powerful analysis applied simultaneously to dynamical systems theory and statistical mechanics. The reader is warned that Ruelle's book is very difficult material; unfortunately, I do not know of any lower-level introductions. The algae problem was invented just as an example for these lectures. While it may have some biological relevance, you should probably consult a real mathematical biologist if you are interested in real algae population dynamics.

FLUID INSTABILITIES

My previous lectures have concentrated on flow *kinematics*. We wrote down specific flow fields, and considered how they transport things like salt, temperature, algae, and magnetic fields. Plenty of complexity occurs even when we know the velocity field exactly.

All too often, we do not know exactly how the fluid is moving. Determining the flow field is the first part of many fluid mechanics problems. But solving the Navier-Stokes equations is hard. Not only are the equations nonlinear, but the solutions can be very complicated functions of space and time.

Turbulence is the name given to fluid flows with plenty of space-time complexity. Turbulence is a tough phenomenon to try to understand directly. Turbulence theorists often study other problems and subsequently argue that the results have some connection with turbulence. Sometimes these other studies turn out to be exciting and challenging in their own right, and become self-contained disciplines. This is where hydrodynamic stability theory comes from.

Suppose we are interested in the motion of water in a saucepan on the stove. The water moves slowly at first, as the small density differences between the warm water below and the cool water above take time to accelerate the fluid. The flow speeds up, and in the process also becomes more complex. By the time the water is boiling, the flow is severely turbulent. Furthermore, the water itself becomes a two-phase material, which adds yet another level of complexity to the situation.

It is hard to understand the development of flow complexity as one long continuous process. It's much easier, although artificial, to imagine a succession of equilibrium states punctuated by discrete transitions. Each transition is presumed to begin with the growth of small disturbances superposed on an existing equilibrium. The transition ends when the disturbances have grown so large that the old equilibrium is destroyed and a new dynamic balance attained. It is tacitly presumed that the initial equilibrium is a simple state, and that successive equilibria are more and more complex.

Instability theory essentially consists of a detailed look at the initial stages of such a transition as described above. In other words, a steady flow is given, and the problem is to determine whether any perturbations to that flow can grow. If so, we would like to know how fast they grow, and what the shape of the fastest-growing mode is. Such information can give us clues about the later stages of the transition. If we are lucky, the instability will also hint at the structure of the next equilibrium.

Today we will look at a few fluid systems and the instabilities that occur in them. The main feature I want to emphasize is the large number of different physical effects that can destabilize fluid systems.

RAYLEIGH-TAYLOR INSTABILITY

Consider a layer of fluid of density ρ_1 and thickness d, lying under a rigid ceiling and above a deep layer of fluid of density ρ_0. This is a perfectly legitimate equilibrium state of two incompressible fluids, even if $\rho_1 > \rho_0$, but it might not be stable.

In other words, if the layer is perturbed so that the upper layer is slightly thicker at one point than at neighboring points, the bulge may continue to sink under its own weight. Eventually it may turn into a big drop of fluid that falls completely away from the ceiling. This is what we expect if the upper fluid is denser than the lower.

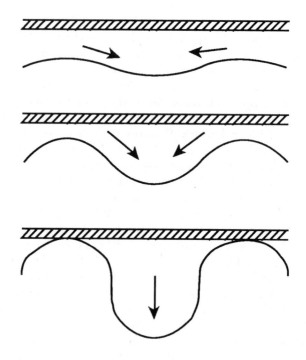

FIGURE 14   Rayleigh-Taylor instability.

Suppose that a small disturbance is imposed on the given equilibrium. "Small" means that quadratic and higher powers of the disturbance can be discarded from the Navier-Stokes equations. The growth rate of the disturbance is then an eigenvalue of the linearized flow equations.

Let us suppose that the surface is perturbed so that the layer thickness is now $d + \epsilon \cos(kx)f(t)$. Here, x is the horizontal coordinate, k the perturbation wavenumber, and ϵ the smallness parameter whose quadratic and higher powers will be discarded. Neglecting surface tension and viscosity, we can derive an equation for the temporal growth function $f(t)$:

$$\ddot{f} = \left[g \left(\frac{\rho_1 - \rho_0}{\rho_1 + \rho_0} \right) k \ \tanh(kd) \right] f . \tag{4.1}$$

If the lower fluid is denser than the upper, Eq. (4.1) predicts that the disturbance will merely oscillate like a wave on a lake. But if the upper is denser than the lower, Eq. (4.1) has solutions that grow unboundedly in time.

When viscous and surface tension effects are included, such a layer can be stabilized by making it thin enough, i.e., by making d small. Those of you who are about to paint their ceilings should contemplate this result.

CAPILLARY INSTABILITY

The phenomenon here is the break-up into drops of the stream of water from a hose or a tap, or any other liquid jet in air. This phenomenon only occurs for liquids, since surface tension is the energy source behind the instability. Let σ denote the surface tension between the liquid and surrounding air.

The air plays almost no role in the phenomenon, so we can pretend the jet is a perfect cylinder of liquid moving uniformly through space. By Galilean relativity, we can study the jet from a co-moving reference frame, from whose point of view we are just looking at a stationary cylinder of liquid.

Again, the stability is investigated by supposing that a very small disturbance is given to the cylinder, and determining its subsequent motion. Suppose the thickness of the perturbed jet is $r(x) = R + \epsilon \cos(kx) f(t)$, where nonlinear powers of ϵ are to be neglected (Figure 15). Then the evolution of f is described by

$$\ddot{f} = \left[\frac{\sigma}{R^3 \rho_1} kR \left(\frac{I_0'(kR)}{I_0(kR)} \right) (1 - k^2 R^2) \right] f \tag{4.2}$$

where I_0 is the Hankel function of the first kind.

It turns out that as long as the surface tension is positive, Eq. (4.2) has growing solutions. This agrees with our experience. At any rate, I have never seen a jet of liquid *not* break up into drops after a few diameters.

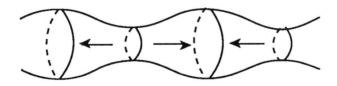

FIGURE 15   Capillary instability: small perturbation analysis.

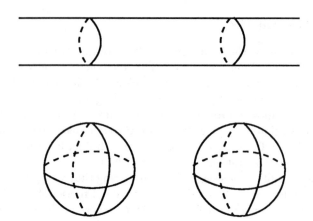

FIGURE 16   Capillary instability: energy stability theory.

Note that surface tension is stabilizing for a plane layer of paint on a ceiling, but destabilizing for a cylinder of liquid. It's worthwhile thinking about why this is. This example warns us that the same physical ingredient can play different roles in different situations.

Some people might object to the analysis sketched above, saying that one ought not need Hankel functions to explain a basic physical process. This is true. The following analysis is unsatisfactory in that it does not give the definite prediction of rapid perturbation growth that we get from Eq. (4.2), but it does highlight the physics.

Consider the total energy stored in a segment of length L of the original cylinder. This is just the surface tension times the surface area of the segment, i.e.,

$$E_{cyl} = 2\pi RL\sigma .\tag{4.3}$$

If this cylinder is deformed into a sphere simply by moving the material around, the sphere will have the same volume as the cylinder, $V = \pi R^2 L$ (Figure 16). The radius of the sphere thus has to be $S = (3R^2 L/4)^{1/3}$. This sphere will have surface energy

$$E_{sph} = 4\pi S^2 \sigma = \pi(6R^2 L)^{2/3}\sigma .\tag{4.4}$$

The transformation of the section of cylinder into a sphere will be energetically favorable if the surface energy of the sphere is less than the surface energy of the original cylinder. This is the case for any L larger than $9R/2$. Hence, there are plenty of available states with lower potential energy than the initial cylindrical configuration.

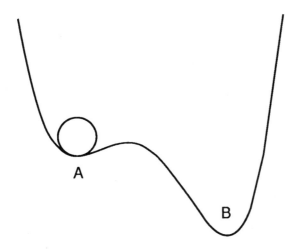

FIGURE 17   Counterexample to overly simplistic energy arguments.

The energy method gives a nice physical explanation of what really drives this instability. However, the fact that a transition from one state to another is energetically favorable does not imply that the first state is unstable. In Figure 17, the ball in trough A has more potential energy than it would in trough B, but A is nonetheless a stable state.

Another shortcoming of the energy method is that it does not predict which of the many lower-energy states the jet will break up into. The small-perturbation analysis predicts that perturbations of a particular wavelength will grow faster than perturbations of any other wavelength. The spacing of drops observed in real jets corresponds remarkably well with this prediction.

SALT FINGERS

The energy method is a powerful tool, especially if applied creatively. Let's use it to investigate the stability of the uppermost few meters of the ocean.

When the sun beats down on the sea, it heats the top layer of water. As the surface warms up, the evaporation rate increases. The result is that the upper layer of water is both warmer and saltier than the lower layer. Typically, the decrease in density due to the warming is greater than the density increase due to the raised salinity, so the heavier fluid lies below the lighter fluid. One would expect such a system to be stable, but it is not.

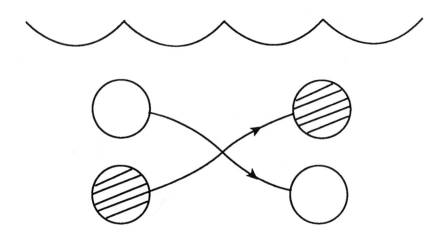

FIGURE 18 Blob interchange in doubly stratified ocean layer.

Imagine swapping two blobs of fluid of equal volumes, whose initial elevations are different (Figure 18). Initially, the hot salty blob will be lighter than its surroundings and tend to rise, while the cold fresh blob will be heavier than its surroundings and tend to sink. The potential energy of the new configuration is greater than the old.

If the blobs are held in their new positions, the cold one will warm up and the hot one will cool down. The blobs will also exchange salt with their new environments. But the diffusivity of salt is so much smaller than the diffusivity of heat (by roughly a factor of 100) that the amount of salt transported by the time the blobs have reached thermal equilibrium is negligible.

Soon after the blobs have been interchanged, therefore, the cold fresh blob has turned into a warm fresh blob, and the hot salty blob to a cold salty blob. The potential energy is now *less* than in the original state. The energy released can be converted into kinetic energy of the blobs, accelerating the instability. The final stage of the salt-finger instability consists of long thin streamers of salty water moving down, intermingled with long thin streamers of fresh water moving up. The term "salt fingers" refers to these long thin channel-like flows.

This instability owes its existence to the competition between two diffusive mechanisms (salt and heat) with different strengths, so it is called *double-diffusive* instability. Diffusion, which is frequently a stabilizing influence, is essential for the instability here.

Double-diffusive instability can occur in systems with many diffusing chemicals. This situation is better described as *multi-diffusive* instability. In chemical manufacturing processes, this kind of instability can be a good thing or an extremely bad thing. Multi-diffusive instabilities are consequently of great interest in chemical engineering and metallurgy research.

PARAMETRIC INSTABILITY

The instabilities discussed so far have obvious meaning. The initial equilibrium is manifestly not the state the fluid wants to be in. The instability is simply the fluid's first step towards a reasonable final state. Not all instabilities are so obvious. Here is one that occurs in many physical systems.

Recall how you played on the swings as a child. Better still, go out to the nearest playground and remind yourself. You can get started with some small-amplitude oscillations just by pushing on the ground. But to get going really high, you need to pump your legs.

The mass at the end of the swing is constant, but the act of pumping changes the effective length of the pendulum. Thus, the angle between the chain and vertical satisfies

$$\ddot{\theta} + \frac{g}{L(t)} \sin\theta = 0, \qquad L(t) = L_0 + L_1(t). \tag{4.6}$$

Here L_0 is the mean effective length of the pendulum, and $L_1(t)$ is the zero-mean perturbation to the effective length. If you pump every T seconds, L_1 will be periodic in time with fundamental period T.

If the natural frequency of the swing $\omega_0 = \sqrt{g/L_0}$ times T is close to an integer multiple of π, the rest state $\theta = \dot{\theta} = 0$ is unstable. That is, an initially small oscillation quickly grows to finite amplitude. This prediction is confirmed by laboratory experiments and also by field studies of elementary-school children in the wild.

This phenomenon is called parametric instability because it is associated with the pendulum length L, which is usually considered a parameter of the system. The most robust parametric instability occurs when $\omega_0 T \approx \pi$. For physiological reasons, we humans tend to pump with $\omega_0 T \approx 2\pi$. This is a more delicate instability, but clearly quite effective when a human with sophisticated feedback control is in charge. Higher-order resonances $\omega_0 T \approx m\pi$ with $m \geq 3$ are tricky to find because the range of unstable pumping frequencies is extremely narrow.

Parametric instabilities occur in containers of fluid subject to vertical oscillations. For example, if you set a loudspeaker on its back, so that the paper cone points upwards, and put a cup of coffee on it, and turn on the sound, you will see this instability. In fact, you will probably see a great many instabilities unless you take the trouble to set up this experiment carefully and not spill the coffee.

When the swing oscillations reach finite amplitude, the period of the unpumped swing lengthens slightly, and human swingers adjust their pumping frequency accordingly. If they did not, the resulting detuning between the natural period and the pumping period would result in chaotic oscillations. Nonlinear swings with constant pumping frequencies are nice examples of real physical systems exhibiting chaos.

SHEAR FLOW INSTABILITIES

Things really get interesting when the energy source for a fluid instability is the kinetic energy of a flow. A simple class of flows susceptible to instabilities are *unidirectional shear flows* (USF's). A USF is a flow in which all fluid particles are moving in straight lines, all in the same direction.

In two dimensions, a USF can be described by specifying the velocity U as a function of z, the coordinate perpendicular to the coordinate x along which the particles are moving. The flow may extend to infinity in either or both directions. Otherwise, the flow may be bounded by one or two planes (lines really, in two dimensions) parallel to the x-axis (see Figure 20).

USF's are good approximations to many real situations. Flow through a pipe or channel is almost a perfect USF. The flow around an aircraft is clearly not a USF, but in the very thin boundary layers adjacent to the structure, the USF approximation is very good. USF's are very convenient to analyze, since almost all problems concerning them can be reduced to scalar ordinary differential equations in z.

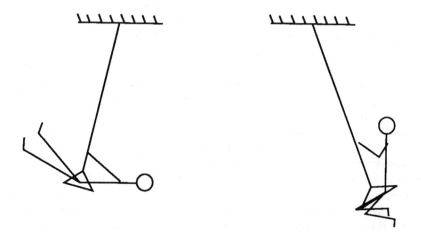

FIGURE 19   Parametric instability.

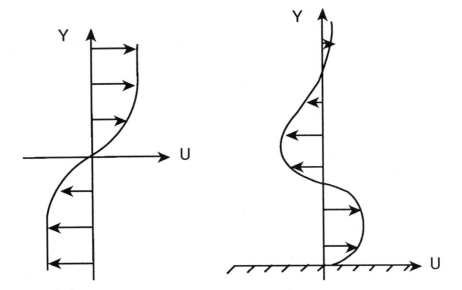

FIGURE 20 Unidirectional shear flows.

The stability of USF's can be investigated by the small-perturbation approach. This was first done by Rayleigh, and he sketched the results that now form the core of shear flow stability theory. The Rayleigh inflection-point theorem states that an internal maximum in the shear, or vorticity, profile is a *necessary* condition for a USF to be unstable. Rayleigh's theorem assumes zero viscosity, which would seem to be a reasonable assumption for flows in which the Reynolds number is very large.

An apparent counterexample to Rayleigh's theorem is furnished by simple two-dimensional flow in a channel with a parabolic velocity profile (Poiseuille flow, Figure 21). This is an exact solution to the viscous Navier-Stokes equations as well as the inviscid equations. This flow is stable at zero viscosity, according to Rayleigh's theorem.

But if the fluid has a small nonzero viscosity, Poiseuille flow is unstable. Viscosity is the destabilizing influence! Consistency with Rayleigh's theorem is restored with the observation that the instability growth rate goes to zero as the viscosity goes to zero.

Energy stability methods have been developed for unidirectional shear flows, and more general two-dimensional flows. These theories are unfortunately not as simple as the salt finger or capillary energy-stability arguments. The shear-flow energy technique makes up for its complexity by having a wonderful mathematical structure that emphasizes some of the most interesting and subtle aspects of fluid mechanics.

The reason for the complexity of energy methods for shear flows is as follows. Since the flows have nontrivial vorticity profiles, we cannot consider transitions from the given state to just any other state; the target state has to have a vorticity distribution consistent with the initial state. Making precise the idea of states with consistent vorticity fields is not trivial! Once this has been done, it is feasible, although not always easy, to characterize manifolds of isocirculational flows and compute relative energies.

General hydrodynamic stability theory is treated well in Drazin and Reid.[4] Instabilities of interest in oceanography are discussed by Pedlosky,[13] and magnetohydrodynamic instabilities important in astrophysics by Parker[12] and Priest.[14] The study of parametric instabilities in fluid systems has been very active recently, and is reviewed by Miles and Henderson.[8] Van Dyke[16] has many beautiful pictures of fluid instabilities developing in simple systems.

TURBULENCE

Turbulence is the dignified senior citizen of complex systems. Turbulence is not a vast, insanely complicated object like the immune system, whose study involves among other things the acquisition of billions of individual facts. Nor is turbulence a fundamental problem like the development of a quantum-relativistic theory of absolutely everything, whose solution is expected to be a fantastically beautiful but totally incomprehensible formula. Turbulence is just an old, everyday problem that has been around for a long time, gently but effectively resisting all attempts to fathom its mysteries.

PHENOMENOLOGY AND KOLMOGOROV SCALING LAWS

The age-old observation is that when a runny fluid flows, the flow has a tendency to be complicated rather than simple. We see complex flows almost every minute of every day, with the result that we often think of flows like milk stirred in tea as being simple. We save our admiration for beautiful pictures like the photographs in Van Dyke's book.[16] But these are really extremely simple flows. For each shot, some dedicated scientist devoted months of his or her life to eliminating extraneous complicating features from the apparatus, so that the resulting picture would deliver an unambiguous message.

As a simple example, consider the flow generated by a fan in a large room (Figure 22).

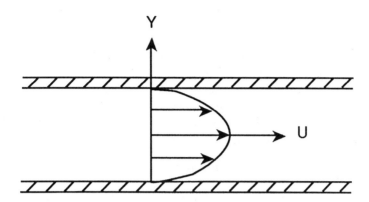

FIGURE 21   Poiseuille flow.

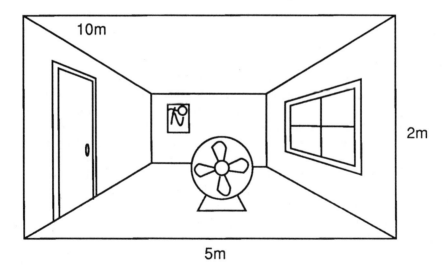

FIGURE 22   Turbulent flow driven by a fan in a room.

The fan drives a large-scale circulation of air around the room. In addition, vortices the size of the fan blades are expected. Measurement of the velocity field everywhere in the room would reveal that there is a continuous spectrum of flow structures ranging from the size of the room itself down to tiny eddies much smaller than any moving part of the fan.

The size of the smallest eddies can be estimated. We know the viscosity of air at room temperature is $0.15cm^2sec^{-1}$, and its density is roughly $1.2 \cdot 10^{-3}gm\ cm^{-3}$. If the fan draws $100\ watts = 10^9 gm\ cm^2 sec^{-3}$, and the room measures $10m \times 5m \times 2m = 10^8 cm^3$, then the power dissipation per unit mass Π is about $8.3 \cdot 10^3 cm^2 sec^{-3}$.

Kolmogorov proposed that the smallest lengthscale be determined solely by the viscosity and the power dissipation per unit mass. If this postulate is accepted, then there is only one possible formula which possesses the right dimensions. This is

$$l_K = \nu^{3/4}\Pi^{-1/4}.\tag{5.1}$$

Plugging in the numbers, we obtain a Kolmogorov length of $2.5 \cdot 10^{-2}$ centimeters as the probable size of the smallest eddies in the room. That's a quarter of a millimeter. That's very small. Still, it's the kind of non-microscopic length scale that we think of as characteristic of our normal experience.

We can use the Kolmogorov length scale in a coarse estimate of the complexity of this fan-driven air flow. If the smallest active objects in the flow have size l_K, the volume occupied by one of them is on the order of l_K^3. The number of Kolmogorov-length sized cells in the room is therefore $V_{room}/l_K^3 = 6.4 \cdot 10^{12}$.

Thus, this turbulent flow has roughly the same complexity as a system with $10^{12} - 10^{13}$ degrees of freedom. It's an interesting reflection that the number of Kolmogorov cubes in the room is the same order of magnitude as the number of air molecules in a Kolmogorov cube.

Although there are plenty of degrees of freedom in fan-driven turbulence in a room, energy is not equipartitioned. We can define the "energy spectrum" $E(k)$ as the contribution to the total flow kinetic energy coming from eddies of size $l_{eddy} \approx 2\pi/k$. $E(k)$ has dimensions $(length)^3(time)^{-2}$.

Kolmogorov proposed that if only the smallest eddies are affected by viscosity, then there is only one possible combination of Π and k that gives the right dimensions for $E(k)$, which is $\Pi^{2/3}k^{-5/3}$. Kolmogorov's prediction that the energy spectrum should be proportional to some dimensionless constant times this dimensional combination is actually verified to within experimental error in real flows. The dimensionless "Kolmogorov" constant is difficult to measure, but is very robustly bounded between 1 and 2.

THE TURBULENCE PROBLEM

Despite this daunting estimate of the complexity of a very mundane turbulent flow, turbulence is simple in many ways. We know from experience almost exactly what will happen when we turn on the fan, and it didn't take very long to build up that experience! The fan blows air along the direction of its rotor axis. The flow is strongest along the centerline of the rotor, and falls off away from this line. There are plenty of fluctuations, but their strength is bounded and their statistics are robust.

This phenomenology of turbulence recalls the situation in the kinetic theory of gases. A macroscopic box of (classical) gas consists of 10^{24} or so molecules. The exact dynamics of all these particles is very complicated, but the average behavior is extremely simple. We can even derive equations for the averaged velocities of the particles and the probability distribution of the velocity fluctuations.

It would seem the most natural thing in the world to try the same thing for turbulence. That is, we would like to obtain deterministic equations for the average flow, and probability distributions for the fluctuations. But the average and fluctuating quantities are strongly coupled in a turbulent flow, in contrast to the weak coupling between molecular and bulk motions in kinetic theory. The result is that no truly satisfactory kinetic theory exists for turbulence. In the absence of turbulence equations derived honestly from the Navier-Stokes equations, theorists have to use models.

One class of theoretical models are the statistical closures obtained by truncating the infinite hierarchy of moment equations (see next section). This procedure is objective and well defined, and requires no adjustable parameters to be introduced into the model. The resulting closures usually give information about the fluctuation statistics in addition to the mean flow.

The other main class of turbulence models are obtained by simply replacing difficult terms by simple ones in the ensemble-averaged Navier Stokes equations. For example, the Reynolds stress tensor (see below) plays a role similar to that of the viscous stress tensor. So we *replace* the Reynolds stress tensor, which we don't know, by a multiple of the ensemble-averaged strain-rate tensor. The multiple is termed the "turbulent viscosity," and can be much larger than the viscosity of the fluid.

We don't know *a priori* what the turbulent viscosity should be in a given flow. We can either postulate a physically reasonable formula, or try a lot of possibilities and choose the one which gives the best agreement with experiments. Indeed, a combination of these methods can give quite good results.

The latter class of turbulence models give predictions only of mean properties of turbulent flows. Often, this is all we want to know, and the fact that we do not have to worry about fluctuations makes the analysis easier than if we used a statistical closure. In the case of the fan in the room, most designers would be completely content to know the non-random large-scale circulation in the room.

The two approaches to turbulence theory are not really antithetical and may be advantageously used together. The Navier-Stokes equations, if creatively applied, give plenty of information about the properties an equation must have in order to be physically consistent. Conversely, some quantities appearing in statistical closures may be better measured in experiments than calculated formally.

ENSEMBLES AND AVERAGES

Turbulent flows look like true random processes. But any flow is really governed by the purely deterministic Navier-Stokes equations, coupled with quite definite boundary and initial conditions. We can only talk about statistics if we do the same experiment over and over, allowing small random changes in the initial or boundary conditions from run to run. If there is some other agency influencing the flow, perhaps an electromagnetic stirring field or gravity waves from a distant pulsar, we should allow for a randomly fluctuating external force field.

Each experimental run is called a *realization* of the experiment. The set of all possible realizations is called the *ensemble* for the experiment. There should be a probability distribution defined on the ensemble, to quantify the likelihood of an experiment being done with certain fluctuations present in the external conditions.

Let us use ω to denote elements of the ensemble. $\mathbf{u}^\omega(\mathbf{x}, t)$ will be the fluid velocity at the point $\mathbf{x}$ at time t in realization ω, $\mathbf{f}^\omega(\mathbf{x}, t)$ the external force field in realization ω, and so on. The velocity and pressure fields are determined by the equations of motion

$$\rho \left(\frac{\partial}{\partial t} + \mathbf{u}^\omega \cdot \nabla \right) \mathbf{u}^\omega = -\nabla p^\omega + \rho \nu \nabla^2 \mathbf{u}^\omega + \mathbf{f}^\omega$$
$$\nabla \cdot \mathbf{u}^\omega = 0$$

(5.1)

together with initial and boundary conditions depending on ω. For simplicity, we are restricting attention to incompressible fluids.

The term $\mathbf{f}^\omega$ represents forces acting within the bulk of the fluid, such as gravity or electromagnetic forces. Researchers often simplify the conceptual ensemble by neglecting the random variations in initial or boundary conditions. The ensemble is then determined by specifying the statistics of the random function $\mathbf{f}^\omega(\mathbf{x}, t)$.

Once the ensemble is defined, we can talk honestly about the statistical properties of turbulence in a given system. The quantities researchers are most often interested in are *ensemble averages* of various quantities. If $P(\omega)$ is the probability distribution for ω, then we can define the ensemble average of $\mathbf{u}^\omega$ as

$$\langle \mathbf{u}(\mathbf{x}, t) \rangle = \int_{ensemble} \mathbf{u}^\omega(\mathbf{x}, t) dP(\omega),$$

the ensemble average of $\mathbf{f}^\omega$ as

$$\langle \mathbf{f}(\mathbf{x}, t) \rangle = \int_{ensemble} \mathbf{f}^\omega(\mathbf{x}, t) dP(\omega),$$

(5.2)

the ensemble-averaged pressure as

$$\langle p(\mathbf{x}, t) \rangle = \int_{ensemble} p^\omega(\mathbf{x}, t) dP(\omega),$$

and so on.

Ensemble averages are attractive things to work with because they are directly measurable in experiments. Indeed, ensemble averages are easy to measure and very reproducible. Practically the only difficulty is taking enough realizations to get good averages. Even that is a minor point for a modern laboratory equipped with computer control capable of automatically repeating experiments many times.

THE CLOSURE PROBLEM

Ensemble-averaging equation (5.1) gives an equation of motion for the ensemble-averaged velocity field:

$$\frac{\partial}{\partial t}\langle \mathbf{u}(\mathbf{x},t)\rangle + \nabla \cdot \langle \mathbf{u}(\mathbf{x},t)\mathbf{u}(\mathbf{x},t)\rangle = -\nabla\langle p(\mathbf{x},t)\rangle + \rho\nu\nabla^2\langle\mathbf{u}(\mathbf{x},t)\rangle + \langle\mathbf{f}(\mathbf{x},t)\rangle$$

$$\nabla \cdot \langle \mathbf{u}(\mathbf{x},t)\rangle = 0 \tag{5.3}$$

The term $\langle \mathbf{u}(\mathbf{x},t)\mathbf{u}(\mathbf{x},t)\rangle$ is a symmetric matrix, or second-rank tensor, whose ij element is

$$\langle u_i(\mathbf{x},t)u_j(\mathbf{x},t)\rangle = \int_{ensemble} u_i^\omega(\mathbf{x},t)u_j^\omega(\mathbf{x},t)dP(\omega)\,.$$

This tensor, more precisely its negative, is called the Reynolds-stress tensor, since it plays the role of a classical stress tensor.

Taking the tensor (or exterior) product of Eq. (5.1) with $\mathbf{u}^\omega(\mathbf{x},t)$, symmetrizing with respect to all vector and tensor indices, and ensemble averaging gives an equation of motion for the Reynolds stress tensor:

$$\frac{\partial}{\partial t}\langle \mathbf{uu}\rangle + \nabla \cdot \langle \mathbf{uuu}\rangle = -\langle \mathbf{u}\nabla p\rangle - \langle \mathbf{u}\nabla p\rangle^T + \rho\nu\langle \mathbf{u}\nabla^2\mathbf{u}\rangle$$

$$+ \rho\nu\langle \mathbf{u}\nabla^2\mathbf{u}\rangle^T + \langle \mathbf{uf}\rangle + \langle \mathbf{uf}\rangle^T\,. \tag{5.4}$$

Equation (5.4) is a matrix equation, and the T superscript on a matrix denotes the transpose of the matrix. The third-rank tensor $\langle \mathbf{uuu}\rangle$ forms part of the so-called energy-transfer tensor.

The equation of motion of $\langle \mathbf{uuu}\rangle$ is obtained by taking the exterior product of Eq. (5.1) with $\mathbf{u}^\omega\mathbf{u}^\omega$, symmetrizing, and ensemble averaging. The resulting equation contains the divergence of the fourth-rank tensor $\langle \mathbf{uuuu}\rangle$.

The alert reader will have guessed that this can go on forever. Each equation in this hierarchy of moment-average equations contains an unknown quantity, which can only be determined by the next higher equation. Hence it's impossible to derive a closed system of equations for any finite subset of them. This difficulty is known as the *closure problem*.

The closure problem arises in any attempt to describe an ensemble of nonlinear deterministic systems statistically. In classical kinetic theory, the closure problem is even more pronounced, because each realization is not only deterministic but reversible. Any statistical description in terms of a finite number of moments would have to reproduce the macroscopically observed irreversibility. But any description derived honestly from the governing equations must preserve reversibility.

STATISTICAL CLOSURES

Boltzmann "solved" the closure problem in kinetic theory dramatically by replacing the double correlation $\langle ff \rangle$ by the product of single correlations $\langle f \rangle \langle f \rangle$. For a discussion of kinetic theory and the Boltzmann equation, including what f is, see any graduate text on kinetic theory.

The corresponding operation on the fluid equations is the replacement of $\langle uu \rangle$ by $\langle u \rangle \langle u \rangle$ in Eq. (5.2). The resulting equation is nothing but the deterministic Navier-Stokes momentum equation. The random force field $\mathbf{f}(\mathbf{x}, t)$ and the initial/boundary conditions are replaced by their ensemble averages.

This system is certainly physically meaningful, but is completely stupid as far as approximating turbulence goes. The whole statistical nature of turbulence has vanished. This approximation could be dignified, if anyone wished, with the name "mean-field" theory of turbulence.

More sophisticated closures can be obtained by playing the replacement game with higher moments. For example, $\langle uuu \rangle$ might be replaced with the appropriate multiple of $\langle u \rangle \langle uu \rangle$, symmetrized with respect to the tensor indices. Or $\langle uuuu \rangle$ could be replaced by a symmetrized linear combination of $\langle uu \rangle \langle uu \rangle$ and $\langle u \rangle \langle uuu \rangle$.

The last proposed, if done carefully, can be motivated by the quite reasonable approximation that the velocity fluctuations have Gaussian, or normal, statistics. It is therefore called the *quasi-normal approximation*. The error made by the replacement of $\langle uuuu \rangle$ by products of lower moments is called the fourth *cumulant* of the velocity distribution; hence, this closure is also referred to as the *zero fourth cumulant* approximation.

The zero fourth cumulant closure has many nice properties. Unfortunately, the squares of some physical quantities are predicted to be negative in this closure! Although the approximation seems good, the small discrepancy results, after enough time, in gross unphysicality.

More sophisticated turbulence closures have been developed in the last thirty-odd years. They perform better than the mean-field and zero fourth cumulant approximations, and, in general, correspond to actual physical models of turbulence. All the models so far contain some crucial ingredient added to the ordinary properties of fluids, explicitly inserted to make the model work. Models which are used with success in many applications are the Direct-Interaction Approximation (DIA), the Eddy-Damped Quasi-Normal Markovian approximation (EDQNM), and a large family of so-called $k - \epsilon$ models.

PHENOMENOLOGICAL CLOSURES

The form of the Reynolds stress tensor $\langle uu \rangle$ is almost identical in appearance to the expression for the fluid stress tensor in kinetic theory. To be sure, the interpretations of this equation is quite different in the two fields. In kinetic theory, the next step is to show that the stress tensor is equal to the viscosity coefficient (determined by the molecular statistics) times the derivative (strain-rate) tensor of the macroscopic velocity field.

We would like to be able to do something similar for turbulent flow, but this cannot be justified by mathematical manipulation of the equations. But the lack of mathematical justification is not so serious to some people as to others. The bold scientist would try it anyway and see what happens, and justify the action by the results.

More precisely, we would like to replace the unknown Reynolds stress tensor with something of the form $\nu_T \nabla \langle \mathbf{u}(\mathbf{x}, t) \rangle$. The "turbulent viscosity" ν_T would be determined by the statistics of the fluctuations, while ∇ acting on the ensemble-average of $\mathbf{u}(\mathbf{x}, t)$ would, of course, be independent of the fluctuations.

In the absence of statistical theory, of course, we do not know how ν_T is really determined. However, the main requirement is that ν_T have dimensions of $(length)^2/(time)$. Any proposed formula giving a dimensionally correct ν_T is worth looking at. Many formulas turn out unsatisfactory in some way or other. The ones that have lasted over the years did so because they are remarkably successful at predicting the mean properties of many different turbulent flows.

Prandtl's *mixing-length* theory assumes that at each point in the flow, one can identify a special length scale L, called the mixing length. The other important property of the mean flow, as far as identifying a viscosity is concerned, is some quantity with dimensions of $(time)^{-1}$, for example, the mean trace-free strain-rate tensor. A reasonable guess for the turbulent viscosity would be some dimensionless constant times the mixing length squared, times the norm of the trace-free strain-rate tensor.

In symbols, the mixing-length equations are

$$\frac{\partial}{\partial t} \langle \mathbf{u}(\mathbf{x}, t) \rangle + \nabla \cdot \langle \mathbf{u}(\mathbf{x}, t) \rangle \cdot \nabla \langle \mathbf{u}(\mathbf{x}, t) \rangle$$
$$= -\nabla \langle p(\mathbf{x}, t) \rangle + \rho \left[\nu + CL^2(\mathbf{x}, t) \parallel \langle \Sigma(\mathbf{x}, t) \rangle \parallel \right] \nabla^2 \langle \mathbf{u}(\mathbf{x}, t) \rangle + \langle \mathbf{f}(\mathbf{x}, t) \rangle \quad (5.5)$$
$$\nabla \cdot \langle \mathbf{u}(\mathbf{x}, t) \rangle = 0 \,.$$

This model assumes that the mixing length $L(\mathbf{x}, t)$ is specified as part of the problem, before we are asked to solve it. For example, in a wall-bounded turbulent flow, a good mixing length prescription would be $L(\mathbf{x}) =$ distance from $\mathbf{x}$ to the nearest boundary. The only way to determine C is to solve Eqs. (5.5) for a range of C values and see which one gives the best agreement with observed flows.

SUMMARY

The way the turbulence problem is usually posed, a theory is often considered lacking if it is not derived systematically from the Navier-Stokes equations. However, most of the current models are extremely successful at predicting some if not all the properties of a given turbulent flow. Although we continue to see turbulence as a fundamental unsolved problem, it is definitely a problem on which progress has been made and about which we know quite a lot.

There are many good books on the phenomenology and theory of turbulence. A good introduction is the book of Landahl and Mollo-Christensen.[6] The two-volume set by Monin and Yaglom[10] was close to a complete exposition of the whole subject of turbulence when it was first published. It is still easily the most comprehensive reference available, although, of course, many new developments are necessarily not included. Van Dyke[16] has many beautiful pictures of turbulent flows, from which much can be learned.

The best reference for turbulence is the natural world. Watch the steam from a kettle or the smoke from a fire, and you will see turbulence in action.

ACKNOWLEDGMENTS

I am indebted to the organizers of the 1990 Complexity Summer School for inviting me to this stimulating course. The Santa Fe Institute staff smoothed my path in many imaginative ways. My work in fluid complexity has been generously supported by the Office of Naval Research, the National Science Foundation, and the Air Force Office of Scientific Research. I would like to acknowledge in particular support from the AFOSR Contract FQ8671- 900589, with the University Research Initiative Program at the University of Arizona, which sponsors the Arizona Center for Mathematical Sciences.

REFERENCES

1. Abramowitz, M., and I. A. Stegun. *Handbook of Mathematical Functions.* New York: Dovers, 1970.
2. Arnold, V. I., and A. Avez. *Ergodic Problems of Classical Mechanics.* New York: Benjamin, 1968.
3. Batchelor, G. K. *Introduction to Fluid Dynamics.* Cambridge, UK: Cambridge University Press, 1967.
4. Drazin, P. G., and W. H. Reid. *Hydrodynamic Stability.* Cambridge, UK: Cambridge University Press, 1981.
5. Halmos, P. R. *Lectures on Ergodic Theory.* New York: Chelsea, 1956.
6. Landahl, M. T., and E. Mollo-Christensen. *Turbulence and Random Processes in Fluids.* Cambridge, UK: Cambridge University Press, 1986.
7. Landau, L. D., and E. M. Lifshitz. *Fluid Mechanics.* New York: Pergamon, 1959.
8. Miles, J. W., and D. Henderson. "Parametrically Forced Surface Waves." *Annual Reviews of Fluid Mechanics* **22** (1990): 143–65.

9. Moffatt, H. K. *Magnetic Field Generation in Electrically Conducting Fluids.* Cambridge, UK: Cambridge University Press, 1978.
10. Monin, A. S., and A. M. Yaglom. *Statistical Fluid Mechanics*, 2 volumes. Cambridge, MA: MIT Press, 1971.
11. Ottino, J. M. *The Kinematics of Mixing, Stretching, Chaos, and Transport.* Cambridge, UK: Cambridge University Press, 1989.
12. Parker, E. N. *Cosmical Magnetic Fields.* New York: Oxford University Press, 1979.
13. Pedlosky, J. *Geophysical Fluid Dynamics.* New York: Springer- Verlag, 1979.
14. Priest, E. R. *Solar Magnetohydrodynamics.* Dordrecht: Reidel, 1982.
15. Ruelle, D. *Thermodynamic Formalism.* Reading, MA: Addison-Wesley, 1978.
16. Van Dyke, M. *An Album of Fluid Motion.* Stanford, CA: Parabolic Press, 1982.
17. Yih, C-S. *Fluid Mechanics.* Ann Arbor, MI: West River Press, 1979.

Wing Yim Tam
Physics Department, University of Arizona, Tucson, Arizona 85721

Pattern Formation in Chemical Systems: Roles of Open Reactors

INTRODUCTION

USF's are good approximations to many real situations. Flow through a pipe or channel is almost a perfect USF. The flow around an aircraft is clearly not a USF, but in the very thin boundary layers adjacent to the structure, the USF approximation is very good. USF's are very convenient to analyze, since almost all problems concerning them can be reduced to scalar ordinary differential equations in z.

Pattern formation is the most fascinating subject in nature.[37] Patterns range from banded structures in rocks to beautiful skin coatings in animals. The formation of these structures is often chemical in origin. The discovery of oscillating chemical reactions early in this century has accelerated the study of chemical pattern formation, which is now a disciplinary area of interest.[22] Chemical patterns are unique and yet they share and exhibit similar constraints and behaviors of nonlinear systems.[19] Nonlinearities are the main ingredients for pattern formation in chemical systems, as well as in other nonlinear systems, and play important roles only when the system is far from thermodynamic equilibrium. In a closed laboratory chemical experiment, nonequilibrium conditions occur only in the beginning of the reaction. As the reagents are being consumed during the reaction, the system evolves irreversibly towards thermodynamic equilibrium. Thus, complex patterns can at most be observed as transients. This is in sharp contrast to living organisms that can sustain themselves by taking in external supplies (food) and excreting unused materials (waste). Another disadvantage of a closed system is the lack of control parameters. These serious restrictions hinder the study of chemical patterns in terms of the well-developed bifurcation theory for complex systems.[11,45,56] Experimental studies in open systems have been common practice in many disciplines, but experiments on chemical patterns were, until recently, conducted in closed systems. In these lectures, I will describe several novel designs for open reactors,[38,52,53,54,55] in which spatiotemporal chemical patterns can be sustained far from thermodynamic equilibrium, and explore phenomena observed in these reactors using oscillating chemical reactions. Due to the limitation of space, I will restrict myself to a few examples and refer interested readers to other references.[14,31,33,40,48]

OSCILLATING CHEMICAL REACTIONS: THE BELOUSOV-ZHABOTINSKII REACTION

Oscillating chemical reaction were discovered as early as 1921,[20,22] but did not attract much attention until the discovery of the now well-known Belousov-Zhabotinskii (BZ) reaction some thirty years later.[10,67,69] The reaction was first discovered by B. P. Belousov[10] in 1958 and later modified by A. M. Zhabotinskii,[69] who brought world attention to the reaction. The reaction has now become the prototype for studies of oscillating chemical reactions even though many different chemical oscillators have been found since then.[22] The reaction involves the oxidation of an organic substrate, e.g., malonic acid, by bromate in a sulfuric acid solution with a metal-ion catalyst, e.g., cerium. Oscillations of the BZ reaction in a homogeneous (well-mixed) system can be observed as periodical concentration variations in certain intermediate species of the reaction, e.g., the bromide ion concentration, and as periodical color changes when a suitable catalyst is used, e.g., from colorless to yellow and back when cerium is used as a catalyst. The visual oscillations are more dramatic when the cerium catalyst is replaced by an iron catalyst, where distinctive red-blue color changes mark the reaction. In addition to temporal oscillations in homogeneous conditions, beautiful spatial patterns of waves, as shown in Figure 1, can be observed in a thin medium of the reagents,[63,64,65,66,68] which will be discussed in another section.

The oscillating phenomena of the BZ reaction can be understood by studying the reaction mechanism[17,69] that was put into its final form by Field, Körös, and Noyes in the early 1970s.[21,23] The mechanism, known as the FKN model,[21] involves 20 different chemical species in a set of 18 reaction steps and has laid the corner stone for the study of the BZ reaction. It is now possible, using simplified versions of the FKN model and inexpensive computing power, to obtain qualitative and sometimes quantitative agreements between numerical models and laboratory

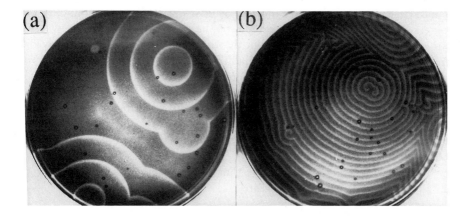

FIGURE 1   Trigger waves in BZ reactions, (a) target waves and (b) spiral waves. Light
regions represent the blue wave fronts and dark regions represent the red background.

experiments.[19] Despite the complexities involved in oscillating reactions, simple
essential criteria must be met in order to sustain oscillations. They are:

1. The system must be far from thermodynamic equilibrium.
2. Autocatalytic feedback mechanisms exist.

The first condition is essential for the second condition to play a role, and could
be met only in the beginning of the reaction in a closed system or in an open reactor
called the continuous-flow stirred-tank reactor (CSTR) where continuous feeds of
chemicals are used to sustain the oscillations. The second condition provides the
nonlinearities needed to produce oscillations and complex behavior. Conditions 1
and 2 are the backbones of the dynamics in oscillating reactions. Oscillations will
stop when the conditions are no longer valid. Thus, chemical oscillations in a closed
reactor are transients, though they may last as long as several hours in some cases.
This limitation can be lifted in a CSTR, which will be discussed later.

CHEMICAL WAVES

The same BZ solution (with iron catalyst) that exhibits oscillations in a homogeneous reactor when left unstirred as a thin layer in a petri dish will also exhibit
spatiotemporal patterns in the form of waves.[1,2,3,12,13,34,35,41,44,46,47,49,62] Beautiful concentric blue wave fronts initiated from some centers will propagate radially
outward in a red background after some global transient oscillations in the layer.
These waves are called trigger waves because they are initiated from centers called

pace makers, which are often gas bubbles or dust particles or the boundary of the container. Figure 1(a) shows an example of the trigger waves. Similar waves can also be observed in excitable BZ media that have chemical receipts slightly different from that of the oscillating BZ media. The blue wave front of the trigger waves has a high concentration of the oxidized iron, while the red background has a high concentration of reduced iron. The concentration profile of the oxidized iron can be mapped out by absorption measurements of monochromatic light at about 488 nm. A schematic diagram of the concentration profile is shown in Figure 2. The wave front has a sharp concentration gradient (about 17 mM/mm of iron concentration) in contrast to the slowly varying (about 1 mM/mm) refractory tail behind the front.[34,35] There are special properties that distinguish these trigger waves from other physical waves, like water waves:

1. There is no attenuation of the wave amplitude; the oxidized iron concentration remains the same at the wave front as the wave propagates.
2. Trigger waves do not interpenetrate each other, but annihilate each other on contact (see the cusp-shaped structures in Figure 1(a)).

The best analogue to trigger waves is fire propagation on grass in the wilderness. The "amplitude" of the fire will be the same as long as there is the same amount of grass ahead of the fire front. The consumed grass behind the front, of course, cannot support another fire front until fresh grass is grown. In the case of the BZ reaction, each wave front consumes only a small percent of the reagents, so that even after a wave front has passed, the medium can support another wave after some refractory time. Thus, successive wave fronts can propagate through the medium many times forming the observed concentric trigger waves. The concentric waves are hence called target waves to label this unique feature, as shown in Figure 1(a).

The origin of the pace makers for the trigger waves has, so far, not yet been settled. Experiments[13] have shown that dust particles, impurities, or bubbles facilitate the formation of pace makers, and that by reducing these heterogeneous centers, waves can be suppressed in an excitable system. But waves can still be initiated in a dust-free oscillating medium. More well-controlled experiments have to be carried out in order to shed light onto the spontaneous origin of the trigger waves.

Target waves are not the only form of chemical waves. When a wave front is disrupted, by physical, chemical, or optical means, propagation of the wave at the perturbed location can be suppressed. The front will then break into two parts. The ends of the broken waves will evolve into a pair of highly regular, spiral-shaped, counter-rotating waves, called spiral waves (Figure 1(b)).[1,2,3,34,35,44,64,65,68] The tips of these waves turn inward around a rotation center while sending waves in the outward direction. The spiral waves, once produced, are self-sustained and do not have pace makers like target waves. The rotation center (spiral core), about 10 μm in diameter, is a singular region where the chemistry remains quasi-stationary. Beyond the core, variations of chemical concentrations resume the full amplitude

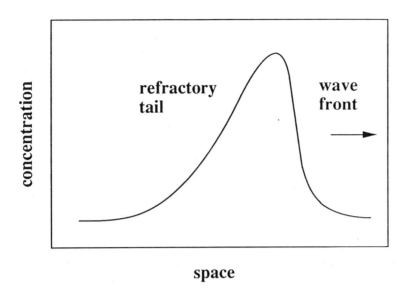

FIGURE 2   Schematic concentration profile of a trigger wave. The wave is traveling to the right as indicated by the arrow.

of the trigger waves. The wave velocity and length (distance between successive wave fronts) of spiral waves are unique for a given medium, in contrast to that of target waves, where waves of different velocities and lengths can be initiated from different pace makers. Detailed measurements[34,35] of the spiral waves have revealed that the shape of the spiral front of isoconcentration level can be approximated by an Archimedean spiral or by an involute of a circle, while numerical results[30] fit closer to the involute. The two curves are asymptotically identical and differ only slightly near the core. Spiral waves are actually quite common in nature, and can be observed in many biological and physical systems. Besides single-armed spirals, "multi-armed" spirals[1] have been reported in BZ reagents. In three-dimensional mediums, fascinating scroll waves have been observed[62,63] and studied in great detail numerically.[66] The wave phenomena in BZ reaction has attracted as much, or even more, interest as in the temporal behavior and deserves further discussion.

FORMATION OF SPATIAL PATTERNS

Spatial patterns in chemical reactions arise mainly from the interaction between reaction kinetics and diffusion of different species. Reactions build up local concentrations to create gradients, while diffusion smooths out spatial variations. It is the competition between the two mechanisms that gives rise to spatial patterns from an initial homogeneous background as first suggested by A. M. Turing in 1950.[58] Turing's work provided the theoretical background for pattern formation in chemical systems, as well as biological systems. The waves in the BZ reactions, though not predicted by Turning, provide a paradigm for the study of spatiotemporal patterns. It has been shown that only two variables are needed to describe the propagation of waves in an excitable medium.[59] The equations that govern the dynamics of chemical waves are

$$\frac{\partial u}{\partial t} = \frac{F(u,v)}{\epsilon} + D_u \nabla^2 u \tag{1}$$

and

$$\frac{\partial v}{\partial t} = F(u,v) + D_u \nabla^2 v \tag{2}$$

where u, v are the independent variables, F and G account for the kinetics of the reaction and are functions of the u, v, and rate constants. D_u and D_v are scaled diffusion coefficients for u and v, respectively. $\epsilon \ll 1$ is a parameter that is proportional to the ratio of the diffusion rate to the chemical reaction rate and plays a crucial role in determining the dynamics of the system. It is easily seen that, because F is scaled by ϵ, u changes much more rapidly than v. The dynamics of Eqs. (1) and (2) are best represented in phase diagrams as shown in Figure 3. Under homogeneous conditions, the equilibrium fixed point (s) of the system corresponds to the intersection of the nullclines of $F(u,v)$ and $G(u,v)$ as shown in Figure 3. The F nullcline has a characteristic S-shaped dependent, while G has a monotonic dependent. If the G nullcline intersects the left branches of the F nullcline, trajectories that start in the neighborhood of the fixed point will flow towards the fixed point quickly, but trajectories that start on the right-hand side of the middle branch of the F nullcline will exhibit a large amplitude excursion before settling back to the fixed point as indicated by the flow of the vector fields in Figure 3(a). This large excursion is a typical property of excitability, where the fixed point is stable to small perturbations, but perturbations about a certain threshold will lead to large amplitude excursions before returning to the fixed point. This excitability when coupled spatially by diffusion is the main ingredient for wave propagation in chemical, as well as biological, systems. Chemicals diffused from the sharp wave front into the neighborhood ahead of the front can excite the quiescent background in that region into a new wave front; thus propagation of the wave front continues. The amplitude and wave velocity depend on the delicate balance between the chemical kinetics and the diffusion. Model studies of Eqs. (1) and (2) with BZ reaction have yielded semi-quantitative arguments with experiments on the dispersion of wave propagation and on the form of spiral waves.[16,30] A side note for Eqs. (1) and (2) is that, when the G nullcline intersects the middle branch of the F nullcline, the fixed point is no longer stable and relaxational oscillations start, as shown in Figure 3(b).

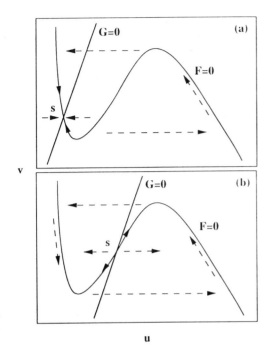

FIGURE 3 Phase diagram of u and v. The solid lines represent the nullclines of F(u,v) and G(u,v) as indicated. s is the fixed point. The arrows indicate the flow fields of the reaction. (a) is excitable and (b) is oscillatory.

NEEDS FOR OPEN REACTORS

The experimental studies of temporal and spatial patterns in chemical reactions discussed so far were conducted in closed systems. In such systems, patterns will envolve irreversibly and uncontrollably toward thermodynamic equilibrium. Any patterns observed are transients and cannot be studied on a long-term basis. Despite the transient nature of patterns in closed systems, detailed studies of these patterns are of fundamental importance and have been studied actively. These studies revealed basic understanding of chemical patterns, but left many questions unanswered, including:

1. Are these patterns stable in an open system?
2. What are the possible patterns?
3. What are the transitions between patterns?

In a closed chemical reactor, one can only set the initial conditions, like the initial concentrations of the reagents. In such systems, there are no tunable parameters that can be varied back and forth; hence, transitions between states cannot be investigated. This puts serious limitations on closed reactors. Thus, it is very desirable to develop open reactors where there are external parameters that one can freely adjust. In the following sections, I will describe some of these open reactors that have been developed over recent years to study temporal and spatial patterns of oscillating chemical reactions.

OPEN TEMPORAL REACTORS
CONTINUOUS-FLOW STIRRED-TANK REACTOR

Chemical processings in open reactors has long been a common practice among the chemical engineering community, but open reactors were first used for the study of oscillating chemical reactions in the mid-1970s by a group of chemists in Bordeaux, France.[20] A schematic diagram of a continuous-flow stirred-tank reactor (CSTR)

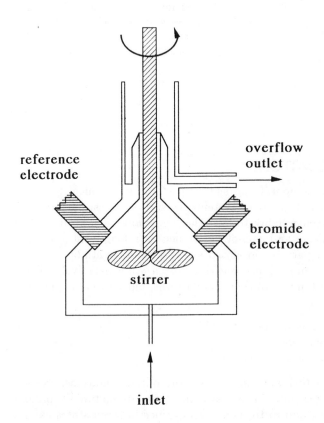

FIGURE 4 Schematic diagram of a continuous-flow stirred-tank reactor.

is shown in Figure 4. Continuous feeding of reagents into the reactor through the inlet is the crucial feature of the CSTR. Peristaltic or precision piston pumps can be used to deliver chemicals into the reactor. A simple overflow mechanism defines the reactor volume and facilitates the removal of chemicals. A rapidly rotating stirrer is used to mix the reagents inside the reactor. Temporal patterns can be monitored or recorded by direct measurement of chemical potentials using ion-sensitive electrodes or optical absorption methods. The temperature of the reactor can be regulated by simply submerging the reactor into a temperature bath. Common control parameters for the CSTR are:

1. initial feed concentrations of reagents,
2. feed rate of chemicals,
3. temperature of the reactor, and
4. rate of mixing in the reactor.

The last control parameter addresses the heterogeneous effect of mixing and has been studied for some mixing sensitive reactions.[36] As for common oscillating reactions, mixing has small effects, and adequate stirring is enough to ensure homogeneous condition. The second control parameter, the feed rate, is the most commonly used. A variation of this parameter is the residence time τ, which is equal to the volume of the reactor divided by the total feed rate. The residence time represents the duration that reagents stay in the reactor. The third control parameter sets the reaction rates of the system and is often held fixed for convenience.

Owing to the above controllable features, the CSTR is an ideal system with which to study temporal patterns of chemical reactions, where nonequilibrium conditions can be maintained and varied to study stabilities and transitions of different temporal states. This tool, together with the techniques developed for the study of dynamics systems, has made tremendous progress in understanding oscillating reactions.[4,5,8,15,28,32,42,43,50,57] Temporal patterns of oscillating reactions observed in CSTR can be summarized as follows:

1. hysteretic transitions between multiple steady states,
2. bifurcations of steady states to limit cycles,
3. birhythmicity—hysteretic behavior between two oscillating states with different frequencies,
4. bifurcations of periodic states to quasi-periodic states with two incommensurate frequencies f_1 and f_2,
5. frequency locking of quasi-periodic states—$f_2 = p/q f_1$, where p,q are integers,
6. period doubling of periodic states—the frequency of the states changes from $f_0 \rightarrow f_0/2 \rightarrow f_0/4 \cdots$,
7. intermittency—irregular bursts of large amplitudes at irregular intervals from a seemingly periodic state, and
8. chaos via several mechanisms—period-doubling cascade, quasi-periodicity, and intermittency.

Model calculations of oscillating reactions in CSTR have reproduced all these findings. The strong evidences from experiments and models have unambiguously demonstrated that complex behavior observed in oscillating chemical reactions in a CSTR is deterministic rather than noisy behavior arising from random driving forces.[4,5,28,42,43,50,57] Moreover, the chaos observed in chemical reactions can be understood in terms of bifurcation theories.

OPEN SPATIAL REACTORS

The study of spatial patterns in open systems, in contrast to temporal patterns, has been lacking. The advancement made with CSTR in temporal studies has not been shared with spatial studies because no similar CSTR reactors, until recently, have been available for spatial patterns. Spatial patterns cannot be retained in a well-mixed environment so that the CSTR is of no use for the study of spatial patterns; any spatial structures will be destroyed by the rapid mixing in the reactor. This drawback of the CSTR is now overcome by novel designs in recently developed open spatial reactors, where patterns can be sustained by diffusion or controllable hydrodynamic flows. In such open spatial reactors, the stabilities and transitions between different spatial patterns can be studied in the same way that temporal patterns are studied in the CSTR's. Three such open spatial reactors will be discussed in this lecture.

RING REACTOR

A schematic diagram of the ring reactor[38] is shown in Figure 5. The main ingredient of the reactor is an annular polyacrylamide gel. The gel suppresses any convective motion generated by concentration gradients or other hydrodynamic instabilities and the formation of bubbles from gases evolved in the BZ reaction,[6] but allows patterns to be formed inside. The patterns formed can be sustained by chemicals diffusing into the gel. The gel is chemically inert to the reaction so that patterns resulting inside the gel are only due to the kinetics of the BZ reaction and the diffusion inside the gel. The gel ring, 1 mm thick, is sandwiched between two fixtures that separate the reactor into two separate compartments such that the gel is in contact with each compartment from either side (see Figure 5). The two compartments serve as reservoirs of chemicals that can be replenished from external feeds.

In the experiment, reagents of the BZ reaction were separated with the sulfuric acid and potassium bromate in the outer compartment, and malonic acid and iron (ferroin) catalyst in the inner compartment. After the start of the experiment, a red front of ferroin could be seen advancing slowly from the inner edge towards the outer edge of the annulus. Then two to three hours later, the circular symmetry of the front started to break, and local irregularities were developed. These irregularities would eventually develop into stable states of an equal number of multiple sources and sinks located around the annular as shown in Figure 6(a) with one source and one sink, where waves are initiated from the source and annihilated at the sink. The source-sink configuration, though it could last as long as the same external feeds are applied, is rather idiosyncratic. A more regular and stable state of rotating waves could be obtained with an appropriate perturbation. This can be done by stimulating waves traveling in two directions, but eliminating waves traveling in one direction, e.g., eliminating the counter-clockwise waves. This will result in uneven waves in the two directions, which will then develop after annihilation of the remaining counter-clockwise waves by the clockwise waves into a state of waves traveling in the clockwise direction. Chemical or optical perturbations can be used to create these one-directional waves. As an example, waves were initiated in the experiment at the pacer (see Figure 5) by adding a solution of higher concentrations of H_2SO_4 and $KBrO_3$ to that location and were inhibited in the counter rotating direction by passing a slow continuous flow of distilled water at the barrier. The perturbations at the pacer and the barrier were then removed after waves traveling in one direction, counterclockwise, were eliminated. The unevenly spaced waves then relaxed to form a state of regular equally spaced waves rotating in the clockwise direction as shown in Figure 6(b) for a state of seven clockwise waves. These highly regular wave patterns, like pinwheels, rotate faithfully around the annulus as long as the same external controls are applied. (The pattern eventually will disrupt due to the shrinking and decomposition of the gel in a strong

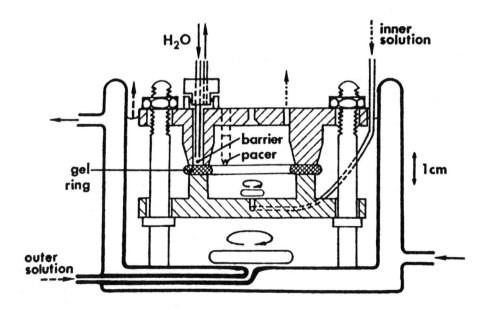

FIGURE 5   Cross section of the ring reactor. See Noszticzius[38] for details.

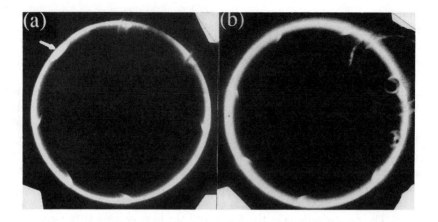

FIGURE 6 Chemical waves in the ring reactor: (a) a state of one source (indicated by
an arrow) and one sink (on the opposite side of the source), (b) a pinwheel state of 7
waves.

acidic medium, but gels of different compositions could be used to solve this minor
problem.) The experiment shows that regular rotating chemical patterns (clockwise
as well as counter clockwise) are stable if the number of waves falls within a certain
interval. This is the first study showing that spatial patterns of chemical reactions
can be sustained in an open reactor. In a recent experiment on chemical pinwheels,
using an improved ring reactor,[31] symmetry breaking and time dependence of the
waves were revealed when the temperature of the reactor was varied. The chemical
pinwheels is the first realization of rotating wave structures predicted for nonlinear
chemical systems in a circular geometry.[37]

An open reactor using the same principle as in the ring reactor has been developed recently by Castets et al.[14] Instead of ring-shaped gel, they use a narrow
rectangular gel. The two opposite long edges are in contact with two chemical reservoirs, which are filled with reagents of a variant of the Chlorite-iodide reaction. The
reagents of Chlorite-iodide reaction are separately distributed in such a way that
neither solution in the reservoirs is active. Reaction takes place only inside the gel.
They observe patterns of stripes parallel to the gel edges in the direction of the
concentration gradient. These stripes then break up into lines of periodic spots of
characteristic size of about 0.2 mm which seems to be nongeometrically related. The
periodic spots are stable and form only for a well-defined range of concentrations.
The pattern exhibits the symmetry-breaking phenomenon in the direction transverse to the imposed gradient and is interpreted as the first experimental evidence
of a genuine Turing structure[58] as discussed in the previous section. Future studies
will certainly verify this extraordinary finding.

CONTINUOUSLY FED UNSTIRRED REACTOR (CFUR)

A cross-sectional view of a CFUR designed by Tam et al.[52] is shown in Figure 7. A disk-shaped, diameter 2.54 cm, polyacrylamide gel (same as in the ring reactor) is sandwiched horizontally between a glass window and a 1 mm thick glass capillary array with evenly spaced 10 μm diameter capillaries packed in a hexagonal matrix. (A nitrocellulose membrane, not shown in Figure 7, is placed between the gel and the capillary array to provide a white background for enhancement of pattern visualization.) The gel is 1 mm thick and is sufficiently thin for two-dimensional wave propagation. The gel communicates through the capillary array with a CSTR below the array; hence, patterns formed in the gel can be sustained by chemicals from the CSTR. The glass capillary array is a crucial feature of the reactor; it provides uniform feed that is in a direction perpendicular to the plane in which patterns can form, and ensures that only vertical mass transport occurs between the gel and the CSTR. If horizontal transport were not suppressed, spatial pattern formation could occur already in the capillary array. This effect is obvious when the capillary array is replaced by a fitted glass disc where mass transport can occur in all directions in the disc.

BZ reaction was used in the experiment. Reagents were delivered into the CSTR by high-pressure piston pumps so that bubble formation from gases produced in the reaction was suppressed by holding 10 atm pressure in the reactor. Bubbles are undesirable because they can block the capillaries and cause inhomogeneous feed to

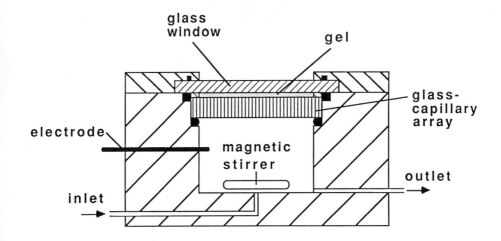

FIGURE 7   Cross-section of the CFUR. See Tam[52] for details.

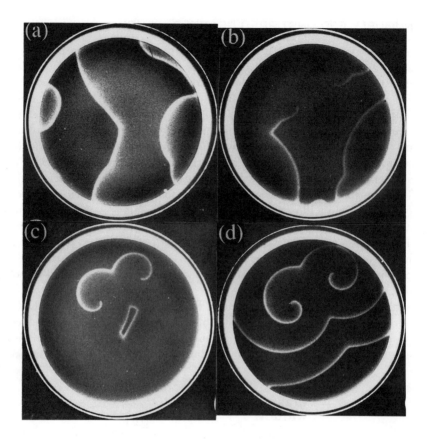

FIGURE 8 States showing the formation of spiral waves in the CFUR by a perturbation as described in the text. (a) irregular waves before the perturbation, and (b) 12 minutes, (c) 24 minutes, and (d) 60 minutes after the perturbation.

the gel. Feeds of fresh reagents were combined into a single line before entering the CSTR. The feed rate and the concentrations of reagents were kept constant, except the concentration of $NaBrO_3$ that was varied over the range 0.01-0.05 M and was used as the control parameter in the experiment. Patterns were digitized using a 512×480 pixel, 8-bit resolution imaging system. The intensity amplitudes of the digitized images are directly related to the concentrations of the oxidized iron.

Homogeneous states of a uniform background are observed for $NaBrO_3$ concentrations below 0.017 M. Spiral patterns similar to those observed in petri dish experiments form quickly after the start of an experiment for $NaBrO_3$ concentrations around 0.018 M, but for concentrations above about 0.025 M, irregular waves form first (see Figure 8(a)). These irregular waves can last a long time until one or more spiral waves appear. The formation of spiral waves from irregular waves can be initiated by a perturbation. A simple method is to stop the stirrer in the CSTR for a short time to create temporary spatial inhomogeneities in the CSTR and, hence, in the feeds to the gel. This perturbation breaks up a wave into two disconnected waves. The tips of the broken waves will then evolve into a pair of spiral cores. Figures 8(b)–(d) show the different stages of the formation of a pair of spiral waves using such perturbation. Spiral waves formed are stable and can last for weeks. Spiral waves can occasionally drift to the walls where they are annihilated. Optical methods can also be used to break the wave front. The transition form homogeneous state to spiral state is shown as intensity amplitudes in Figure 9 as a function of the $NaBrO_3$ concentration. The insert in Figure 9 clearly shows a small hysteretic loop with location and size reproducible within a few percent in runs with different gels. Figure 9 is the first study of the transition of spatial patterns of oscillating reactions in an open reactor. The hysteretic behavior observed in Figure 9 could simply be due to the excitability of the BZ reaction, but a detailed calculation may shed more light into the nature of the transition.

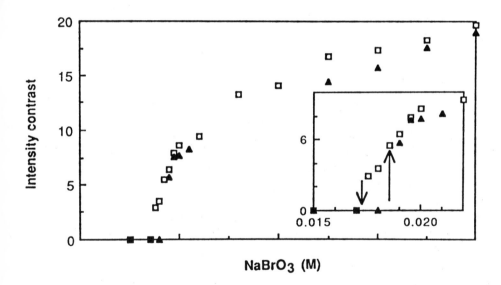

FIGURE 9 The intensity contrast, which is the difference between the maximum intensity of a spiral wave front (the oxidized state) and the minimum intensity (the reduced state), is shown as a function of $NaBrO_3$ concentration. ($\square$, decreasing concentration; $\triangle$, increasing concentration). The appearance (disappearance) of the spiral waves with increasing (decreasing) concentration is indicated by $\uparrow$ ($\downarrow$) in the insert.

The CFUR is a very useful tool in the study of two-dimensional chemical waves. An example is to study the dynamics of a single spiral wave. The tip of a spiral wave is known from experiments and numerical simulations to rotate in a circle about a core region. But, meandering tips of spiral waves have also been observed in BZ reagents.[2,7,9,25,29,48]

The meandering effect could have been caused by boundary effects, hydrodynamics, or inhomogeneities. Recent studies of Jahuke, Skaggs, and Winfree[29] on chemical waves in gels loaded with BZ reagents have revealed that the meandering motion is not an artifact, but is inherent in the chemistry used. They observed circular tip orbits for some chemistries, as well as epicycle-like (compound) orbits for others in both experiment and model calculation. Recent detailed studies by Skinner and Swinney[48] on the motion of spiral tip using a CFUR has brought new understanding in the tip motion. They found supercritical transition from simple rotation (one frequency) to compound rotation (two frequencies). (The control parameter used in that experiment was the feed $NaBrO_3$ concentration.) The supercritical transition is consistent with recent model studies using a simple kinetic for the BZ reaction.[9] Skinner and Swinney[48] conjectured that the refractory tail of the wave plays a dominant role in controlling the tip motion. This can be verified by future model studies.

COUETTE REACTOR

The transport of reagents in the ring reactor and the CFUR is through molecular diffusion, which is about 10-5 cm^2/s. Spatial structures resulting from reaction-diffusion mechanisms have a characteristic size proportional to the square root of the product of the diffusion coefficient and the averaged reaction time. Thus, spatial structures of BZ reactions in gels have a typical size of above few millimeters. Large spatial structures can be obtained by enhancing the mass transport that can be easily modeled in simulations, but not in experiments. The Couette reactor[27,40,53] is designed to have the unique property of controllable mass transport. A schematic diagram of the Couette reactor[53] is shown in Figure 9. It consists of a fluid filling the gap between two concentric cylinders with the inner cylinder rotating and the outer cylinder at rest. This arrangement is exactly the Couette-Taylor system[26] used in hydrodynamic studies of shear flows; but instead of nonreacting fluid, the fluid used in the Couette reactor is now the BZ reagent. Experiments[18,51] indicate that the transport arising from the hydrodynamic flow at sufficiently high cylinder speed can be modeled as a one-dimensional diffusion process in the axial direction. (The effective diffusion has a simple power dependence on the rotation rate of the inner cylinder and can be orders of magnitude larger than molecular diffusion.) Thus, the reactor is effectively a one-dimensional reaction-diffusion system where the transport, effective diffusion, can be controlled by varying the rotation rate. Patterns in the Couette reactor can be sustained by feeds at the ends as shown in Figure 9.

A dual substrate glucose-acetone BZ reaction was used in the experiment because the reaction produces an insignificant amount of gas compared to the usual bubble-producing, malonic acid BZ system.[39] Bubbles are undesirable in the Couette reactor because they interfere with the flow state and change the effective diffusion coefficient. Another advantage of the dual substrate system is that the reaction exhibits only steady and simple periodic states in homogeneous systems; thus, any complex spatiotemporal behavior observed in the Couette reactor must arise from an interplay between the simple local dynamics and effective diffusion. In the experiment, the oxidizer $(KBrO_3)$ was fed into the cylindrical annulus at one end, $Z = 0$, while the reducers (glucose and acetone) were fed at the other end, $Z = 1$. The rate of removal of chemicals at each end was carefully adjusted to match the feed rate at that end so that there was no net axial flow. The feed rate α_0 of the oxidizer was varied from 0-35 ml/h, while the feed rate α_1 of the reducer was fixed at 10 ml/h (H_2SO_4 was fed at both ends, while $MnSO_4$ (catalyst) was fed only at $Z = 1$). Three rotation rates of 6, 9, and 12 Hz, corresponding to an effective diffusion coefficient of 0.12, 0.16, and 0.22 cm^2/s estimated from Tam,[51] respectively, were studied in the experiment. The concentration of one of the intermediate species, bromide ion, was measured at 16 locations with ion-selective electrodes $(Ag - AgBr)$ spaced equally along the axial extent of the reactor. Time series, power spectra, and phase portraits were used to identify the spatiotemporal patterns for each α_0. Figures 10(a)–(c) show bifurcation sequences obtained as the feed rate α_0 is varied at the rotation rates of 6, 9, and 12 Hz, respectively. Data were obtained with both increasing and decreasing α_0 and no hysteresis was observed.

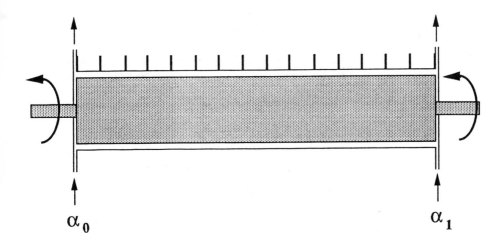

$$\alpha_0 \qquad\qquad \alpha_1$$

FIGURE 10   The Couette reactor. α_0 and α_1 are the feed rates at Z=0 and Z=1, respectively. The vertical lines show the location of the 16 evenly spaced Ag-AgBr electrodes.

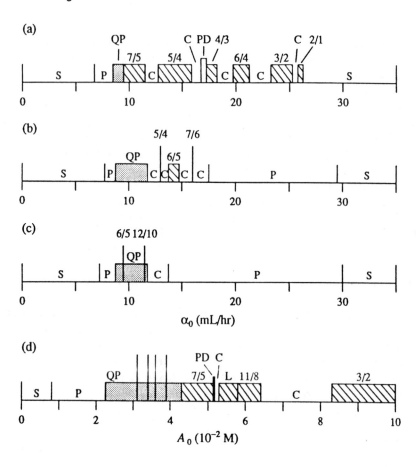

FIGURE 11 (a)-(c) Bifurcation sequences observed in the Couette reactor as a function of α_0 for cylinder rotation rates of 6, 9, and 12 Hz, respectively. (d) Bifurcation sequence for a reaction diffusion model as a function of bromate feed concentration A_0 (for D = 0.078 cm2/s).[53,61] The labels for the different dynamical regines are: steady (S), periodic (P), quasi-periodic (QP), frequency-locked (p/q is the frequency locking ratio, where p and q are integers), period-doubled (PD), and chaotic (C); the region labelled L in (d) corresponds to many different frequency-locked states, and the four long verticle lines in (d) are very narrow windows of frequency-locked states.

At low α_0, only time-independent states with spatial bromide concentration variations are observed. The first transition is from a time-independent state to an oscillating state with increasing α_0 for all rotation rates. Figures 11(a) and 12(b) show the power spectrum and the phase portrait of an oscillating state, respectively. The periodic oscillations are neither traveling waves nor standing waves—the phase has a non-monotonic spatial dependence. The amplitudes of the oscillations are not the same everywhere along the reactor; they are large near the $Z = 0$ end and small near the $Z = 1$ end. At higher α_0 after the first transition, a second transition to a quasi-periodic state occurs. The frequencies obtained from time series and power spectra (Figure 11(b)) of this state are incommensurate and localized; α_0 (the frequency of the original periodic state) is dominant near $Z = 0$, while the new frequency ω_1 is dominant near $Z = 1$. A phase portrait of a quasi-periodic state is shown in Figure 12(b). No third frequency is observed with increasing α_0 instead the two frequencies of the quasi-periodic state can become locked together at some integer ratio to form a frequency-locked state. Figure 11(c) and 12(c) show the power spectrum and the phase portrait of a frequency-locked state, respectively. Further increase in α_0 leads to a chaotic state, then (except at 12 Hz) to many frequency-locked and chaotic states as shown in Figures 10(a)–(c). Chaotic states are identified

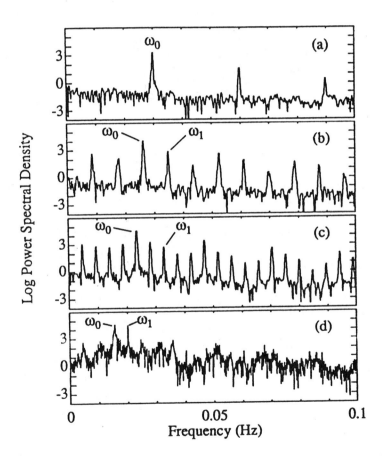

FIGURE 12   (a)-(d) Power spectra determined from the time series data for, respectively, periodic (α_0 = 8.0 mL/h), quasi-periodic (α_0 = 9.0 mL/h), frequency-locked (α_0 = 10.0 mL/h), and chaotic states (α_0 = 16.0 mL/h), obtained in each case from a bromide electrode located at Z=0.2 with rotation rate of 6 Hz and α_1 = 10.0 mL/h.

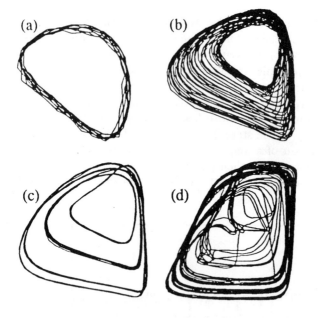

FIGURE 13 (a)-(d) Phase portraits of attractors constructed from time series data[24] for periodic, quasi-periodic, frequency-locked, and chaotic states, respectively. The same time series data giving the power spectra in Figure 12 are used in the construction of the attractors.

by their power spectra (Figure 11(d)) and phase portraits (Figure 12(d)). In one special case, labeled PD in Figure 10(a), the transition to chaos is via a period-doubling of a frequency-locked state. Presumably other observed transitions from periodic states to chaotic states also proceed through a period-doubling cascade, but the parameter range is too small to detect. At very high α_0, only steady states are observed as shown in Figures 10(a)–(c). The bifurcation sequences are similar for all rotation rates, except that, at high rotation rates, complex behavior occurs for a smaller range in α_0. This behavior is due to the fact that the higher the mass transport rate (effective mixing), the closer the system is towards homogeneous conditions as in a CSTR where only steady and simple periodic states are observed.

Results at lower rotation rates have revealed more complicated dynamics, but at that rotation rate, the effective diffusion model for mass transport is no longer applicable and interpretation of the results is not trivial. The bifurcation sequences in Figures 10(a)–(c), except for a shift of a few percent in the transition α_0 values, are reproducible from run to run. Recent model calculations of a one-dimensional reaction-diffusion system capture the observed phenomena.[53,60,61] Figure 10(d) shows a bifurcation sequence obtained from a model calculation with only two species for the local chemistry and a single effective diffusion of 0.078cm2/s (close to 6Hz rotation rate) for both species. The correspondence between the numerical results and the experimental results at 6 Hz is remarkable. Model calculations at higher rotation rates are also similar to the experimental results at high rotation rates.[61] The similarity between the model and the experiment is to some degree fortuitous. The boundary conditions and control parameters are different, having flow rate in the experiment and feed concentration in the model. Furthermore, the two-species model is for the malonic acid BZ system, and the relation to the glucose-acetone system is unknown. Despite this, the qualitative agreements between experiment and model suggest that similar bifurcation sequences can be observed in other chemistries in a one-dimensional reactor with opposing oxidation and reduction gradients.

DISCUSSION

The study of chemical patterns has reached another height since the discovery of the BZ reaction. This progress is possible primarily due to the adoption of open reactors to study chemical reactions and the advance in the studies of dynamical systems. The CSTR has proven to be a critical tool in the study of temporal patterns. The success in the CSTR can now be seen in the open spatial reactors. Sustained spatiotemporal patterns, transitions and bifurcations sequences of spatial patterns are, for the first time, obtained using these reactors. With these newly developed open spatial reactors, it should now be straightforward to study sustained spatial patterns in other geometries and boundary conditions, and new phenomena are expected to be discovered. The study of chemical patterns in open systems may shed light in understanding the complexity of pattern formation in living things.

ACKNOWLEDGMENTS

I thank Professor Dan Stein and Professor Lynn Nadel for the invitation to lecture at SFI. I am indebted to Z. Noszticzius, J. A. Vastano, W. Horsthemke, W. D. McCormick, H. L. Swinney, Q. Ouyang, and P. DeKepper for their collaborations in the works reported in these lectures.

REFERENCES

1. Agladze, K. I., and V. I. Krinsky. "Multi-Armed Vortices in an Active Chemical Medium." *Nature* **296** (1982): 424–426.
2. Agladze, K. I., A. V. Panfilov, and A. N. Rudenko. "Nonstationary Rotation of Spiral Waves: Three-Dimensioal Effect." *Physica* **29D** (1988): 409–415.

3. Agladze, K. I., V. I. Krinsky, and A. M. Pertsov. "Chaos in the Non-Stirred Belousov-Zhabotinsky Reaction is Induced by Interaction of Waves and Stationary Dissipative Structures." *Nature* **308** (1984): 834–835.

4. Argoul, F., A. Arneodo, P. Richetti, J. C. Roux, and H. L. Swinney. "Chemical Chaos: From Hints to Confirmation." *Acc. Chem. Res.* **20** (1987): 436–442.

5. Argoul, F., A. Arneodo, P. Richetti, and J. C. Roux. "From Quasiperiodicity to Chaos in the Belousov-Zhabotinskii Reaction I. Experiment." *J. Chem. Phys.* **86** (1987): 3325–3338.

6. Avnir, D., and M. Kagan. "Spatial Structures Generated by Chemical Reactions at Interfaces." *Nature* **307** (1984): 717–721.

7. Barkey, D. "A Coupled-Map Lattice for Simulating Waves in Excitable Media." *Nonlinear Structures in Physical Systems*, edited by L. Lam and H. C. Morris. New York: Springer, 1990.

8. Barkley, D. E. "Studies of the Complex Dynamics of The Belousov-Zhabotinskii Reaction." Ph.D. thesis, University of Texas at Austin, 1988.

9. Barkley, D., M. Kness, and L. S. Tuckerman. "Spiral-Waves Dynamics in a Simple Model of Excitable Media: The Transition from Simple to Compound Rotation." *Phys. Rev. A* (1990): to appear.

10. Belousov, B. P. "A Periodic Reaction and Its Mechanism." In *Sbornik Referatov po Radiatsionni Meditsine*, 145. Moscow: Medgiz, 1958.

11. Bergé, P., Y. Pomeau, and C. Vidal. *Order within Chaos*. New York: Wiley, 1984.

12. Bodet, J. M., J. Ross, and C. Vidal. "Experiments on Phase Diffusion Waves." *J. Chem. Phys.* **86** (1987): 4418–4424.

13. Bodet, J. M., C. Vidal, A. Pacault, and F. Argoul. "Experimental Study of Target Patterns Exhibited by the B. Z. Reaction." *Nonequilibrium Dynamics in Chemical Systems*, edited by C. Vidal and A. Pacault, 102–106. Heidelberg: Springer-Verlag, 1984.

14. Castets, V., E. Dulos, J. Boissonade, and P. DeKepper. "Experimental Evidence of a Sustained Standing Turing-Type Nonequilibrium Chemical Pattern." *Phys. Rev. Lett.* **64** (1990): 2953–2956.

15. Coffman, K. G., W. D. McCormick, Z. Noszitezius, R. H. Simoyi, and H. L. Swinney. "Universality, Multiplicity and the Effect of Iron Impurities in the Belousov-Zhabotinskii Reaction." *J. Chem. Phys.* **86** (1987): 119-129.

16. Dockery, J. D., J. P. Keener, and J. J. Tyson. "Dispersion of Traveling Waves in the Belousov-Zhabotinskii Reaction." *Physica D* **30** (1988): 177–191.

17. Edelson, D., R. J. Field, and R. M. Noyes. "Mechanistic Details of the Belousov-Zhabotinskii Oscillations." *Int. J. Chem. Kinet.* **7** (1975): 417–432.

18. Enokida, Y., K. Nakata, and A. Suzuki. "Axial Turbulent Diffusion in Fluid Between Rotating Coaxial Cylinders." *AIChE J.* **35** (1989): 1211–1214.

19. Epstein, I. R. "Chemical Oscillations and Nonlinear Chemical Dynamics." *1989 Lectures in Complex Systems*, edited by E. Jen. Santa Fe Institute Studies in the Sciences of Complexity, Lec. Vol II, 213–269. Redwood City, CA: Addison-Wesley, 1990.

20. Epstein, I. R., K. Kustin, P. DeKepper, and M. Orban. "Oscillating Chemical Reactions." *Sci. Am.* **248** (1983): 112–123.
21. Field, R. J., E. Koros, and R. M. Noyes. "Oscillations in Chemical Systems. Part 2. Thorough Analysis of Temporal Oscillations in the Ce-BrO3-Malonic Acid System." *J. Am. Chem. Soc.* **94** (1972): 8649–8664.
22. Field, R. J., and M. Burger. *Oscillations and Travelling Waves in Chemical Systems.* New York: Wiley, 1985.
23. Field, R. J., and R. M. Noyes. "Oscillations in Chemical Systems. Part 4. Limit Cycle Behavior in a Model of a Real Chemical Reaction." *J. Chem. Phys.* **60** (1974): 1877–1884.
24. Fraser, A. M., and H. L. Swinney. "Independent Coordinates for Strange Attraction from Mutual Information." *Phys. Rev. A* **33** (1986): 1134–1140.
25. Gerhardt, M., H. Schuster, and J. J. Tyson. "A Cellular Automaton Model of Excitable Media Including Curvature and Dispersion." *Science* **247** (1990): 1563–1566.
26. Gorman, M., and H. L. Swinney. "Spatial and Temporal Characteristics of Modulated Waves in the Circular Couette System." *J. Fluid Mech.* **117** (1982): 123–142.
27. Grutzner, J. B., E. A. Patrick, P. J. Pellechia, and M. Vera. "The Continuously Rotated Cellular Reactor." *J. Am. Chem. Soc.* **110** (1988): 726–728.
28. Hudson, J. L., M. Hart, and D. Marinko. "An Experimental Study of Multiple Peak Periodic and Nonperiodic Oscillations in the Belousov-Zhabotinskii Reaction." *J. Chem. Phys.* **71** (1979): 1601–1606.
29. Jahnke, W., W. E. Skaggs, and A. T. Winfree. "Chemical Vortex Dynamics in the Belousov-Zhabotinsky Reaction and in the 2-Variable Oregonator Model." *J. Phys. Chem.* **93** (1989): 740–749.
30. Keener, J. P., and J. J. Tyson. "Spiral Waves in the Belousov-Zhabotinskii Reaction." *Physica D* **21** (1986): 307–324.
31. Kreisberg, N., W. D. McCormick, and H. L. Swinney. "Symmetry Breaking in a Chemical Pinwheel." *J. Chem. Phys.* **91** (1989): 6532–6533.
32. Maselko, J., and H. L. Swinney. "Complex Periodic Oscillations and Farey Arithmetic in the Belousov-Zhabotinskii Reaction." *J. Chem. Phys.* **85** (1986): 6430–6441.
33. Maselko, J., and K. Showalter. "Chemical Waves on Spherical Surfaces." *Nature* **339** (1989): 609–611.
34. Müller, S. C., T. Plesser, and B. Hess. "Two-Dimensional Spectrophotometry of Spiral Wave Propagation in the Belousov-Zhabotinskii Reaction." *Physica D* **24** (1987): 87–96.
35. Müller, S. C., T. Plesser, and B. Hess. "The Structure of the Core of the Spiral Waves in the Belousov-Zhabotinskii." *Science* **230** (1985): 661–663.
36. Nagypal, I., and I. R. Epstein. "Fluctuations and Stirring Rate Effects in the Cholorite-Thiosulfate Reaction." *J. Phys. Chem.* **90** (1986): 6285–6292.
37. Nicolis, G., and I. Prigogine. *Self-Organization in Nonequilibrium Systems,* 153–156. New York: Wiley, 1977.

38. Noszticzius, Z., W. Horsthemke, W. D. McCormick, H. L. Swinney, and W. Y. Tam. "Sustained Chemical Waves in an Annular Gel Reactor: A Chemical Pinwheel." *Nature* **329** (1987): 619–620.

39. Ouyang, Q., W. Y. Tam, P. DeKepper, W. D. McCormick, Z. Noszticzius, and H. L. Swinney. "Bubble-Free Belousov-Zhabotinskii-Type Reactions." *J. Phys. Chem.* **91** (1987): 2181–2184.

40. Ouyang, Q., J. Boissonade, J. C. Roux, and P. DeKepper. "Sustained Reaction Diffusion Structures in an Open Reactor." *Phys. Lett. A.* **134** (1989) 282–286.

41. Ross, J., S. C. Müller, and C. Vidal. "Chemical Waves." *Science* **240** (1988): 460–464.

42. Roux. J. C., J. S. Turner, W. D. McCormick, and H. L. Swinney. "Experimental Observations of Complex Dynamics in a Chemical Reaction." *Nonlinear Problems: Present and Future*, edited by A. R. Bishop, D. K. Campbell, and B Nicolaenko, 409–422. Amsterdam: North-Holland, 1982.

43. Roux, J. C., R. H. Simoyi, and H. L. Swinney. "Observation of a Strange Attractor." *Physica D* **8** (1983): 257–266.

44. Saul, A., and K. Showalter. "Propagating Reaction-Diffusion Fronts." *Oscillation and Travelling Wave in Chemical System*, edited by R. J. Fields and M. Burger, 419–439. New York: Wiley, 1985.

45. Schuster, H. G. *Deterministic Chaos*. Weinheim: Physik-Verlag, 1984 .

46. Showalter, K. J. "Trigger Waves in the Acidic Bromate Oxidation of Ferroin." *J. Phys. Chem.* **85** (1981): 440–450.

47. Showalter, K., R. M. Noyes, and H. Turner. "Detailed Studies of Trigger Waves Initiation and Detection." *J. Am. Chem. Soc.* **101** (1979): 7463–7469.

48. Skinner, G. S., and H. L. Swinney. "Periodic to Quasiperiodic Transition of Chemical Spiral Rotation." *Physica D*, to appear.

49. Smoes, M. L. "Chemical Waves in the Oscillatory Zhabotinskii System. A Transition from Temporal to Spatio-temporal Organization." In *Dynamics of Synergetic Systems*, edited by H. Haken, 80–96. Berlin: Springer, 1980.

50. Swinney, H. L., and J. C. Roux. "Chemical Chaos." In *Nonequilibrium Dynamics in Chemical Systems*, edited by C. Vidal and A. Pacault, 124–140. Berlin: Springer, 1984.

51. Tam, W. Y., and H. L. Swinney. "Mass Transport in Turbulent Couett-Taylor Flow." *Phys. Rev. A* **36** (1987): 1374–1381.

52. Tam, W. Y., W. Horsthemke, Z. Noszticzius, and H. L. Swinney. "Sustained Spiral Waves in a Continuously Fed Unstirred Chemical Reactor." *J. Chem. Phys.* **88** (1988): 3395–3396.

53. Tam, W. Y., J. A. Vastano, H. L. Swinney, and W. Horsthemke. "Regular and Chaotic Chemical Spatiotemportal Patterns." *Phys. Rev. Lett.* **61** (1988): 2163–2166.

54. Tam, W. Y. "Pattern Formation in Chemical Systems." In *Nonlinear Structures in Physical Systems*, edited by L. Lam and H. C. Morris. New York: Springer, 1990. (Figure 1 on p. 89 should read: Continuously Fed Unstirred Reactor (CFUR).)

55. Tam, W. Y., and H. L. Swinney. "Spatiotemporal Patterns in a One-Dimensional Open Reaction-Diffusion System." *Physica. D*, to appear.
56. Thompson, J. M. T., and H. B. Stewart. *Nonlinear Dynamics and Chaos.* New York: Wiley, 1986.
57. Turner, J. S., J. C. Roux, W. D. McCormick, and H. L. Swinney. "Alternating Periodic and Chaotic Regimes in a Chemical Reaction-Experiment and Theory." *Phys. Lett. A* **85** (1981): 9–12.
58. Turing, A. M. "The Chemical Basis of Morphogenesis." *Phil. Trans. R. Soc.* **B237** (1952): 37–72.
59. Tyson, J. J., and P. C. Fife. "Target Patterns in a Realistic Model of the Belousov-Zhabotinskii Reaction." *J. Chem. Phys.* **73** (1980): 2224-2236.
60. Vastano, J. A. "Bifurcation in Spatiotemporal Systems." Ph.D. thesis, The University of Texas at Austin, 1988.
61. Vastano, J. A., T. Russo, and H. L. Swinney. "Bifurcation to Spatiotemporal Chaos in a Reaction-Diffusion System." Submitted to Physica D, 1990.
62. Welsh, B. J., J. Gomatam, and A. E. Burgess. "Three-Dimensional Chemical Waves in the Belousov-Zhabotinskii Reaction." *Nature* **304** (1983): 611–614.
63. Winfree, A. T. "Organizing Centers for Chemical Waves in Two and Three Dimensions." In *Oscillation and Travelling Wave in Chemical Systems*, edited by R. J. Field and M. Burger, 441–472. New York: Wiley, 1985.
64. Winfree, A. T. "Spiral Waves of Chemical Activity." *Science* **175** (1972): 634–636.
65. Winfree, A. T. "Rotating Chemical Reactions." *Sci. Am.* **230** (1974): 82–95.
66. Winfree, A. T., and S. H. Strogatz. "Organizing Centers for Three-Dimensional Chemical Waves." *Nature* **311** (1984): 611–615.
67. Winfree, A. T. "The Pre-history of the Belousov-Zhabotinskii Reaction." *J. Chem. Educ.* **61** (1984): 661–663.
68. Zaikin, A. N., and A. M. Zhabotinsky. "Concentration Wave Propagation in Two-Dimensional Liquid-Phase Self-Oscillating System." *Nature* **225** (1970): 535–537.
69. Zhabotinskii, A. M. "Periodic Processes of the Oxidation of Malonic Acid in Solution (Study of the Kinetics of Belousov's Reaction)." *Biofizika* **9** (1964): 306–311.

Sidney R. Nagel
The James Franck Institute and the Department of Physics, The University of Chicago,
Chicago, Illinois 60637

Experimental Analysis of Disordered Systems

A SUSCEPTIBILITY ANALYSIS OF THE GLASS TRANSITION

Disordered systems present a special challenge to anyone interested in understanding their underlying physical properties. We can see one aspect of this quite vividly if we compare an amorphous solid with a crystal or periodic system. Suppose you are given a material and are told that it has a particular crystal structure with a specified lattice constant (i.e., nearest-neighbor spacing). Once you are given the position of one atom and the orientation of the crystal, you will know the *exact* equilibrium position of every atom in the sample. A tremendous amount of information, the position of roughly 10^{23} particles, has been conveyed in just a few sentences. It is not surprising that, using mathematical tools, this information can be used efficiently to calculate many of the material's important physical properties. For an amorphous solid, on the other hand, although there may be some short-range order (such as that every atom is surrounded by a well-defined coordination shell of neighbors), there is no long-range order. All that we can say about the material is that at large distances an atom is surrounded by an *average* density of other particles. A knowledge of this average density is much less impressive than a knowledge of the exact positions of all the particles and it provides much less information with which to construct a theory.

Likewise, from an experimental point of view, it is inherently more challenging to work with disordered materials than crystals. Whereas in a simple crystal all sites are identical, in an amorphous solid each atom exists in a different environment. Even though the local order may be quite similar from site to site, the surrounding disorder can lead to long-range effects such as strains in the structure which may introduce distributions in almost every measured quantity. For example, in a perfect crystal, all atoms will have the same potential energy, whereas in an amorphous solid, there will be a wide distribution of energies—a distribution which is essential for the phenomenon of electron localization. Another property affected by disorder, and which is more pertinent to the subject of this lecture, is the relaxation time, τ, for a spin or dipole to change orientation. In a disordered medium, the relaxation times for such processes are often spread over many orders of magnitude. We can see the origin of this effect if we observe that the frequency, $\omega = \tau^{-1}$, for a particle to jump over a barrier of energy E is given by the Arrhenius or activated form: $\omega = \omega_0 e^{-E/k_B t}$. (Here k_B is Boltzmann's constant, T is the temperature, and ω_0 is a constant representing the attempt frequency for the system to cross the barrier.) The strain energies contribute to the energy barrier appearing in the exponential and thus have a profound effect on the relaxation times themselves: a small distribution of energies will lead to a very wide distribution of times. The single relaxation time in the ideal crystal has been replaced by an enormously wide distribution. The study of these distributions can become a formidable experimental task. We are no longer interested solely in measuring the time constant itself but in determining the entire *shape* of the distribution as well. The experimentalist is then faced with the objective of identifying not just a number but a continuous function. The task of measuring a broad response function is generally necessary in most experimental investigations of disordered systems.

In the first half of this lecture, I will focus on one long-standing problem in the physics of disordered materials, namely how a liquid, when supercooled sufficiently far below the equilibrium crystallization temperature, can freeze into an amorphous solid which we call a "glass." We will see that this question will lead us naturally to a study of relaxation-time distributions of the kind discussed in the last paragraph. These distributions are particularly well studied by susceptibility techniques.

Surprisingly little is known about the ubiquitous process of glass formation despite the long history and obvious technological importance of these materials.[1] One question that immediately presents itself is whether this "transition" from a liquid to a glass is a true thermodynamic phase transition or whether it is simply a gradual slowing down, or kinetic freezing, of all motion without an intrinsic thermodynamic singularity. Careful diffraction measurements[18] comparing the atomic structure of the liquid and the glass have shown that the structures of these two "phases" are indistinguishable. There is no indication in the structure that the sample has changed from a liquid to a frozen glass. Thus, the glass transition is a more

---

[1]See Wong and Angell[94] for an excellent review of much of the literature on the glass transition before 1976. For a more recent review of many aspects of the dynamics of glass-forming liquids, see *Dynamics Aspects of Structural Change in Liquids and Glasses*.[3]

subtle phenomena than we might at first expect. The most conspicuous physical change in the material is that the molecules themselves can no longer diffuse so that the liquid is unable to flow. We are therefore forced to concentrate our investigations on the dynamics of the liquid. Even if there is an underlying phase transition, it is nevertheless clear that kinetic effects are responsible for much of what is seen experimentally.[1] It is the subtle intertwining of kinetics with thermodynamics that makes this problem so interesting and complicated.

The study of critical phenomena over the past several decades has motivated the search for universal aspects in all transitions. Even if the change from a liquid to a glass cannot be regarded as a true phase transition, there is nevertheless the anticipation that something universal, independent of material, is underlying the process by which this change is made.

CRYSTALLIZATION AND GLASS FORMATION

Normally most liquids will freeze into a crystalline material. Crystallization, however, is a first-order transition which takes place via a nucleation and growth mechanism.[79,87] For crystallization to occur, a nucleus of the crystalline phase must be formed spontaneously and grow to envelop the entire system. The stability of a nucleus is determined by two contributions to its total free energy: the bulk energy which favors the nucleus since the crystalline state is the thermodynamically preferred state and an unfavorable surface term analogous to a surface tension. Since the former varies as the volume and the latter varies as the surface area, the criterion for the stability of a nucleus involves a critical radius where the two contributions to the total energy are equal in magnitude but opposite in sign. When the nucleus is greater than this critical size, it will continue to grow; when it is smaller than that size, it will tend to shrink. If the nucleus grows as a result of single particles either joining or leaving its surface, then it will take many attempts, and therefore a long time, before a nucleus of critical size is attained.

The critical size depends sensitively on the temperature. At T_m, the melting (or freezing) temperature, the bulk liquid is in equilibrium with the bulk crystal, and consequently there is no positive bulk contribution to the nucleus energy. With decreasing temperature, the crystal becomes energetically more favorable so that the contribution of the bulk free energy becomes greater and the size of the critical nucleus decreases. This occurs exponentially with the amount of supercooling.

There is another process important for calculating the nucleation rate. As the liquid is supercooled, its viscosity increases rapidly so that it becomes progressively more difficult to rearrange the particles. Therefore, at temperatures much lower than the freezing point, even though the critical size of a nucleus can be small, it will nevertheless take a very long time to form such a nucleus because the dynamics for rearrangement is so slow. At low enough temperatures, the rise in viscosity dominates the temperature dependence of the nucleation rate so that the nucleation rate will actually decrease as the temperature is lowered. The nucleation rate therefore grows rapidly as the temperature is lowered past T_m, reaches a peak at some intermediate degree of supercooling, and then decreases rapidly as the temperature is lowered even further.

In order to form a glass, it is essential to bypass the nucleation process. This is possible if the liquid is cooled rapidly past the temperature where there is a maximum in the nucleation rate. Upon further cooling, nucleation events become progressively less likely to occur. In this temperature region, it then becomes possible to decrease sufficiently the rate of cooling so that we can make physical measurements on the supercooled liquid without fear of the sample crystallizing and interrupting the experiment.

The most common example of a glass is silicon dioxide, the transparent material that covers our windows. It is less commonly known that many different kinds of liquids can also be formed into a glass.[22] It is plausible that all materials are potential glass formers if only the liquid could be cooled at a sufficiently rapid rate.[24,85,89] Thus there are, aside from the covalently bonded network glasses such as SiO_2, ionic glasses, glasses made from liquids held together by van der Waal's or hydrogen bonds, and metallic glasses. The rates for supercooling necessary for glass formation vary enormously. Some materials, such as As_2S_3, are difficult to form into a crystal and always become glasses in laboratory experiments.[4] Only on geological times scales have crystals of this mineral orpiment been made. For the fabrication of metallic glasses, on the other hand, cooling rates greater than a million degrees per second are often necessary.[85] In computer simulations,[13,42,46,70,73,78,92] it has been possible to make a glass out of a liquid of Lennard-Jones atoms which represents liquid argon. Such a glass has not been produced in the laboratory, but because of the enormously faster cooling rates achievable in a computer simulation, it can be observed in a computer experiment.

KAUZMANN ENTROPY AND COOLING RATES

Can a liquid be supercooled to indefinitely low temperatures and still not form a glass? A powerful argument[61] against such a possibility was given by Kauzmann in 1948. It rests on the calculation of the entropy, S, for a supercooled liquid. Since this state is not in equilibrium, it is not clear that the entropy can always be defined. In the case where there is an underlying crystalline phase, however, we can calculate the entropy as a function of the temperature, T, from the measured heat capacity, $C_p = \partial Q / \partial T \mid_p$, in the following way. From the thermodynamic relation

$$S(T_f) - S(T_0) = \int_{T_0}^{T_f} \frac{C_p(T)}{T} dT, \qquad (1.1)$$

we can use the heat capacity, measured from zero temperature to T_m in the crystal phase, to determine the entropy of the crystal just below the melting point. As the crystal melts, we can measure the latent heat and obtain the entropy of the liquid. Now, as we recool the liquid below the freezing temperature T_m, we can again measure the heat capacity and use Eq. (1.1) to determine the entropy of the supercooled liquid.

In Figure 1, I show a schematic plot of how the heat capacity and entropy evolve in the various regions of interest: crystal, liquid, supercooled liquid, and glass.[72] The heat capacity of the liquid is, in general, much larger than that of the crystal, which is approximately the classical value of 3R per mole of sample. The supercooled liquid is a smooth continuation of this liquid behavior to lower temperature. At a temperature T_g which depends on the cooling rate, C_p of the liquid drops to a value close to that of the crystal. Below this point, the heat capacity of the glass and the crystal are similar. The slower the cooling rate, the lower will be the temperature T_g. In the figure, therefore, the dashed curve is cooled at a faster rate.

As outlined above, the entropy, S, can be obtained from this heat capacity curve. As shown in the figure, the slope of S versus T is largest in the liquid and supercooled liquid because the C_p is largest in these regions. As the temperature decreases, the supercooled liquid looses entropy faster than a crystal does at the

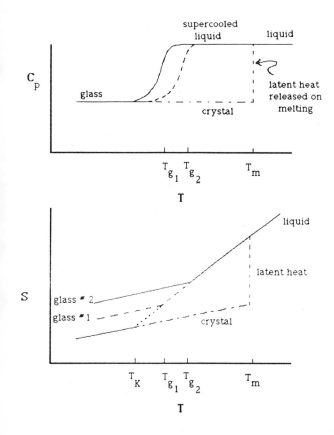

FIGURE 1   A schematic diagram of the heat capacity, C_p, and entropy, S, of a crystal, liquid, supercooled liquid, and glass as a function of temperature. Glass #1 and glass #2, with apparent glass transition temperatures T_{g1} and T_{g2} respectively, are obtained for different rates of cooling. In the upper figure, the dashed curve represents the result of a cooling rate which is faster than that used to produce the solid curve.

same temperature. Thus, although the entropy of the liquid near the melting temperature is higher than that of the crystal (because of the release of latent heat), the entropies of the two phases quickly approach one another because of the difference of slopes. If this behavior continued to an arbitrarily low temperature, then the entropy of the liquid would cross that of the crystal at the "Kauzmann temperature," T_K. This is apparently a paradoxical result since one would not normally expect to have the liquid with a smaller entropy than that of the ordered crystal. To prevent this situation from occurring, the glass transition must intervene so that the entropy of the glass remains higher than that of the crystal. Although counter-intuitive, this argument does not by itself violate any law of thermodynamics. Nevertheless, if we were to continue this argument further and let the liquid entropy continue to decrease below T_K without any change in slope, we soon would have a violation of the third law since S would become negative before $T = 0$.

I have noted that the slower one cools the liquid, the lower will be the value of T_g. The Kauzmann argument effectively puts a limit on the lowest value of T_g: $T_g \geq T_K$. Thus, no matter how slowly one cools the liquid, there should be a heat capacity anomaly (i.e., a drop in the value of C_p) at a temperature above the Kauzmann temperature.

TIME SCALES, THE FALL FROM EQUILIBRIUM, AND THE DEFINITION OF A GLASS

The curves shown above for the heat capacity are not in equilibrium. There are two ways in which this is true. First, as soon as we have cooled below T_m, the equilibrium phase is the crystal and the supercooled liquid phase is metastable. This presents some experimental difficulties, as I mentioned in a previous section, but it does not really concern us if we are studying the glass transition. In effect, we are studying the behavior of the liquid *as if a crystalline state did not exist.* Theoretically this essentially means that we have removed the very small region of phase space where the crystal exists without disturbing the rest of the liquid phase space. Although it is a plausible to ignore the existence of a crystal phase in this manner, it is not completely satisfactory. I do not know of a better solution.

For the purposes of the present discussion, however, there is a more important manner in which the curves in the figure are out of equilibrium. If we follow the dashed curve in the upper figure, which was cooled at a high cooling rate, we realize that, as soon as it starts to deviate from the high-temperature liquid behavior, we are no longer in equilibrium with the liquid regions of phase space. Clearly, if we had only been a little more patient in our cooling, we would have obtained the solid curve which remains in the liquid regime to lower temperatures. Thus, whenever we see a deviation from liquid behavior, it implies that we are out of equilibrium.[14]

The fall from equilibrium that appears at different temperatures depending on the cooling rate indicates that there is an inherent relaxation time, τ, in the system. As the temperature is lowered, this time increases rapidly and crosses the experimental time scale, that is, the time necessary to do the experiment. For slow cooling rates, the experimental time scale is longer so that τ does not cross it until a lower temperature is reached.

It should be no surprise that a relaxation time is important for glass transition phenomena. After all, the most obvious physical change in going from a liquid to a glass is the increase in viscosity, η; the liquid becomes progressively more sluggish as the glass transition is approached.[45] Since $\eta = G_\infty \tau$ (where G_∞ is the infinite frequency shear modulus and τ is the Maxwellian relaxation time[64]), then the rapid increase in the viscosity implies a similarly rapid increase in τ.

The rise in viscosity occurs very rapidly as the temperature is lowered. Indeed one experimental definition[88] of where the glass transition occurs is when h reaches the value 10^{15} poise (1 poise = 1 dyne sec/cm^2). This value corresponds to a relaxation time on the order of a day for the rearrangement of a molecule. This is assumed to be the typical limit to the patience of an experimenter. (For comparison the viscosity of water or alcohol at room temperature is 10^{-2} poise.) In Figure 2, the viscosity data for salol, an organic glass-forming liquid, is shown in the vicinity of the glass transition.[66] From this figure, one can clearly see that, as the glass transition is approached, the temperature dependence of the viscosity (or equivalently the relaxation time) departs from Arrhenius behavior, $\eta = \eta_0 \exp(A/T)$ (which would have been a straight line on this kind of plot). The deviation is in the direction of a greater temperature dependence.

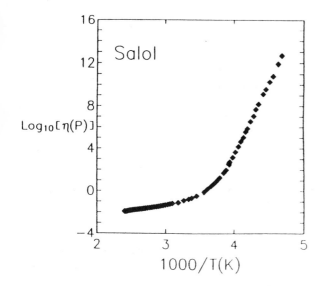

FIGURE 2 The logarithm of the viscosity of salol as a function of the inverse temperature. This form for plotting the data is chosen since a straight line would indicate Arrhenius, or activated, behavior. The data is taken from Laughlin[66] and Cukierman.[28]

The temperature dependence of η, at least near the glass transition, can often be fit quite well with the Vogel-Fulcher form[91]:

$$\eta = \eta_0 \exp\left(\frac{A}{T - T_0}\right). \tag{1.2}$$

Although this is one of the most commonly used fitting forms, it does not by any means fit the data over the entire temperature range. Another form that has recently been used[12,29,43,63,67] to fit this kind of data is power law behavior motivated by analogies with critical phenomena,

$$\eta = \eta_0\left(\frac{T - T_0}{T_0}\right)^{-\alpha}. \tag{1.3}$$

Many other forms have been tried,[11,71] but there is no clear evidence that any one form is better than the others. In particular, none of the forms fit the data over the entire range of temperature.

One question naturally presents itself at this point: do different probes (such as viscosity and specific heat) measure the same relaxation time or are there different time scales which become important for different experiments in the supercooled liquid? As we shall see later in this lecture, it appears that there is only one relaxation time that is important for all the different experiments which we have been able to perform. In this case, then, since the Vogel-Fulcher form works well for the viscosity (at least near the glass transition temperature), it can also be used to describe the temperature dependence of the relaxation times as measured by the different experiments.

We can already see that there are a plethora of possible definitions of the glass transition temperature:

1. There is the temperature where $\eta = 10^{15}$ poise.
2. There is the temperature where the heat capacity drops from a liquid-like to a solid-like value.
3. There is the Kauzmann temperature where the entropy of the liquid crosses that of the crystal.
4. There is the temperature T_0 in the Vogel-Fulcher equation (or other chosen fitting form) where the relaxation times diverge.

The problem with the first definition is that the value $\eta = 10^{15}$ poise, where the time scale is one day, is purely arbitrary and there is no unanimity in its choice; indeed, the value $\eta = 10^{13}$ poise is often chosen instead.[2] The problem with the second one is that the value chosen depends on the cooling rate and is not intrinsic to the sample. The third and fourth definitions rely on extrapolations over a very large temperature range. Depending on what relaxation function one chooses, one will get very different values for the glass transition temperature. Nevertheless, despite this essentially experimental difficulty of getting close enough to T_0 or T_K, these last two definitions at least have a chance of giving a temperature which is intrinsic to the equilibrium sample. To use them, we need good experimental data to help guide us in choosing the correct extrapolation function.

STRONG AND FRAGILE LIQUIDS

As mentioned above, the Vogel-Fulcher form only fits a portion of the viscosity data and cannot be used over the entire range of temperature for most materials. Apparently, there are a variety of different kinds of liquids which can be quenched into the liquid state. Angell[2] has classified these liquids into two categories: strong and fragile. He plots the logarithm of the viscosity versus inverse-scaled temperature. This graph is similar to what we have done in Figure 2 except that, for each sample, the temperature is divided by the glass transition temperature T_g where the viscosity reaches 10^{13} poise. Thus, the data for all the liquids coincide at $T_g/T = 1$. At high temperatures $T_g/T = 0$, the data again approach a common value since, in the non-viscous limit, liquids have an approximately sample independent value for the viscosity $\eta \approx 10^{-2}$ poise. Different glasses show different amounts of curvature on such a plot. We show such a plot, taken from Angell,[2] in Figure 3. Those glasses that are characterized as "strong," such as the covalently bonded network glass SiO_2, show no curvature and therefore have an Arrhenius dependence of the viscosity on temperature. At the other limit, the glasses which show the most curvature, such as the associated liquids salol and o-terphenyl, are deemed to be "fragile." The viscosity of these liquids tends to decrease rapidly above the glass transition temperature so that they flow easily at slightly elevated temperatures. This is the rationale for the nomenclature "fragile" and "strong."

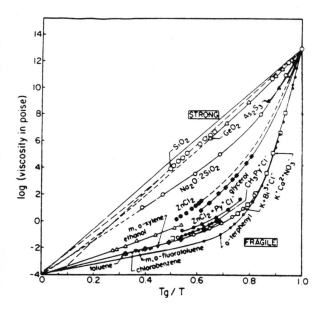

FIGURE 3   A plot of the viscosity versus temperature developed by Angell to classify glass-forming liquids. The temperature is scaled by the glass transition temperature for each sample, where $\eta = 10^{13}$ poise. The data are taken from Angell.[2]

For a given value of the viscosity, the fragile glasses are closer to their value of T_g than are the strong ones. Thus, if one's aim is to investigate the behavior near the glass transition, it is natural to choose to study the fragile glass formers. In addition, the Kauzmann "paradox" is more important for these fragile glass formers.[2] In these liquids, the difference between the liquid and crystalline specific heat is relatively large so that the Kauzmann temperature is often not far below the experimentally accessible temperature range. In a strong liquid like SiO_2, the liquid and solid specific heats are nearly the same so that the extrapolated value of T_K is close to $T = 0K$. Since, as I argued above, the Kauzmann temperature puts a limit on how far a liquid can be supercooled without undergoing a transition, the fragile liquids are again the systems of choice for experiments seeking to investigate the nature of the glass transition. Most of the data that I will discuss in the remainder of this lecture will be on the fragile liquids.

SUSCEPTIBILITY PROBES OF RELAXATION PHENOMENA

Many different experimental probes have been used to study relaxation phenomena in glasses. An important class of experiments has to do with measuring the frequency-dependent susceptibility of the material to a small perturbing field. For example, dielectric susceptibility[30] can detect the reorientation of polar molecules due to the presence of a small oscillating electric field, specific heat spectroscopy[14,15,23] measures the enthalpy relaxation due to a small perturbation of the temperature and ultrasonic attenuation[31] measures the response to a density perturbation. In all of these experiments, the response of the sample is measured as the frequency of the perturbation is varied. I will present in this section some of the theory which is common to all these techniques and which will allow us to see how these experiments probe the relaxation phenomena in which we are interested. I will restrict consideration to the case where the response is linearly proportional to the magnitude of the perturbation.

In order to be concrete, I will treat the case of the dielectric susceptibility, although the same results will be applicable to any other linear susceptibility. In the case of dielectric susceptibility, the perturbation is a small electric field E and the response is the electric displacement vector D. The displacement vector $D(t)$ at time τ is the sum of the system's response to the electric field, $E(t')$, at all earlier times t':

$$D(t) = \int_{-\infty}^{\infty} \epsilon(t - t')E(t')dt' \qquad (1.4)$$

where $\epsilon(t - t')$ is the dielectric constant. Since there cannot be a response until the field is present, $\epsilon(t - t') = 0$ for $(t - t') \leq 0$. We can take the Fourier transform of both sides of this equation in order to find the response to a perturbation varying sinusoidally with frequency ω. We find:

$$D(\omega) = \int_{-\infty}^{\infty} D(t)e^{i\omega t}dt$$

$$= \int_{-\infty}^{\infty} \epsilon(t - t')e^{i\omega(t-t')}d(t - t') \int_{-\infty}^{\infty} E(t')e^{i\omega t'}dt' \qquad (1.5)$$

$$= \epsilon(\omega)E(\omega)$$

If $E(t')$ is a step function, then the response will, in general, be a complicated function $D(t) = D_0\phi(t)$. I will begin with the Debye model which is the response of a system with only a single relaxation time, τ. For such a simple system, the response to a step function perturbation will be an exponential decay from the initial to the final state. Thus, if

$$E(t) = \begin{cases} E_0 & t < 0 \\ 0 & t \geq 0, \end{cases}$$

then

$$D(t) = \begin{cases} D_0 = \epsilon_0 E_0 & t < 0 \\ (\epsilon_0 - \epsilon_\infty)E_0\phi(t) = (\epsilon_0 - \epsilon_\infty)E_0 e^{-1/\tau} & t \geq 0 \\ \quad + D_0\phi(t) = (\epsilon_0 - \epsilon_\infty)E_0 e^{-t/\tau} \end{cases}$$

Here ϵ_0 is the dielectric constant evaluated at $\omega = 0$ and ϵ_∞ is the dielectric constant evaluated at the high frequency limit $\omega \to \infty$. Note that, unless $\epsilon_\infty = 0$, there is an instantaneous jump in $D(t)$ at $t = 0$ when the field is turned off. In order to calculate $D(\omega)$ in Eq. (1.5), we need to know the response to a delta-function field at time t'. This can be immediately obtained from $\phi(t)$ which in Eq. (1.6) is the response to a step function: $\epsilon(t) = \epsilon_\infty\delta(t) - (\epsilon_0 - \epsilon_\infty)D\phi(t)/dt$. The response is given by

$$D(\omega) = E(\omega)\left(\epsilon_\infty + (\epsilon_0 - \epsilon_\infty)\int_0^\infty e^{i\omega t}\frac{d}{dt}\{-e^{-t/\tau}\}dt\right)$$

$$= E(\omega)\left(\epsilon_\infty + (\epsilon_0 - \epsilon_\infty)\frac{1}{1 - i\omega\tau}\right) \qquad (1.7)$$

$$= E(\omega)\left(\epsilon_\infty + (\epsilon_0 - \epsilon_\infty)\left(\frac{1}{1 + \omega^2\tau^2} + i\frac{\omega\tau}{1 + \omega^2\tau^2}\right)\right).$$

The dielectric constant for the Debye (single relaxation time) model is therefore

$$\epsilon(\omega) = \epsilon_\infty + (\epsilon_0 - \epsilon_\infty)\left(\frac{1}{1 + \omega^2\tau^2} + i\frac{\omega\tau}{1 + \omega^2\tau^2}\right). \qquad (1.8)$$

There is both a real and an imaginary part to $\epsilon(\omega)$. The real part, ϵ', indicates the in-phase response, while the imaginary part, ϵ'', indicates the out-of-phase response of the system to the applied perturbation. The imaginary part of the response is symmetric when plotted versus $\log_{10}\omega$ and has a full width at half maximum of $W_D = 1.14$ decades. The angular frequency, $\omega_p \equiv 2\pi\nu_p$, at which ϵ'' is a maximum occurs when $\omega = 1/\tau$. Thus, a susceptibility measurement which measures $\epsilon(\omega)$ over a range of frequency can determine the value of τ for each temperature at which the experiment is performed. As I discussed above, τ approximately obeys Eq. (1.2), the Vogel-Fulcher equation, as the temperature is lowered towards the glass transition. Thus, τ increases rapidly, so that ω_p and the susceptibility curves move to lower frequency, with decreasing temperature. In Figure 4, I have shown some typical dielectric data for salol,[35] a fragile glass-forming liquid, taken at several different temperatures. The characteristic shift of the curves to lower frequency as the temperature is lowered can be seen.

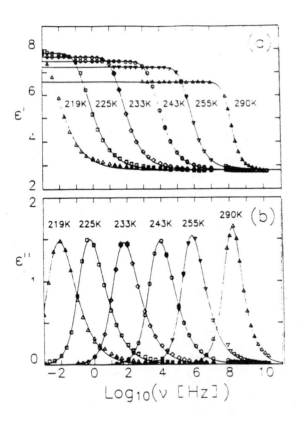

FIGURE 4　The real and imaginary parts of the dielectric susceptibility of salol as a function of $\log_{10}\mu$. As marked in the figure, each curve represents a different temperature. The solid lines are the best-fit curves using the stretched exponential form. The data is taken from Dixon.[35]

In any real experiment, it is quite rare to find the simple Debye relaxation function. Instead, the response function is often significantly broader than that given by Eq. (1.8). This can be seen in the data shown in Figure 4, where the width becomes noticeably larger as the temperature is lowered. The most natural explanation for a width broader than that given by the Debye model is that there is more than one relaxation time in the material. The total response is therefore a sum of the responses from all the different relaxation mechanisms, each with its own characteristic relaxation time. Thus, we can take $\phi(t) = \int_0^\infty \rho(\tau)e^{-t/\tau}d\tau$ where $\rho(\tau)$ is the distribution of relaxation times. This translates in the frequency domain to

$$\phi(\omega) = \epsilon_\infty + (\epsilon_0 - \epsilon_\infty)\int_0^\infty \rho(\tau)\frac{1}{1 - i\omega\tau}d\tau$$

$$= \epsilon_\infty + (\epsilon_0 - \epsilon_\infty)\int_0^\infty \rho(\tau)\frac{1}{1 + \omega^2\tau^2}d\tau + i(\epsilon_0 - \epsilon_\infty)\int_0^\infty \rho(\tau)\frac{\omega\tau}{1 + \omega^2\tau^2}.$$
$$(1.9)$$

This is a general way to fit any broad susceptibility peak.

Susceptibility data taken by different probes appear to have certain similarities with one another. For example, not only are the peaks in ϵ'' broader than those given by a Debye spectrum ($W = 1.14$ decades), but, as in the data shown in Figure 4, they generally tend to be asymmetric with a tail to high frequencies. This is true not only of data from dielectric susceptibility but is also true of data from ultrasonic attenuation,[56] specific heat spectroscopy,[14,15,23] shear modulus,[57] and other probes. The similarities are so prevalent that it is unsatisfying to express the relaxation in terms of a distribution of relaxation times that in general would be expected to vary from one sample to another. Instead, a number of phenomenological forms for the relaxation have been suggested. Although these forms lack a solid theoretical justification, they have been moderately successful at fitting the data over limited ranges of frequency. One of the most often quoted of these forms was suggested by Davidson and Cole[30]:

$$\epsilon(\omega) = (\epsilon_0 - \epsilon_\infty)\frac{1}{(1 - i\omega\tau)^\beta} \qquad \text{for } 0 \leq \beta \leq 1. \qquad (1.10)$$

This form, which equals the Debye form when $\beta = 1$, becomes broader and develops an asymmetric high frequency tail as β decreases. A second form due to Cole and Cole[25] is also broader than Debye but is symmetric in shape:

$$\epsilon(\omega) = (\epsilon_0 - \epsilon_\infty)\frac{1}{1 + (i\omega\tau)^{1-\alpha}} \qquad \text{for } 0 \leq \alpha \leq 1. \qquad (1.11)$$

Again, when a = 0, this reduces to the Debye case. In recent years the stretched-exponential form (also known as the Kohlrausch-Williams-Watts form[62,93]) has become increasingly popular. This form cannot be expressed analytically as a function of frequency but must be numerically integrated from its form in the time domain:

$$\phi(t) = \phi_0 e^{-(t/\tau)^{\beta}} \qquad \text{for } 0 \leq \beta \leq 1. \qquad (1.12)$$

As with the Davidson-Cole form, in the frequency domain it is broader than Debye with an asymmetric tail to high frequencies. In Figure 4, the solid lines through the data points are the best fits to this stretched-exponential form. There have been a number of theoretical derivations of this form for the relaxation.[10,19,26,33,59,76] I will not review this literature here, since some of these approaches have been reviewed in a previous volume in this lecture series.[77]

SUSCEPTIBILITY DATA OF GLASS-FORMING LIQUIDS

In Figure 4, I have shown the dielectric susceptibility data for the fragile, glass-forming liquid salol which is typical of many other materials. One of the first questions that one might ask is whether the relaxation times obtained from this dielectric susceptibility data are the same relaxation times as would be determined by other probes. Since these curves are wider than the simple Debye form, the frequency $\nu_p = \omega_p/2\pi$, where ϵ'' has a peak at any given temperature, does not measure the unique relaxation frequency of the liquid but rather determines an average value for the distribution of relaxation frequencies. If we plot $-\log \nu_p$ versus $1/T$, we obtain a curve reminiscent of that shown in Figure 2.

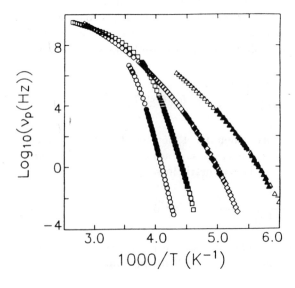

FIGURE 5 The peak frequency in the susceptibility curves versus the inverse temperature for four liquids. The circles are for o-terphenyl, the squares are for salol, the diamonds are for glycerol, and the triangles are for propylene glycol. The open symbols are taken from dielectric susceptibility measurements and the closed symbols are taken from specific heat spectroscopy data. Data is taken from Wu et al.[95]

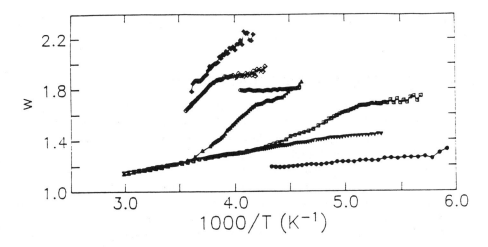

FIGURE 6   The normalized width of $\epsilon''(\omega)$, $w = W/W_D$, plotted versus temperature for several glass-forming liquids: glycerol (∇), propylene glycol (solid circle), salol (solid triangle), dibutyl-phthalate (open square), α-phenyl-o-cresol with 13% o-terphenyl impurity (open circle), and o-terphenyl with 9% (solid diamond) and 33% (open diamond) o-phenylphenol impurity. Data is taken from Dixon et al.[35]

In Figure 5, I show a plot for four different liquids (glycerol, salol, propylene glycol, and o-terphenyl) of $\log_{10}(\nu_p)$ versus $1/T$ obtained by two different susceptibility probes.[95] The open symbols are from dielectric susceptibility measurements and the solid symbols are taken from specific-heat spectroscopy measurements. The dielectric susceptibility data were taken over a very wide range of frequency (in some cases 14 decades), whereas the specific-heat data could only be obtained over a narrower range (roughly 6 decades). Nevertheless, where they overlap, the two different probes give nearly identical values for ν_p. Thus, the average relaxation time at any temperature determined from the specific heat, which couples to *all* relaxing modes in the liquid, is the same as that obtained from dielectric susceptibility, which couples predominantly to dipolar reorientation. Similar comparisons have been made[16] with data taken from ultrasonic attenuation, specific heat spectroscopy, and the frequency-dependent shear modulus. Again all the probes gave essentially the same average relaxation time.

As we saw in Figure 4, the width, W, of the peak in ϵ'' varied as a function of temperature. In Figure 6, I show the width of the spectra normalized to the value given by the Debye model: $w = W/W_D$. At high temperatures[35] (which also means high peak frequencies), w approaches 1. Thus, a single relaxation-time model appears to fit the relaxation function when ν_p approaches a typical phonon frequency. Although for all the different probes the average values for the relaxation times are the same, the widths of the distributions are not. Thus, for some samples, such as glycerol, the width determined from dielectric susceptibility is narrower than that obtained from specific heat data; for other samples, such as salol[35] and o-terphenyl,[34] it is the other way around.

UNIVERSAL CURVES FOR DIELECTRIC RESPONSE

It is clear that the behavior of these samples, as shown by the susceptibility measurements, have many features in common. They are broader than would be expected from a single relaxation-time model, they are all slightly asymmetric, they seem to approach the Debye form at high temperatures (and frequencies), and the average relaxation times appear to diverge (or at least vary faster than Arrhenius behavior) at a finite temperature. Nevertheless, it is not clear whether these curves demonstrate any truly *universal* features. For example, in Figure 4 we see that, when the data is fit by a stretched exponential form, the high frequency tails do not lie on the data. One may ask whether these deviations are unique to this sample or whether they are generic to all the samples at all temperatures. In order to demonstrate such generic behavior, we would want to be able to place the data, appropriately scaled, from one sample directly on top of that from another. Before this can be achieved, we need, for a single sample, to be able to take the data for all temperatures and place it on a single master curve.

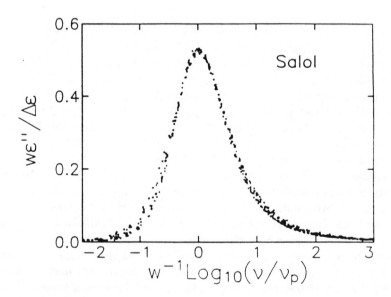

FIGURE 7 An unsuccessful attempt, using the most natural scaling variables, at making a master curve for the salol dielectric susceptibility data. The data is taken from Wu.[95]

In order to make such a master curve, we might expect that all that needs to be done is to align the peaks' positions and then take account of the variation in width. This can be achieved by plotting the data versus $1/W \log_{10}(\nu/\nu_p)$. The division by ν_p brings the peaks into alignment and the division by w makes all the half widths the same. In order to get the height to be uniform, the values of ϵ'' should be divided by $\Delta\epsilon = \epsilon_0 - \epsilon_\infty$ and multiplied by w. (From Kramers-Kronig analysis, we know that the area under $\epsilon''(\omega)$ plotted logarithmically in frequency is just $\Delta\epsilon$. Therefore $\Delta\epsilon/w$ should be proportional to $\epsilon''(\omega_p)$.) In Figure 7, I show an attempt to make a master curve in this manner for salol.[95] The result is not very successful. The data from different temperatures do not lie on top of one another in the tails of the curve. A more sophisticated scaling procedure must be used if a single master curve is to be found for this data.

In Figure 8, I show a much more successful scaling procedure which does collapse all the salol data onto a single curve.[36] The abscissa is now $1/w(1 + 1/w)\log_{10}(\nu/\nu_p)$ and the ordinate is $1/w \log_{10} \epsilon''\nu_p/\nu\Delta\epsilon$. The division of $\epsilon''/\Delta\epsilon$ by the normalized frequency, ν/ν_p, in the logarithm tilts the curve and decreases its slope on a logarithmic plot so that it is monotonically decreasing. The multiplication by $(1 + 1/w)$ makes the high frequency slopes the same for all temperatures. The division by w on both axes shifts the curves so that they lie on top of one another for all temperatures. This division by w on both axes is reminiscent of the multifractal fitting procedure[48,49,58] done in the analysis of chaos data. At low frequency, $\nu < \nu_p$, the curves are all horizontal and lie on top of one another indicating that ϵ'' is proportional to ν in this region.

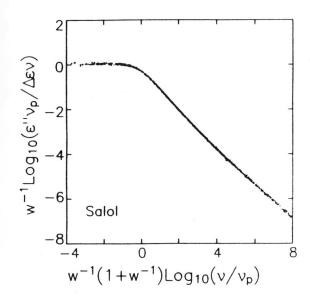

FIGURE 8   A successful attempt at making a master curve for the salol dielectric susceptibility data using the rather unconventional scaling axes $1/w(1 + 1/w)\log_{10}(\nu/\nu_p)$ and $1/w \log_{10} \epsilon''\nu_p/\nu\Delta\epsilon$. The data is taken from Dixon et al.[36]

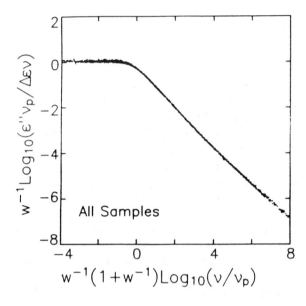

FIGURE 9 A master curve of all the dielectric data at all temperatures on several different samples. The data is taken from Dixon et al.[36]

If we now plot the susceptibility data for different samples on the same graph, we find that we do get the desired result: all the data for all temperatures on all samples lie on a single "universal" curve. This is shown in Figure 9.

We can now claim that there is universal behavior in the susceptibility data on glass-forming liquids, albeit not of the simple kind that we might at first have expected. It only appears when the data is scaled in this unconventional way which is motivated to some extent by the multifractal scaling of chaos data. It is also clear from this scaling procedure that none of the forms that have been suggested so far in the literature can fit the data over its entire range. I also note here that the dielectric data that was used in these fits covers an exceptionally large range of frequency—in some cases over 14 decades. Thus, this universal behavior is found over almost the entire range that is experimentally accessible. It will be difficult to extend the range by many more decades.

CONCLUSIONS

I have presented in this lecture a susceptibility study of the dynamics of supercooled liquids as they are about to undergo the glass transition. It should be apparent from this discussion that some of the fundamental questions regarding the nature of the glass transition remain unanswered. For example, the experimental evidence so far does not reveal whether or not the glass transition is a true phase transition. It is difficult to tell whether the relaxation times deduced from susceptibility data diverge at a finite temperature or whether they only do so at zero temperature. The data shows curvature on a plot of $\log \tau$ versus $1/T$. This indicates that perhaps a scaling, Vogel-Fulcher, or one of the other fitting forms may be a good fit to the data. However, it is apparent that the curvature decreases as the temperature is lowered. This has previously been interpreted as the system reverting to Arrhenius behavior[28,66] at low temperatures. A careful analysis of the susceptibility data, however, has shown that the there is still curvature visible for the relaxation time in all the samples down to the lowest value of temperature and frequency measured.[35,36,34] Although this is suggestive that there is a phase transition, it is certainly not conclusive evidence of its existence.

A more convincing experiment would be one which measures an increasing correlation length as the temperature is lowered towards the glass transition. Several attempts have been made to observe such a length scale related to the order parameter of the glass. One of these, in analogy with a measurement which coupled to the order parameter in spin-glasses, was to measure the nonlinear dielectric susceptibility of a glass.[96] This experiment showed no evidence of a diverging correlation length. Monroe[69] has suggested that such a nonlinear susceptibility probe may not measure a correlation length in the same simple way for a glass as it does for a spin glass. A second measurement attempted to see a difference between the viscosity when measured on different microscopic lengthscales.[37] This experiment also gave a null result and indicated that, if a diverging correlation length does exist, it does not couple to the viscosity. A series of computer simulations[39] of the glass transition has also searched for a divergent lengthscale in properties which would not be easily accessible to a laboratory experiment. In these calculations, correlation functions associated with bond orientational order as well as with the density correlations were examined. These correlation functions did not show any signature of an order parameter growing as the temperature was lowered.

These experiments must certainly caution us not to jump too readily to the conclusion that there is a true underlying phase transition as a liquid is cooled into a glass. However, they do not rule out that possibility, since the order parameter associated with the glass may be very subtle and may therefore be difficult to measure directly. Without an adequate theory to help us identify the order parameter, the experimental attempts at measuring the order parameter may simply be probing the wrong observables.

One important outcome of the susceptibility measurements reported in this lecture has been the demonstration that the spectrum of relaxation times for many different samples measured at different temperatures do fall on a single universal curve. Thus, what I indicated in the introduction to this lecture might be true of disordered systems generally and is certainly true here: the *shape* of a distribution function, as distinct from simply its average value, has become one of the most interesting aspects of the glass transition. Why do the susceptibility curves have the shape that they have, why is it universal, and why is it necessary to use the rather non-intuitive scaling variables as the natural variables for the problem? Does the fact that there is a single master curve for all this data indicate that there is an underlying universality similar to the universality appearing in phase transitions? Does the fact that both axes in Figure 9 are multiplied by w^{-1} indicate that the physics of the glass transition is dominated by multifractal scaling? If so, what does this type of scaling tell us about the nature of the transition? The susceptibility data has opened up a new set of questions that one can ask about the glass transition. These are questions that need to be addressed by any theory that hopes to describe the dynamics of supercooled liquids. Although these types of experiments may not have answered many of the questions we originally had about the glass transition, it has perhaps given us some new ways to think about the problem.

AVALANCHES AND EARTHQUAKES[2]

Sand in a sandbox, snow on a hillside, screes at the bottom of a mountain, salt piles at the side of a road, and sugar in a bowl all form piles with a non-trivial shape. These, in common with other examples of granular materials, will generally be found in mounds with an irregular and non-horizontal upper surface. This behavior is in sharp contrast to what is found in most normal fluids where the upper surface is invariably flat. Although granular materials are stable if the upper surface rests at a sufficiently gentle angle, they nevertheless are unstable if the slope (i.e., the angle the top surface makes with respect to the horizontal) becomes too great. In that case, an avalanche will occur as the material begins to flow. The flow will continue until the pile is again stable, at which point the surface is at "the angle of repose." The physics of granular materials, and of avalanches in particular, is not well understood. Nevertheless, sandpiles have often been used as a metaphor for other phenomena in physics such as the motion of vortices in superconductors[32] or the behavior of spin glasses.[84]

In 1773, Coulomb[27] made the connection between the angle of repose, θ_r, and the static internal friction of a given material. He postulated a friction coefficient between the grains in a pile which related the shear and the normal stresses within the material. This allowed a phenomenological understanding of the static non-zero angle of repose. However, to understand the behavior of avalanches, we need to know not only the static shape of a pile but also how a granular material flows when the pile is tilted past the point of stability. Reynolds,[51,80] in 1885, provided one of the crucial ingredients to our understanding of flow in these materials by introducing the notion of dilation; in order to flow, or even deform at all, a granular material must expand. You can observe this phenomenon of dilation every time you walk on a beach and see the wet sand around your foot dry out as you put your weight on it. The deformation caused by the pressure of your foot causes the sand to expand and allows the water to drain away. In 1954, Bagnold[5,6] initiated the study of rapidly flowing granular material. Subsequently, much effort has been expended

---

[2]Much of what appears in this first section on avalanches is a modified version of another review paper. See Jaeger et al.[97]

in order to measure and calculate the various stresses that are built up in such systems. An excellent review of this work has been given by Savage.[81] Although much is now known about the time-independent bulk flow properties of granular materials, the instabilities and fluctuations that occur in these flow patterns are less well understood. These instabilities can be important in developing and controlling many of the widely used industrial processes that employ granular materials.

In order to get a qualitative understanding of an avalanche, it is appealing to think of a generalized phase transition occurring at the angle of repose, θ_r: for slopes $\theta < \theta_r$, no flow of sand can occur and the pile appears to be solid, whereas for $\theta > \theta_r$, the sand will flow freely downhill. Bak, Tang, and Wiesenfeld[7] introduced the idea of "self-organized criticality" which relies on an analogy with a second-order phase transition. This idea may be of general applicability but is particularly easy to express it in terms of a model of how a model sandpile might be expected to behave. They argued that, when the slope is small, any motion in the pile should be due to local rearrangements. The size of the effected region should grow as the slope is increased. Above some critical value, however, a global avalanche should occur as the system relaxes back to the angle of repose. They suggested that the angle of repose could, in the language of phase transitions, be thought of as "critical": the size of an avalanche would increase with the slope of the pile until it spanned the entire system at θ_r. However, in order to reach this critical behavior, no parameters would need to be adjusted by the experimentalist: the system *organizes itself* in a critical state. A "self-organized critical state," characterized by long-range spatial and temporal correlations, is therefore predicted to occur at the steady-state angle (i.e., the angle where the current of particles added into the system equals the current of particles leaving the system through avalanches). Such correlations could give rise to many types of scaling behavior, such as power law or "1/f" power spectra of fluctuations around the steady-state particle flow. Self-organized critical behavior was found in cellular-automata computer simulations[7] which modeled the behavior of avalanches. The scaling properties of these cellular-automata computer models[58,74] have been studied in detail. In general, the behavior is more complicated than is found in conventional critical phenomena. In one dimension, it was found that a multifractal analysis[48,49] gave a better fit to the simulation data than did a simple finite-size scaling hypothesis. In two, three, and four dimensions, scaling seemed to work better but was still slightly inferior to the multifractal fits.

Do *real* sand piles behave in the manner predicted by the ideas of self-organized criticality? The avalanches in a sand pile can be studied in a number of different experimental configurations.[53] Schematic diagrams of three such experiments are shown in Figure 10. In Figure 10(A), sand is sprinkled randomly onto the top surface of a pile confined by three walls and with one open side. As sand is added randomly to the top surface, the amount of sand falling off the platform at the open boundary is measured by a capacitor underneath the platform's edge. Capacitance changes correspond directly to changes in the flow of particles. The resolution was sufficient to detect the passage of only a single particle. The time sequence of capacitance changes, ΔC, was fed into a spectrum analyzer to obtain its

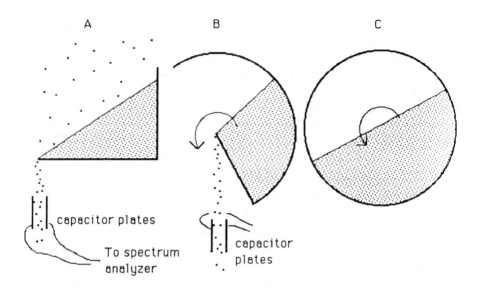

FIGURE 10 A schematic diagram of the side view of three experimental configurations: (A) an open box with sand added from the top; (B) a rotating semicircular
drum; and (C) a closed cylindrical drum. In the first two, the sand falls through a pair
of capacitor plates. The signal from the capacitor is sent to a spectrum analyzer.

power spectrum. In the middle experiment, the sand was placed in a semi-cylindrical
drum which was rotated at a slow rate. As in the first configuration, whenever an
avalanche occurred, the sand leaving the system fell through a pair of capacitor
plates. The sizes and shapes of the particles were varied in both experiments. The
results obtained in both experiments were qualitatively the same.

In Figure 11, I show a typical time trace and its corresponding power spectrum
obtained[53] with the drum rotating at $\Omega = 1.3°$/min using the spherical glass beads.
Each spike in the signal corresponds to an avalanche which spans the entire system
and consists of thousands of particles. Each avalanche occurs after a rather well
defined interval of time has elapsed since the preceding event. The avalanches start
at an angle $\theta_m \equiv \theta_r + \delta$ and return the pile to its stable state at θ_r. The models of
self-organized criticality would predict the occurrence of avalanches of all different
sizes. The experiment shows that there were no avalanches that were local and
which did not span the entire system. The nearly uniform spacing of avalanches,
$\langle \Delta t \rangle = \langle \Delta \theta \rangle / \Omega$, corresponds to the broadened peak at $f = 1/\langle \Delta t \rangle$ in the power
spectrum shown in Figure 11. The roll-off at high frequencies above $f = 1/\langle \tau \rangle$ is
due to the finite width τ of the individual avalanches. Its average power law decay,
close to f^{-3}, is due to variations between rectangular and triangular pulse shapes
of individual events. In the relevant frequency range, between $1/\langle \Delta t \rangle$ and $1/\langle \tau \rangle$,
the spectrum appears to be approximately frequency independent.

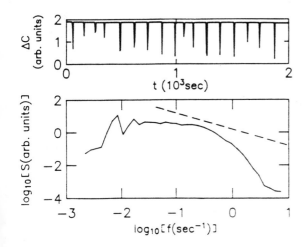

FIGURE 11 a) The time trace of a typical sequence of avalanches in a rotating drum filled with spherical glass beads. ΔC is the change in capacitance due to the flow of beads between the plates. b) The power spectrum, $S(f)$, for the above trace. Dashed line shows a $1/f$ power spectrum for comparison. Data from Jaeger et al.[53]

These findings are inconsistent with the ideas of self-organized criticality which, for the power spectrum, predicted a power law behavior, $f^{-\phi}$, with $\phi \cong 1$. Such behavior would have followed from a power law distribution of widths, τ, implying events over a wide range of time scales. Instead, we find that the power spectrum is indicative of a linear superposition of global, system-spanning avalanches with a narrowly peaked distribution of time scales. There are thus two characteristic angles for a sand pile, not one, as was implicitly assumed in the models of Bak, Tang, and Wiesenfeld. $(\theta_m - \theta_r)$ is on the order of a few degrees and is the product of the interval between spikes in Figure 11 and the rotation speed of the drum. In the interval between θ_m and θ_r, the stationary pile is metastable; only if it is perturbed will an avalanche occur even though its slope is greater than θ_r. This study suggests that the behavior is self-organized but *not* critical. The mechanism for initiating an avalanche is more akin to a nucleation and growth phenomenon than to critical fluctuations and the data resemble more what one would expect from a first-order rather than a second-order phase transition.

A recent experiment was performed on small piles of sand in which individual grains were added one at a time.[52] After each addition, any avalanche that occurred was given time to completely subside before the next particle was added. The distribution of sizes for these avalanches was measured. A qualitative difference in the nature of the avalanche distributions was found as the size of the system was varied. For large systems, the distribution looked similar to what I have already described: only large, system-spanning events occurred. If, on the other hand, the radius of the base of the pile was small, less than approximately 30 bead diameters, then the distributions appeared broad with events occurring with many different sizes. Finite-size scaling could then be used for these small systems to lay the distribution from one pile on top of another.

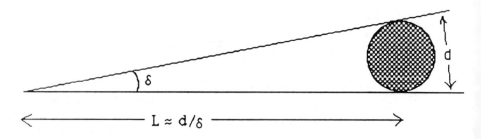

FIGURE 12   Illustration of the criterion for when finite size effects should become important in a sandpile. If the radius of the pile is smaller than $L \approx d/\delta$, then the pile can no longer distinguish between θ_r and θ_m.

I will present here a simple argument[68] of how finite-size effects should enter into the behavior of a sandpile. As shown above, the difference between the angle of maximum stability and the angle of repose, $\delta = (\theta_m - \theta_r)$, is on the order of a few degrees. The criterion that the system look "large" is that a single particle can exist totally within the space subtended by that angle, as shown in Figure 12. If the base of the pile is too small, even if the slope of the top surface starts off below θ_r where the pile is stable, the addition of a single particle will bring it into the unstable region where the angle is greater than θ_m. If we take $\delta \approx 2°$, then $L \approx 30d$ (where d is the diameter of a single grain). This corresponds to the value observed in the experiment where the system crosses over from large to small system size behavior. This argument would suggest that the behavior in the small size regime is not critical. The fluctuations that occur appear to be due simply to the system being too small for it to show the two distinct angles θ_r and θ_m that determine the behavior in the infinite system. In particular, this data on small size systems does not appear to me to support the application of self-organized critical behavior to the behavior of sandpiles.

In order to see whether critical behavior can be restored when there is only one characteristic angle in the pile, vibrations[40,41,53,65] were introduced to the pile so that, whenever the angle rises above θ_r, there would be enough of a perturbation to make it collapse. Even in this case the power spectrum does not resemble a simple power law. What happens, instead, is that the high-frequency cut-off broadens as the vibration intensity is increased until it covers the entire dynamic range of the experiment. This is shown in Figure 13.

When there are vibrations present in the system, the slope of a pile of sand relaxes in time. Some data for this type of relaxation is shown in Figure 14. For large intensity, the decay is logarithmic over several decades of time. For lower intensity, the relaxation seems to saturate at long times. We can understand the logarithmic decay in terms of a simple single-particle model assuming that the vibrations act like an effective temperature, T_{eff}, and that the particles move by activated hopping over barriers created by their neighbors. The heights of the barriers themselves depend on θ and increases as θ decreases. The barrier height, U, disappears at $\theta = \theta_m$ since, above that angle, the pile is unstable even when $T_{eff} = 0$. We, thus, approximate U by $U \cong U_0 + U_1(\theta_r - \theta)$. We relate the particle flow, $j = d\theta/dt$, to the driving force, proportional to θ and obtain:

$$\frac{d\theta}{dt} = -A\theta e^{\beta(\theta - \theta_r)} \tag{2.1}$$

where $\beta = U_1/kT_{eff}$ and A is a constant. For $\beta\theta \gg 1$, the solution to this equation is approximately

$$\theta - \theta_r - \frac{1}{\beta}\ln(\beta A\theta_r t + 1) \tag{2.2}$$

which is the form needed to fit the data for large vibration intensity in Figure 14.

By now it should be apparent that granular materials have a number of intriguing properties. In some ways they behave like a liquid and in other ways they behave like a solid. For example, the fact that there exists an angle of repose which allows the formation of a mound of material is reminiscent of a solid where complicated shapes can be formed readily. On the other hand, an angle of maximum stability, θ_m, is suggestive of liquid behavior since above that angle the material will flow easily. Between θ_r and θ_m, there is a region of bi-stability where, depending on the initial conditions, either stationary or flowing behavior is observed. I have

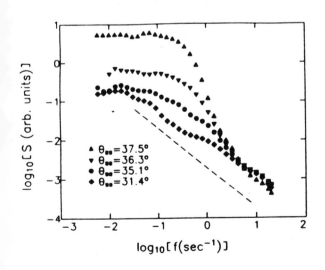

FIGURE 13 The power spectra for avalanches of aluminum oxide particles in a rotating drum for different vibration intensities. The intensities are greatest for the bottom curves. The vibration intensity is parametrized by the θ_{ss} which is the average, or steady-state, angle of the pile when the drum is rotated with the given value of Ω (in this case $\Omega = 1.3°/\text{min}$). The dashed line has slope -1 to show a $1/f$ power spectrum for comparison. Data from Jaeger et al.[53]

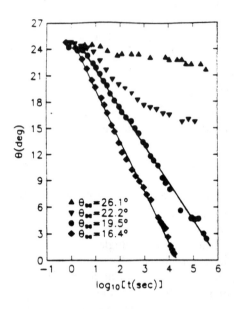

FIGURE 14 The relaxation of the slope in a stationary drum with glass beads. Vibration intensities increase from top to bottom. Straight lines indicate $\log(t)$ behavior. Data from Jaeger et al.[98]

suggested that this is similar to a first-order phase transition where hysteresis occurs and a nucleation and growth phenomenon may take place. Another point that I have not yet mentioned is that, when the material does flow, it does so in a very non-Newtonian manner: the flowing region is confined to a boundary layer and does not penetrate into the bulk of the material. Thus, when an avalanche occurs in a sandpile all the motion takes place in a shallow layer near the upper surface. In what follows, I would like to present a model for the friction in granular materials that can explain these phenomena in a qualitative manner.

We start[55] by writing down Newton's equation of motion for the time-averaged position $\langle r \rangle$ of a grain of mass m:

$$m\langle \ddot{r} \rangle = \sigma - F. \tag{2.3}$$

σ is the shear or driving force and F is the friction force due to dissipation. F will, in general, be a function of the relative velocity of the particle with respect to its neighbors. If $\dot{\gamma}$ is the shear rate and d is the diameter of a grain, then this relative velocity is $\dot{\gamma}d$. If we are concerned only with finding the steady-state behavior, we can set $\langle \ddot{r} \rangle = 0$ so that $\sigma = F$.

F can be found from the average energy dissipated by the grain per unit distance travelled in the forward, x, direction: $F = \langle dE/dx \rangle$. There are at least two contributions to F: one is from the kinetic energy lost due to forward motion with constant relative speed, $\dot{\gamma}d$, and the other is from the extra energy a grain loses between collisions by falling in the gravitational potential.

FIGURE 15 Path of a particle following the terrain below it at a low shear rate (dashed line). At higher shear rate (solid line), it leaves the underlying surface.

The first contribution is

$$F_{ke} = \frac{1}{2\lambda} m(\dot{\gamma}d)^2 \qquad (2.4a)$$

where λ is the mean distance a particle travels before its excess energy compared to that of its neighbors is dissipated in collisions. Typically λ will be on the order of several times the particle diameter, d. In dimensionless units using $\beta \equiv d/2\lambda$ and $\tilde{\dot{\gamma}} \equiv \dot{\gamma}(d/g)^{1/2}$,

$$\tilde{F}_{ke} \equiv \frac{F_{ke}}{mg} = \beta \tilde{\dot{\gamma}}^2 . \qquad (2.4b)$$

β depends on the coefficient of restitution of the particles.

Several groups[5,6,50,82] have measured the relation between stress, σ, and shear rate, $\dot{\gamma}$, in a variety of configurations. At high shear, the stress varies quadratically with $\dot{\gamma}$. These studies have focused on the behavior at large $\dot{\gamma}$ but did not explore the phenomena when $\dot{\gamma}$ is small.

The second contribution to the frictional force occurs because a colliding particle has typically just fallen some distance in the gravitational potential. Its average relative motion is therefore larger than $\dot{\gamma}d$. Since all the collisions are inelastic, a collision with the layer beneath it will dissipate a fraction of its total kinetic energy. In effect, this leads to a dissipation of some of the gravitational energy from the last fall. We assume the grain falls into depressions between the particles on the layer above which it is moving as shown in Figure 15. The typical size of such a depression (and the typical distance between layers) is d. For large relative velocities, the particle does not have a chance to fall very far before it reaches the other side of the hollow. The faster it goes, the less it will drop and the less potential energy will be lost. This contribution to the friction will *decrease* as the velocity is increased. For large shear, we find

$$F_{pe} = \alpha_a \frac{mg^2}{2d} \dot{\gamma}^{-2} = \alpha_a \frac{mg}{2} \tilde{\dot{\gamma}}^{-2} \qquad \tilde{\dot{\gamma}} \gg 1 . \qquad (2.5a)$$

In the limit of small relative velocity, the particle can only fall to the bottom of the well, that is, approximately a distance d, so that the potential energy lost per unit distance approaches a constant. Thus,

$$F_{pe} = \alpha\, mg \qquad\qquad \dot{\tilde{\gamma}} \ll 1 \qquad\qquad (2.5b)$$

where α and α_a are constants. These are the asymptotic expressions for high and low shear rates. An expression which interpolates between these two limiting forms is

$$\tilde{F} \equiv \frac{F_{pe}}{mg} \approx \frac{\alpha}{1 + \alpha_1 \dot{\tilde{\gamma}}^2} \; . \qquad\qquad (2.6)$$

The total frictional force in dimensionless form is:

$$\tilde{F} = \tilde{F}_{pe} + \tilde{F}_{ke} = \frac{\alpha}{1 + \alpha_1 \dot{\tilde{\gamma}}^2} + \beta \dot{\tilde{\gamma}}^2 \; . \qquad\qquad (2.7)$$

The constants α_a, α, and β depend on the properties of the granular material such as the coefficient of restitution, the density, and the detailed nature of the geometrical arrangements of the particles. The first term, which is due to the extra velocity that the particles pick up due to falling in the gravitational field, may dominate the frictional force at low relative velocities whereas the second term, which accounts for the collisions in the direction of *average* motion, dominates at large $\dot{\gamma}$. In the low $\dot{\gamma}$ regime, the vertical velocity picked up from falling can be larger than the horizontal relative velocity.

Figure 16 shows the behavior of Eq. (2.7). A significant feature of the curves is that F can have a minimum at a non-zero value of $\dot{\gamma}$. At values of $\dot{\gamma}$ much larger than $\dot{\gamma}_0$, the value of $\dot{\gamma}$ at which the minimum occurs, F increases quadratically. This agrees with experiment.[50,82]

I would now like to demonstrate that the friction curve shown in Figure 16 can account qualitatively for some of the unusual behavior observed in the flow of granular materials that I mentioned above. For steady-state solutions, we want to find those value of $\dot{\gamma}$ which satisfy $\sigma = F$. Since $\sigma = mg \sin\theta$ is a constant set only by the angle of the pile, all the dependence on $\dot{\gamma}$ appears in F. For all values of θ for which σ is less than $F(\dot{\gamma}_0)$, there is only one steady-state solution which occurs at $\dot{\gamma} = 0$. Thus, the pile is stable up to an angle of repose, given by $mg \sin\theta_r = F(\dot{\gamma}_0)$. Even at θ_r the pile will remain stationary since, if it is perturbed away from $\dot{\gamma} = 0$, the acceleration in Eq. (2.3) is negative. This will decrease the motion until the stationary state is again obtained. Only if the grains are already moving with relative velocity $\dot{\gamma}d \geq \dot{\gamma}_0 d$ will the final steady-state solution be at the non-zero value $\dot{\gamma}_0 d$. There is also only one steady-state solution when σ is greater than $F(\dot{\gamma} = 0)$. Above this value, the pile is always unstable and flow must occur. Therefore, the value of θ_m is determined from the condition $mg \sin\theta_m = F(\dot{\gamma} = 0)$.

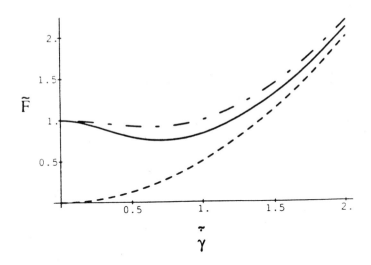

FIGURE 16 A plot of Eq. (2.7) for two choices of the constants: $\alpha = 1, \alpha_1 = 1, \beta = 0.5$ and $\alpha = 1, \alpha_1 = 2, \beta = 0.5$. For comparison we show the curve for simple quadratic dissipation ($\alpha = 0, \beta = 0.5$).

More complicated behavior occurs in the region where there is more than one solution to $F = \sigma$; that is in the region where $F(\dot{\gamma}_0) < \sigma < F(\dot{\gamma} = 0)$. This corresponds to the range of angles: $\theta_r < \theta < \theta_m$. In this range there are three solutions: one with $\dot{\gamma} = 0$, one with $\dot{\gamma}$ in the region of negative slope, and one with $\dot{\gamma}$ on the rising part of the curve. Only the first and last of these are stable solutions. The middle one, where the slope is negative, is unstable against any small perturbation of the shear rate: any perturbation around this solution produces an acceleration in Eq. (2.3) which drives the system to one of the other two solutions. This picture accounts for the hysteresis observed in granular flow. The "negative resistance" region in the frictional force introduces complexity to the behavior in a number of ways. Even apart from the case where the shear rate lies in this unstable region, the existence of two stable steady-state solutions over much of the interesting range implies that different results may be found depending on the history of any experiment.

This model has consequences, as well, for how a static pile of sand responds to vibrations or "noise." If the noise produces characteristic velocities of magnitude larger than the value of the unstable solution, then we would expect the pile to start to flow. As the slope increases, the amount of noise necessary to create an avalanche decreases until it approaches zero at θ_m. This is seen qualitatively in experiments.[53]

Finally, the fact that the material flows in a boundary layer can also be explained by this model of friction. Any layer of material which is flowing at a relative velocity less than $\dot{\gamma}_0 d$ (but not zero) will be unstable. The relative velocity will either slow down and become zero, or else it will grow to some value greater than $\dot{\gamma}_0 d$. If the top layer is flowing at a fixed velocity, there can only be a finite number of layers which have a relative velocity greater than or equal to $\dot{\gamma}_0 d$. That is, if v_{top} is finite and $\dot{\gamma}_0 d$ is also non-zero, the number of layers in which flow occurs must be of order $n \approx v_{top}/\dot{\gamma}_0 d$. All other layers must be at rest.

I have argued that granular material does not have any obvious critical behavior and cannot be modeled by the ideas of self-organized criticality. However, it should be clear that granular flow is interesting in its own right. The onset of flow in these systems is reminiscent of a first-order transition with nucleation and growth phenomena. In the metastable regime, flow will start only if there is a "nucleus" of sufficiently high velocity; velocities less than a critical size will decay back to zero and values greater than that size will grow to envelope the entire system. The question still remains whether there are other physical systems which display critical behavior of the type suggested by Bak, Tang, and Wiesenfeld. The phenomena of earthquakes may be such a case. I will discuss this briefly in the next section. A question that is still not understood in any depth is why do the cellular-automata models of avalanches,[7,58,86] which seem to show signs of self-organized critical behavior, behave so differently from the real systems. The analysis of friction that I have presented above for the granular materials is entirely different from what is included in the computer simulations. It appears likely that the omission of dissipation may play an important role in the qualitative behavior of these simulations. Recent experiments[9] have shown that arching may also be an important effect in the bulk properties of flowing granular material. How this kind of effect is included in the equations of motion has also not been addressed.

EARTHQUAKES

It is fortunate that very large earthquakes occur only infrequently. However, the sporadic appearance of these devastating events makes it all the more imperative that we learn how to predict the next major incident. At present, although we do not know how to predict reliably the time and place of an event, we do know a lot about the statistics for the rates of earthquakes of different magnitudes. The Gutenberg-Richter law embodies one of the statistical laws governing earthquake production: big events do not happen often, but small ones occur more frequently with a distribution of events which follows a simple power law. It has never been adequately explained why the Gutenberg-Richter law works for data from faults all over the globe, and in particular why it shows roughly the *same* power law exponent for all these events. It is therefore important to find the origin of this behavior and quite challenging to try to understand the mechanisms which can produce earthquakes with such an enormous variation in both rate and magnitude of events.

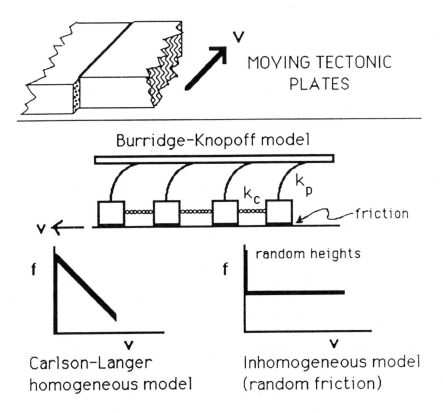

FIGURE 17 The top panel shows the motion of two tectonic plates moving with respect to one another at constant velocity. This motion is what is believed to be the underlying cause of earthquake production. The middle panel shows a schematic diagram of the Burridge-Knopoff model. The blocks are coupled to one another by springs of constant k_c and are coupled to the top support by springs of constant k_p. The bottom plate moves at a constant velocity V with respect to the top support. The bottom panel shows two possible forms for the friction between the blocks and the bottom plate. The one at the left is the one used in the homogeneous model studied by Carlson and Langer.

Along a single geological fault, there will be earthquakes of many different sizes. The empirical Gutenberg-Richter law[47] relates ρ, the rate of occurrence of an earthquake, to its seismic moment, m:

$$\rho(m) = Am^{-b-1} \qquad (2.8)$$

where A and b are constants and m is given by

$$m = \int_s \xi ds' . \qquad (2.9)$$

Here s is the area of the active region during an earthquake and ξ is the local displacement of the two sides of the fault. All available data indicate that b appears to have a nearly "universal" value that lies between 0.8 and 1.1. The "Richter magnitude" of the earthquake is defined,[38,60] at least for small-sized quakes, as $M \equiv \log_{10} m$. Deviations from this behavior occur for very large magnitudes.

A variety of models for earthquakes have been introduced. Their success has been judged by whether or not they are able to reproduce the Gutenberg-Richter law with the experimentally observed value of b. Burridge and Knopoff[17] introduced a model consisting of blocks coupled to one another by one set of springs and to an upper support by a second set of springs. The blocks were in frictional contact with a surface that moved at a constant velocity, V, relative to the top support. The configuration is shown in Figure 17. The friction was taken to be of a highly nonlinear "stick/slip" form. By assuming the existence of intrinsic spatial irregularities in the model, they were able to reproduce many of the characteristics of real earthquake data. More recently, Carlson and Langer have shown that, even in the absence of inhomogeneities, this same model produces results that still have many of the qualitative properties of real earthquakes. In particular, this model reproduces the Gutenberg-Richter law for a large class of parameters. In a different cellular automata model, Bak and Tang have produced data which can also be described by a scaling law.[8]

It is interesting to note that the stick-slip friction law used by Carlson and Langer is similar to the one we derived in the last section for granular materials. In both cases, the friction is finite at zero velocity (or shear rate) and then decreases as the velocity is increased. This behavior, in particular the region of negative slope, was important for producing the metastability, the hysteresis, and the boundary layers in the flow of granular material. This particular form for the friction is also what makes the earthquake models so rich in their behavior. Without the region of negative slope, there would be no power law distribution of earthquake sizes in this model.

As I have mentioned various models with quite different assumptions have succeeded in giving a power law behavior for the distribution of earthquake magnitudes. Moreover, they have also been able to give values for the exponent b close to those seen experimentally, i.e., $b \approx 1$. Since the value of b appears to be insensitive to the assumptions in these models, one might expect there to be some underlying argument for why this value for the exponent is ubiquitous. An argument based on a sum rule has been proposed[90] which purports to give such an demonstration. The sum rule is simply that on average each segment of a fault must move at the same constant velocity V with which the two tectonic plates are moving:

$$\frac{1}{s} \int_0^S \rho(s)m(s)ds = V \, . \tag{2.10}$$

Here S is the total area of the fault and $\rho(s)$ and $m(s)$ are respectively the distribution and magnitude of earthquakes which have area s. The gist of the argument is that, if scaling laws are assumed for both $\rho(s)$ and $m(s)$, then we can get some bounds for the exponent b itself. Thus if:

$$\rho(s) = Ws^{-\omega}$$
$$m(s) = M_0 s^{\mu}$$

(2.11)

then $b = (\omega - 1)/\mu$. As long as there are earthquake events occurring down to arbitrarily small size, then, since the integral in Eq. (2.10) must not diverge, it is easy to show that $b < 1$. In addition, if the scaling of $\rho(s)$ and $m(s)$ exists up to S, the total area of the fault, then, since the right-hand side of the integral in Eq. (2.10) does not depend on S, we can conclude[3] that $b \approx 1$. This argument does not explain why the Gutenberg-Richter law exists in the first place; that is, why $\rho(s)$ and $m(s)$ have the form given in Eq. (2.11). What it does show instead is that, in those cases where scaling does exist, the exponent b must be close to the experimental value, i.e., $b \approx 1$.

The behavior of earthquakes, as distinct from what is seen in avalanches, does seem to show critical, or scaling, properties. In both types of phenomena, the way energy is dissipated and the functional form for the friction are crucial ingredients in determining the overall behavior. It is perhaps surprising that, for both earthquakes and avalanches, quite similar friction laws have been proposed which nevertheless lead to qualitatively different behavior. I do not find it obvious to see how one would have predicted, prior to knowing the results of the experiment, which of these two forms of catastrophic events would lead to a scaling behavior and which would lead to an essentially first-order transition. Much work, both experimental and theoretical, remains to be done to clarify this issue.

ACKNOWLEDGEMENTS

I would like to thank Norman Birge, Shobo Bhattacharya, Linda Busse, John Carini, Paul Dixon, Richard Ernst, Gary Grest, Yoon Jeong, N. Menon, Kevin O'Brien, Bruce Williams, Heinrich Jaeger, Chu-heng Liu, Leo Kadanoff, Maria de Sousa Vieira, Giovani Vasconcelos, Thomas Witten, Sumin Zhou, and Lei Wu who were my collaborators on the various experiments discussed in this lecture. I am grateful to Austen Angell, Morrel Cohen, Raymond Goldstein, Gene Mazenko, Richard Palmer, James Sethna, Daniel Stein, and Jan Tauc for many stimulating discussions on the subject of the glass transition. I would also like to thank Robert Behringer, Jean Carlson, James Langer, Jorge Lomnitz-Adler, and Bruce Shaw for many stimulating discussions on avalanches and earthquakes. This work was supported in part by NSF DMR 88-02284 and the NSF–Materials Research Laboratory.

[3] Another assumption necessary here is that W is an approximately extensive parameter, that is, $W = W_0 S^{1-\epsilon}$ where ϵ is small.

REFERENCES

1. Angell, C. A., and W. Sichina. *Annals of the New York Academy of Sciences* **279** (1976): 53.
2. Angell, C. A. In *Proceedings of the Workshop on Relaxations in Complex Systems*, edited by K. L. Ngai and G. B. Wright, 3. Springfield, VA: U.S. Dept. of Commerce, 1984.
3. Angell, C. A., and M. Goldstein, eds. *Dynamic Aspects of Structural Change in Liquids and Glasses*. Annals of the New York Acad. of Sciences, Vol. 484. New York: New York Academy of Sciences, 1986.
4. Bagley, B. G. In *Amorphous and Liquid Semiconductors*, edited by J. J. Tauc, 4. London: Plenum Press, 1974.
5. Bagnold, R. A. *Proc. R. Soc. Lond. A* **225** (1954): 49.
6. Bagnold, R. A. *Proc. R. Soc. Lond. A* **295** (1966): 219.
7. Bak, P., C. Tang, and K. Wiesenfeld. *Phys. Rev. Lett.* **59** (1987): 381.
8. Bak, P., and C. Tang. *J. Geophys. Research* **94** (1989): 15635.
9. Baxter, G. W., R. P. Behringer, T. Fagert, and G. A. Johnson. *Phys. Rev. Lett.* **62** (1989): 2825.
10. Bendler, J. T., and M. F. Shlesinger. *J. Mol. Liquids* **36** (1987): 37.
11. Bendler, J. T., and M. F. Shlesinger. *J. Stat. Phys.* **53** (1988): 531.
12. Bengtzelius, U., W. Gotze, and A. Sjolander. *J. Phys. C* **17** (1984): 5915.
13. Bernu, B., J. P. Hansen, Y. Hiwatari, and G. Pastore. *Phys. Rev. A* **36** (1987): 4891.
14. Birge, N. 0., and S. R. Nagel. *Phys. Rev. Lett.* **54** (1985): 2674.
15. Birge, N. 0. *Phys. Rev. B* **34** (1986): 1631.
16. Birge, N. 0., Y. H. Jeong, and S. R. Nagel. *Dynamic Aspects of Structural Change in Liquids and Glasses*, edited by C. Austen Angell and M. Goldstein, 101. Annals of the New York Acad. of Sciences, Vol. 484. New York: New York Academy of Sciences, 1986.
17. Burridge, R., and L. Knopoff. *Bull. Seismol. Soc. Am.* **57** (1967): 341.
18. Busse, L. E. *Phys. Rev. B* **29** (1984): 3639.
19. Campbell, I. A., J.-M. Flesselles, R. Jullien, and R. Botet. *Phys. Rev. B* **37** (1988): 3825.
20. Carlson, J. M., and J. S. Langer. *Phys. Rev. Lett.* **62** (1989): 2632.
21. Carlson, J. M., and J. S. Langer. *Phys. Rev. A* **40** (1989): 6470.
22. Chen, H. S. *Rep. Prog. Phys.* **43** (1980): 353.
23. Christensen, T. *J. Phys. (Paris) Colloq.* **46** (1985): C8-635.
24. Cohen, M. H., and D. Turnbull. *J. Chem. Phys.* **31** (1959): 1164.
25. Cole, K. S., and R. H. Cole. *J. Chem. Phys.* **9** (1941): 341.
26. Cohen, M. H., and G. S. Grest. *Phys. Rev. B* **24** (1981): 4091.
27. Coulomb, C. *Memoires de Mathematiques et de Physique Presentés à l'Academie Royale des Sciences par Divers Savants et Lus dans les Assemblées*, (1773): 343.

28. Cukierman, M., J. W. Lane and D. R. Uhlmann. *J. Chem. Phys.* **59** (1973): 3639.
29. Das, S. P., G. F. Mazenko, S. Ramaswamy, and J. J. Toner. *Phys. Rev. Lett.* **54** (1985): 118.
30. Davidson, D. W., and R. H. Cole. *J. Chem. Phys.* **19** (1951): 1484.
31. Davis, T. D., and T. A. Litovitz. *Physical Acoustics*, Vol. 2B. New York: Academic Press, 1965.
32. de Gennes, P. G. *Superconductivity of Metals and Alloys*, 83. New York: W. A. Benjamin, 1966.
33. Degiorgio, V., T. Bellini, R. Piazza, F. Mantegazza, and R. E. Goldstein. *Phys. Rev. Lett.* **64** (1990): 1043.
34. Dixon, P. K., and S. R. Nagel. *Phys. Rev. Lett.* **61** (1988): 341.
35. Dixon, P. K. *Phys. Rev. B* **42** (1990): 8179.
36. Dixon, P. K., L. Wu, S. R. Nagel, B. D. Williams, and J. P. Carini. *Phys. Rev. Lett.* **65** (1990): 1108.
37. Dixon, P. K., S. R. Nagel, and D. Weitz. To be published.
38. Ekstrom, G., and A. M. Dziewonski. *Nature* **332** (1988): 319.
39. Ernst, R. M., S. R. Nagel, and G. S. Grest. To be published.
40. Evesque, P., and J. Rajchenbach. *Phys. Rev. Lett.* **62** (1989): 44.
41. Faraday, M. *Trans. Roy. Soc. London* **52** (1831): 299.
42. Fox, J., and H. C. Andersen. *J. Phys. Chem.* **88** (1984): 4019.
43. Fredrickson, G. H., and H. C. Andersen. *Phys. Rev. Lett.* **53** (1984): 1244.
44. Fulcher, G. S. *J. Am. Ceram. Soc.* **77** (1925): 3701.
45. Goldstein, M. *J. Chem. Phys.* **51** (1969): 3728.
46. Grest, G. S., and S. R. Nagel. *J. Phys. Chem.* **91** (1987): 4916.
47. Gutenberg, B., and C. F. Richter. *Ann. Geofis.* **9** (1956): 1.
48. Halsey, T. C., P. Meakin, and I. Procaccia. *Phys. Rev. Lett.* **56** (1986): 854.
49. Halsey, T. C., M. H. Jensen, L. P. Kadanoff, I. Procaccia, and B. I. Shraiman. *Phys. Rev. A* **33** (1986): 1141.
50. Hanes, D. M., and D. L. Inman. *J. Fluid Mech.* **150** (1985): 357.
51. Harr, M. E. *Mechanics of Particulate Media*, 258. New York: McGraw Hill, 1977.
52. Held, G. A., D. H. Solina II, D. T. Keane, W. J. Haag, P. M. Horn, and G. Grinstein. *Phys. Rev. Lett.* **65** (1990): 1120.
53. Jaeger, H. M., C.-h. Liu, and S. R. Nagel. *Phys. Rev. Lett.* **62** (1989): 40.
54. Jaeger, H. M., C.-h. Liu, S. R. Nagel, and T. A. Witten. *Relaxation in Complex Systems and Related Topics*, edited by I. A. Campbell and C. Giovannella, 235. London: Plenum Presss, 1990.
55. Jaeger, H. M., C.-h. Liu, S. R. Nagel, and T. A. Witten. *Europhysics Letters* **11** (1990): 619.
56. Jeong, Y. H., S. R. Nagel, and S. Bhattacharya. *Phys. Rev. A* **34** (1986): 602.
57. Jeong, Y. H. *Phys. Rev. A* **36** (1987): 766.
58. Kadanoff, L. P., S. R. Nagel, L. Wu, and S.-M. Zhou. *Phys. Rev. A* **39** (1989): 6524.

59. Kakalios, J., R. A. Street, and W. B. Jackson. *Phys. Rev. Lett.* **59** (1987): 1037.
60. Kanamori, H. In *Earthquakes, Observation, Theory and Interpretation*, edited by H. Kanamori and E. Boschi, 596. Amsterdam: North Holland, 1983.
61. Kauzmann, W. *Chem. Rev.* **43** (1948): 219.
62. Kohlrausch, R. *Pogg. Ann. Phys.* **91** (1854): 198.
63. Kirkpatrick, T. R. *Phys. Rev. A* **31** (1985): 939.
64. Landau, L. D., and E. M. Lifshitz. *Theory of Elasticity*, 3rd Edition, Ch. 36. Oxford: Pergamon, 1986.
65. Laroche, C., S. Douady, and S. Fauve. *J. Phys. France* **50** (1989): 699.
66. Laughlin, W. T., and D. R. Uhlmann. *J. Phys. Chem.* **76** (1972): 2317.
67. Leutheusser, E. *Phys. Rev A* **29** (1984): 2765.
68. Liu, C.-H., and S. R. Nagel. To be published.
69. Monroe, D. "The Electron Glass vs. Ising Spin Systems—Implication for Experiments, preprint, 1990.
70. Mountain, R., and D. Thirumalai. *Phys. Rev. A* **36** (1987): 3300.
71. Movaghar, B., M. Grunewald, B. Ries, H. Bassler, and D. Wurtz. *Phys. Rev. B* **33** (1986): 5545.
72. Moynihan, C. T., A. J. Easteal, and J. Wilder. *J. Phys. Chem.* **78** (1974): 2673.
73. Nosé, S., and F. Yonezawa. *J. Chem. Phys.* **84** (1986): 1803.
74. O'Brien, K. P., L. Wu, and S. R. Nagel. *Phys. Rev. B*, in press.
75. Oxtoby, D. W. In *Annals of the New York Acad. of Sciences*, Vol. 484, 26. New York: New York Academy of Sciences, 1986.
76. Palmer, R. G., D. Stein, E. Abrahams, and P. W. Anderson. *Phys. Rev. Lett.* **53** (1984): 958.
77. Palmer, R. G., and D. L. Stein. *Complex Systems*, edited by D. Stein. Santa Fe Institute Studies in the Sciences of Complexity, Lect. Vol. I. Reading, MA: Addison-Wesley, 1988.
78. Rahman, A., M. J. Mandell, and J. P. McTague. *J. Chem. Phys.* **64** (1976): 1564.
79. Ramakrishnan, T. V., and M. Yussouff. *Phys. Rev. B* **19** (1979): 2775.
80. Reynolds, O. *Phil. Mag., Ser. 5* **20** (1885): 469.
81. Savage, S. B. *Advances in Applied Mechanics* **24** (1984): 289.
82. Savage, S. B., and M. Sayed. *J. Fluid Mech.* **142** (1984): 391.
83. Sornette, A., and D. Sornette. *Europhys. Lett.* **9** (1989): 197.
84. Souletie, J. *J. Physique* **44** (1983): 1095.
85. Spaepen, F., and D. Turnbull. "Rapidly Quenched Metals." *Second International Conference*, edited by N. J. Grant and B. C. Giessen, Section 1 205. Cambridge: MIT Press, 1976.
86. Tang, C., and P. Bak. *Phys. Rev. Lett.* **60** (1988): 2347.
87. Turnbull, D., and J. C. Fisher. *J. Chem. Phys.* **17** (1949): 71.
88. Turnbull, D. *Contemp. Phys.* **10** (1969): 473.
89. Turnbull, D., and M. H. Cohen. *J. Chem. Phys.* **52** (1970): 3038.
90. Vasconcelos, G. L., M. deSousa Vieira, and S. R. Nagel. To be published.

91. Vogel, H. *Phys. Z.* **22** (1921): 645.
92. Wendt, H. R., and F. F. Abraham. *Phys. Rev. Lett.* **41** (1978): 1244
93. Williams, G., and D. C. Watts. *Trans. Faraday Soc.* **66** (1970): 80.
94. Wong, J., and C. A. Angell. *Glass: Structure by Spectroscopy.* New York: Dekker, 1976.
95. Wu, L., P. K. Dixon, S. R. Nagel, B. D. Williams, and J. P. Carini. *J. Non-Cryst. Solids.* In press.
96. Wu, L. In press.

J. F. Traub† and H. Woźniakowski†‡

†Department of Computer Science, Columbia University and ‡Institute of Informatics, University of Warsaw

Theory and Applications of Information-Based Complexity

1. OVERVIEW

A very much simplified view of an important scientific paradigm is the following. A mathematical formulation of a natural phenomenon is created. Computations stemming from the mathematical formulation lead to predictions.

An instance is provided by the following familiar example. The phenomenon is planetary motion. The mathematical formulation is given by Newton's equations. The failure of Newton's equations to correctly predict the motion of Uranus leads to the discovery of Neptune.

The relation between a real world phenomenon, a mathematical formulation, and the computation are schematized in Figure 1.

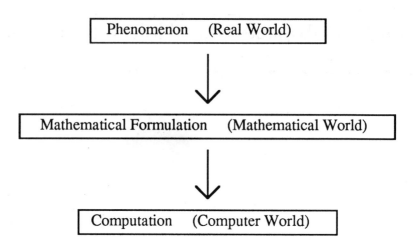

FIGURE 1 Three worlds.

We comment on this tripartite relation.

i.   At one time computations were done by hand or by mechanical calculations. Now, scientific and engineering calculations push the frontiers of supercomputing.

ii. Mathematical formulations of scientific problems are often infinite dimensional and *continuous*. They are expressed, for example, as systems of ordinary or partial differential equations, integral equations, optimization problems, or path integrals.

Information-based complexity (IBC) studies the complexity of infinite-dimensional continuous problems. It is a branch of computational complexity which studies the minimal computer resources (typically, time or space) needed to solve a mathematically posed problems. Thus, a problem's complexity measures its *intrinsic* difficulty. Readers should note that the term computational complexity serves double duty both as the name of a field and as a crucial invariant characteristic of a problem. The invariant is sometimes abbreviated to complexity.

What are the implications of studying the complexity of infinite-dimensional continuous problems? In a typical application of IBC such as numerical approximation of a multivariate integral, the input is a multivariate function. But a digital computer can handle only finite sets of numbers. Thus, referring to Figure 1, the multivariate function exists in the mathematical world, but it must be represented in the computer world by only a finite set of numbers. This leads to the following observations:

i.   The mathematical input is infinite dimensional. The computer input is a finite set of numbers. Hence, the information available for the calculation is *partial*.

ii. Since the information is partial, the original mathematical problem can be only approximately solved.

iii. We wish to determine the complexity of approximately solving the original mathematical problem.

iv. Because the information is partial, we are able to obtain lower bounds on a problem's computational complexity using adversary arguments at the information level. That is why this branch of computational complexity is called information-based complexity.

Although the focus of IBC is on infinite-dimensional problems, there has been some work on finite-dimensional problems. Examples are large linear systems and eigenvalue problems (see Traub et al.,[25] Chapter 5, Sections 9 and 10) as well as the discrete problem of synchronizing cloaks in a distributed system.[30]

As we discussed above, the information is partial. Typically information is also *contaminated* and *priced*. Thus, IBC is characterized by partial, contaminated, and priced information. A comprehensive treatment of IBC and an extensive bibliography may be found in Traub et al.[25] Expository material that emphasizes particular aspects of IBC and also presents a variety of open questions may be found in Packel and Traub,[14] Packel and Woźniakowski,[15] Traub and Woźniakowski,[28] and Woźniakowski.[33]

We contrast IBC with two other branches of computational complexity. Blum, Shub, and Smale[2] have studied the complexity of continuous problems defined by a finite number of parameters. An example of such a problem is 4-satisfiability; that is, does a system of n real polynomials of degree at most four have a real zero? Since the mathematical problem can be entered into the computer (modulo round-off errors), the mathematical and computer inputs are the same and the information given to the computer is complete.

When the information is complete for continuous or discrete problems, there are, of course, no information-level adversary arguments. Complexity bounds must then be obtained using combinatorial arguments. Currently, these arguments can't establish the problem complexity but can only provide the complexity class to which a problem belongs. A class of particular interest is the class NP (non-deterministic polynomial).[5] The hardest problems in NP are said to be NP-complete.

A major result of Blum et al.[2] is that 4-satisfiability is NP-complete over the reals. This is the first NP-completeness result over the reals.

Both IBC and Blum-Shub-Smale complexity use the real number model as an abstraction of finite-precision, floating-point numbers. The interest in finite-precision floating point is that most scientific computing is done in this system (see, e.g., Traub et al.,[24] Section 5.2, and Traub et al.,[25] Chapter 3, Section 2.3, for a specification of the real number model).

Both IBC and Blum-Shub-Smale complexity are branches of continuous computational complexity, for infinite- and finite-dimensional problems, respectively. This may be contrasted with discrete complexity which studies the complexity of discrete problems.

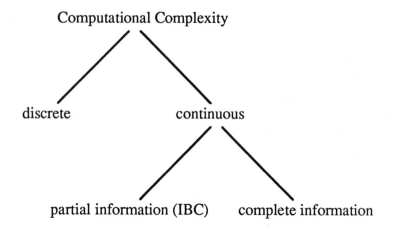

FIGURE 2   Structure of computational complexity.

Discrete complexity generally uses the bit model (Turing machine model). In this model the cost of an arithmetic operation depends on the size of the operands. But the cost of a floating-point operation is essentially independent of the operand size and this is reflected in the real number model.

The complexity can depend critically on the model of computation. For example, linear programming has polynomial complexity in the bit model but is conjectured not to have polynomial complexity in the real number model.[27]

Figure 2 schematizes the relations among the branches of computational complexity discussed above while Table 1 indicates the model of computation and information used by each.

We summarize the remainder of this chapter. To make this chapter self-contained, we provide a brief introduction to the basic concepts of IBC in Section 2. Many continuous problems have been proven noncomputable or intractable. In Section 3 we discuss whether noncomputability and intractability can be broken using randomization or the average-case setting and pose four open problems.

The noncomputability and intractability results are certainly beautiful theoretical computer science. But do they indicate limits for the foundations of physics and for supercomputing? We speculate on this in Section 4. In the following section we report on the application of IBC to ill-posed problems, partial differential equations, and integral equations. In the concluding section we show how number theory results were recently used to obtain the average case complexity of multivariate integration.

TABLE 1 Three branches of computational complexity.

	Discrete Complexity	Blum-Shub-Smale Complexity	IBC
Model of Computation	Bit Model	Real Number Model	Real Number Model
Information	complete exact free	complete exact free	partial contaminated priced

2. BASIC CONCEPTS OF IBC

To make this chapter self-contained, we provide a brief introduction to the basic concepts of IBC. We present an abstract formulation illustrated by a simple example. A proof technique which leads to tight complexity bounds for some problems will also be indicated. A comprehensive treatment of IBC and an extensive bibliography may be found in Traub et al.[25] Let

$$S: F \to G,$$

where F is a subset of a linear space and G is a normed linear space over the real or complex field. We wish to approximate $S(f)$ for all f from F.

Let $U(f)$, where $U: F \to G$, denote a computed approximation to $S(f)$ for $f \in F$. We now explain how the approximation U can be constructed. To do this we need to discuss the concept of information.

The basic assumption of IBC is that, in general, we do not have full knowledge of an element f since typically f is an element from an infinite-dimensional space and it cannot be represented on a digital computer. Instead, it is assumed that we can gather some knowledge about f by computations of the form $L(f)$, where $L: F \to H$ for some set H.

Let Λ denote a class of permissible information operations L. That is, $L \in \Lambda$ iff $L(f)$ can be computed for each f from F. For example, if F is a set of functions, then Λ is sometimes taken as a set of L consisting of function evaluations, $L(f) = f(x)$ for some x from the domain of f.

Let

$$N(f) = [L_1(f), L_2(f), \dots, L_n(f)], \qquad L_i \in \Lambda, \quad \forall f \in F, \qquad (2.1)$$

be the computed information about f. We stress that the L_i as well as the number n can be chosen adaptively. That is, the choice of L_i may depend on the already computed $L_1(f), L_2(f), \dots, L_{i-1}(f)$. The number n may also depend on the computed $L_i(f)$. (This permits arbitrary termination criteria.)

$N(f)$ is called the information about f, and N the information operator. In general, N is many-to-one, and thus, knowing $y = N(f)$, it is impossible to recover the element f. For this reason, the information N is called *partial*.

The approximation $U(f)$ is constructed by combining the computed information $N(f)$. That is, $U(f) = \phi(N(f))$, where $\phi: N(F) \to G$. A mapping ϕ is called an algorithm. The approximation U can thus be identified with the pair (N, ϕ), where N is an information operator and ϕ an algorithm that uses the information N.

We illustrate these concepts by an example.

EXAMPLE: INTEGRATION. Let F be a class of functions $f: [0, 1] \to \mathbf{R}$ that satisfy a Lipschitz condition with constant q,

$$|f(x) - f(y)| \le q\,|x - y|, \quad \forall\, x, y \in [0, 1].$$

Let $G = \mathbf{R}$ and

$$S(f) = \int_0^1 f(t)\,dt.$$

The class Λ is a collection of $L: F \to \mathbf{R}$, such that for some x from $[0, 1]$, $L(f) = f(x)$, $\forall f \in F$. The information N is given by

$$N(f) = [f(x_1), f(x_2), \ldots, f(x_n)]$$

with the points x_i and the number n adaptively chosen. The approximation U is now of the form $U(f) = \phi(N(f)) = \phi(f(x_1), f(x_2), \ldots, f(x_n))$. An example of an algorithm ϕ is a linear algorithm given by $U(f) = \phi(N(f)) = \sum_{i=1}^n a_i f(x_i)$ for some numbers a_i. ∎

We now present a model of computation. It is defined by two postulates:

1. We are charged for each information operation. That is, for every $L \in \Lambda$ and for every $f \in F$, the computation of $L(f)$ costs c, where c is positive and fixed, independent of L and f.

2. Let Ω denote the set of permissible combinatory operations including the addition of two elements in G, multiplication by a scalar in G, arithmetic operations, comparison of real numbers, and evaluations of certain elementary functions. We assume that each combinatory operation is performed exactly with unit cost.

In particular, this means that we use the real number model, where we can perform operations on real numbers exactly and at unit cost. Modulo round-offs and the very important concept of numerical stability, this corresponds to floating-point arithmetic widely used for solving scientific computational problems.

We now discuss the cost of the approximations $U(f) = \phi(N(f))$. Let $\mathrm{cost}(N, f)$ denote the cost of computing the information $N(f)$. Note that $\mathrm{cost}(N, f) \ge cn$, and the inequality may occur since adaptive selection of L_i and n may require some combinatory operations. If $N(f)$ can not be computed by using n information operations and a finite number of operations from Ω, then $\text{cost}(N, f) = +\infty$.

Knowing $y = N(f)$, we compute $U(f) = \phi(y)$ by combining the information $L_i(f)$. Let $\text{cost}(\phi, y)$ denote the number of combinatory operations from Ω needed to compute $\phi(y)$. If $\phi(y)$ can not be computed by using a finite number of operations from Ω, then $\text{cost}(\phi, y) = +\infty$.

The cost of computing $U(f)$, $\text{cost}(U, f)$, is given by

$$\text{cost}(U, f) = \text{cost}(N, f) + \text{cost}(\phi, N(f)).$$

We now define the concepts of error and cost of the approximation U. The definitions of error and cost depend on the setting. We first discuss three settings: worst case, average case, and probabilistic. Then we turn to a randomized setting.

In the worst-case setting, the error and cost of U are defined as

$$e(U) = \sup_{f \in F} \|S(f) - U(f)\|,$$

$$\text{cost}(U) = \sup_{f \in F} \text{cost}(U, f).$$

In the average-case and probabilistic settings, we assume that the set F is equipped with a probability measure μ. In the average-case setting, the error and cost of U are defined as

$$e(U) = \left(\int_F \|S(f) - U(f)\|^2 \mu(df) \right)^{1/2},$$

$$\text{cost}(U) = \int_F \text{cost}(U, f) \, \mu(df).$$

In the probabilistic setting, we assume that we are given a number $\delta \in [0, 1]$, and the error and cost of U are defined as

$$e(U) = \inf\{ \sup_{f \in F - A} \|S(f) - U(f)\| : A \text{ such that } \mu(A) \leq \delta\}.$$

$$\text{cost}(U) = \sup_{f \in F} \text{cost}(U, f).$$

We now discuss a randomized setting. In this setting the approximation U is defined by a random selection of information and algorithm. More precisely, let ρ be a probability measure on a set T. Then, for each $t \in T$, we select information N_t and an algorithm ϕ_t, and compute $U_t(f) = \phi_t(N_t(f))$. Here t is a random variable distributed according to the measure ρ. Random information N_t is of the form (2.1) with randomly chosen L_i and n. A random algorithm is $\phi_t: N_t(F) \rightarrow G$. The approximation U can now be identified as the 4-tuple, $U = (N, \phi, T, \rho)$.

The error of U in the randomized setting is defined as

$$e(U) = \sup_{f \in F} \int_T \|S(f) - U_t(f)\| \, \rho(dt).$$

In the randomized setting, the cost of $U_t(f) = \phi_t(N_t(f))$ is defined as above, and then the cost of U is defined as

$$\text{cost}(U) = \sup_{f \in F} \int_T \text{cost}(U_t, f)\, \rho(dt).$$

We illustrate the randomized setting by continuing the integration example.

EXAMPLE (CONTINUED). Consider the classic Monte Carlo algorithm

$$U_t(f) = \frac{1}{n} \sum_{i=1}^{n} f(t_i),$$

with uniformly distributed points t_i. Here $t = [t_1, t_2, \ldots, t_n] \in T = [0,1]^n$ and ρ is the uniform distribution over the unit n-dimensional cube. In this case,

$$N_t(f) = [f(t_1), f(t_2), \ldots, f(t_n)]$$

is random information with randomly chosen points t_i and deterministically chosen n. The algorithm ϕ_t is deterministic and equal to $\phi_t(y_1, y_2, \ldots, y_n) = \frac{1}{n} \sum_{i=1}^{n} y_i$. The error of U is proportional to $n^{-1/2}$ and the cost of U is proportional to n. ∎

We are ready to define the computational complexity of IBC problems The basic notion of ε-complexity is defined as the minimal cost among *all* U with error at most ε,

$$\text{comp}(\varepsilon) = \inf\{\text{cost}(U) :\ U \text{ such that } e(U) \leq \varepsilon\}.$$

(Here we use the convention that the infimum of the empty set is taken to be infinity.) Depending on how $e(U)$ and $\text{cost}(U)$ are specified this defines ε-complexity for each of the four settings discussed above.

We stress that we take the infimum over *all* possible U for which the error does not exceed ε. In the worst-case, average-case, and probabilistic settings, U can be identified with the pair (N, ϕ), where N is the information and ϕ is the algorithm that uses that information. This means that we take the infimum over *all* information N consisting of information operations from the class Λ, and over *all* algorithms ϕ that use N such that (N, ϕ) computes approximations with error at most ε. In the randomized setting, U can be identified with the 4-tuple (N, ϕ, T, ρ) and we take the infimum over *all* random information N_t and *all* random algorithms ϕ_t, where $t \in T$ is distributed accordingly to an arbitrary probability measure ρ. Sometimes we write

$$\text{comp}^{\text{wor}}(\varepsilon), \ \text{comp}^{\text{avg}}(\varepsilon), \ \text{comp}^{\text{prob}}(\varepsilon, \delta), \ \text{and} \ \text{comp}^{\text{ran}}(\varepsilon)$$

to emphasize the setting and the dependence on the parameter δ in the probabilistic setting. If we want to stress that we use one of the deterministic settings, we then say, for example, the worst-case deterministic setting or the average-case deterministic setting and write $\text{comp}^{\text{wor}-\text{det}}(\varepsilon)$ or $\text{comp}^{\text{avg}-\text{det}}(\varepsilon)$.

EXAMPLE (CONTINUED). For the integration problem, the model of computation assumes that one function evaluation costs c, and each arithmetic operation, comparisons of real numbers, and evaluations of certain elementary functions can be performed exactly at unit cost. Usually $c \gg 1$.

The worst-case ε-complexity for the class F of Lipschitz functions with constant q is

$$\mathrm{comp}^{\mathrm{wor}}(\varepsilon) = \Theta\left(c\,\frac{q}{\varepsilon}\right), \quad \text{as } \varepsilon \to 0.$$

We use big Θ notation which means that there exist positive constants α_1, α_2 and α_3 such that $\alpha_1\, cq/\varepsilon \leq \mathrm{comp}^{\mathrm{wor}}(\varepsilon) \leq \alpha_2\, cq/\varepsilon$ for $\varepsilon \in [0, \alpha_3]$.

For the average-case and probabilistic settings, assume that μ is a truncated classical Wiener measure placed on the first derivatives. Then, in the average-case setting, we have

$$\mathrm{comp}^{\mathrm{avg}}(\varepsilon) = \Theta\left(c\left(\frac{1}{\varepsilon}\right)^{1/2}\right), \quad \text{as } \varepsilon \to 0.$$

In the probabilistic setting for $q \gg \ln(1/\delta)$, we have

$$\mathrm{comp}^{\mathrm{prob}}(\varepsilon, \delta) = \Theta\left(c\left(\frac{\sqrt{\ln(1/\delta)}}{\varepsilon}\right)^{1/2}\right), \quad \text{as } \varepsilon \to 0.$$

Finally, in the randomized setting, we have

$$\mathrm{comp}^{\mathrm{ran}}(\varepsilon) = \Theta\left(c\left(\frac{1}{\varepsilon}\right)^{2/3}\right), \quad \text{as } \varepsilon \to 0.$$

The complexity of integration in different settings has been studied for various classes of functions by many researchers. Recent surveys may be found in Novak[13] and Traub et al.[25] ∎

Among the goals of IBC is to find or estimate the ε-complexity, and to find an ε-complexity optimal U, or equivalently, an ε-complexity optimal pair (N, ϕ) (or an ε-complexity optimal 4-tuple (N, ϕ, T, ρ) in the randomized setting). By ε-*complexity optimality* of U we mean that the error of U is at most ε and the cost of U is equal to, or not much greater than, the ε-complexity. For a number of problems this goal has been achieved due to the work of many researchers.

We briefly indicate a proof technique sometimes used to obtain tight bounds on computational complexity of IBC problems. In what follows we restrict ourselves to the worst-case setting although a similar approach is used in other settings.

As already explained, the approximation $U(f)$ is computed by combining information operations from the class Λ. Let $y = N(f)$ denote this computed information. In general, the operator N is many-to-one, and therefore the set $N^{-1}(y)$ consists of many elements from F which are indistinguishable from f. Then the set $SN^{-1}(y)$ consists of all elements from G which are indistinguishable from $S(f)$.

Since $U(f)$ is the same for any f from the set $N^{-1}(y)$, the element $U(f)$ must serve as an approximation to any element g from the set $SN^{-1}(y)$. It is clear that the quality of the approximation $U(f)$ depends on the "size" of the set $SN^{-1}(y)$.

The radius of the set $A = SN^{-1}(y)$ is defined as the smallest radius of the ball which contains the set A,

$$\text{rad}(A) = \inf_{g \in G} \sup_{a \in A} \|a - g\|.$$

The *radius of information* $r(N)$ is then defined as the maximal radius of the set $SN^{-1}(y)$ for $y \in N(F)$,

$$r(N) = \sup_{y \in N(F)} \text{rad}(SN^{-1}(y)).$$

Clearly, the radius of information $r(N)$ is a sharp lower bound on the worst-case error of any U. We can guarantee an ε-approximation iff $r(N)$ does not exceed ε (modulo a technical assumption that the corresponding infimum is attained).

The cost of computing $N(f)$ is at least $c\,n$, where c is the cost of one information operation, and n, called the *cardinality* of N, denotes the number of information operations in N. By the ε-*cardinality number* $m(\varepsilon)$, we mean the minimal number n of information operations for which the information N has radius $r(N)$ at most equal to ε,

$$m(\varepsilon) = \min\{n: \text{there exists } N \text{ of cardinality at most } n \text{ such that } r(N) \le \varepsilon\}.$$

From this we obtain a lower bound on the ε complexity,

$$\text{comp}^{\text{wor}}(\varepsilon) \ge c\,m(\varepsilon).$$

For many problems it turns out that it is possible to find an information operator N_ε consisting of $m(\varepsilon)$ information operations, and a mapping ϕ_ε such that the approximation $U(f) = \phi_\varepsilon(N_\varepsilon(f))$ has error at most ε and $U(f)$ can be computed with cost at most $(c + 2)\,m(\varepsilon)$. (For examples of such problems see Traub et al.,[25] Chapters 5 and 7). This yields an upper bound on the ε-complexity,

$$\text{comp}^{\text{wor}}(\varepsilon) \le (c + 2)\,m(\varepsilon).$$

Since usually $c \gg 1$, the last two inequalities yield the almost exact value of the ε-complexity,

$$\text{comp}^{\text{wor}}(\varepsilon) \simeq c\,m(\varepsilon).$$

This also shows that the pair $(N_\varepsilon, \phi_\varepsilon)$ is almost ε-complexity optimal.

In each setting of IBC, one can define a radius of information such that we can guarantee an ε-approximation iff $r(N)$ does not exceed ε. This permits one to sometimes obtain tight complexity bounds in other settings.

The essence of this approach is that the radius of information as well as the ε-cardinality number $m(\varepsilon)$ and the information N_ε do not depend on particular algorithms, and they can often be expressed entirely in terms of well-known mathematical concepts. Therefore, we can sometimes obtain tight complexity bounds by drawing on powerful mathematical results.

3. NONCOMPUTABILITY AND INTRACTABILITY

3.1. FUNDAMENTAL CONCEPTS

Many important problems have been proven noncomputable or intractable in the worst-case deterministic setting. Since these are complexity results, we can only circumvent them by changing the setting. We can replace deterministic by randomized or we can replace worst case by average case. (Another possibility is to replace worst case by probabilistic case but we shall not pursue that here.) That is, we might be able to decrease the complexity by weakening the worst-case deterministic guarantee to a randomized or average-case guarantee.

We will give examples where noncomputability or intractability are broken and other examples where they are not. We will pose four open problems regarding the power of randomization or the average-case setting.

Given an error tolerance $\varepsilon > 0$, it has been shown for many problems that, in the *deterministic worst-case setting*, the computational complexity is of order $(1/\varepsilon)^{d/r}$, where d is the problem dimension and r measures the smoothness.[25] That is,

$$\text{comp}^{\text{wor}-\text{det}}(\varepsilon) = \Theta\left(\left(\frac{1}{\varepsilon}\right)^{d/r}\right), \quad \text{as } \varepsilon \to 0.$$

An example of a problem with such complexity is multivariate integration of function defined on the unit cube D in d dimensions. Thus, we wish to approximate

$$S(f) = \int_D f(t)\,dt$$

for functions for which $f^{(r)}$ is continuous and uniformly bounded in the sup norm. See Section 6 where multivariate integration is discussed in detail.

Other problems with such complexity are[25] approximation, nonlinear optimization, systems of nonlinear equations, linear elliptic differential equations, and Fredholm integral equations. The latter two will be discussed in Section 5.

A numerical example illustrates the implications of the complexity function. If one wants eight-place accuracy in computing triple integrals of functions that are once differentiable, then $\varepsilon = 10^{-8}, d = 3, r = 1$, and $(1/\varepsilon)^{d/r} = 10^{24}$. Even if we generously assume the existence of a sequential computer that performs 10^{10} function evaluations per second, the computation will take on the order of 10^{14} seconds or over 3 million years. Observe also that if the dimension d, that is, the number of variables, is increased by one, then the complexity is increased by a factor of 10^8. Even if ε is quite large, say, $\varepsilon = 10^{-2}$, the complexity is increased by a factor 100 when the dimension is increased by one.

Note two important differences between a typical NP-complete problem such as the traveling salesman problem and the multivariate problems discussed above. Exponential complexity for the traveling salesman problem (and many other problems) is currently only conjectured, but exponential complexity has been established for the indicated problems in information-based complexity. Secondly, the role of input size in combinatorial complexity is, in information-based complexity, often taken by the dimension d. We observe that problems of high dimensionality occur in such diverse disciplines as physics and nonlinear economic modeling.

If the complexity is of order $(1/\varepsilon)^{d/r}$, then for fixed smoothness r and increasing dimension d the problem is exponential in d and we say it is *intractable*.

An even more serious impediment than intractability to solving a problem may occur; a problem may be noncomputable. We say a problem is *noncomputable* if it is *impossible* to compute an ε-approximation at finite cost. That is, there exists ε_0 such that

$$\text{comp}^{\text{wor}-\text{det}}(\varepsilon) = \infty, \quad \forall \varepsilon \leq \varepsilon_0.$$

The motivation for this definition is that if the complexity is infinite, there does not exist an algorithm for computing even an approximate solution which uses a finite number of information and combinatory operations.

We give a simple example of a noncomputable problem. We wish to compute an ε-approximation to a definite integral in the deterministic worst-case setting. It is enough to consider the scalar case. Assume that we know the class of integrands consists of continuous and uniformly bounded functions. This problem is noncomputable.

This is a very special case of a general situation. We observed above that often the complexity is of order $(1/\varepsilon)^{d/r}$ for positive r. If the functions are only continuous and uniformly bounded, that is, if $r = 0$, then generally the complexity is infinite for all these problems.

We emphasize that these negative results concerning intractability and non-computability are instrinsic. There is no way around them by being clever.

The intractability and noncomputability results stated above are for the *worst-case deterministic setting*. This is the same setting in which a problem in combinatorial complexity is NP-complete. If we wish to mitigate these negative conclusions, we are forced to consider other settings.

3.2. DOES RANDOMIZATION HELP?

One possible way to break intractability or noncomputability is to permit randomization. Physicists have long known about the power of randomization. For example, classical Monte Carlo methods may be regarded as using randomized information to make the number of function evaluations needed to compute an ε-approximation to a multivariate integral independent of dimension. More recently, computer scientists have used randomized algorithms to solve problems with complete information such as primality testing (see, for example, Rabin[19]).

We emphasize that, whenever randomization is introduced, it weakens the assurance we can offer regarding the error. Even if we are willing to live with this weaker assurance, will randomization decrease complexity? As we shall discuss below, there are both negative and positive results. Obviously,

$$\text{comp}^{\text{wor-ran}}(\varepsilon) \leq \text{comp}^{\text{wor-det}}(\varepsilon), \quad \forall \varepsilon \geq 0,$$

because determinism is a special case of randomization.

Since we are trying to break noncomputability or intractability, we are only interested in the situation where there is a big win, that is, when $\text{comp}^{\text{wor-ran}}(\varepsilon)$ is much less than $\text{comp}^{\text{wor-det}}(\varepsilon)$. Specifically, we are interested when a problem which is intractable (or even noncomputable) in the worst-case deterministic setting is tractable (or computable) in the worst-case setting with randomization.

It turns out that randomization may help a great deal or it may not help at all. A problem where randomization helps is multivariate integration[1,13] (these results may also be found in Traub et al.,[25] Chapter 11). As already mentioned this problem is *intractable* for the class of r-times differentiable functions defined on the d-dimensional unit cube with uniformly bounded derivatives in the sup norm,

$$\text{comp}^{\text{wor-det}} = \Theta\left(\left(\frac{1}{\varepsilon}\right)^{d/r}\right), \quad \text{as } \varepsilon \to 0.$$

On the other hand,

$$\text{comp}^{\text{wor-ran}}(\varepsilon) = \Theta\left(\left(\frac{1}{\varepsilon}\right)^{2/(1+2r/d)}\right), \quad \text{as } \varepsilon \to 0.$$

Since the exponent is bound by 2,

$$\text{comp}^{\text{wor-ran}}(\varepsilon) = O\left(\left(\frac{1}{\varepsilon}\right)^{2}\right), \quad \text{as } \varepsilon \to 0,$$

and the problem is *tractable*. Now let $r = 0$. Then there exists ε_0 such that

$$\text{comp}^{\text{wor-det}}(\varepsilon) = \infty, \quad \forall \varepsilon \leq \varepsilon_0,$$

and the problem is *noncomputable*. On the other hand,

$$\text{comp}^{\text{wor-ran}}(\varepsilon) = \Theta\left(\left(\frac{1}{\varepsilon}\right)^{2}\right), \quad \text{as } \varepsilon \to 0,$$

and the problem is computable and tractable. Indeed, this bound is achieved by the classical Monte Carlo method.

Thus for multivariate integration, randomization helps enormously and

$$\text{comp}^{\text{wor-ran}}(\varepsilon) \ll \text{comp}^{\text{wor-det}}(\varepsilon), \quad \text{as } \varepsilon \to 0.$$

On the other hand, for certain approximation problems(see Traub et al.,[25] pp. 423–425) randomization does not help and

$$\text{comp}^{\text{wor}-\text{ran}}(\varepsilon) = \Theta(\text{comp}^{\text{wor}-\text{det}}(\varepsilon)), \quad \text{as } \varepsilon \to 0.$$

We only have examples where randomization does or does not break intractability and noncomputability. There is no theory which characterizes for which problems randomization can help. Furthermore, there is no theory which sets limits on how much randomization can help. This suggests the following questions.

QUESTION 1. Characterize those problems for which randomization helps significantly. In particular, for which problems does randomization break noncomputability or intractability?

QUESTION 2. What are the limits on how much randomization can help?

3.2.1. THE EFFECT OF PSEUDO-RANDOMIZATION.

As we have discussed above, randomization can break intractability for a problem such as multivariate integration. Such problems are solved on digital computers using pseudo-random number generators. If we show that a problem is tractable with randomization, is it still tractable with pseudo-randomization? Although there is a huge literature on tests for pseudo-random number generators (see, for example, Knuth,[7] Chapter 3, and Niederreiter[11]), these tests do not answer our question. In Traub and Woźniakowski[29] we study this question for multivariate integration. We summarize the conclusion here.

The good properties of the classical Monte Carlo algorithm that we would like to preserve are:

i. The cost needed to compute an ε-approximation is proportional to ε^{-2}, independent of the dimension d.

ii. This holds even if the class of integrals is a ball in $\mathbf{L}_2$.

The conclusion of our paper may be partitioned into bad news and good news. The bad news is that:

i. The cost result does not hold generally.

ii. The cost result holds only for a smaller class of integrals.

The good news is that:

i. If care is taken with the pseudo-random number generator, then the cost result holds.

ii. It holds for, say, the class of Lipschitz or smoother integrands.

Thus, with a little extra smoothness of integrands and if a little care is taken with pseudo-random number generators, we can preserve the essential properties of the Monte Carlo algorithm and maintain tractability of multivariate integration with pseudo-randomization.

3.3. DOES THE AVERAGE CASE HELP?

As we have seen, randomization may break intractability or noncomputability. Another possibility is to replace the worse-case setting by an average setting. That is, we weaken a worst-case assurance to an average assurance. For simplicity of exposition we limit ourselves here to the average-case deterministic setting.

We will limit our discussion to *linear* problems. In the worst-case settings, a problem is said to be linear if S is a linear operator and F is a ball of finite radius. In the average-case setting, we additionally assume that the measure is a truncated Gaussian measure over F (see Traub et al.,[25] Chapters 4 and 6, for more details). Examples of linear problems include multivariate integration, multivariate approximation, and linear partial differential equations, with appropriate measure.

For linear problems, and ignoring multiplicative constants of order 1 which are independent of ε, we have the following basic relations among the three settings defined above,

$$\text{comp}^{\text{avg}-\text{det}}(\varepsilon) \leq \text{comp}^{\text{wor}-\text{ran}}(\varepsilon) \leq \text{comp}^{\text{wor}-\text{det}}(\varepsilon), \ \forall \varepsilon \geq 0.$$

The inequality

$$\text{comp}^{\text{avg}-\text{det}}(\varepsilon) \leq \text{comp}^{\text{wor}-\text{ran}}(\varepsilon)$$

will be mathematically used in Section 6. The inequality

$$\text{comp}^{\text{wor}-\text{ran}}(\varepsilon) \leq \text{comp}^{\text{wor}-\text{det}}(\varepsilon)$$

was discussed before. That leaves the inequality

$$\text{comp}^{\text{avg}-\text{det}}(\varepsilon) \leq \text{comp}^{\text{wor}-\text{det}}(\varepsilon).$$

Can this lead to a big win? It turns out that settling for an average assurance may help a great deal or it may help only a little. A problem where an average assurance helps a great deal is integration of continuous functions which may or may not be uniformly bounded. This will be discussed in Section 6. There are also problems where an average assurance does not help.[25,28]

As in the case of randomization we lack a theory which characterizes for which linear problems the complexity in the average-case setting is significantly less then for the worst-case setting.

Our discussion suggests the following questions.

QUESTION 3. Characterize linear operators and measures for which the average-case setting helps significantly. In particular, for which problems and measures does the average-case setting break noncomputability or intractability?

QUESTION 4. Extend the analysis of average case to nonlinear problems; in particular, for such nonlinear problems as nonlinear equations, optimization, and nonlinear differential equations.

4. COMPLEXITY AND SCIENCE

In Section 3 we saw that many problems are noncomputable or intractable. What does this imply for scientific problems?

This is a highly speculative section on whether intrinsic limits on what can be achieved through computation limit what can be known in science.

Our goal is more to suggest directions for thought and to pose questions than to answer them. This section draws heavily on Traub[23] to which the reader is referred for additional material on this subject.

4.1. SCIENCE AND COMPUTATION

A number of researchers have commented on the relation between science and computation. Although the following quotations are primarily about the foundations of physics, the same relation holds for other sciences.

Pagels[16] put it succinctly: "To know a mathematical truth you must be able to compute it" (p. 44).

White[32] writes, "Foundational Physics seeks predictive theories, and that predictive process must be a computational algorithm executed on a real physically limited computer, a relationship termed the Wheeler cycle."

Feinberg[4] states, "I am not saying that theoretical physicists who work on fundamental questions, such as the behavior of subatomic particles, do not perform calculations. Doing calculations is what makes them theoretical physicists."

Geroch and Hartle[6] require that a physical "theory be such that its predictions can be extracted not merely in principle, possibly by ever higher levels of skill and sophistication, but also in practice, mechanically."

Landauer[10] summarizes the issue eloquently: "What computers can do will define the ultimate nature of the laws of physics. After all, the laws of physics are algorithms for the processing of information, and will be futile, unless these algorithms are implementable in our universe, with its laws and resources. 'What computers can do' refers not to the forseeable technological future, but rather to that which can be done in principle with the resources available in our actual physical universe."

In fairness, we must tell that one of us (JFT) recently spoke to L. D. Faddeev, the Director of the Steklov Institute in Leningrad and a world-famous mathematical physicist. Faddeev's view is that there is no connection between the foundations of physics and computation. If he's right, that's the end of this story, but we believe he's mistaken.

What intrinsic impediments are there to "what computers can do"? We mention three of them here.

i.   *Chaos.* Chaos is extreme sensitivity to initial conditions. Since the precise initial conditions are either not known or, even if known, cannot be exactly entered into a digital computer, the behavior of a chaotic system cannot be predicted.

ii. *Physical Limits.* There is an extensive literature on physical limits to computation. See, for example, Bremmerman[3] and Landauer.[9]

iii. *Intractablity.* If a problem is *intractable*, then there can never be sufficient computer resources for its solution. Of the various impediments to computation, we will confine ourselves here to intractability.

4.2. INTRACTABILITY

Many continuous problems are intractable in dimension in the worst-case setting. Furthermore, there is no theory as to when intractability can be broken using randomization or the average-case setting.

Noncomputability might be another barrier. For simplicity we confine ourselves to intractability.

What is known about the complexity of discrete problems? The intractability of many discrete problems has been *conjectured* but not proven. A discrete problem is intractable if it is exponential in input size. What is known is that there exist numerous problems, hundreds of problems, that all have essentially the same complexity. They are all intractable or all tractable, and the common belief among experts is that they are all intractable. For technical reasons these problems are said to be NP-complete.[5] One of the great open questions in discrete complexity theory is if P$\neq$NP or equivalently, whether the NP-complete problems are indeed intractable.

The above is for the worst-case deterministic setting. There are examples of problems which are tractable in randomized or average-case settings but, as with continuous problems, there is no theory which characterizes for which problems this is true.

4.3. STRONGLY INTRACTABLE PROBLEMS

To date, research in computational complexity has been for particular settings such as worst case, randomized, average, probabilistic, and asymptotic. It seems desirable to have a notion of intractability that is independent of the setting.

We will introduce the concepts of *strong noncomputability* and *strong intractability*. The description of these concepts will be rather vague; how they might be made precise will be discussed elsewhere. We will describe strong intractability; the description of strong noncomputability is analogous.

Consider a problem which is intractable in the worst-case deterministic setting. Assume that as we back off from the deterministic worst-case assurance to randomized, average, probabilistic, or asymptotic assurances, the problem remains intractable. (These are examples of assurances and not a comprehensive list.) Furthermore, this is to hold for every "fair" measure.

Such a problem is *strongly intractable*. If a problem is strongly intractable, the computer resources will never be found to solve the problem, even for weakened assurances.

Do strongly intractable problems exist? The answer depends obviously on what exactly is meant by setting and "fair" measure. For instance, let us restrict ourselves to worst-case, randomized, average-case, probabilistic, and asymptotic settings. For the settings which require a measure (average case, probabilistic) define a "fair" measure as one that maximizes the complexity of the problem. Then there are problems which are strongly intractable since they are intractable in all these five settings. An example of such a strongly intractable problem is approximation of periodic smooth functions of d variables.[8,25] Of course, for a different set of settings and a different definition of "fair" measure, the existence of strongly intractable problems may be open.

The concept of strong intractability is new and this question has not been previously posed. What the existence of strongly intractable problems implies for what is scientifically knowable will be explored in the concluding section.

4.4. WHY MIGHT WE EXPECT DIFFICULTIES WITH SUPERCOMPUTING AND WITH THE FOUNDATIONS OF PHYSICS?

Now that we have discussed the known or conjectured results regarding intractability from theoretical computer science, we will very briefly summarize why intractability might be an impediment to what is knowable in science.

To repeat, in the worst-case deterministic setting, the complexity of many discrete problems is conjectured to grow exponentially with the number of objects while the complexity of many continuous problems is known to grow exponentially with dimension. Furthermore, although some problems are tractable in the randomized or average settings, there are others which remain intractable. Indeed, we believe there are problems which are strongly intractable, that is, intractable in all settings.

This might indicate intractability of many supercomputing problems, and for the foundations of physics. For example, computational chemistry, computational design of pharmaceuticals, and computational metallurgy involve computation with huge numbers of particles. Economic modeling can involve a large number of variables. Path integrals, which are of great importance in physics, are infinite dimensional, and therefore invite high-dimensional approximations.

4.5. DO INTRACTABILITY RESULTS REALLY LIMIT WHAT IS KNOWABLE IN SCIENCE?

We have reviewed what is known about intractability for both discrete and continuous problems and introduced the concept of strong intractability. The intractability results and conjectures are certainly daunting. They suggest that many problems involving a large number of objects or high dimensionality might be impossible to solve.

Nevertheless, in this section we advance some reasons why the intractability results might not prevent us from solving scientific problems. We limit ourselves here to three reasons. See Traub[23] for a discussion of whether other modes of computations, such as neural networks, massively parallel computers, cellular automata, or reversible computation might enable us to circumvent the existing intractability results.

A SCIENTIFIC PROBLEM MAY HAVE MANY MATHEMATICAL FORMULATIONS. Intractability results are for a particular mathematical formulation. However, a scientific problem need not have a unique formulation. The following example is due to M. Kalos. Consider the flow of an incompressible fluid past an obstacle. The following are formulations of this scientific problem.

i.   Navier Stokes equation.

ii. A coupled system of linear ODE's which expresses Newton interactions between a large number of particles.

iii. Cellular automaton.

Although a particular formulation of a scientific problem may be intractable, it may be possible to find another one that is tractable. We believe this issue is important and we shall devote the concluding section to its discussion.

THE PROBLEM INSTANCE MAY BE EASY. Complexity theory tells us about the computational complexity of a class of problem instances. It does not tell us anything about a particular problem instance. Two examples will illustrate this observation.

The Traveling Salesman Problem is NP-complete and therefore probably intractable. However, that does not imply that a particular instance of this problem is difficult.

As a second example, consider integrating a function of d variables of smoothness r. For r fixed, the complexity grows exponentially with d. This tells us nothing about the difficulty of integrating a particular integrand.

In the probabilistic setting of information-based complexity, we wish to be assured of a small error except on sets of small measure. That is, we wished to be assured of a small error on "most" problems. Even in the probabilistic setting, the fact that a problem is intractable does not mean that a particular problem instance is difficult.

Even if a problem is strongly intractable, we cannot draw an inference about a particular problem instance. To what extent does nature give us easy problem instances?

THE GÖDEL EFFECT. Gödel's theorem states that in a sufficiently rich mathematical system, there are theorems that cannot be proven. Although the theorem has had the most profound effects on logic and the foundations of mathematics, it has had little or no impact on the work of mathematicians. (However, Smale[22] observes that Gödel's theorem has changed the agenda of mathematics.) Since a thorough discussion would take us too far afield, we will only indicate a couple of the possible reasons for the lack of impact

Mathematicians do not prove arbitrary theorems. They prove theorems that arise from areas of interest to mathematicians and in a historical context. It has been observed that mathematicians prove theorems they can prove.

Note an analogy between Gödel's theorem and computational complexity. Gödel's result does not permit us to infer that a particular theorem is undecidable, just as computational complexity does not permit us to conclude that a particular problem instance is hard. (Note, however, an important difference between Gödel's theorem and computational complexity: Gödel's theorem is negative whereas computational complexity sometimes gives us positive results by telling us certain problems are computable or tractable.)

This raises a number of questions. Will it turn out that noncomputability and intractability will have the same relation to science that Gödel's theorem has had with mathematicians? That is, are the results profound but insignificant to the practice of science? On the other hand, might intractability change the agenda of science?

Of course, mathematicians have broad latitude in their choice of problems. How much latitude do scientists have in their choice of scientific problems and mathematical formulations? We return to this question in the last section.

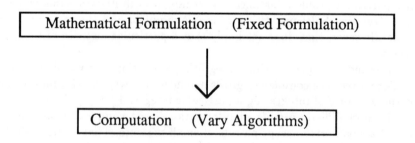

FIGURE 3 Computational complexity.

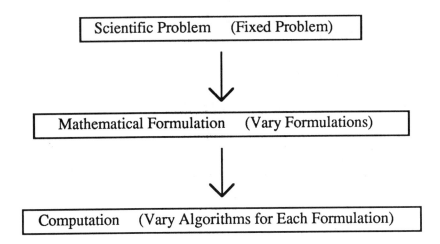

FIGURE 4   Computational complexity of scientific problems

4.6. COMPLEXITY OF SCIENTIFIC PROBLEMS

Computational complexity studies the complexity of a particular mathematical problem. As we pointed out, a scientific problem need not have a unique mathematical formulation. These formulations should meet twin desiderata.

i.   They should capture the essence of the science.

ii.   These should be computationally tractable.

This suggests the need for a theory of computational complexity of scientific problems. We contrast such a theory with computational complexity. In computational complexity the mathematical formulation is given and we vary algorithms for its solution. The resources required by the optimal algorithm is the complexity. This situation is diagrammed in Figure 3.

In the proposed theory of the computational complexity of scientific problems, a scientific problem is given. Mathematical formulations are then varied to see if any meet the twin desiderata above. For each mathematical formulation the complexity is obtained to check tractability. This situation is diagrammed in Figure 4.

A central question in such a theory would be: "Do there exist scientific problems such that every mathematical formulation is strongly intractable?" We believe that there are such scientific problems and that they are *unknowable*.

5. APPLICATIONS

In this section we sample some applications of IBC for three problems: ill-posed problems, partial differential equations, and integral equations. Most of the results are due to Arthur G. Werschulz. The readers who are interested in proofs and a more detailed discussion of these subjects are referred to the monograph of Werschulz.[31]

5.1. ILL-POSED PROBLEMS

We review complexity results for ill-posed problems in the worst and average-case settings for the class Λ_1 of all linear functionals.

By an ill-posed problem we mean that

$$S: D \subset F_1 \to G$$

is a linear *unbounded* operator. Here F_1 is a normed linear space and D is an infinite-dimensional linear subspace of F_1. We wish to approximate $S(f)$ for f from

$$F = \{f \in D : \|f\| \leq 1\}.$$

An example of an ill-posed problem is the solution of a Fredholm problem of the first kind for which the operator S is the inverse of a compact operator.

Not surprisingly, in the worst-case setting the problem is noncomputable.[31] To overcome this negative result, we must change some assumptions. For instance, we may change the error criterion. That is, instead of the absolute error criterion

$$\|S(f) - U(f)\|,$$

we switch to the *residual criterion* which is defined by

$$\|f - S^{-1}(U(f))\|.$$

If S is the inverse of a compact operator M, then the residual criterion means that we try to guarantee that

$$\|f - M(U(f))\| \leq \varepsilon, \quad \forall f \in F.$$

It turns out that the residual criterion for ill-posed problems leads to positive results which are related to the approximation problem in the worst-case setting.[31]

Another option to overcome the noncomputability of ill-posed problems under the absolute error criterion is to consider a different setting. Suppose we switch to the average-case setting. We need to choose a probability measure μ on the domain D of the operator S. Assume that μ is Gaussian.

Very recently Kon, Ritter, Vakhania, and Werschulz (private communications) showed that for a Gaussian measure the second moment of any linear unbounded operator is finite,

$$\int_D \|S(f)\|^2 \mu(df) \ < \ +\infty.$$

It can be shown[31] that this implies that the average-case complexity of an ill-posed problem is *finite* for all positive ε,

$$\text{comp}^{\text{avg}}(\varepsilon) \ < \ +\infty, \quad \forall \varepsilon > 0.$$

Hence, the ill-posed problem, which is *noncomputable* in the worst-case setting, becomes *computable* in the average-case setting with a Gaussian measure.

Although the average-case complexity is finite, it can tend to infinity arbitrarily fast as ε goes to zero. More precisely, the average-case complexity depends on the eigenvalues λ_i of the finite trace correlation operator C_ν of the Gaussian measure $\nu = \mu S^{-1}$. We have

$$\text{comp}^{\text{avg}}(\varepsilon) \ \simeq \ c \min \left\{ n: \sum_{i=n+1}^{\infty} \lambda_i \leq \varepsilon^2 \right\}.$$

For example, if $\lambda_i = i^{-p}$ with $p > 1$, then

$$\text{comp}^{\text{avg}}(\varepsilon) = \Theta \left(c \left(\frac{1}{\varepsilon} \right)^{2/(p-1)} \right).$$

Thus, for p close to 1 the average-case complexity is huge. For instance, if F_1 is a space of functions of d variables, and if $p = 1 + 1/d$, then the ill-posed problem is intractable even on the average. On the other hand, if p is not too close to 1, say, $p \geq 2$, then the average case complexity is at most proportional to ε^{-2}, and the ill-posed problem is tractable on the average.

5.2. LINEAR PARTIAL DIFFERENTIAL EQUATIONS

We summarize the complexity results of Werschulz for linear elliptic differential equations. For simplicity we restrict ourselves only to the worst-case setting. Complexity results for different settings may be found in Werschulz.[31]

Let F be the unit ball of the Sobolev space $H^r(D)$ for a bounded, simply connected C^∞ region D of $\mathbf{R}^d$. Let $G = H^m(D)$. Let A be a $2m$th-order elliptic linear partial differential operator.

The operator $S: F \to G$ is defined as the variational solution of the elliptic differential equations with the operator A and zero boundary values. That is, $S(f) = A^{-1}f$.

We discuss two classes of information operations. As in Section 5.1, the class Λ_1 consists of all linear functionals, and let the class Λ_2 consist of function evaluations. Obviously, $\Lambda_2 \subset \Lambda_1$.

For the class Λ_1, the worst-case complexity is given by

$$\text{comp}^{\text{wor}}(\varepsilon) = \Theta\left(c\left(\frac{1}{\varepsilon}\right)^{d/(r+m)}\right), \quad \text{as } \varepsilon \to 0.$$

Furthermore, the finite-element information and finite-element algorithm of degree $k \geq 2m - 1 + r$ are nearly ε-complexity optimal.

For the class Λ_2, if $r \leq d/2$, then the problem is noncomputable, whereas if $r > d/2$, then

$$\text{comp}^{\text{wor}}(\varepsilon) = \Theta\left(c\left(\frac{1}{\varepsilon}\right)^{d/r}\right), \quad \text{as } \varepsilon \to 0.$$

For $r > d/2$, the (modified) finite-element information and (modified) finite-element algorithm of degree $k \geq m - 1 + r$ are nearly ε-complexity optimal.

As we see, the worst-case complexity in the class Λ_2 of function evaluations is infinite or goes to infinity (as ε tends to zero) much faster than the corresponding worst-case complexity in the class Λ_1 of all linear functionals. This means that function evaluations supply less information about linear elliptic differential equations than linear functionals.

Finite-element information and algorithms are classical ways of solving elliptic problems. It turns out that if the degree of the finite-element algorithm is properly chosen reflecting the order of the elliptic operator and the smoothness of right-hand functions, then they are also nearly ε-complexity optimal.

5.3. INTEGRAL EQUATIONS

We review the complexity results of Werschulz for integral equations in the worst-case setting for the two classes Λ_1 and Λ_2 introduced in Sections 5.1 and 5.2. Analyses of different settings may be found in Werschulz.[31]

As in Section 5.2, let F be the unit ball of the Sobolev space $H^r(D)$ for a bounded, simply connected C^∞ region D of $\mathbf{R}^d$. Assume that $r > 1/2$. Let $G = \mathbf{L}_2(D)$. Let $K: \mathbf{L}_2(D) \to \mathbf{L}_2(D)$ be a compact integral operator with $\|K\| < 1$.

The operator $S: F \to G$ is defined as the variational solution of the Fredholm problem of the second kind. That is, $S(f) = (I - K)^{-1}f$.

It turns out that the worst-case complexity of integral equations for the two classes Λ_1 and Λ_2 is roughly the same and equal to

$$\text{comp}^{\text{wor}}(\varepsilon) = \Theta\left(c\left(\frac{1}{\varepsilon}\right)^{d/r}\right), \quad \text{as } \varepsilon \to 0.$$

Furthermore, the finite-element information and finite-element algorithm of degree $k \geq r - 1$ (appropriately defined for the class Λ_1 and Λ_2) are nearly ε-complexity optimal.

In contrast to the elliptic problem of Section 5.2, the integral equations problem is an example of a problem for which function evaluations and linear functionals are roughly of the same power. Similarly as for the elliptic problem, if the degree of the finite-element algorithm is appropriately chosen, then the classical way of solving integral equations by finite-element information and finite-element algorithm turns out to be nearly ε-complexity optimal.

6. MULTIVARIATE INTEGRATION

In this section we discuss the complexity of multivariate integration. We draw heavily on Woźniakowski.[35] The approximate computation of multivariate integrals has been extensively studied in many papers[11,12,13,25,26] for hundreds of references.

For a given class F of functions $f: D \to \mathbf{R}$ defined on the d-dimensional unit cube $D = [0, 1]^d$, let $S: F \to G = \mathbf{R}$ be

$$S(f) = \int_D f(t)\, dt, \quad \forall\, f \in F.$$

We briefly indicate the complexity results in the worst-case setting. These are known for many classes F. For example, if $F = W_p^{r,d}$ is the Sobolev class of functions whose rth distributional derivatives exist and are bounded in the $\mathbf{L}_p$ norm by one, then for $p\,r > d$ we have

$$\mathrm{comp}^{\mathrm{wor}}(\varepsilon) = \Theta\left(c\left(\frac{1}{\varepsilon}\right)^{d/r} \right), \quad \text{as } \varepsilon \to 0$$

(see Novak[13] for a recent survey).

For d large relative to r, the worst-case complexity is huge for even moderate ε. For r fixed, multivariate integration is intractable.

To overcome intractability of multivariate integration in the worst-case setting, we switch to the average-case setting. In contrast to the enormous literature for the worst-case setting, the average-case setting for multivariate integration has been studied in relatively few papers; (see Novak[13] for a recent survey).

We restrict ourselves to the class $F = C_d$ of continuous functions defined on D and equipped with the classical Wiener sheet measure w.[1] That is, w is Gaussian

---

[1] A uniform bound on the sup norm of continuous functions is not essential in the average-case setting with a Gaussian measure. Complexity results are essentially the same with or without such an uniform bound; see Woźniakowski[34] for details.

with mean zero and covariance kernel

$$R(s,t) \overset{\text{def}}{=} \int_{C_d} f(s)\,f(t)\,w(df) = \min(s,t) \overset{\text{def}}{=} \prod_{j=1}^{d} \min(s_j, t_j)$$

for any vectors $s = [s_1, \ldots, s_d]$ and $t = [t_1, \ldots, t_d]$ from D.

We now show that the average-case complexity for the class C_d is at most proportional to ε^{-2},

$$\text{comp}^{\text{avg}}(\varepsilon) = O\left(c\,\varepsilon^{-2}\right), \quad \text{as } \varepsilon \to 0, \quad \forall\, d \geq 1. \tag{6.1}$$

The proof of Eq. (6.1) can be provided by using a well-known error estimate of the Monte Carlo algorithm in the randomized setting. We stress that we use a different setting only to get an estimate in the average-case setting. The Monte Carlo algorithm computes

$$U(f) = \frac{1}{n} \sum_{i=1}^{n} f(x_i) \tag{6.2}$$

with independently and uniformly distributed points t_i over D. The expected error of the Monte Carlo algorithm with respect to the points t_i satisfies the following inequality

$$\int_{D^n} (S(f) - U(f))^2 \, dx_1 \cdots dx_n = \frac{1}{n} \left(\int_D f^2(t)\,dt - \left(\int_D f(t)\,dt \right)^2 \right)$$

$$\leq \frac{1}{n} \int_D f^2(t)\,dt.$$

We integrate both side of the above inequality with respect to the Wiener measure w. Since

$$\int_{C_d} \int_D f^2(t)\,dt\, w(df) = \int_D \left(\int_{C_d} f^2(t)\, w(df) \right) dt = \int_D t_1 \cdots t_d\, dt = 2^{-d},$$

we have

$$\int_{C_d} \int_{D^n} (S(f) - U(f))^2 \, dx_1 \cdots dx_n\, w(df) \leq \frac{1}{n} 2^{-d}. \tag{6.3}$$

Denote by $e^{\text{avg}}(U; x_1, \ldots, x_n)$ the average-case error of Eq. (6.2) for fixed x_i,

$$e^{\text{avg}}(U; x_1, \ldots, x_n) = \left(\int_{C_d} \left(S(f) - \frac{1}{n} \sum_{i=1}^{n} f(x_i) \right)^2 w(df) \right)^{1/2}$$

Then Eq. (6.3) can be rewritten as

$$\int_{D^n} e^{\text{avg}}(U; x_1, \ldots, x_n)^2 dx_1 \cdots dx_n \leq \frac{1}{n} 2^{-d}. \tag{6.4}$$

Applying the mean value theorem to the left-hand side of Eq. (6.4), we conclude that there exist points $x_1^*, \ldots, x_n^*$ such that

$$e^{\text{avg}}(U; x_1^*, \ldots, x_n^*) \leq \frac{1}{\sqrt{n}} 2^{-d/2}. \tag{6.5}$$

Thus, for $n = \lceil 2^{-d} \varepsilon^{-2} \rceil$, the average-case error of $U(f) = n^{-1} \sum_{i=1}^{n} f(x_i^*)$ is at most ε. Since the cost of $U(f)$ is $(c + 1)n = \Theta(c \varepsilon^{-2})$, Eq. (6.1) is proven.

Observe that Eq. (6.1) means that multivariate integration on the average is tractable since the average-case complexity is at most proportional to ε^{-2} no matter how large the dimension d. The proof of Eq. (6.1) is not constructive since we use the mean value theorem. Thus, we do not know how to construct points x_i^* for which the estimate (6.5) holds.

To obtain the average-case complexity, one needs to find optimal sample points, i.e., points of function evaluations which lead to minimal average-case error. It is known that grid points are not optimal sample points.[18,36]

The lower bound on the average-case complexity[17] is

$$\text{comp}^{\text{avg}}(\varepsilon) = \Omega\left(c\, \varepsilon^{-1}\right), \quad \text{as } \varepsilon \to 0.$$

Hence, the average-case complexity is at least proportional to ε^{-1} and at most proportional to ε^{-2}. To find its actual dependence on ε, we will construct optimal sample points.

We now show that the optimal sample points are related to $\mathbf{L}_2$ *discrepancy* which is defined as follows (see Niederreiter[11,12] where the background, history, and about 500 references concerning discrepancy may be found).

For $t = [t_1, \ldots, t_d] \in D = [0, 1]^d$, define $[0, t) = [0, t_1) \times \cdots \times [0, t_d)$. Let $\chi_{[0,t)}$ be the characteristic (indicator) function of $[0, t)$. For $z_1, \ldots z_n \in D$, define

$$R_n(t; z_1, \ldots, z_n) = n^{-1} \sum_{k=1}^{n} \chi_{[0,t)}(z_k) - t_1 t_2 \cdots t_d$$

as the difference between the fraction of the points z_i in $[0, t)$ and the volume of $[0, t)$. The $\mathbf{L}_2$ *discrepancy* of $z_1, \ldots, z_n$ is defined as the $\mathbf{L}_2$ norm of the function $R(\cdot; z_1, \ldots, z_n)$. Roth[20,21] proved that

$$\inf_{z_1, \ldots, z_n} \left(\int_D R_n^2(t; z_1, \ldots, z_n)\, dt \right)^{1/2} = \Theta\left(n^{-1} (\log n)^{(d-1)/2}\right). \tag{6.6}$$

The points $z_1^*, \ldots, z_n^*$ with $\mathbf{L}_2$ discrepancy of order $n^{-1} (\log n)^{(d-1)/2}$ are related to *Hammersley* points. Let $p_1, p_2, \ldots, p_{d-1}$ be the first $(d - 1)$ prime numbers. Any integer $k \geq 0$ can be uniquely represented as $k = \sum_{i=0}^{\lceil \log k \rceil} a_i p_j^i$ with integers $a_i \in [0, p_j - 1]$. The radical inverse function ϕ_{p_j} is given as

$$\phi_{p_j}(k) = \sum_{i=0}^{\lceil \log k \rceil} a_i p_j^{-i-1}.$$

The sequence $\{u_k\}$ of $(d-1)$-dimensional points for $k = 0, \pm 1, \pm 2, \ldots$ is defined by

$$u_k = [\phi_{p_1}(k), \phi_{p_2}(k), \ldots, \phi_{p_{d-1}}(k)], \quad k = 0, 1, \ldots, M - 1,$$

with $M = (p_1 p_2 \cdots p_{d-1})^{\lceil \log n \rceil}$, and by $u_{k+M} = u_k$, $\forall k$. Then there exists a real number t^* such that the d-dimensional points $z_1^*, \ldots, z_n^*$ are obtained by adding one component to the $(d-1)$-dimensional points u_k,

$$\{z_1^*, \ldots, z_n^*\} = \left\{ [(k + t^*)n^{-1}, u_k] \; : \; 0 \le k + t^* < n \right\}. \tag{6.7}$$

For $t^* = 0$, they are Hammersley points. Thus, the points z_i^* are obtained by adding t^*/n to the first component of Hammersley points.

We are ready to show how the results on $\mathbf{L}_2$ discrepancy can be used to obtain optimal sample points and average-case complexity.

Let $n = \Theta(\varepsilon^{-1} (\log \varepsilon^{-1})^{(d-1)/2})$ be chosen such that the $\mathbf{L}_2$ discrepancy of the function $R_n(\cdot, z_1^*, \ldots, z_n^*)$ is at most ε. Letting $\vec{1} = [1, 1, \ldots, 1]$, we define

$$x_k^* = \vec{1} - z_k^*, \quad k = 1, 2, \ldots, n. \tag{6.8}$$

We approximate the integral of f from C_d by the arithmetic mean of its values at x_k^*,

$$U^*(f) = n^{-1} \sum_{k=1}^{n} f(x_k^*), \quad \forall f \in C_d. \tag{6.9}$$

Clearly the cost of computing $U(f)$, is $(c+1)\,n$. To show that the average-case error of U^* is at most ε, we use

$$\int_{C_d} \left(I(f) - U^*(f) \right)^2 w(df) = \int_D R_n^2(t; z_1^*, \ldots, z_n^*)\, dt.$$

This formula can be checked by direct calculation. Due to the choice of n and the construction of z_i^*, the average-case error of U is at most ε. This implies that the average-case complexity is bounded by the cost of U^*,

$$\mathrm{comp}^{\mathrm{avg}}(\varepsilon) = O(c\,\varepsilon^{-1}(\log \varepsilon^{-1})^{(d-1)/2}). \tag{6.10}$$

It turns out that the bound equation (6.10) is sharp.[35] This means that function evaluations at the points x_k^* given by Eq. (6.8), and the algorithm U^* given by Eq. (6.9) are nearly ε-complexity optimal and

$$\mathrm{comp}^{\mathrm{avg}}(\varepsilon) = \Theta(c\,\varepsilon^{-1}(\log \varepsilon^{-1})^{(d-1)/2}).$$

We complete this section by a number of remarks.

■ The definition (6.8) of the points x_k^* is not fully constructive due to the unspecified constant t^* in Eq. (6.7). If one takes the sample points as the Hammersley points

$$z_k = [kn^{-1}, \phi_{p_1}(k), \ldots, \phi_{p_{d-1}}(k)], \quad k = 1, \ldots, n,$$

then the $\mathbf{L}_2$ discrepancy of the points z_k is of order $n^{-1}(\log n)^{d-1}$ (see for example Niederreiter[11]). Then the approximation $U(f) = n^{-1}\sum_{k=1}^{n} f(x_k)$ with $x_k = \vec{1} - z_k$ has average-case error at most ε provided that $n = \Theta(\varepsilon^{-1}(\log \varepsilon^{-1})^{d-1})$. This leads to a slight increase in the cost. More precisely, instead of the minimal average-case cost $\Theta(c\varepsilon^{-1}(\log \varepsilon^{-1})^{(d-1)/2})$, we approximate multivariate integrals at average-case cost $O(c\varepsilon^{-1}(\log \varepsilon^{-1})^{d-1})$ using the sample points x_k.

- The average-case complexity is specified to within a constant which may depend on dimension d. We believe it is difficult to find this constant even asymptotically.
- It would be interesting to extend the analysis of multivariate integration for smoother classes of functions equipped with folded Wiener sheet measures, or for other problems such as approximation of functions of d variables equipped with a classical or folded Wiener sheet measure.

ACKNOWLEDGMENTS

This research was supported in part by the National Science Foundation.

REFERENCES

1. Bakhvalov, N. S. "On Approximate Computation of Integrals" (in Russian). *Vestnik MGV, Ser. Math. Mech. Astron. Phys. Chem.* **4** (1959): 3–18.
2. Blum, L., M. Shub, and S. Smale. "On a Theory of Computation and Complexity over the Real Numbers: NP-Completeness, Recursive Functions and Universal Machines." *Bull. Amer. Math. Soc.* **21** (1989): 1–46.
3. Bremmerman, H. J. "Complexity and Transcomputability." In *The Encyclopedia of Ignorance*, edited by R. Duncan and M. Weston-Smith. Oxford, UK: Pergammon, 1977.
4. Feinberg, G. "On the Method of Theoretical Physics (Redux)." Unpublished manuscript of lecture presented at Conference on Method in Science and Philosophy, Columbia University, New York City, December, 1988.
5. Garey, M. R., and D. S. Johnson *Computers and Intractability: A Guide to the Theory of NP-Completeness.* New York: W. H. Freeman, 1979.
6. Geroch, R., and J. B. Hartle. "Computability and Physical Theories." *Foundations Phys.* **16** (1986): 533–550.
7. Knuth, D. E. *The Art of Computer Programming*, vol. 2, 2nd edition. Reading, MA: Addison Wesley, 1981.
8. Kowalski, M. A., and W. Sielski. "Approximation of Smooth Periodic Functions in Several Variables ." *J.Complexity* **4** (1988): 356–372.
9. Landauer, R. W. "Wanted: A Physically Possible Theory of Physics." *IEEE Spectrum* **4** (1967): 105–109.
10. Landauer, R. W. "Computation and Physics: Wheeler's Meaning Circuit?" *Foundations Phys.* **16** (1986): 551–564.
11. Niederreiter, H. "Quasi-Monte Carlo Methods and Pseudo-Random Numbers." *Bull. Amer. Math. Soc.* **84** (1978): 957–1041.
12. Niederreiter, H. "Quasi-Monte Carlo Methods for Multidimensional Numerical Integration." In *Numerical Integration III*, edited by H. Braß and G. Hämmerlin. International Series of Numerical Mathematics, vol. 85, 157–171. Basel: Birhäuser Verlag, 1988.
13. Novak, E. *Deterministic and Stochastic Error Bounds in Numerical Analysis.* Lectures Notes in Math, vol. 1349. Berlin: Springer Verlag, 1988.
14. Packel, E. W., and J. F. Traub. "Information-Based Complexity." *Nature* **328** (1987): 29–33.
15. Packel, E. W., and H. Woźniakowski. "Recent Developments in Information-Based Complexity." *Bull. Amer. Math. Soc.* **17** (1987): 9–36.
16. Pagels, H. R. *The Dreams of Reason: The Computer and the Rise of the Sciences of Complexity.* New York: Simon and Schuster, 1988.
17. Papageorgiou, A. "On Average Case Complexity." Ph. D. thesis, Department of Computer Science, Columbia University, 1989.
18. Papageorgiou, A., and G. W. Wasilkowski. "On the Average Complexity of Multivariate Problems." *J. Complexity* **6** (1990): 1–23.

19. Rabin, M. O. "Probabilistic Algorithms." In *Algorithms and Complexity: New Directions and Recent Results*, edited by J. F. Traub, 21–39. New York: Academic Press, 1976.
20. Roth, K. F. "On Irregularities of Distribution." *Mathematika* **1** (1954): 73–79.
21. Roth, K. F. "On Irregularities of Distribution, IV." *Acta Arith.* **37** (1980): 67–75.
22. Smale, S. Private communication, 1990.
23. Traub, J. F. "What is Scientifically Knowable?" *Proceedings of the 25th Anniversary Symposium, Computer Science Department, Carnegie-Mellon University.* Reading, MA: Addison Wesley, 1991.
24. Traub, J. F., G. W. Wasilkowski, and H. Woźniakowski. *Information, Uncertainty, Complexity.* Reading, MA: Addison-Wesley, 1983.
25. Traub, J. F., G. W. Wasilkowski, and H. Woźniakowski. *Information-Based Complexity.* New York: Academic Press, 1988.
26. Traub, J. F., and H. Woźniakowski. *A General Theory of Optimal Algorithms.* New York: Academic Press, 1980.
27. Traub, J. F., and H. Woźniakowski. "Complexity of Linear Programming." *Oper. Res. Lett.* **1** (1982): 59–61.
28. Traub, J. F., and H. Woźniakowski. "Information-Based Complexity: New Questions for Mathematicians." *Mathematical Intelligencer* (1991), to appear.
29. Traub, J. F., and H. Woźniakowski. "The Monte Carlo Algorithm with a Pseudo-Random Generator." Report, Computer Science Department, Columbia University, 1989; Mathematics of Computation (1992), to appear.
30. Wasilkowski, G. W. "A Clock Synchronization Problem with Random Delays." *J. Complexity* **5** (1989): 1–11.
31. Werschulz, A. G. *The Computational Complexity of Differential and Integral Equations.* Oxford: Oxford University Press, 1991.
32. White, I. "The Limits and Capabilities of Machines—A Review." *IEEE Trans. Systems, Man and Cybernetics* **18** (1988): 917–938.
33. Woźniakowski, H. "A Survey of Information-Based Complexity." *J. Complexity* **1** (1985): 11–44.
34. Woźniakowski, H. "Average Complexity for Linear Operators over Bounded Domains." *J. Complexity* **3** (1987): 57–80.
35. Woźniakowski, H. "Average Case Complexity of Multivariate Integration." *Bull. AMS* (new series) (1991): 185–194.
36. Ylvisaker, D. "Designs on Random Fields." In *A Survey of Statistical Design and Linear Models*, edited by J. Srivastava, 593–607. Amsterdam: North-Holland, 1975.

Andrew G. Barto
Department of Computer and Information Science, University of Massachusetts, Amherst, MA 01003

Some Learning Tasks from a Control Perspective

1. INTRODUCTION

Progress in understanding complex systems strongly depends on choices made in defining the problems to be studied. This process of abstraction requires both simplification sufficient to make solutions feasible and the retention of enough complexity to make solutions relevant to the real systems being studied. These aspects of modeling methodology have been discussed extensively, but in specific cases it is easy to overlook the advice of philosophers and system theorists by forgetting that the solution to a problem abstraction may not be the ultimate goal. There is a strong tendency to commit what Alfred North Whitehead called the "Fallacy of Misplaced Concreteness"[40]: we tend to mistake abstractions for reality, especially if these abstractions are successful. This admonition applies to the study of learning because so many researchers are currently studying such a small fraction of the issues relevant to learning. The purpose of this chapter is to describe a collection of learning tasks in order to place the most commonly studied learning tasks into a broad context and to outline a range of tasks that are important despite the fact that they are receiving relatively little attention.

Artificial intelligence researchers often distinguish between a system's "environment," "performance element," and "learning element."[15] Dietteric[15] referred to these elements to make a useful statement about learning:

> The task of the learning element can be viewed as the task of bridging the gap between the level at which the information is provided by the environment and the level at which the performance element can use the information to carry out its function. (p. 328)

One can consider a spectrum of possibilities for the extent of this gap. At one extreme, the environment provides reliable, explicit, and detailed specification of desired actions. In this case, the learning element simply has to remember what it is told and employ an appropriate indexing scheme for accessing this information. A somewhat wider gap occurs if the environment can provide desired actions for only a subset of the situations that might occur in the future. Learning in this case requires storing information and forming an indexing scheme that can generalize beyond the specific instances used in training. This is usually called a supervised learning task and is being widely studied in the context of artificial neural networks. It is possible to consider much wider gaps in which the environment provides less information, and the learning element has to perform more sophisticated processing. For example, the environment may provide unreliable and infrequent assessments of the consequences of the performance element's actions. In this case, the learning element must somehow use a problem-solving method to discover what is "worth" remembering and combine it with a storage mechanism and an appropriate indexing scheme so that this knowledge can be used to improve future performance.

Widely recognized in the field of artificial intelligence is the need to consider learning in the context of problem solving. For example, there is a trend toward developing comprehensive "learning architectures."[25,36] In the field of artificial neural networks, or connectionist networks, there is also recognition of the need to study comprehensive learning architectures, but most neural network research addresses only a few subproblems of more general learning tasks. Although restricting attention to subproblems is necessary for making initial progress in understanding more complete systems, and solutions to subproblems can be of practical use in their own right, it is useful to reassess progress in light of more realistic frameworks. In this chapter, instead of the problem-solving framework of artificial intelligence, we adopt the framework of *control*. In addition to its closer ties to traditional engineering problems, a control framework more realistically emphasizes real-time interaction with reactive environments.

This chapter only addresses learning *tasks*; methods for solving these tasks are not addressed at all. Learning tasks are defined in terms of the task's objectives, the nature of the learning system's environment, and the nature of the information available to the learning system. Although the methods applicable to a particular task depend on the task's characteristics, there is not a one-to-one correspondence between tasks and methods, despite the fact that tasks and methods are often given the same names. Most methods can be applied to several classes of tasks, and most tasks can be solved by several different methods. By discussing only tasks, we hope to provide a sound basis for understanding the methods applicable to learning, while avoiding the confusion invariably produced by confounding tasks and methods.[1] Hinton[19] provides a good overview of learning methods developed for artificial neural networks, and Barto[2] discusses some of these methods in the context of control.

We describe several classes of learning tasks within a control framework designed to highlight the limitations of each class and the relationships between them. Because each class of tasks is defined by specializing this framework in specific ways, it is possible to see what additional complexities are encompassed by the more elaborate tasks, and what simplifying assumptions reduce the more elaborate tasks to the simpler ones. Some of the tasks are rather degenerate examples of control tasks, but viewing them as such serves our purpose of exposing issues they *do not address*. Our goal in using this framework is not to provide an exhaustive categorization or a "unified theory" of learning tasks. The subject of learning is so broad, with so many domain-specific features, that a useful comprehensive theory probably does not exist. Our framework does nothing to illuminate some interesting and significant classes of learning tasks, and it avoids many subtleties. Further, for the most part we avoid mathematical formalism even though some of the discussion could benefit from the additional precision that exists in extensive mathematical literature pertinent to each class of tasks. The reader may be able to suggest improvements to this framework and how we embed tasks within it, but we believe it is adequate for showing that there is more to learning than is encompassed by the tasks most frequently studied.

2. LEARNING CONTROL TASKS

The philosopher Daniel Dennett[16] expressed the root idea of control in common language as follows:

> A *controls* B if and only if the relation between A and B is such that A can *drive* B into whichever of B's normal range of states A *wants* B to be in. (p. 52)

For our purposes, an even broader definition is better: A controls B if and only if the relation between A and B is such that A can cause B to behave as A wants B to behave. Some of the behavior in which we are interested is not readily expressible in terms of a range of states. *Learning control* tasks involve system A

---

[1] The distinction between tasks and methods is similar to the distinction between Marr's[28] computational and algorithmic levels. Jordan and Rumelhart[22] provide additional insight into this distinction as it applies to issues in learning.

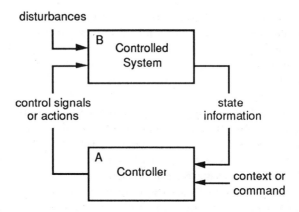

FIGURE 1   A Basic Control Loop. The controller, A, receives information about
the state of the controlled system, B, as well as context signals or commands. The
behavior of B is influenced by control signals from A and unpredictable disturbances.

learning through experience how to make a system B, and systems similar to B, do
what A wants them to do.

Figure 1 shows systems A and B in a basic control loop. System A is the controller, system B is the controlled system, and the signals that the controller sends
to the controlled system are the control signals. We refer to these signals simply as
A's actions. Other inputs to the controlled system are classified as "disturbances,"
which are unpredictable signals whose influence on the controlled system is to be
counteracted by the controller. The input signals to the controller, on the other
hand, typically provide information about the current state of the controlled system as well as specification of commands to the controller, which one can regard
as the "context" of the control task. Commands may indicate, for example, a desired output of the controlled system (e.g., the set point of a thermostat), a desired
time course of system variables (e.g., the desired trajectory of robot manipulator
positions), or more general commands (e.g., "make the controlled system do X").

The range of learning tasks described in this chapter requires elaborating Figure
1 by adding an additional component and the interconnections as shown in Figure
2. We focus on the controller, A, as both the learning element and the performance
element,[2] and we assume that its input can be divided into three categories. First,
A receives information describing the state of B. This information is usually incomplete: B only provides "clues" to A about B's current state. The second category

---

[2] We find it less useful to distinguish between a learning element and a performance element than
do researchers following a symbolic approach to machine learning. In connectionist networks—and
in biological systems—learning mechanisms are distributed throughout performance systems.

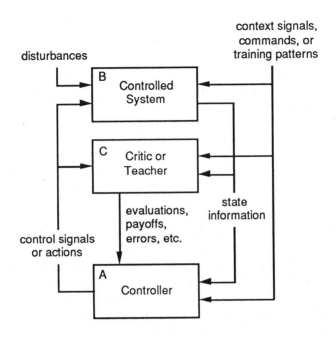

FIGURE 2 A General Schema for Describing a Range of Learning Tasks. The basic control loop of Figure 1 involving the controller, A, and the controlled system, B, is augmented with a critic or teacher, C, and an input to all components labeled "context, command, or training pattern" which sets the context for the learning task.

of information A receives comes from a component, labeled C, which we call a "critic" or a "teacher," depending on the type of task being considered. The critic provides information to A that guides learning. C has access to clues about the state of B and also has access to the actions of A. It implements some kind of evaluative procedure that assesses A's actions in light of B's current state according to a measure of "correctness," "goodness," "utility," etc. Sometimes the critic provides "payoffs" to A. The gap referred to above between the information provided by a learning system's environment and the information needed to accomplish some task is largely determined by the nature and quality of the information provided to A by C. Significant differences between learning tasks depend on the nature of this information.[3] The third category of information component A receives "sets the context" for the learning task. This information corresponds to the context or command information A receives in Figure 1, but in the expanded schema of Figure 2 it includes the training patterns required for some learning tasks. This

[3] We do not consider the case of multiple critics because the single-critic case is sufficiently complex for our purposes; many significant issues not discussed in this chapter come into play in the multi-critic case.

information is available to all components via the line labeled "context signals, commands, or training patterns," but we assume that the source of the signals on this line is outside of the system and "beyond the control" of components A, B, and C.

In general, components A, B, and C are dynamic systems. This means that the current output of any of these components can depend on an internal state, which changes over time, in addition to a current input. Consequently, the current output of a component can depend on the past history of its inputs in addition to its current input. Component A *must* be a dynamic system if it is to be capable of learning from its past experience, but for the learning tasks we consider here, some or all of the other components are assumed to be "memoryless," i.e., their current output only depends on their current input. We assume throughout that C is deterministic and memoryless, an assumption whose consequences we discuss in Section 3.

In placing particular tasks into the schema shown in Figure 2, what constitutes the components A, B, and C, and what corresponds to the context signals, commands, or training patterns are not intrinsic properties of the mechanisms and phenomena under consideration. Some divisions of reality into such boxes, with certain communication channels designated as input and output channels, may have obvious advantages over others in terms of explanatory power, but these divisions are best seen as the result of the process of constructing models for specific purposes. Many different ways of matching specific cases to this schema can be useful. In particular, we point out two aspects of the schema shown in Figure 2 that are misleading if taken too literally. First, the controller, A, receives three categories of information via three distinct input channels. This is a useful abstraction that may appear differently in real situations. For example, there may be no identifiable separation of A's input channels into three categories because all of this information might be represented in a distributed code involving overlapping sets of input channels. A second misleading conclusion is that the controller, A, corresponds to something like an entire adaptive organism. This identification might make sense in some circumstances (e.g., for very simple adaptive organisms), but it is better to avoid this interpretation because parts of some or all of the components shown in Figure 2 may reside inside the organism (e.g., B might include the organism's respiratory system, and C might represent the organism's highly evolved preference structure).

Because we frequently refer to the components of Figure 2 only by their labels A, B, and C, the following summary should be helpful in remembering their meaning:

A. *Controller*: the learning system; combines learning and performance elements.

B. *Controlled System*: behavior can be influenced by the controller and is evaluated by the critic or teacher; not present in all tasks.

C. *Critic* or *Teacher*: implements evaluative process that assesses A's actions in light of B's current state and the current context, command, or training pattern according to a measure of "correctness," "goodness," "utility," etc.

3. PERFORMANCE MEASURES AND GRADIENTS

Although it is notoriously difficult to give a definition of learning that is both adequate and precise, most attempts to do so involve the idea of a system improving performance over time according to some performance measure. If one thinks of the performance measure as a function defined over the set of possible external behaviors of the learning controller (being, for the moment, purposefully vague about what the controller's external behavior is), and if one visualizes this function as a surface, then any given behavior of the controller corresponds to a point on this surface. For a controller to improve performance over time, the point corresponding to its behavior has to move to successively higher points on this performance surface; or, assuming continuous changes in behavior, the point has to move uphill across this surface.

One critical aspect of most interesting learning tasks is the problem of measuring the learning system's performance according to a performance measure that reflects the true objectives of the task. The "true" performance measure might evaluate the overall performance of the learning system over its lifetime, but it is impossible to determine this *during* the system's lifetime when learning has to occur. In practice, one has to devise measures that are closely correlated with the true performance and yet can be measured more easily and more quickly. The problem of specifying accessible performance measures that can act as surrogates for basic but hard-to-measure criteria is a central issue in most realistic learning tasks.

Given this preamble, we can provide the rationale for our view of the critic as a memoryless system. Referring to Figure 2, the critic, C, provides the controller, A, with information pertinent to the *immediate consequences* of the action just taken by A on the controlled system, B, as assessed within the current context. The performance measure embodying the task's true objective, however, is not necessarily revealed in the immediate consequences of the controller's actions. The true performance measure usually evaluates behavior over longer periods of time, and although C's signals *over time* provide information relevant to A's true performance, C does not directly implement this measure. Hence, restricting discussion to memoryless critics emphasizes the problem of improving performance according to a measure that is not directly accessible. One might think of C as the "primary critic" which is part of a task's specification and provides the most basic evaluative information. Some learning methods involve "secondary critics" which provide performance information that is more accurate and more timely than the information provided by C. Research exists on how useful secondary critics can be constructed by means of learning processes (e.g., Barto, Sutton, and Anderson,[7] Sutton,[34,35] Watkins,[37] Werbos[39]), but within the framework presented here, these "adaptive critic" methods are considered to be implemented within the controller A.[4]

---

[4] Alternatively, if one focused only on the learning task faced by an adaptive critic, the task would appear as a special kind of adaptive prediction task within the class of supervised learning tasks discussed in Section 5.

In some tasks there is no conflict between short-term performance, as revealed by the critic's instantaneous signals, and long-term performance, as determined by the task's true performance measure. In these tasks, performing each action to produce the best immediate response from the critic coincides with the task's true objective. This occurs, for example, when the training experiences to which A is exposed at each time step are statistically independent and the true performance measure is the time average of the critic's outputs. In these cases, acting to optimize each immediate signal of the critic coincides with the objective of optimizing the critic's average signal over time. In other tasks, however, dependencies over time may make it necessary to sacrifice short-term performance in order to produce improvement with respect to the true performance measure. This occurs when A's actions not only influence the critic's immediate evaluation but also influence a dynamic process which plays a part in determining future evaluations. Section 6.3 discusses tasks with this property, called *sequential decision tasks*. All the other tasks we discuss are defined to avoid conflict between short-term and long-term performance.

One can distinguish two basic types of learning tasks on the basis of the local characteristics of the performance surface about which the critic's instantaneous signals provide information. We call a task a *supervised learning task* if each instantaneous signal of the critic provides information pertinent to the *gradient* of the performance surface at the point corresponding to the controller's current external behavior. Recalling that the gradient of a surface at a point is a vector pointing in the direction of steepest ascent, a supervised learning task is characterized by the availability from C of *directed* information about how the controller should change its behavior in order to improve performance (at least locally). This gradient information often (but not always) takes the form of an error vector giving the difference between the controller's action and some desired, or target, action. Although each error vector (based on a single training pattern) is not the gradient of the true performance measure (a measure of the error over all training patterns), it provides useful information about the true gradient. In other cases, the critic's signals may lack the magnitude information present in error vectors and only provide directional information; for example, only the signs of errors may be provided. Although less informative than other kinds of gradient information, this is still gradient information according to our view, and such tasks are supervised learning tasks.[5] The key point is that in a supervised learning task, the learning system is provided with information about (1) whether or not a local improvement is possible (is the gradient zero?), and (2) if improvement is possible, *how* (what direction) behavior should be changed to achieve improvement.

In other learning tasks, which we call *reinforcement learning tasks*, instead of directly providing gradient information during learning, the instantaneous output of C gives the controller information about the current *value* of the performance measure. In contrast to gradient information, information about the performance

---

[5] The signs of the components of a gradient vector instruct a direction of local improvement but not the direction of most rapid local improvement.

measure's value does not itself indicate how the learning system should change its behavior to improve performance; it is not *directed* information. A system facing a reinforcement learning task has to concern itself with estimating gradients based on information about performance measure values provided by C over time. Although C's outputs are usually scalars in a reinforcement learning task, whereas they are vectors in supervised tasks, the distinction between scalar and vector-valued information from C is not central. For example, although we do not discuss them in this chapter, in multi-criterion reinforcement learning tasks the critic signals are vectors consisting of the values of multiple performance measures. The key point is that each output of the critic in a reinforcement learning task evaluates behavior but does not in itself indicate if improvement is possible or *direct* changes in behavior.

The distinction between supervised and reinforcement learning tasks is only one of many distinctions between learning tasks that one can make. We use this distinction as an organizing principle in our view of learning tasks because it corresponds to a significant division of existing theoretical traditions and their literatures, and it separates tasks according to issues that have been treated as almost orthogonal: *generalization* is central in most supervised learning tasks, whereas *exploration* is central in reinforcement learning tasks. Addressing these issues separately provides a good basis for understanding the more complex tasks discussed below which combine them. However, in practice it may be more difficult than we have suggested to distinguish between these classes of tasks even in relatively simple cases. For example, it may have occurred to the reader that whether a task's objective is a global or local extremum of the performance measure is a factor that complicates the situation. Just because directional information is immediately available in a task does not mean that the task's objective is best served by always following it. Doing so would result in a local extremum, whereas a global extremum may be the objective. A task such as this would match our definition of a supervised learning task in a local sense, but it would be more like a reinforcement learning task when viewed globally because exploration is a key issue for global optimization.

4. UNSUPERVISED LEARNING TASKS

Unsupervised learning tasks are concerned with recoding information into forms that are more compact, better organized, or otherwise better suited for understanding or further processing. Usually, the objective is to recode while retaining as much of the original signal's information as possible. The framework used in this chapter has not been designed to do justice to unsupervised learning tasks. However, if one were to place these tasks within our control framework, they would appear as tasks in which there is no controlled system, B, and the signals that A must learn to recode are the context, command, or training patterns. We might regard the critic, C, as embodying the principle by which the recoding is to take place. Unlike all the other tasks discussed below, however, this critic would be *fixed* and completely *known* for each type of unsupervised learning task.[6] Consequently, a better view might omit the separate critic altogether and regard its function as being built into the learning system, A, from the start. In this case, there is no closed-loop interaction between any of the components, and the term "controller" for A is highly inappropriate. For example, in one type of unsupervised learning task known as a clustering task, the objective might be to separate the input data into disjoint classes (clusters) so as to maximize the ratio of measures of the between-cluster and within-cluster scatters. All the details of this performance measure are known in advance and can be incorporated into the clustering algorithm.

In contrast, the learning tasks discussed below have the property that important aspects of the performance measure are unknown and can vary with each task instance. For example, in a supervised learning task, the desired pairings of training patterns and target actions depend on the specific task instance (Section 5). In a reinforcement learning task, even less is known about the performance measure (Section 6). Despite the rather degenerate form they take within the framework used here, the category of unsupervised learning tasks is appropriate for addressing issues of information representation that are essential in all the more elaborate learning control tasks that one can define. Although we consider them important, these issues are orthogonal to the ones we have chosen to address in this chapter. Barlow[1] provides a good philosophical overview of unsupervised learning tasks.

5. SUPERVISED LEARNING TASKS

A critic capable of providing gradient information about a performance measure is usually called a "teacher." Following Jordan and Rumelhart,[22] we distinguish between supervised learning tasks with "proximal" and "distal" teachers. We discuss the case of the proximal teacher first because it is the prototypical supervised learning task.

PROXIMAL TEACHER. In a supervised learning task with a proximal teacher, the performance measure is defined in terms of a set of input vectors to A, called *training patterns*, and the actions that A should produce in response to them. These correct actions are sometimes called *target actions*. In practice, training patterns and target actions are represented as bit vectors or vectors of real numbers. Within the framework of Figure 2, the controlled system, B, plays no role, and the training patterns arrive from outside the system as context signals. Based on each action

---

[6]For the other types of tasks we discuss, the critic is fixed, and possibly completely known, for each task *instance*, e.g., a specific supervised pattern recognition task. In contrast, the critic is fixed and known for each *entire class* of unsupervised learning tasks, e.g., all tasks in which principal components are to be extracted.

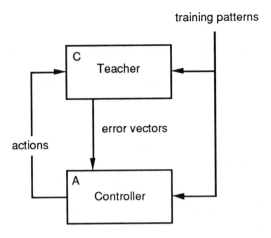

FIGURE 3   The basic components of a supervised learning task with a proximal teacher. It differs from Figure 2 in that there is no role for the controlled system, B.

of A and knowledge of the target action for each training pattern, C provides an error vector to A for each training pattern, where the error vector is the vector difference between A's actual action and the target action for the current training pattern. In an alternative view of these tasks, C simply provides A with the target action for each training pattern instead of an error vector. In this case, A itself has to compute the error vector (and the term critic is not very descriptive of C). But whether C provides A with error vectors or target actions, each instantaneous output of C tells A something about how it should change the action it produces in response to each training pattern. In what follows, we regard C as providing error vectors. Figure 3 is the variant of Figure 2 appropriate for the supervised learning tasks most commonly studied. It differs from Figure 2 in that the controlled system, B, is absent.

The objective of a supervised learning task with a proximal teacher is for A to learn to produce the correct action (i.e., the target action) in response to each training pattern and to generalize correctly to patterns not presented during training. Stating this somewhat more precisely, the objective is for A to form a mapping from the set of possible input patterns to actions that would produce error vectors of small magnitude on all possible sets of training patterns. Exactly what constitutes the magnitude of an error vector can be defined in a variety of ways which involve issues that we do not address here. Supervised learning tasks, as described above, provide the simplest framework permitting the study of generalization, and applicable methods can be useful in a wide range of practical problems. In artificial intelligence, learning in this type of task is called "learning from examples."[15]

By assuming that the training patterns arrive from outside of the system, we incorporate a key feature of supervised learning tasks as they are usually studied: the selection of training patterns is not influenced by the behavior of the controller,

A. There is no closed-loop interaction between A and the source of training patterns. The teacher, C, and controller, A, interact in a closed loop because the error vectors supplied by C depend on A's actions, but the generator of training patterns is not in this loop. The only real learning control that occurs is that A is learning how to control C in the context of each training vector: it wants to make C produce error vectors equal to zero.

A number of special cases of supervised learning tasks with a proximal teacher have been studied separately. Rote memory storage is the variant of this task in which the issue of generalization is omitted. For example, a standard computer random access memory solves the version of this task in which each training pattern is an address and each target action is the bit vector to be stored at the given address. Extending rote memory to include some form of generalization leads to more general associative memory systems, which have been extensively studied in the form of artificial neural networks (e.g., Hinton and Anderson[20] and Kohonen[23,24]). If the target actions are elements of a finite set, then supervised learning tasks are sometimes called adaptive pattern classification tasks where the target actions are the class labels (e.g., Duda and Hart[17]). If the target actions are real numbers, then these tasks are function approximation problems in which generalization involves interpolation and extrapolation.

Open-loop system identification[18,27] is also a supervised learning task with a proximal teacher. Open-loop system identification is the task of modeling an incompletely known system by observing how it responds to a set of inputs. It is open loop because there is no simultaneous attempt to control the system being identified. The inputs to the system being identified and its corresponding outputs play the roles, respectively, of training patterns and target actions. This fits into the schema of Figure 3 by letting C contain the system being identified. Training patterns are the inputs to the system being identified and also inputs to A. The error vectors provided to A during learning are computed by comparing A's actions with the outputs of the system being identified. As the magnitudes of the error vectors are reduced, A's input/output behavior more closely matches that of the system being identified. According to this formulation, the controller, A, is not engaged in controlling the system being identified: it is attempting to control C, in the sense described above, and the task is arranged so that it has to identify the unknown system in order to do this.

This view of the open-loop system identification task does not do justice to many of its features, especially the use of delay-coordinate representations of system input and output, but it does correctly characterize the nature of the learning task according to the distinctions we are making. Similarly, open-loop adaptive prediction tasks appear as tasks in which the role of the critic is played by the system whose behavior is to be predicted. Storage buffers must be introduced for storing training patterns until the target actions become available. Barto[2] discusses these tasks in somewhat more detail; Widrow and Stearns[41] and Goodwin and Sin[18] provide views of these tasks from the perspectives, respectively, of adaptive signal processing and adaptive control. Adaptive critic methods (e.g., Sutton[35]) are methods for solving open-loop adaptive prediction tasks in which the objective is to predict some statistic about the critic's signals over the future (such as their expected sum over the future) instead of the critic's signal at a prespecified time in the future.

Some tasks that might be called supervised learning tasks do not conform to this open-loop view. For these tasks, the generation of training patterns is not "beyond the control" of component A as we have specified. The training process may incorporate pedagogic principles which alter the order, frequency, etc. of training patterns based on A's behavior. For example, to accelerate the learning process, training may concentrate on the patterns on which large errors have been made, or training "queries" can be generated to test hypotheses[12]. These techniques, as well as others such as "fading" and "shaping" in which the training information changes over time depending on A's behavior,[7] require that both A and C influence the source of training patterns. These techniques are not often studied from a theoretical perspective and do not fit into Figure 3, but they can be important for improving the rate and/or quality of learning. Similarly, the closed-loop system identification task requires identification to proceed at the same time that the system being identified is being controlled. Such tasks are more properly viewed as examples of adaptive sequential decision tasks discussed in Section 6.3.

DISTAL TEACHER. The basic supervised learning task described above can be generalized to what Jordan and Rumelhart[22] call "supervised learning with a distal teacher." The gradient information provided by a distal teacher is not expressed in the same coordinate system as are A's actions. Jordan and Rumelhart[22] use the example of learning to move a robot arm to specified locations in the arm's workspace. For this problem, A's actions specify joint angles, but C provides error vectors that are differences between desired spatial positions of the end effector and the positions actually attained. Here, A is working in joint space, but C is providing spatial training information. Learning methods must incorporate some means for deducing from the distal teacher's training information what a proximal teacher would specify if one were provided. Jordan and Rumelhart[22] discuss methods that effectively construct an appropriate proximal teacher based on the distal teacher's training information.

Figure 4 is an elaboration of Figure 3 illustrating the components and connections required for a supervised learning task with a distal teacher. The controlled system, B, is placed in the pathway from the controller, A, to the teacher, C. Component B transforms the space of A's actions to the space in which C is able to provide error vectors. As in other supervised learning tasks, training patterns, assumed to be generated outside of the system, are available to both A and C.

---

[7] These terms describe techniques used to train animals.[21] Fading refers to the alteration of stimulus patterns over time as a function of the animal's behavior, whereas shaping refers to the alteration over time of the criteria by which the delivery of reinforcement is determined. Although these techniques more properly relate to reinforcement learning than to supervised learning, both might be thought of as alterations of the pattern/target pairs as a function of the animal's behavior.

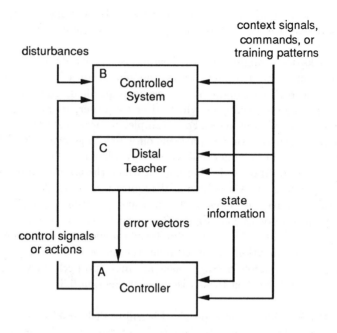

FIGURE 4 The basic components of a supervised learning task with a distal teacher. A controlled system, B, is placed in the pathway connecting the controller, A, with the distal teacher, C.

If B is the identity transformation, then the schema shown in Figure 4 reduces to that shown in Figure 3. For the arm movement example, B is the forward kinematic transformation of the arm, i.e, the transformation from joint angles (the actions of A) to the corresponding spatial coordinates of the end effector (the outputs of B). Training patterns are coded descriptions of the target spatial positions which play the role of commands to move the end effector to these positions. During training, a series of such commands is given to A, so that A can learn how to generate appropriate actions for each commanded position, and to C, so that C can properly assess the arm's actual movement. Although the arm, B, does not require access to these commands in the example of Jordan and Rumelhart,[22] in general these signals may include context information that influences B's behavior; for example, movement may have to occur in the context of different loads.

The distal teacher, C, provides error vectors to A based on discrepancies between B's outputs and target outputs for B for each context signal, training pattern, or command. In the arm movement task, for each training pattern, C knows how to determine an error vector in spatial coordinates. Component A somehow has to figure out how to alter its actions in response to each training pattern to reduce the magnitude of these error vectors. This task can be nontrivial, and a variety of methods can be used. The availability of B's outputs to A, as provided by the link from B to A in Figure 4, may be useful to A depending on what learning strategy it employs. In tasks such as this, there is a closed loop of influence passing from A, through B and C, and back to A. We can say that A is trying to control C, and the task is arranged so that in the process A must also learn how to control B. One of the issues that arises in these tasks is that there may be many actions of A that can achieve a distally specified target (for example, there may be many arm configurations that place the end effector at a target spatial location). In these cases, the objective may be to achieve any one of the possible solutions, or to achieve a solution that satisfies additional constraints.

The generalization of supervised learning tasks to include distal teachers, i.e., to cases in which B is not the identity transformation, is much more widely applicable than suggested by the arm control example. In fact, when details of the internal structure of the learning component are taken into account, supervised learning tasks *always* involve a distal teacher. To understand this, recall that in Section 3 we characterized supervised learning tasks by the availability of gradient information pertaining to the performance surface defined over the *external* behavior of the controller A. Because we are discussing learning tasks and not learning algorithms, we have said nothing about how gradient information pertinent to external behavior can be translated into gradient information pertinent to the *internal* parameters of the learning component that are to be adjusted during learning. The controller, A, is being treated as a black box.

But if one focuses on an internal adjustable component of A (such as a weight of a neural network), and shifts perspective so that this internal component is now the learning component A of Figure 4, then this A faces a task with a distal teacher. This task can be viewed in terms of Figure 4 by regarding component B as the process that translates the internal component's actions into the external actions of the overall controller. How the learning task of such an internal component can be solved depends on how much knowledge is available about the transformation implemented by B. For example, in the case of a layered artificial neural network, if A is a hidden unit, component B represents the transformation implemented by the part of the network between A and the network's output units. Because the details of this transformation are known, the appropriate translation of gradient information from distal to proximal coordinate systems can be accomplished, in this case, by error back-propagation[33] (which evaluates the Jacobian transpose of B at the current operating point). If the transformation performed by B is not known in detail sufficient to permit the required translation of gradient information, B can be identified via supervised learning (as discussed above) so that the required translation can be approximated using the resulting model.[22]

In the examples of supervised learning with a distal teacher given above, the controlled system B was treated as being memoryless (arm dynamics were ignored in the version of the arm movement task discussed). If B is a dynamic system, i.e., if its output depends on an internal state in addition to input, then the tasks we have classified as supervised learning tasks with a distal teacher are the same as adaptive control tasks in which the objective is to cause B's behavior to track a desired reference trajectory, where the trajectory is specified either at all time instants or only at selected time instants. If the reference trajectory is specified at all time instants, the context or command signals directly provide the target trajectory, so that the distal teacher can provide the controller with a tracking error at each time instant. If the reference trajectory is not specified at all time instants, then the task's objective is to achieve the specified behavior at the designated times while perhaps satisfying other constraints in the intervening time intervals. An example of the latter type of task occurs when the context or command signals specify a target output to be produced by B at a later time but do not specify a target trajectory for B's states or outputs over the intervening time interval. Jordan and Rumelhart[22] refer to the teacher in such tasks as being "distal in time." Although in such tasks the process intervening between A's actions and the relevant output of B extends over time, the logic is the same as for other tasks with distal teachers: somehow, knowledge about B must be used to deduce what a proximal teacher would specify at each time instant if one were present.

Supervised learning tasks with teachers that are distal in time are related to the adaptive sequential decision tasks we discuss in Section 6.3 as examples of reinforcement learning tasks. In fact, it may not be very useful to draw sharp distinctions between these tasks because they can both require learning to produce sequences of actions without the immediate availability of training information. However, within the framework we are using in this chapter, no matter how infrequent the training information, if it occurs in the form of target outputs of B, then the task is supervised because C can provide gradient information by comparing B's actual outputs with the targets. In a reinforcement learning task, on the other hand, the objective is to make B's behavior improve according to a performance measure that does not necessarily involve target outputs. But the situation is more complicated than this because learning with a teacher distal in time usually involves satisfying trajectory constraints that also are not specified in terms of target outputs. We leave these subtleties for a more discriminating framework and turn now to a discussion of reinforcement learning tasks.

6. REINFORCEMENT LEARNING TASKS

In a supervised learning task, the teacher, C, supplies information to A in the form of error vectors. Each error vector provides information about the gradient of an underlying performance measure at the current operating point. The gradient vector points in the direction in which the controller's behavior should be changed to yield the best local improvement. In a reinforcement learning task, in contrast, C provides A with information pertinent to just the value of the performance measure for the current behavior. Accordingly, for reinforcement learning tasks we call C's outputs "evaluations." An evaluation does not tell A how it should change its behavior, or even whether improvement is possible or not. We distinguish between reinforcement learning tasks according to what information is available to A in addition to C's evaluations. In *nonassociative* tasks, A receives only C's evaluation, whereas in *associative* tasks, A has access to information that can influence its actions via associative relationships.

6.1 NONASSOCIATIVE REINFORCEMENT LEARNING TASKS

In a nonassociative reinforcement learning task, the only inputs A receives are the evaluations provided by C. Each of A's actions is separately evaluated by C, and the objective is for A to find the best action (or one of possibly many best actions) as evaluated by C. This basic picture is complicated by the possibility that C's evaluations may not depend in a simple way on A's actions. Evaluations may depend on a complex, and possibly stochastic, dynamic system intervening between A's actions and C, as well on other factors that A's actions cannot influence. To accommodate these possibilities within the general schema of Figure 2, we use all of the components and delineate special cases by assigning specific properties to them.

Figure 5 shows the basic arrangement of these components for nonassociative reinforcement learning tasks. Because neither the context signals nor the outputs of B are available to it, A cannot select actions conditionally on this information even though the evaluations it receives may depend on this information as well as on

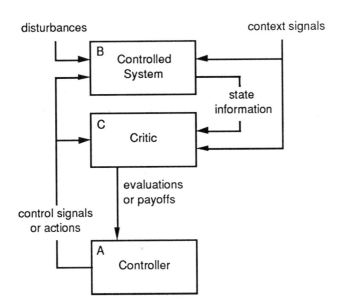

FIGURE 5   The basic components of a nonassociative reinforcement learning task. The only input A receives is the evaluative input from C.

its actions. We use the term nonassociative because A cannot form an associative mapping from this information to actions. Because C evaluates the consequences of A's actions on B, C may not need access to the actions themselves. However, in some cases C's evaluation depends on the actions themselves as well as on their consequences (for example, the actions may have different "costs"), and thus we provide the direct connection from A to C in Figure 5 (which is present for the same reasons in all remaining figures depicting reinforcement learning tasks).

If the output of B is a deterministic function of A's actions (that is, if there are no disturbances and B is memoryless and insensitive to any changes in context signals) and the performance measure concerns only the instantaneous behavior of A, then the task is a function optimization task. If C implements a real-valued function of B's output, as is often assumed, then B and C together assign a real number to each action of A. The objective is for A to determine which of its actions maximizes this number, or, since this is generally impossible, to find an action that is a local maximum. Within the framework being used in this chapter, this is a relatively simple *type* of task, although specific instances can be extremely difficult to solve due to the nature of the functions implemented by B and C.

If B's output is not a deterministic or memoryless function of A's actions, then the evaluation A receives from C in response to an action may not be the same each time that action is performed. This can also happen if B and/or C are sensitive to multiple context signals. In these cases, the performance measure has to take this variability into account. It often makes sense to model the overall evaluation process in terms of probability distributions, and to define the performance measure as the average over time of the evaluations produced by C. Two cases are distinguished depending on whether the probability distributions are stationary or nonstationary over time. A stationary probabilistic model is appropriate when B is memoryless and B and C are insensitive to multiple contexts. In this case, the action that yields the optimal average evaluation stays the same over time. A nonstationary model is appropriate when B is not memoryless and/or there is sensitivity to multiple context signals. Here, an action that is best at one time may not be best at another. Because A is blind both to B's state and to any context signal, to A it appears that C is changing its evaluation criteria over time.

Stationary tasks of this kind have been extensively studied as "bandit problems"[13] or addressed by means of the theory of learning automata.[29] For example, learning automata theorists have studied the following problem. The controller, A, has n actions $a_1, a_2, \ldots, a_n$. Immediately after A generates an action, C sends to A a signal indicating either "success" or "failure." This signal is determined from A's action by some complex nondeterministic process implemented by components B and C of Figure 5. Whatever this process may be, it is modeled simply as a collection of success probabilities $d_1, d_2, \ldots, d_n$, where d_i is the probability of returning "success" given that A has produced action a_i. Each d_i can be any number between 0 and 1 (the d_i's do not have to sum to one), and one assumes that A has no initial knowledge of these values. The objective is for A to eventually maximize the probability of receiving "success," which is accomplished when A always performs the action a_j such that $d_j = \max\{d_i | i = 1, n\}$. In a variant of this task, the success probabilities are known, but it is not known which corresponds to which action.

In the nonstationary version of this task, the success probabilities, d_i, change over time. A reasonable objective under these circumstances is for A to maintain a high level of performance over time by continuing to change its action in an attempt to track the maximum of the nonstationary evaluation process. It can do this by attempting to construct a predictive model of the evaluation process, but this is difficult to do only on the basis of the information provided by C, or it may be impossible due to unpredictable shifts of context. More realistically, if A can adjust its behavior quickly enough to track the nonstationary evaluation process, then it can maintain good performance over time without constructing a predictive model. The point, however, is that even though evaluations of A's actions may depend on the context and on the state of B, it is not possible for A to select actions conditionally on this information because A is completely blind to this information.

To prevent too narrow an interpretation of nonassociative reinforcement learning tasks, note that the following qualifies as an example. Suppose each of A's actions is an entire control rule which can be applied to a complex stochastic dynamic system. Let the transformation B assign to each such action (now an entire control rule) some description of the dynamic system's behavior under the control of the corresponding control rule. For example, B might produce a description of the state the dynamic system reaches after some time interval. Based on this state description, C provides to A an evaluation of the choice of control rule.[8] The solution sought in this task is a control rule optimal in a sense determined by the critic. This formulation of learning control does not have much to recommend it due to the size and complexity of the sets and systems involved, but it is an example of a nonassociative reinforcement learning task.

Deeper aspects of reinforcement learning tasks exist when the objective is not just to discover *eventually* which action is optimal, but also to perform the optimal action as frequently as possible during the discovery process or, more generally, to maintain performance throughout the discovery process at as high a level as possible. The issue arises of how to balance the requirement for maintaining a high level of performance over time with the requirement for estimating the relative worth of the actions. Two factors must influence each action selection: (1) the desire to use what is already known about the relative merits of the actions, and (2) the desire to acquire more knowledge about actions' consequences to make better selections in the future. These two factors ordinarily conflict: the best decision according to one is not best according to the other. This is called the conflict between control and identification. It is present in its simplest form in the stochastic success/failure task described above, which is one of the simplest adaptive optimal control tasks.[31]

[8] Biological evolution is sometimes viewed this way: each of A's actions is the genetic material of an organism, and C provides a measure of the reproductive success of that organism over its lifetime.

Despite the limitations imposed by restricting the information to which component A has access, nonassociative reinforcement learning tasks clearly illustrate differences between reinforcement learning tasks and supervised learning tasks. Obviously, when C directly provides A with gradient information, as it does in a supervised learning task, the learning algorithm implemented by A does not have to estimate the gradient of the performance measure from a set of evaluations. It is effectively told by C how to change its behavior. In a reinforcement learning task, however, the learning component has to *do something* to estimate the gradient. What it does depends on the specifics of the learning algorithm, but in all cases it must perform some form of exploratory behavior and must manipulate the resulting set of evaluations to determine how to change the rule for generating behavior. In a supervised learning task, as described in Section 5, all the responsibility for exploratory behavior is relegated to whatever process generates training patterns, and this process is usually fixed and simple (e.g., it repeatedly cycles through a finite set of training patterns). Consequently, the conflict between control and identification—the issue that emerges in its most stark form in nonassociative reinforcement learning tasks—is not present in supervised learning tasks unless the generation of training examples can be influenced by A's actions.

6.2 ASSOCIATIVE REINFORCEMENT LEARNING TASKS

Associative reinforcement learning tasks have all of the properties of nonassociative reinforcement learning tasks except that A has access to information other than C's evaluations. Consequently, A can take advantage of information that can help it perform better than can its blind counterpart facing a nonassociative version of the same task. In Figure 6, the controller, A, has access to information about the state of B and, if present, context or command information.

One of the simplest associative reinforcement learning tasks is a generalization of the "success/failure" nonassociative task described in Section 6.1. Suppose that at any time, there can be one of several context signals and that the success probabilities depend on the context signal as well as the action of A. Specifically, suppose d_{ix} is the probability of C sending "success" to A given that A produced action a_i while the context signal was x. To achieve the best rate of success, whenever A receives context signal x, it should select the action a_j such that $d_{jx} = \max\{d_{ix}|i = 1, n\}$. Clearly, if A selects actions independently of the context signals—as it would be forced to do if facing a nonassociative task—then in general it could, at best, achieve a lower success rate than if its actions could depend on the context signals. A's sensitivity to context signals eliminates nonstationarity by allowing the nonstationary task to be solved as multiple stationary tasks. Tasks of this kind are discussed by Barto, Sutton, and Brouwer,[8] Barto and Anandan,[5] Barto,[3,4] and Williams.[42,43]

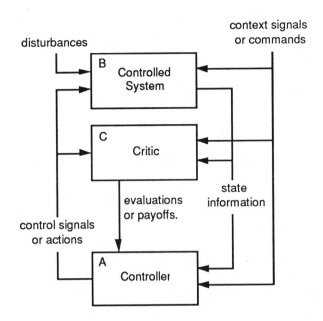

FIGURE 6   Basic components of an associative reinforcement learning task. The critic provides evaluative information to A, which also has access to information about the state of B and the context signal or command.

In other associative reinforcement learning tasks, the controller, A, can take advantage of information about B's current state if this state influences C's evaluation of the current action. For example, B's current state may act as context for the evaluation process in the same way that the context signals do in the example just described, except that in this case, A's actions can influence how these context signals change over time. Consequently, A's actions not only can influence the immediate evaluation signal produced by C, but they can also influence future evaluations by influencing B's future behavior. However, by an associative reinforcement learning task, we mean a task in which the underlying performance measure is such that the the objective is to maximize only the *immediate* evaluation at each step. More complex tasks, called adaptive sequential decision tasks, require learning how to perform actions so as to maximize a measure of long-term performance by manipulating B's long-term behavior. In these tasks, it can make sense to forego a favorable immediate evaluation in order to achieve a better evaluation in the future. We discuss adaptive sequential decision tasks in the next section.

Associative reinforcement learning tasks combine the important aspects of nonassociative reinforcement learning tasks with aspects of supervised learning tasks, while avoiding the additional complexities present in sequential tasks. For example, the context signals in the associative version of the "success/failure" task just described correspond to pattern vectors in a supervised learning task. The objective is for A to respond to each pattern with the action that is optimal for that pattern, but it must learn how to do this on the basis of success/failure feedback instead of error vectors. All the issues concerning generalization which are important in supervised learning are important here, but there is a conflict between control and identification as well.

6.3 ADAPTIVE SEQUENTIAL DECISION TASKS

Adaptive sequential decision tasks are characterized by performance measures that evaluate the controller's behavior over extended periods of time, but unlike some of the tasks discussed above—which are special cases—the full framework for sequential tasks takes into account the possibility of the controller influencing the long-term behavior of the controlled system, B. Whereas in associative reinforcement learning tasks, the controller, A, has to discover what actions produce the most favorable immediate evaluation from C, in sequential tasks A has to learn how to select the actions that maximize some cumulative measure of the evaluations it receives over time. This is more difficult than merely trying to achieve the best immediate evaluation. Some actions may be useful in producing a high immediate evaluation, but these same actions may cause B to enter states from which *later* high evaluations are unlikely or impossible. Hence, performing these actions would result in worse performance over the long-term than might be possible otherwise. Conversely, some actions may produce low evaluations in the short-term but are necessary to set the stage for better evaluations in the future. The controller's decision-making method must somehow account for both the short- and long-term consequences of actions. By a task's horizon, we mean the duration of the time interval into the future over which the consequences of current actions are significant.

Many tasks of practical significance can be modeled as sequential decision tasks. Finding the least-cost route from one place to another is perhaps the most generic example. Choice points along a route correspond to the states of B, A's actions determine the path taken and the next place reached, and the evaluation from C in response to an action is related to the cost of traveling the path. In this example, it is clear that an action influences the immediate evaluation (the cost of the path immediately taken) as well as the evaluations possible over the future (determined by the place reached). More complex tasks involving resource allocation, investment, gambling, and foraging for food are also examples of sequential decision tasks, as are many of the planning and problem-solving tasks studied by artificial intelligence researchers. Unlike most of the tasks described above, solving sequential decision tasks can be difficult even if one knows all the details about the systems and performance measures involved. Bertsekas[14] provides a good account of sequential decision tasks and their solution using computational methods known as dynamic programming, which are applicable when there are accurate models of the systems and performance measures involved.

By an *adaptive* sequential decision task, we mean a sequential decision task in which one does not have the accurate models required for applying these solution methods. These tasks are obviously very difficult to solve: solving them requires learning, and they can simultaneously involve the difficulties present in all of the other learning tasks we have discussed. The associative reinforcement learning tasks discussed in Section 6.2 are special cases of adaptive sequential decision tasks whose horizons are reduced so that only the immediate consequences of actions are significant. Some approaches to solving adaptive sequential decision tasks are discussed by Barto, Sutton, and Watkins[9,10] and Barto and Singh.[11]

One type of sequential decision task receiving attention is known as a Markov decision task.[32] In this kind of task, the controller, A, interacts with a controlled system, B, which is a discrete-time, finite-state, stochastic dynamic system. At each time step t, A observes the system's current state, x_t, and selects an action, a_t. After the action is performed, A receives a certain amount of payoff, r_t, from the critic, C, that depends on a_t and x_t, and B makes a transition from state $x_t = x$ to state $x_{t+1} = y$ with probability $P_{xy}(a_t)$. Upon observing state x_{t+1}, the controller A chooses another action, a_{t+1}, and continues in this manner for a sequence of time steps. The objective of the task is for A to form a rule to use in selecting actions, called a decision policy, that maximizes a measure of the total amount of payoff accumulated over time.

One commonly studied measure of cumulative payoff is the expected infinite-horizon discounted return. This is defined by using a discount factor, γ, $0 \le \gamma < 1$, to weight future payoffs less than immediate payoffs. Specifically, the expected infinite-horizon discounted return for a given policy and state, x, is

$$E\left[\sum_{t=0}^{\infty} \gamma^t r_t | x_0 = x\right], \qquad (1)$$

where x_0 is the initial system state, and E is the expectation assuming that A uses the given policy. The objective of the decision task is to form a policy (there may be many) that maximizes the expected infinite-horizon discounted return defined by Eq. 1 for each system state x. This sequential decision task reduces to the "success/failure" associative reinforcement learning task described in Section 6.2 if there are only two possible payoff values and the discount factor, γ, equals zero. In this case, one can identify the larger payoff value with the "success" signal of the associative reinforcement learning task, and the smaller value with the "failure" signal of that task. When γ equals zero, Eq. 1 becomes simply $E[r_0|x_0 = x]$, which is maximized when the probability of immediate success is maximized for each state x. If, in addition, B has only one state, this task further reduces to the "success/failure" nonassociative reinforcement learning task described in Section 6.1.

The most general adaptive sequential decision tasks require all the components and connections shown in Figure 2, with the exception of the pathway for context signals, commands, or training patterns. If such signals were present, they could indicate which of several different adaptive sequential decision tasks was present at any particular time. For example, in the context of "thirst," a foraging animal's task would be different than in the context of "hunger." The critic's evaluation criteria would depend on the context, and the animal's foraging strategy, implemented by A, may come to depend on the context also. We might call such tasks "associative adaptive sequential decision tasks" in analogy with the associative reinforcement learning tasks discussed above. Solving them would involve learning associative relationships between context signals and control rules effective in different contexts, and generalization among different sequential decision tasks could be important. This observation brings us in a full circle back to the issues addressed by supervised learning tasks, except that here the gap between what must be learned and the kind of training information available is wide indeed.

7. DISCUSSION

In this chapter we described a collection of learning tasks in order to place the tasks most commonly studied into a broad context and to describe other types of learning tasks that are receiving relatively little attention. We restricted discussion to learning *tasks*—defined in terms of the task's objectives, the nature of the learning system's environment, and the nature of the information available to the learning system—and did not discuss methods for solving these tasks. By placing all the tasks within a control framework we attempted to illuminate the limitations of each type of task and the relationships between them, even though some of the tasks are rather degenerate kinds of control tasks. We reiterate that our aim has not been to provide an exhaustive categorization or a unified theory of learning tasks.

We distinguished two broad classes of tasks on the basis of the type of training information available during learning. In supervised learning tasks, information about the gradient of the performance measure is directly available to guide learning. By gradient information we mean information that directly tells the system if local improvement in behavior is possible and, if so, specifies how the behavior should be changed, i.e, in what direction a change should be made. The central issue in supervised learning tasks is *generalization*: how can a complete mapping be inferred from a sample of examples? In contrast, the training signals in reinforcement learning tasks do not directly contain directional information. These signals evaluate behavior but do not directly indicate if local improvement is possible or how the behavior should be changed for improvement. Directional information must be obtained from a collection of signals evaluating different behaviors, but the learning system itself has to perform this integrative process. It has to probe the environment—perform some form of *exploration*—to obtain information about how to change its behavior. In so doing, a system in a reinforcement learning task can encounter a conflict between how it has to change behavior in order to obtain directional information, and how the resulting directional information tells it to change its behavior for improvement. This is known as the conflict between control and identification and is absent in supervised learning tasks. Although many tasks involve aspects of both supervised learning and reinforcement learning tasks—so that the distinction between these classes of tasks may not be as sharply drawn in practice as we have suggested—this distinction has significant consequences for the design of learning methods.

Following Jordan and Rumelhart,[22] we discussed supervised learning involving proximal and distal teachers. Whereas a proximal teacher provides gradient information in the coordinate frame of the learning system's actions, a distal teacher provides this information in a different coordinate frame. Solving an example of the latter task requires transforming distal training information into the information a proximal teacher would provide if one were present. Tasks with distal teachers involve control in more substantive ways than do other supervised learning tasks. To learn to achieve distal targets implies that the learning system must learn to control aspects of the system that transforms its actions into the distal coordinate frame (component B in Figure 4). If this is a dynamic system, these tasks correspond to adaptive control tasks in which the objective is to control a system so that its output tracks a desired reference trajectory.

We also emphasized a distinction between nonassociative and associative learning tasks. In a nonassociative task, the learning system tries to find a single optimum action, whereas in an associative task, it tries to construct an associative mapping from inputs providing state and/or context information to optimal actions. We discussed this distinction only in terms of reinforcement learning tasks. Nonassociative reinforcement learning tasks are the simplest tasks involving the issues of exploration and the conflict between control and identification. Supervised learning tasks, on the other hand, must be associative in order to be interesting (except, perhaps, in cases involving distal training information). A key feature of supervised learning is the possibility for forming associative mappings that generalize appropriately to novel inputs, and in nonassociative cases there are no novel inputs.

For any kind of learning task, access to state and/or context information is important because it may allow nonstationary nonassociative tasks to be transformed into stationary associative tasks. In the nonstationary case, the critic's evaluation criteria may appear to vary over time because information relevant to performance is unavailable to the learning system; in the stationary case, context signals allow the learning system to alter its behavior according to which of a collection of stationary tasks is being faced at any time. Given the ability to make its behavior conditional on relevant state and/or context information, a learning system has the potential for dramatically improved performance over what it could achieve if it were blind to this information. It is obviously important, therefore, for a learning system to have access to information that it can use to reduce or eliminate nonstationarity by means of associative processes.

In associative learning tasks, state and context signals play similar roles in supplying information that can be used to reduce nonstationarity. In the framework used in this chapter, the difference between state and context signals is that the former have the potential for being influenced by the learning system, whereas we assumed that the latter were beyond the learning system's control. We let the context signals play the role of the training patterns in supervised learning tasks because in most studies of supervised learning applied to pattern classification or function approximation, the sequence of training patterns is not influenced by the learning system's behavior. But it should be clear that signals regarded as context in one task formulation may become state signals in another formulation that extends the learning system's influence.

Finally, we discussed sequential decision tasks in which it can make sense to forgo short-term performance in order to achieve better performance over the long-term. Solving a task having this property requires extensive planning that can be difficult even if all the details of the task are known in advance. When these details are not all known in advance, performance over the long-term can be improved by learning, and we used the term adaptive sequential decision task for these cases. The other reinforcement learning tasks discussed in this chapter are special cases of adaptive sequential decision tasks. Reducing the horizon so that only the immediate consequences of actions influence the performance measure yields associative reinforcement learning tasks. If, in addition, the learning system is not permitted access to information other than the critic's evaluations, the task becomes a nonassociative reinforcement learning task. Consequently, all the issues we have discussed in this chapter must be considered in solving adaptive sequential decision tasks.

Although the framework adopted in this chapter for comparing and contrasting learning tasks is one of many that could have been created, the exercise of embedding a range of tasks within it demonstrates that there are many factors relevant to learning beyond those addressed in the most commonly studied learning tasks. Conducting this exercise while avoiding discussion of learning methods eliminates the confusion arising when aspects of tasks and methods are confounded. When studying learning methods against this background, a clearer view of their capabilities and limitations is possible. For example, one can consider separately two different categories of limitations on the capabilities of a learning method.[9] A method may be limited in its ability to solve tasks *of a given type* that are more complex than a certain level, and a method may be limited because it can only solve a *certain type of task*. For example, an artificial neural network consisting of a single layer of linear threshold units using the perceptron learning rule can learn to implement only linear discriminate functions. A very different limitation of such a network is that even if it could learn to implement arbitrarily complex discriminant functions (for example, by using the error back-propagation method[26,30,33,38]), it would still require an environment capable of providing a specific type of training information.

Although solution methods for difficult nonlinear pattern classification, function approximation, and function optimization tasks will have important roles in sophisticated learning systems, it seems to us that sophisticated learning behavior can result from a system designed to solve many interrelated learning tasks of different types, where each task is a relatively simple example of its type given the available prior knowledge. One key to designing useful learning systems may be to

[9] This point was made by Barto and Sutton.[6]

design systems applicable to realistic models of the tasks actually faced by animals instead of abstract tasks isolating only a few features of realistic learning tasks.

8. ACKNOWLEDGMENTS

The author gratefully acknowledges Charles Anderson, Michael Jordan, Harry Klopf, and Richard Sutton for the many hours of discussion which have contributed to the material outlined here; Vijaykumar Gullipalli, James Houk, and Richard Yee for helpful comments on drafts of this chapter; and the Air Force Office of Scientific Research, Bolling AFB, for support through grant AFOSR-89-0526. Some of the observations made here were also made by Barto and Sutton.[6]

REFERENCES

1. Barlow, H. B. "Unsupervised Learning." *Neural Computation* **1** (1989): 295–311.
2. Barto, A. G. "Connectionist Learning for Control: An Overview." In *Neural Networks for Control*, edited by T. Miller, R. S. Sutton and P. J. Werbos. Cambridge, MA: MIT Press. To appear.
3. Barto, A. G. "Learning by Statistical Cooperation of Self-Interested Neuron-like Computing Elements." *Human Neurobiology* **4** (1985): 229–256.
4. Barto, A. G. "From Chemotaxis to Cooperativity: Abstract Exercises in Neuronal Learning Strategies." In *The Computing Neuron*, edited by R. Durbin, R. Maill and G. Mitchison, 73–98. Reading, MA: Addison-Wesley, 1989.
5. Barto, A. G., and P. Anandan. "Pattern Recognizing Stochastic Learning Automata." *IEEE Transactions on Systems, Man, and Cybernetics* **15** (1985): 360–375.
6. Barto, A. G., and Sutton, R. S. "Goal Seeking Components for Adaptive Intelligence: An Initial Assessment." Technical Report AFWAL-TR-81-1070, Air Force Wright Aeronautical Laboratories/Avionics Laboratory, 1981.
7. Barto, A. G., R. S. Sutton, and C. W. Anderson. "Neuronlike Elements That Can Solve Difficult Learning Control Problems." *IEEE Transactions on Systems, Man, and Cybernetics* **13** (1983): 835–846. Reprinted in *Neurocomputing: Foundations of Research*, edited by J. A. Anderson and E. Rosenfeld. Cambridge, MA: MIT Press, 1988.
8. Barto, A. G., R. S. Sutton, and P. S. Brouwer. "Associative Search Network: A Reinforcement Learning Associative Memory." *IEEE Transactions on Systems, Man, and Cybernetics* **40** (1981): 201–211.

9. Barto, A. G., R. S. Sutton, and C. Watkins. "Learning and Sequential Decision Making." In *Learning and Computational Neuroscience*, edited by M. Gabriel and J. W. Moore. Cambridge, MA: MIT Press, to appear.

10. Barto, A. G., R. S. Sutton, and C. Watkins. "Sequential Decision Problems and Neural Networks." In *Advances in Neural Information Processing Systems 2*, edited by D. S. Touretzky, 686–693. San Mateo, CA: Morgan Kaufmann, 1990.

11. Barto, A. G., and S. P. Singh. "On the Computational Economics of Reinforcement Learning." In *Proceedings of the 1990 Connectionist Models Summer School*, edited by D. Touretzky, G. Hinton and T. Sejnowski. San Mateo, CA: Morgan Kaufmann, to appear.

12. Baum, E. B. "Neural Net Algorithms that Learn in Polynomial Time from Examples and Queries." *IEEE Transactions on Systems, Man, and Cybernetics*, to appear.

13. Berry, D. A., and B. Fristedt. *Bandit Problems*. London: Chapman and Hall, 1985.

14. Bertsekas, D. P. *Dynamic Programming: Deterministic and Stochastic Models*. Englewood Cliffs, NJ: Prentice-Hall, 1987.

15. Cohen, P. R., and E. A. Feigenbaum, eds. *The Handbook of Artificial Intelligence*, Vol. 3. Stanford, CA: HeurisTech Press, 1982.

16. Dennett, D. C. *Elbow Room. The Varieties of Free Will Worth Wanting*. Cambridge, MA: MIT Press, 1985.

17. Duda, R. O., and P. E. Hart. *Pattern Classification and Scene Analysis*. New York: Wiley, 1973.

18. Goodwin, G. C., and K. S. Sin. *Adaptive Filtering Prediction and Control*. Englewood Cliffs, NJ: Prentice-Hall, 1984.

19. Hinton, G. E. "Connectionist Learning Procedures." *Artificial Intelligence* 40 (1989): 185–234.

20. Hinton, G. E., and J. A. Anderson. *Parallel Models of Associative Memory*. Hillsdale, NJ: Erlbaum 1981.

21. Honig, W. K., and J. E. R. Staddon. *Handbook of Operant Behavior*. Englewood Cliffs, NJ: Prentice Hall, 1977.

22. Jordan, M. I., and D. E. Rumelhart. "Forward Models: Supervised Learning with a Distal Teacher." Submitted for publication.

23. Kohonen, T. *Content-Addressable Memories*. Berlin: Springer-Verlag, 1980.

24. Kohonen, T. *Self-Organization and Associative Memory*. Berlin: Springer-Verlag, 1984.

25. Laird, J. E., P. S. Rosenbloom, and A. Newell. "Chunking in SOAR: The Anatomy of a General Learning Mechanism." *Machine Learning* 1 (1986): 11–46.

26. le Cun, Y. "Une Procedure d'apprentissage pour Reseau à Sequil Assymetrique [A Learning Procedure for Asymmetric Threshold Network]." *Proceedings of Cognitiva* 85 (1985): 599–604.

27. Ljung, L., and T. Söderstrom. *Theory and Practice of Recursive Identification*. Cambridge, MA: MIT Press, 1983.

29. Narendra, K., and M. A. L. Thathachar. *Learning Automata: An Introduction*. Englewood Cliffs, NJ: Prentice Hall, 1989.
30. Parker, D. B. "Learning Logic." Technical Report 47, Massachusetts Institute of Technology, 1985.
31. Polderman, J. W. "Adaptive Control and Identification: Conflict or Conflux?" Amsterdam, The Netherlands: Centrum voor Wiskinde en Informatica, 1987.
32. Ross, S. *Introduction to Stochastic Dynamic Programming*. New York: Academic Press, 1983.
33. Rumelhart, D. E., G. E. Hinton, and R. J. Williams. "Learning Internal Representations by Error Propagation." In *Parallel Distributed Processing: Explorations in the Microstructure of Cognition*, Vol.1: *Foundations*, edited by D. E. Rumelhart and J. L. McClelland. Cambridge, MA: Bradford Books/MIT Press, 1986.
34. Sutton, R. S. *Temporal Credit Assignment in Reinforcement Learning*, Amherst, MA: University of Massachusetts Press, 1984.
35. Sutton, R. S. "Learning to Predict by the Methods of Temporal Differences." *Machine Learning* **3** (1988): 9–44.
36. Sutton, R. S. "Integrating Architectures for Learning, Planning, and Reacting based on Approximating Dynamic Programming." In *Proceedings of the Seventh International Conference on Machine Learning*, 216–224. San Mateo, CA: Morgan Kaufmann, 1990.
37. Watkins, C. J. C. H. *Learning from Delayed Rewards*. Cambridge, England: Cambridge University, 1989.
38. Werbos, P. J. *Beyond Regression: New Tools for Prediction and Analysis in the Behavioral Sciences*. Cambridge: Harvard University Press, 1974.
39. Werbos, P. J. "Building and Understanding Adaptive Systems: A Statistical/Numerical Approach to Factory Automation and Brain Research." *IEEE Transactions on Systems, Man, and Cybernetics*, 1987.
40. Whitehead, A. N. *Science and the Modern World (Lowell Lectures, 1925)*. New York: The Macmillan Company, 1925.
41. Widrow, B., and S. D. Stearns. *Adaptive Signal Processing*. Englewood Cliffs, NJ: Prentice-Hall, 1985.
42. Williams, R. J. "Reinforcement Learning in Connectionist Networks: A Mathematical Analysis." Technical Report ICS 8605, University of California at San Diego, 1986.
43. Williams, R. J. "Reinforcement-Learning Connectionist Systems." Technical Report NU-CCS-87-3, Northeastern University, 1987.

Z. Hasan
Departments of Physiology and of Electrical and Computer Engineering, University of Arizona, Tucson, AZ 85724

Moving a Human or Robot Arm with Many Degrees of Freedom: Issues and Ideas

INTRODUCTION

We will address the problems involved in the seemingly simple task of moving a set of linked segments (e.g., upper arm, forearm, hand) from one position to another. As will be seen, the complexity of these problems is such that one cannot help marvelling that animals and humans are able to carry out limb movements at all! Some of the problems are common to biological motor control and robotics, whereas others are specific to the biological system. Many ideas have been advanced as to how the problems might be solved. Because of their biological flavor, some of the proposed neural-net approaches will be given more than proportionate attention. No definitive answers have emerged, however, and this field remains full of opportunities. For an overview of the issues, several references[10,18,22,25,48,52] are accessible to those without an extensive background in biology.

One of the obstacles in reading the literature on movement control is lack of uniformity in the terminology and notation used by different authors. In the present article, an attempt is made to employ a uniform system of notation, which has necessitated recasting the work of some of the authors in a form they may not find satisfying. The presentation, moreover, would seem to some experts as mixing up what they consider the elementary and the advanced issues, and to be merely skimming over some of them. These liberties are taken in the interest of students who approach this topic for the first time, but who are not naive about other complex systems.

THREE BLACK BOXES

Given the specification of the task, for example, move the fingertip 10 cm to the right, certain temporal patterns of activity should somehow be generated for each of the dozens of muscles of the arm, so as to accomplish the task. Many authors (e.g., Atkeson,[2] Soechting[48]) break this down into a sequence of three component problems. (See Figure 1.)

- *Problem I:* The task needs to be translated into the requisite kinematic details. Specifically, one needs to select the joint angles, $\theta_n(t)$, as functions of time t (where n indexes the N kinematic degrees of freedom), which would be appropriate, for example, to move the fingertip 10 cm to the right.
- *Problem II:* Given the desired $^D\theta_n(t)$, (where ^{D} stands for desired), one needs to determine the torques, $\mu_n(t)$, which, when applied to the joints, would cause the motions $^D\theta_n(t)$.
- *Problem III:* Given the desired $^D\mu_n(t)$, the commands $\alpha_m(t)$ to the muscles (where m indexes the M muscles; $M > N$) have to be specified in light of the complex geometrical arrangement and mechanical properties of the muscles. (For example, a muscle often contributes to more than one μ_n; the torque contribution of a muscle is always movement dependent.)

In robotics, *Problem III* usually does not arise, since motors can be situated in one-to-one correspondence with the kinematic degrees of freedom, and the mechanical (impedance) properties of the motors do not vary. *Problem II* is usually avoided by employing "stepper" motors (or, at some risk of instability, position servomotors), which respond directly to the $^D\theta_n(t)$ commands. Such a motor overcomes, by brute force, any flopping that would have resulted from the motions of

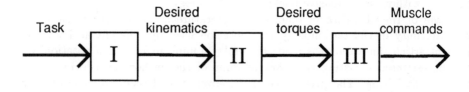

FIGURE 1   A scheme for movement control, with three successive transformations.

the other linked segments. Faster and more efficient robots, however, are being designed using d.c. torque motors, taking into account the inertial interactions, which makes it necessary to solve this problem.[41] Nevertheless, it is *Problem I* that has attracted the greatest attention in robotics.

Does this dissection of one black box into three black boxes correspond to biological reality? We don't know. We will proceed on the assumption that it does, and will explore how each of the black boxes might be implemented neurally. Later, we will consider some alternatives to this general scheme.

BLACK BOX I: FROM TASK TO JOINT ANGLES

The task in the example of moving the fingertip 10 cm to the right was specified in a body-centered frame of reference. In contrast, the task specification for reaching a visually sensed target begins in a retina-centered frame. The two frames are obviously not equivalent, as the eyes can move with respect to the head and the head can move with respect to the body. Since there is no apparent difference in the way arm movements are performed for these types of tasks, it is generally assumed that the machinery exists for transforming all movement tasks into the same frame of reference.[48] (But see Pellionisz and Peterson.[46]) Indeed, there is evidence that certain neurons in the motor areas of the cerebral cortex code for the direction of impending movement in a body-centered frame.[11] This frame, which encompasses three-dimensional space and can be represented by Cartesian coordinates, is traditionally called "extrinsic," despite the fact that it is linked to the body. In contrast, the "intrinsic" frame encompasses the degrees of freedom of the body segments; it can be represented by θ_n as coordinates.

Assuming that the task has been specified in the extrinsic frame, the problem to be solved by *Black Box I* is its transformation into the intrinsic frame ("inverse kinematics" problem). To start with, we consider this in two parts: first, how to transform a stationary target position into the constants $^T\theta_n$ where ^{T} stands for target, and second, given $^T\theta_n$ and the initial angles $^I\theta_n$, how to generate the functions $^D\theta_n(t)$.

FROM TARGET POSITION TO FINAL JOINT ANGLES. A model for this transformation is provided by elementary trigonometry, but in order for this to be useful computationally, one must have accurate knowledge of segment lengths, and one must commit to a particular choice of coordinates for the extrinsic frame. (Each choice is associated with certain unsolvable positions where the Jacobian vanishes.) Usually, the nonlinear, multi-valued nature of inverse trigonometric functions leaves no alternative but successive approximation to find the angles.

Kuperstein and Rubinstein[37] have proposed a network-learning algorithm to implement the transformation from stationary target position to joint angles. In their scheme, the target is located on a topographic map (TV camera image), rather than described in Cartesian or spherical coordinates. No knowledge of segment lengths is assumed. In the learning phase, the robot's gripper holds the target, moves it successively by random changes in joint angles and, based upon the camera signals, updates the parameters ("weights") of its sensory-motor map. This is unsupervised learning that relies on achieving self-consistency between the (random) output signals and the resultant camera signals. As in an infant's acquisition of visually guided behavior, the relevant visual signals in the learning phase are elicited by self-produced movement. Once this phase is completed, the presentation of the target, which is no longer in the robot's grip, elicits from the sensory-motor map the outputs necessary for the gripper to correspond to the target. (The accuracy, however, is not impressive, but it can be made impressive by fine tuning the map using another type of learning, described later.) It is instructive to see what it takes to implement this mapping, which works reasonably well without knowledge of segment lengths or of analytical models. A brief description derived from Kuperstein and Rubinstein[37] follows.

The robot has four degrees of freedom (plus one for the gripper) in three-dimensional space. The target is a cylinder, whose position and orientation are registered by two TV cameras fixed in extrinsic space. From the camera images, 18 uni-dimensional input arrays s_k (each with 640 elements) are constructed. Six of these correspond to the two dimensions of the images from each of the two cameras and from their stereo disparity. Twelve correspond to a discretized set of four orientations of the cylinder's image, again for each camera and for the disparity. (Note that the Cartesian coordinates of the target are never computed.) The sensory-to-motor mapping is described by:

$$^M\theta_n = \sum_k \sum_i s_{k,i} w_{k,i,n} ,$$

where ^{M} stands for output of the map, $s_{k,i}$ is the ith element of the array s_k ($k = 1$ to 18, $i = 1$ to 640), and w represents the weights. Here, s describes the partially digested sensory input, $^M\theta$ the output, and w the mapping.

In the learning phase the weights are modified according to:

$$\Delta w_{k,i,n} = \sigma s_{k,i}(\theta_n - {}^M\theta_n) ,$$

where θ_n are randomly chosen angles. Note that only the weights corresponding to active sensory inputs are modified, since $s_{k,i}$ is used as a multiplier. All weights are initially set to zero. The constant σ determines the learning rate.

After about 1200 learning trials, the sensory-motor map settles. It is then good enough to achieve a position accuracy of 3% of the length of the arm, and orientation accuracy (between the cylinder and the gripper) of 6 deg solid angle. The errors can be made undetectable, at least in a local volume region, by adding two to four learning trials using a completely different algorithm. This algorithm, which is based on sensory error, employs an additional set of s arrays that capture the position of the gripper rather than the target. The computed difference between the output signals elicited from the sensory-motor map by the s arrays for the gripper and for the real target, is now used to modify the weights. This sensory-error-based learning can do wonders for fine tuning the weights, at least in a local region, but cannot be made to converge with arbitrary starting weights. The first learning scheme, based on self-consistency, can start with arbitrary weights, but its asymptotic accuracy is modest.

Thus, a carefully designed sequence of learning algorithms can establish an accurate sensory-motor map that allows one to determine the appropriate joint angles $^T\theta_n$ for any given stationary target. This determination, of course, is independent of initial joint angles $^I\theta_n$, and leaves open the question of how the angles should change as functions of time.

FROM FINAL JOINT ANGLES TO KINEMATICS. One could simply choose for $^D\theta_n(t)$ an interpolation, linear or otherwise, between the initial and final values, as is often done in stepper-motor robots. But this would fly in the face of known features of the kinematics of human arm movements. For example, consider a movement for which the initial and final positions of the wrist both lie on the arc of a circle centered on the elbow. Since the requisite final shoulder angle is the same as the initial, any interpolation scheme for θ_n would keep the shoulder from moving at all, and thus movement of the wrist will be along an arc. For such initial and final positions of the wrist, however, a human subject will show a far straighter path than an arc, which requires that the shoulder should move and then reverse direction. This contradicts simple interpolation schemes.

Bullock and Grossberg[4] attempt to account for some of the kinematic details in a model called "Vector Integration To Endpoint." In a slightly simplified form of this model, the command $^D\theta_n(t)$ is generated by integrating the product of two terms:

$$^D\theta_n(t) = \int G(t)v_n(t)dt\,,$$

where $G(t)$ is a nonspecific "Go" signal, and $v_n(t)$ represents a low-pass version of the difference between the target angle and the currently commanded angle:

$$v_n(t) + k\frac{dv_n(t)}{dt} = {}^T\theta_n - {}^D\theta_n(t)$$

where k is a constant.

Clearly, for positive G, $^D\theta_n(t)$ will approach $^T\theta_n$ in time. By choosing an appropriately shaped $G(t)$, the model can generate reasonable $^D\theta_n(t)$, and can exhibit the experimental phenomenon that subjects asked to move at the same speed but to a more distant target in fact increase their speed. Thanks to the postulated increase of G with time, a change in target position after movement has started elicits a faster response than does the initial target. This too conforms to experiment. However, in this model each joint is independent of the others; the model will not work in situations where a joint angle changes away from the target value and then comes back to it.

In discussing *Black Box I*, we have so far focused on the joint-angle commands $^D\theta_n(t)$ that emerge from this box, and have made no mention of the actual joint angles $\theta_n(t)$. Stepper-motor enthusiasts have no need even to make this distinction. Bullock and Grossberg[4] do distinguish between the two, and they propose a learning algorithm that calibrates the commands with the actual angles. This algorithm operates only during externally imposed movements ($G = 0$) to alter the calibration memory trace. Note that the feedback of actual joint angle information is utilized only for calibration purposes during pauses in active movement. Unlike the case of a servomechanism, the feedback does not serve to correct errors during active movement, which obviates any problems of instability.

Houk et al.[27] have proposed a cerebellum-inspired neural network for generating $^D\theta_n(t)$, given $^T\theta_n$. The signals corresponding to $^T\theta_n$ as well as the current $\theta_n(t)$ all converge via modifiable synapses onto nodes P_i. (For those with some knowledge of neuroanatomy, these represent synapses of "parallel fibers" on "Purkinje cells.") The node P_i has bistable properties (with hysteresis), so that P_i is always either zero or one. These nodes, in turn, project inhibition to internal positive-feedback loops (involving "cerebellar" and "red" nuclei). The output of each loop is either zero, or, when it is triggered by a "Go" signal, a value that is interpreted as $^D\dot{\theta}_n(t)$. Thus, in the simplest case, when a loop has been triggered,

$$^D\theta_n(t) = \omega_n \int \left(1 - \sum_{i=1}^{I} \frac{P_i}{I}\right) dt\,,$$

where ω_n are constants.

The P_i depend on the $^T\theta_n$ and $\theta_n(t)$ inputs, with random weights to start with. The weights w are modified during successive movements according to:

$$\Delta w_{ij} = [\sigma_1(1 - C)(1 - P_i) - \sigma_2 C P_i]\, e_{ij}\,,$$

where j indexes the various inputs to the node P_i. C (representing "climbing fiber" activity) is zero most of the time, but it transiently becomes 1 after a movement that falls short of the target; e_{ij} is a measure of the "eligibility" of the synapse for modification (i.e., a trace of local input activity); and σ_1, σ_2 are constants.

After many attempts the weights settle; the network can then produce the $^D\theta_n(t)$ for any initial and final positions. The target and current position inputs can, in fact, preselect the P_i, with no immediate effect on the output before the occurrence of the "Go" trigger, since the internal positive-feedback loops are then in the zero state. (There is no need to postulate a "Go" signal of tailor-made form to produce the temporal changes in the outputs.) The feedback of $\theta_n(t)$, moreover, does not cause instability because it affects only the timing of the zero-one transitions in P_i, not the magnitude. This scheme is well founded on available neurophysiological data, though questions remain as to whether the $^T\theta_n$ are actually determined and represented in the brain.

OTHER KINEMATIC FEATURES. A good deal of information is available about the kinematics of human movements, which should exert a selection pressure on hypotheses, but this does not always happen. Each hypothesis gives prominence only to some of the known facts. Our ignorance about *Black Box I* is underscored by lack of explanations for certain readily observed kinematic features of human movement. One such feature is that the changes in joint angles, for example, during a pointing movement, do not all start (or end) at the same time.[29] It has been pointed out that a strategy of interpolation of θ_n, but with staggered onset times for different joints, can sometimes give rise to a more straight path in extrinsic space.[26] However, this is not always so, and in any case, no general rules have been put forward regarding the amount of stagger. (Later, we will argue that the apparent stagger may emerge from kinetic rather than kinematic rules.)

Often, there are fundamental features of a task that are difficult to describe in either extrinsic or intrinsic frames. An example is the task of pinching the thumb against a finger, when the point of skin contact is important, but the location of the point with respect to other segments is not important. In such tasks there may be great variability in the details of $\theta_n(t)$, and yet the task is performed successfully on each trial.[5] This makes one wonder if the transformation to intrinsic coordinates, which is the role of *Black Box I,* is ever actually performed in such tasks.

Certain kinematic features of real movements are simpler to describe in extrinsic rather than intrinsic coordinates. For example, in pointing movements, the speed of the end-effector (e.g., hand) in the extrinsic frame is generally a fairly symmetrical, bell-shaped function of time, irrespective of which region of the workspace is traversed, whereas joint angular velocity profiles may be quite complicated (and even include reversals). Another such feature, well known from studies of freehand drawing, is the speed-curvature tradeoff: regions of high curvature of the path in extrinsic space are traversed at lower speed.[38] This may seem entirely "natural" because of our experience with centrifugal forces in steering an automobile, but the data do not support constancy of centrifugal acceleration. (Also, unlike the automobile example, the center of mass is not even close to the path.) In order to elaborate on this, we need to define path variables.

Let p represent length measured *along* the path, and η the orientation of the tangent to the path with respect to some fixed direction in extrinsic space. The velocity vector has the magnitude dp/dt, and is directed along η. (Note that p and η, which change along the path, would be called "intrinsic coordinates" in analytical geometry, but in order to avoid confusion with $\{\theta_n\}$ we will not use this terminology.) The radius of curvature at a point on the path is defined by $r = dp/d\eta$. (Thus, $dp/dt = r d\eta/dt$.) It is an experimental fact[38] based on studies of handwriting and various other movements of relatively small excursion that:

$$d\eta/dt = kr^{-2/3} ,$$

where k is constant, at least for a single "stroke" of the movement.

This last statement is equivalent to: $(dp/dt)^3/r = $ constant. In other words, high curvature (small r) is associated with low speed. (Note that the centrifugal acceleration, given by $(dp/dt)^2/r$, is not constant.) The statement is also equivalent, in Cartesian coordinates, to:

$$\left(\frac{d^2y}{dt^2}\right)\left(\frac{dx}{dt}\right) - \left(\frac{d^2x}{dt^2}\right)\left(\frac{dy}{dt}\right) = \text{constant}.$$

Why such a strange kinematic relationship? If, for some reason, $x(t)$ and $y(t)$ were both sinusoidal functions with the same frequency (but any amplitudes and phases), not necessarily covering a complete cycle, it is easy to show that the "strange" relationship would in fact obtain. The path (a Lissajous figure) would then be an arc of an ellipse. (A straight line represents the special case of a collapsed ellipse.) Thus, if the actual path were a concatenation of elliptical arcs, each of which can be produced readily by driving the *extrinsic*, Cartesian coordinates sinusoidally, the observed speed-curvature tradeoff would be an automatic consequence.

By asking subjects to draw large ellipses freehand in three-dimensional space, Soechting and Terzuolo[50] found that the appropriate *intrinsic* coordinates varied approximately sinusoidally. Since the intrinsic-to-extrinsic geometric transformation is nonlinear, the Cartesian coordinates exhibited a departure from sinusoidal variation in certain cases, and the ellipses in extrinsic space then showed up as distorted. The distortion, which depended on the plane of the figure, was predicted correctly based upon the assumption that the extrinsic-to-intrinsic transformation (implemented by *Black Box I*) was only a (linear) approximation to the geometrically correct one. If the correct transformation had been implemented, the intrinsic coordinates should not have varied sinusoidally in order to produce the presumably desired sinusoidal variation of the extrinsic coordinates. Instead, the desired variation was transformed linearly, resulting in sinusoidal variation of intrinsic coordinates and consequent distortion of the figure in extrinsic space. It is difficult to understand why a neural network would prefer to implement a linear rather than a nonlinear transformation. At present, however, there is no alternative explanation of the consistent distortions and errors that have been described experimentally in extrinsic space.

PREFERRED INTRINSIC COORDINATES AND REDUNDANCY. We introduced the idea of an intrinsic space in terms of the N joint angles θ_n, where N is the number of degrees of freedom. The organism, however, does not float free in outer space, and therefore some of its degrees of freedom correspond to movements with respect to external supports. For arm movements with the trunk fixed, the degrees of freedom associated with the shoulder have to be defined in relation to the fixed chair supporting the trunk. They are nevertheless intrinsic, inasmuch as the angles could be measured by sensors located entirely under the skin. Having defined the intrinsic space as one for which the set $\{\theta_n\}$ is a basis, we can transform the coordinate system, and describe the same N-dimensional intrinsic space using a different set $\{\phi_n\}$ as coordinates. The transformation effected by *Black Box I* could very well have as its output a set $\{\phi_n\}$ different from $\{\theta_n\}$.

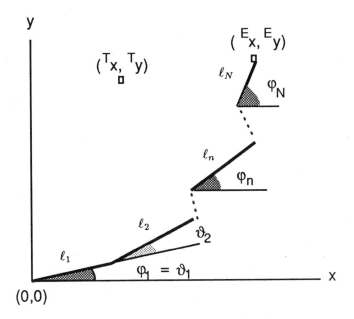

FIGURE 2 Diagram of a planar linkage of N hinged segments. Thick lines denote the segments, θ_n are the joint angles, and ϕ_n the segment orientations.

To fix ideas, and yet not be lost in trigonometric details, consider the case of a planar linkage of N straight, hinged segments. (See Figure 2.) The segment lengths are ℓ_n and the joint angles are θ_n. (Note that θ_1, as in the example of the shoulder, is defined with respect to a fixed direction along the x-axis.) For subsequent use, we define ϕ_n by:

$$\phi_n = \sum_{i=1}^{n} \theta_i .$$

Clearly, ϕ_n is the orientation of the nth segment with respect to the fixed x-axis; yet $\{\phi_n\}$, being an invertible transformation of $\{\theta_n\}$, can serve as a coordinate system for intrinsic space. The extrinsic space is two-dimensional, and we have arbitrarily chosen Cartesian coordinates (x, y) to describe it. The proximal end of segment #1 is hinged at $(0,0)$, and the distal end of segment #N is the "end effector," at $(^Ex, ^Ey)$, that is required to reach the given target at $(^Tx, ^Ty)$.

Psychophysical data (in fact, for three-dimensional space) indicate that subjects are more accurate in their awareness of absolute orientations (such as ϕ_n) than of the joint angles between segments.[49] (Moreover, the "appropriate intrinsic coordinates" mentioned earlier in connection with ellipse-drawing movements, were in fact absolute orientations; the joint angles did not show certain regularities that the orientation angles did.) The data seem to indicate that even though the biological sensors span joints and thus respond to θ_n, the awareness of position is related to ϕ_n.

From Figure 2, the application of elementary trigonometry leads to the following relationship between extrinsic and intrinsic coordinates.

$$^Ex = \sum_{n=1}^{N} \ell_n \cos \phi_n, \quad ^Ey = \sum_{n=1}^{N} \ell_n \sin \phi_n .$$

For given $(^Ex, ^Ey)$, there is an infinite number of solutions for $\{\phi_n\}$ when $N > 2$. (The same is true for $\{\theta_n\}$.) This is a problem of redundancy we have not addressed so far: the intrinsic coordinates for a given target position in extrinsic space may not be unique. Experimentally, it is found for large excursions that the intrinsic coordinates at the end of the movement are different for different starting arm positions, even though the target position in extrinsic space remains the same.[6] This suggests that, instead of first determining the target angles and then choosing the temporal pattern, the system may choose increments in the angles based upon their current values, according to some rules that are yet to be explicated. The rules might be such as to strike a compromise between achieving a straight-line path in extrinsic space, and keeping the joint angles within limits.

One possibility, not proposed before, is to choose the desired $^D\phi_n(t)$, without knowledge of the final values $^T\phi_n$, as follows:

$$^D\phi_n(t) = \int \sum_{j=1}^{N} \ell_j a_{j,n} \cos(\phi_j - \zeta)dt ,$$

where ϕ_j represents a current value. It can be shown that if $a_{j,n}$ is any arbitrary anti-symmetric array $(a_{n,j} = -a_{j,n})$, this choice would result in the velocity of the end-effector being directed along the orientation ζ in extrinsic space. The redundancy, of course, is captured here in the arbitrariness of a. Perhaps the rules for choosing the kinematics pertain to $a_{j,n}$, rather than directly to the $^D\phi_n(t)$.

BLACK BOX II: FROM KINEMATICS TO TORQUES

The torques necessary to achieve the desired kinematics in the intrinsic frame can be calculated by using equations derived from classical mechanics ("inverse dynamics" computation). These "equations of motion" relating the torques $\mu_n(t)$ to intrinsic variables, however, can be tedious to derive, cumbersome to apply, and not be useful at all without accurate knowledge of a large number of parameters (e.g., mass, center-of-mass location, principal axes, and moments of inertia, for each segment).[3,41,52] Moreover, some of the parameters change when loading conditions are changed.

A SLIGHTLY TEDIOUS DERIVATION. We derive the equations of motion for the planar system of Figure 2, using the Lagrangian method. (Readers unfamiliar with this method may choose to skip this section.) The kinetic energy, E, of the system is given by:

$$E = \frac{1}{2} \sum_{n=1}^{N} \left[m_n(\dot{x}_n^2 + \dot{y}_n^2) + I_n \dot{\phi}_n^2 \right] ,$$

where m_n is the mass, (x_n, y_n) the location of the center of mass (c.m.), and I_n the moment of inertia (about the c.m.), for the nth segment.

In order to express (x_n, y_n) in terms of $\{\phi_j\}$, we denote by b_n the fixed distance between the c.m. and the proximal hinge of the nth segment, and define:

$$u_{ij} = \ell_j \text{ if } j < i, \qquad u_{ij} = b_i \text{ if } j = i, \qquad \text{and } u_{ij} = 0 \text{ if } j > i .$$

Then, from trigonometry,

$$x_n = \sum_{j=1}^{N} u_{nj} \cos \phi_j, \qquad y_n = \sum_{j=1}^{N} u_{nj} \sin \phi_j .$$

Differentiating these and substituting in the expression for E, we obtain:

$$E = \frac{1}{2} \sum_{n=1}^{N} \left[m_n \sum_{i=1}^{N} \sum_{j=1}^{N} u_{ni} u_{nj} \cos(\phi_i - \phi_j) \dot{\phi}_i \dot{\phi}_j \right] + \frac{1}{2} \sum_{n=1}^{N} I_n \dot{\phi}_n^2 ,$$

which expresses the kinetic energy in terms of intrinsic variables alone. It is convenient to define H, a symmetric matrix with constant elements, by:

$$H_{ij} = \sum_{n=1}^{N} m_n u_{ni} u_{nj} + I_j \delta_{ij} ,$$

where δ_{ij} is the Kronecker symbol ($\delta_{ii} = 1, \delta_{i \neq j} = 0$). Then:

$$E = \frac{1}{2} \sum_{i=1}^{N} \sum_{j=1}^{N} H_{ij} \dot{\phi}_i \dot{\phi}_j \cos(\phi_i - \phi_j) .$$

From this expression, after some manipulation, one can deduce that:

$$\frac{d}{dt} \left(\frac{\partial E}{\partial \dot{\phi}_i} \right) - \frac{\partial E}{\partial \phi_i} = \sum_{j=1}^{N} H_{ij} \left[\ddot{\phi}_j \cos(\phi_i - \phi_j) + \dot{\phi}_j^2 \sin(\phi_i - \phi_j) \right] .$$

Now, according to the Lagrangian formulation, this last expression must equal $-\partial U / \partial \phi_i$, where U is the potential energy. Ignoring gravity, $-\partial U / \partial \theta_i = \mu_i$. It follows that $-\partial U/\partial \phi_i = (\mu_i - \mu_{i+1})$, where μ_{N+1} is defined to be zero. Using the trivial relation

$$\mu_n = \sum_{i=n}^{N}(\mu_i - \mu_{i+1}),$$

we obtain finally

$$\mu_n = \sum_{i=n}^{N}\sum_{j=1}^{N} H_{ij}\left[\ddot{\phi}_j\cos(\phi_i - \phi_j) + \dot{\phi}_j^2\sin(\phi_i - \phi_j)\right].$$

These are the N equations of motion (for $n = 1$ to N). Writing them in terms of ϕ rather than θ resulted in a compact expression. Traditionally, the equations of motion are written in terms of θ. Our $\dot{\phi}^2$ terms, then, give rise to many terms in products of angular velocities, and $\ddot{\theta}$ terms appear with a multiplying "inertia matrix" whose elements are functions of θ. Here, the matrix H is independent of position or movement. Its elements depend only on the length and inertia parameters (ℓ_n, b_n, m_n, I_n), which are $4N$ in number.

SPECIAL CASES. For the special case of a planar, two-joint system ($N = 2$), the equations of motion, after substituting $\phi_1 = \theta_1$, $\phi_2 = \theta_1 + \theta_2$ and expanding the summation symbols, are given below.

$$\mu_1 = (H_{11} + H_{22} + 2H_{12}\cos\theta_2)\ddot{\theta}_1 + (H_{22} + H_{12}\cos\theta_2)\ddot{\theta}_2 - H_{12}\dot{\theta}_2(\dot{\theta}_2 + 2\dot{\theta}_1)\sin\theta_2,$$
$$\mu_2 = (H_{22} + H_{12}\cos\theta_2)\ddot{\theta}_1 + H_{22}\ddot{\theta}_2 + H_{12}\dot{\theta}_1^2\sin\theta_2,$$

where the H_{ij}, from their definition, are given by:

$$H_{11} = m_1 b_1^2 + m_2\ell_1^2 + I_1, \quad H_{22} = m_2 b_2^2 + I_2, \quad H_{12} = H_{21} = m_2\ell_1 b_2.$$

Thus, only three combinations of the eight parameters are needed in the equations. Note that the angular acceleration at each joint depends on the angular velocity and acceleration at the other joint. Thus, flopping can occur as a result of the motions of the other joint.

Another interesting special case (for later reference) is one where the entire mass of the N-segment system is concentrated at the end-effector. In this case, $H_{ij} = m_N\ell_i\ell_j$, and the equations of motion can be reduced to:

$$\mu_n = m_N\sum_{i=n}^{N}\ell_i(a_y\cos\phi_i - a_x\sin\phi_i),$$

where a_x, a_y are the components of the acceleration of the end-effector. In this special case, the knowledge of the desired trajectory of the end-effector in extrinsic space (in particular, knowledge of desired a_x, a_y), combined with information about current angles, is sufficient to generate the requisite torques. There is no need to specify $\{^D\phi_n(t)\}$. The kinematic redundancy resolves itself, as if by magic! (This is because the massless segments would be driven to infinite angular acceleration by any other $\{\mu_n\}$.)

In the general case, when the mass is not concentrated at the tip, if we were nevertheless to choose μ_n according to the last equation, the end-effector would follow some trajectory (call it $\mathfrak{S}$) in extrinsic space, though of course its acceleration would *not* be given by (a_x, a_y). It can be shown that $\{\phi_n(t)\}$ in this situation will be the same as would have been obtained with $\mu_n = 0$ if the end-effector had been dragged along the trajectory $\mathfrak{S}$ by an appropriate external force. This suggests that if there was some way of choosing $a_x(t)$ and $a_y(t)$ (which in the general case are not components of acceleration of the end-effector), such that the resultant $\mathfrak{S}$ would coincide with the desired trajectory in extrinsic space, then kinematic redundancy could be resolved without ever having to specify $\{^D\phi_n(t)\}$. The resolution of the redundancy would be a "natural" one, in that the more massive segments would move less. Also, the computation of the torques would be a relatively simple matter. However, it is not at all obvious how one should choose $a_x(t)$ and $a_y(t)$. We will return to this issue in connection with equilibrium-point hypotheses.

INVERSE DYNAMICS WITH KNOWLEDGE OF THE EQUATIONS. If the equations of motion are known, and there are good estimates of the parameters, the determination of the torques $\{^D\mu_n(t)\}$ required for achieving $\{^D\phi_n(t)\}$ is straightforward, speaking mathematically, but it is a considerable problem computationally. For robotics applications, however, elegant computational techniques have been designed for fast calculation of torques, as well as for refining the estimates of the parameters on the basis of feedback of force and torque from the distal joint.[3,34] (For a review, see Atkeson.[2]) Provided the degrees of freedom are not too many, present-day computers can just about perform the calculations for real-time applications.

Kawato et al.[33] present a learning algorithm for generating the torques, in which no *a priori* knowledge of the parameters is assumed. Certain expressions derived from the equations of motion, however, are computed by a set of "subsystems." The scaling of these expressions is what is modified based on performance errors.

The state $S = \{\theta_n, \dot{\theta}_n\}$ is monitored, and compared with the desired state ^DS at every moment of time. The torques are generated according to the following equation (in vector notation):

$$^D\mu = G(^DS - S) + f(^DS),$$

where G is a feedback-gain matrix and $f(\ )$ is the output of the internal model. G determines the proportional and derivative gains; the latter are made significant only in the vicinity of the target position. The internal model produces simply a weighted sum of the outputs of the subsystems Q:

$$f_n(^DS) = \sum_j w_{nj} Q_{nj}(^DS).$$

Of the 26 different Q's actually used, an example is:

$$\sin {}^D\theta_2 \cos({}^D\theta_2 + {}^D\theta_3){}^D\dot{\theta}_1 {}^D\dot{\theta}_2 \,.$$

The weights w, initially set to zero, are modified according to :

$$\frac{dw_{nj}}{dt} = \sigma Q_{nj}({}^DS)[\mu_n - f_n({}^DS)] \,,$$

where σ is a constant. (It is remarkable that for certain conditions on σ, the convergence of the weights has been proved analytically.) As in all such models, the torques on the first attempt are feedback generated, but as learning proceeds, the feedback becomes less and less important.

This scheme was tested on a three-segment, nonplanar manipulator. For a particular ${}^DS$, some 40 repetitions made the system produce perfect responses. A change in ${}^DS$ without additional learning produced a modestly good performance. But if the desired movements were fast, the performance could be disastrous. A change in loading required some additional 40 attempts for full adaptation.

INVERSE DYNAMICS WITHOUT KNOWLEDGE OF THE EQUATIONS. The modeling of a complicated, nonlinear system of unknown dynamics presents a serious challenge even with distributed-processing approaches. We describe one example of implementing inverse dynamics by a table look-up scheme in which the table builds itself.

In the scheme of Miller et al.,[41] all possible values of the state $S = \{\theta_n, \dot{\theta}_n, \ddot{\theta}_n\}$ are assigned integer identifiers (q). Note that angular accelerations are part of the state here. A "function generator" module produces the output f when presented with q:

$$f(S) = \sum_{i=1}^{I} w_{qi} z_i \,,$$

where z_i is an element of a memory array of size I, and w is a connection matrix, whose elements are 0 or 1, assigned randomly, but subject to the condition that for each q, there are only K nonzero elements. At each control cycle, *Black Box I* is assumed to produce not only the desired state ${}^DS$ for the current cycle, but also the desired state ${}^DS'$, for the next cycle. The torques are generated by *Black Box II* according to:

$$^D\mu = G({}^DS - S) + f({}^DS') \,,$$

where G is a feedback-gain matrix, S is the current state as reported by peripheral sensors, and $f()$ is calculated (by the addition of K memory elements) as described above. After each control cycle the memory elements z_i are updated according to:

$$\Delta z_i = \sigma w_{qi} \frac{{}^D\mu - f(S)}{K} \,.$$

Here, $^D\mu$ is the same as in the previous equation, q and $f(S)$ now refer to the current state (not the desired state), and σ is a constant. Note that w_{qi} is *not* modified. All z_i are initially set to zero.

This learning algorithm was used for controlling a planar, two-joint system. For desired movements that were rather complicated, the correct torques were generated after only a few attempts, using $I = 50000$, $K = 60$, and $\sigma = 0.6$. With a much smaller memory ($I = 500$), however, instability occurred for such high values of σ. (No convergence proofs are available.) The parameter most critical to performance seemed to be K. Small values did not allow sufficient generalization from one set of states to another, and large values resulted in deleterious generalization. The system adapted to changes in loading or in desired kinematics, though only after several attempts.

BACK TO BIOLOGY. If *Black Box II* is implemented by a network of real neurons, we should be able to find it. One possibility is that the relevant neurons may be located in the upper regions of the spinal cord (rather than the brain): selective damage in this region can lead to poor performance in reaching movements of the arm, sparing the performance of grasping movements.[1]

Karst and Hasan[30] have approached the issue from the output side. We recorded muscle activity associated with the initiation of pointing movements from various initial to final positions when the arm was constrained to be a planar, two-joint system; shoulder and elbow rotations were allowed in the horizontal plane. By focusing on initiation (when $\dot{\phi}_j = 0$), we could obtain at least a rough indication of the initial torques. We found that each of the two torques varied systematically, not with the direction of the target in extrinsic space (ξ), but rather with ($\xi - \phi_2$). This finding hints that yet another frame of reference, which is linked to the distal segment, may be relevant to neural motor control.

Although there were small differences in the timing of activation of shoulder and elbow muscles, the difference was not correlated with the apparent stagger in the onsets of shoulder and elbow rotations.[32] Figure 3A gives an example involving hand movement to the left and toward the trunk, in which the shoulder flexion (represented by increase in θ_1) seems to start later than the elbow flexion, despite the fact that shoulder flexor (pectoralis major) activity occurs early. A simulation, depicted in figure 3B, which was based on the equations of motion, shows that even if the torques at the two joints are turned on at the same instant, the joint rotation onsets appear staggered. (The magnitudes of the torque steps for the simulation were chosen in light of the data for widely varying values of $[\xi - \phi_2]$.) This can be interpreted in two ways.

i.   *Black Box I* chooses the $^D\theta_n(t)$ to have staggered onsets, and the stagger is such that the requisite torques have almost coincident onsets.

ii. The system applies nearly coincident torques, and the rotational onset difference simply emerges from the mechanics of the arm.

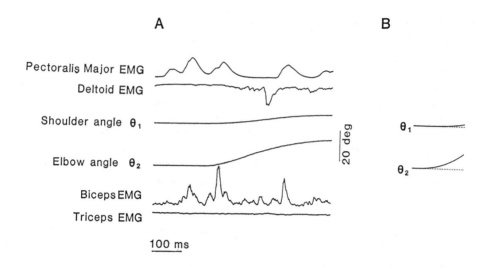

FIGURE 3 A. Records associated with a two-joint, pointing movement of the human arm in the horizontal plane. The filtered, absolute value of the electromyographic activity (EMG) of four muscles is shown. (Pectoralis major = shoulder flexor; Deltoid = shoulder extensor, whose activity is plotted increasing downward; Biceps = elbow/shoulder flexor; and Triceps = elbow extensor.) The joint angles are plotted with flexion upward. (Observations of G. M. Karst and Z. Hasan.) B. Results of a simulation in which torque steps are applied at the same time to the shoulder and elbow; yet the change in shoulder angle (θ_1) has the appearance of starting later than the change in elbow angle (θ_2).

For reasons of parsimony, we prefer the second interpretation. However, if we are right in this preference, then there is little basis for demarcating the boundary between *Black Boxes I* and *II*.

BLACK BOX III: FROM TORQUES TO MUSCLE ACTIVATIONS

A muscle at fixed length L will produce a force F for a given neural command α. For the same α, however, F will be different for different L. Muscles also exhibit damping, in that F depends on dL/dt (in a highly nonlinear way), for the same α and L. The torque arising from muscle activity is the product of F and the moment arm; the latter varies with joint angle, in a non-monotonic fashion. A muscle, moreover, may span more than one degree of kinematic freedom, in which case L depends on all the spanned angles, and so does each of the moment arms. All that the nervous system can do is to determine $\alpha_m(t)$, where m indexes the muscles. This recital of muscle properties should convince the reader that it is a far cry from $\alpha_m(t)$ to $\mu_n(t)$, hence the need for *Black Box III*.

MUSCLE REDUNDANCY. Typically, the number of muscles (M) that can possibly affect the relevant joints exceeds the number of kinematic degrees of freedom (N); in other words, for a desired $\{{}^D\mu_n\}$, the set $\{\alpha_m\}$ is not unique. The resolution of this redundancy has been studied mostly in tasks requiring the exertion of forces against fixed, external supports, i.e., in isometric conditions.

We will focus on only two of the degrees of freedom of the arm: (i) flexion/extension at the elbow, and (ii) supination/pronation of the forearm, i.e., rotation of the forearm about its long axis. (The second is, technically, a radio-ulnar rather than elbow or wrist movement.) There are nine muscles that can exert torques to affect these movements, but three of them have negligible effects.[14] Among the six major players, the biceps exerts strong elbow-flexor as well as forearm-supinator torques; two other muscles also have effects on both degrees of freedom. Given the desired amounts of flexion/extension and supination/pronation forces, there is an infinite number of ways of choosing the forces of the six muscles. A possible solution[14] is outlined below.

Consider a two-dimensional space with orthogonal coordinates, which represent forces in the flexion/extension and supination/pronation directions. The muscle m, when maximally activated, can then be represented by a two-dimensional vector f_m, whose components are the force contributions in the two directions. Define the symmetric tensor:

$$d_{ij} = |f_i||f_j| \cos \psi_{ij}; \qquad i,j = 1 \text{ to } 6 \,,$$

where ψ_{ij} is the angle between the vectors f_i and f_j. The tensor d_{ij} has six eigenvalues, all real, two of which are different from zero. Let V_1, V_2 represent the eigenvectors corresponding to the non-zero eigenvalues. (Note that V_1, V_2 are six-dimensional vectors.) For any given desired force vector ${}^D f$ (which is two-dimensional), we choose the six muscle activations as components of the six-dimensional vector $(aV_1 + bV_2)$, where a and b are unique in that the combination produces ${}^D f$. In essence, we have resolved the redundancy by discarding four eigenvectors. It is as if the relative activation of the six muscles could follow one of only two rules, or a combination thereof. The two rules, however, are not imposed arbitrarily; they reflect the geometry of the muscles' arrangement and their relative strengths.

A similar procedure has been applied[46] to the far more complicated arrangement of 30 muscles involved in rotating the head of a cat in three dimensions. The muscle activations induced reflexively by imposed rotations about different axes were predicted and compared with experiment. The axis for which any given muscle was maximally active corresponded well to the prediction. Agreement was poorer for target-following movements of the head.

A problem with this approach is that some of the components of the eigenvectors usually turn out to be negative, whereas a muscle cannot be activated to generate pushing force. (Muscles are unidirectional in their force capability; the only way to change the sign of the torque about a joint is to use different muscles.) Jongen et al.[28] have proposed a variant of the tensor method in which negative activations do not arise, and have applied it to the two degrees of freedom of the arm discussed earlier. Instead of the M-dimensional eigenvectors (V_1, V_2), they construct a number of two-dimensional "principal direction" vectors (in fact, four in number); then, for given ^Df, they pick the two principal directions that are closest to ^Df, and express ^Df as a linear combination of these. Each principal direction corresponds to a unique set of relative activations of the muscles. These principal directions (and the associated relative muscle activations) are defined as those for which an imposed perturbation elicits a collinear force, assuming that only the muscles stretched by the perturbation contribute forces. (In effect, for each conceivable direction, a pair of eigenvectors is formed, but only some of the muscles are included; a different direction may involve a different set of muscles and a different pair of eigenvectors.) The muscle redundancy is resolved as in the previous scheme, but here the system has learned to avoid negative activations, though at the cost of having to store more principal directions. (Learning of this type can be demonstrated in a network consisting of a circular array with modifiable weights.[7]) Predictions from this theory have been compared with experimental data regarding the recruitment of single motoneurons in tasks requiring different combinations of flexion/extension and supination/pronation forces under isometric conditions.[28] The predictions were reasonably accurate.

MECHANICAL PROPERTIES OF MUSCLES. When actual movement is allowed (i.e., in non-isometric conditions), the relative activation of muscles is not necessarily the same as in isometric conditions.[15,42] Most likely, this is due primarily to the difference in the use of the muscles that span more than one joint ("multiarticular muscles").

At any given instant, according to the equations of motion, the mechanical power $\mu_n \cdot \dot{\theta}_n$ does not, in general, have the same sign for all n. Therefore, if each muscle spanned only one joint, then some of the muscles could be providing energy which is dissipated simultaneously by muscles at other joints! Multiarticular muscles can confer efficiency on the system by avoiding this dissipation[15]: a multiarticular muscle that shortens during the movement provides net positive work to the mechanical system, yet it can allow different signs of $\mu_n \cdot \dot{\theta}_n$ at different joints. Moreover, even in the situation where all the requisite $\mu_n \cdot \dot{\theta}_n$ are positive, a multiarticular muscle may sometimes be better suited to producing the torques, compared to separate uniarticular muscles. This can happen when, with rotations of two joints in opposite directions, the shortening speed of a biarticular muscle is small, whereas the shortening speed of uniarticular muscles is high. (For all muscles, the greater the shortening speed, the less the force.) How *Black Box III* may make these choices among uni- and multiarticular muscles remains unclear, however.

One aspect of the mechanical properties of muscle that is clear is that an active muscle, when forcibly stretched, exhibits some peculiar properties, including a drop in force during lengthening. The nervous system appears to compensate for this "yielding" by reflexively increasing the activation of the muscle.[45] This fast-acting reflex (which is also manifested in the familiar knee jerk), however, is not the only reflex response, as we shall discuss shortly.

ALTERNATIVES TO THE THREE BLACK BOXES

The general scheme that we have followed so far (Figure 1) might have seemed a matter of common sense when it was introduced at the beginning, but by now the patient reader is probably having second thoughts about the wisdom of accepting it! Indeed, dissenting opinions exist, and we will briefly discuss two of them.

MUSCLE SYNERGIES

This is an ill-defined, overused, and therefore justifiably maligned term. For our purposes we define a *muscle synergy* simply as a coordinate for the space of all *observed* muscle activation patterns $\{\alpha_m(t)\}$. Thus, in the static case, the synergies may be identified with the "principal directions" introduced earlier. The concept is useful only if all conceivable patterns are not observed: the observed set can then be expressed in terms of a finite number of synergies. Note that this set would typically consist of an infinite number of patterns. (Some authors imply by synergy that only a finite number of patterns should be observed, which is not how it is defined here.) The task specification determines the combination of the synergies that will be elicited, and the actual elicitation is contingent on either an external or an internal trigger. This idea is tantamount to collapsing all three *Black Boxes* into one, with no computation of either $^D\phi_n(t)$ or $^D\mu_n(t)$.

Support for this idea comes from experiments with sudden perturbation of upright posture; in this situation the subject has no time to go through the three-black-box computations if he is not to topple. It was found that in response to several types of perturbations, the activity of the various muscles exhibited only a few different patterns,[44] which, moreover, were the same as when the subject voluntarily mimicked the perturbation-evoked corrective movements.[43] Hence, the idea arose of a discrete set of muscle activity patterns. (These patterns cannot be generated simply by activating each stretched muscle in response to the perturbation; there are long-latency effects in stretched as well as unstretched muscles.[13,39]) More recently, however, it has been found that if the repertoire of perturbations is expanded, so is the number of observed muscle activity patterns,[40] which negates the concept of synergy as a fixed pattern, but does not necessarily negate what is defined here. The concept of synergies has been developed mostly in the context of imposed perturbations, where the presumed task is to resist the perturbation. It should be recognized, however, that due to the complexity of the equations of motion, this task may not be accomplished if each muscle resists only its own stretch, unless every stretched muscle were to become rigid.[19] In fact, the response of each muscle seems to depend on the "body scheme," i.e., the perceived $\{^I\phi_n\}$, irrespective of the correctness of the perception.[16]

We have already cited some evidence that the concept of synergies might also be applicable to voluntary movements. Recently, Koshland et al.[36] have reported a temporal similarity between the voluntary pattern and the perturbation-elicited pattern (both dependent on $\{^I\phi_n\}$) in a situation in which the latter was *inappropriate* for resisting the motion at one joint caused by the perturbation of the other joint. This gives credence to the view that synergies, although they have been studied mostly with perturbations, may be designed for voluntary movement. It is possible that the synergies (but not the plethora of observed muscle activity patterns) are learned in childhood through trial and error. Babies, for instance, commonly activate simultaneously the opposing muscles at a joint, whereas adults do it only transiently during most movements, but they too resort to sustained co-contraction when the task is unfamiliar. There is no excuse for such profligacy according to the three-box scheme.

If it is true that the nervous system does not compute $^D\phi_n(t)$ or $^D\mu_n(t)$, but selects directly a combination of synergies, then how does one account for the fact that a change in loading conditions (and therefore in the parameters of the equations of motion) does not cause a completely wrong movement? Data of Karst and Hasan[31] suggest that when the inertial load is changed, the pattern of initial muscle activity is not modified appropriately (even though the subject has had some experience with the new load), and thus the initial movement is indeed in a wrong direction. The direction, however, is corrected fairly quickly, which presumably shows the efficacy of reflexes in correcting the direction of motion (rather than correcting for position, speed, force, or other variables that have been proposed as "regulated"). Recent experiments have shown that patients who lack a sense of position (and muscle reflexes) make consistent errors, despite the presence of vision, in multijoint pointing movements; these errors are related to the direction of movement.[12] Thus, a point of view is emerging in which, unlike the three-box scheme, a combination of muscle synergies is chosen for the given task on the basis of long experience, and movement is initiated; veering motions and other effects arising from the mechanics of the arm, then, are corrected on the fly, using feedback information. There is, however, no model (which incorporates realistic feedback delays) to lend precision and detailed testability to this point of view.

Related ideas that involve synergies at the kinematic rather than muscle level have been proposed by Kelso et al.[35] Certain links in the movements of the two arms in bimanual tasks, for instance, can be made comprehensible in this approach, whereas there is no mechanical reason for such a linkage.

PULL OF EQUILIBRIUM

One way that the N-segment system of Figure 2 could be made to transport the end-effector to the target is to tie a spring between the end-effector and the target, such that the force in the spring would be zero only when the former reaches the latter. With some damping in the spring, it is obvious that this method will work: the end-effector will be pulled by the spring and will eventually reach the target, though possibly along a wild trajectory. (A force in the target direction does not cause acceleration in that direction, since the end-effector is not the center of mass.)

In this method there is no need to know anything about the equations of motion or their parameters, and the kinematic redundancy is resolved automatically.

If the force exerted on the end-effector has components $(-f_x, -f_y)$, it can be shown from *statics* that the torques at the joints necessary to balance this force are given by:

$$\mu_n = \sum_{i=n}^{N} \ell_i (f_y \cos \phi_i - f_x \sin \phi_i).$$

(Not surprisingly, this equation is of the same form as the *dynamics* equation, presented earlier, for the special case when the mass of the arm is concentrated at the end-effector.) Now, tying the above-mentioned spring is equivalent to choosing the torques according to the last equation, with $f_x = \sigma(^T x - {}^E x)$, $f_y = \sigma(^T y - {}^E y)$, where σ represents the stiffness of the equivalent spring. Thus, given the target position, and the current $\{\phi_n\}$, one can use a simple, statics-based method of choosing the torques, and be assured by including some damping that the end-effector will reach the target and will stop there, no matter how complicated the dynamics. If the mass were mostly at the end-effector, the path resulting from the application of this method will be nearly straight.

The fact, however, is that the mass is very far from being concentrated at the tip, yet the path of the tip is not very far from straight. (Recall the earlier discussion about choosing $a_x(t), a_y(t)$.) That the central nervous system (CNS) does not follow the equivalent-spring strategy is further confirmed by the data of Hasan and Karst,[20] who show that even the sign of μ_n from this equation does not tally with experiment.[31]

The basic idea, nevertheless, is attractive, because it obviates the necessity of choosing $\{^D \phi_n(t)\}$, resolves the kinematic redundancy, and avoids inverse-dynamics computation altogether. Consequently, a number of modifications and refinements of this idea have been proposed, all of which go under the rubric of "equilibrium-point" hypothesis. Let us briefly consider two, both of which posit a gradual shift of the equilibrium point, instead of assuming that the final target position is the only equilibrium point throughout the movement.

ANISOTROPIC STIFFNESS OF THE END-EFFECTOR. In the strategy just described, the force exerted by the end-effector would always be directed toward the equilibrium position in extrinsic space. In the light of experimental results showing that, in general, the force is *not* so directed when the arm is perturbed from a postural position of equilibrium, Hogan[24] generalized the concept of "spring-like" beyond the simple spring we postulated. Namely, if the force vectors can be described as gradients of a scalar (potential) function U, then the system is spring-like. Experimentally, this is very nearly the case for the forces at the end-effector elicited by perturbations in postural conditions. The stiffness, though, is not isotropic: the magnitude of the restoring force is different for different directions of perturbation. The function U can therefore be imagined as a bowl with elliptical rather than circular isopotential contours. The bottom of the bowl corresponds to the equilibrium point in extrinsic space. The force observed in response to perturbation from the equilibrium point has the direction of the steepest descent along the bowl, and the magnitude of the force equals the steepness of the slope.

For a voluntary movement the whole bowl moves toward the target, dragging the end-effector with it, as the latter, figuratively speaking, keeps falling into the bowl. The path of the end-effector, though it is governed by the equations of motion, cannot veer too far from the bottom of the bowl. What creates the bowl and what moves it toward the target position? Its creation is a matter of letting the torques μ_n be dependent on current joint angles, such that the across-joint effects are reciprocal. The dependence on current angles can come about partly through the mechanical properties of the muscles and partly through reflex effects. (Similarly for the dependence on $\{\dot{\theta}_n\}$, which is necessary to damp oscillations.) What moves the bowl toward the target is a change in the (non-reflex) activation of the muscles, which defines a new position of equilibrium; this position changes continuously. Simulations[9] have found it necessary to assume a higher magnitude of stiffness than revealed in postural conditions. We simply do not know what the function U looks like during a voluntary movement.

MUSCLE ACTIVATION BASED ON EQUILIBRIUM POINT. A somewhat different version of the equilibrium-point hypothesis has been presented by Feldman et al.[8] In this version, which we describe for the planar, two-joint case without an external force, the equilibrium point does move in a straight line and at constant speed toward the target in extrinsic space, though of course the end-effector does not. The activation for each muscle is based on the difference between two variables: the actual muscle length, and the length that the muscle would have to have in order for the arm to be in the currently commanded equilibrium position. (Actually, the muscle activation is postulated to depend also on speed of length change, which is consonant with reflex properties, and makes the temporal pattern of predicted activation during voluntary movement more akin to the electromyographic observations.) Note that it is the equilibrium position of the end-effector in extrinsic space, and not the muscle activation, that is the command variable according to this hypothesis. Thus, feedback of actual muscle length is crucially important for determining muscle activation. The hypothesis allows for the observed possibility of, for instance, a flexor muscle initiating an extension movement at one of the joints, but only when the extension is small. Issues of kinematic and muscle redundancy require additional assumptions for their resolution.

All versions of the equilibrium-point idea strive to present a unified account of movement and posture. It remains controversial, however, whether this unification is even desirable in the light of some of the observations that were mentioned in the section on muscle synergies, e.g., difference in relative activation of muscles for posture and for movement, response to perturbation that is inappropriate for correction. (Hasan and Stuart[21] cite other examples where the CNS seems to assist rather than resist a perturbation, and discuss how this might help in coordination.) It is disconcerting that many discussions of the equilibrium-point idea convey the flavor of circular argumentation. After all, given any observed trajectory, one can always construct a motion of an equilibrium point such that the observed trajectory would be obtained. The situation with regard to the testability of these constructs is improving, however, as the newer data go beyond kinematics by digging into the neural and electromyographic levels. In any case, the cognizance taken of muscle properties and of reflex effects makes the equilibrium-point ideas more physiological than the stepper-motor-oriented schemes from robotics.

OTHER NOTIONS

We have not discussed any of the optimization approaches to the understanding of movement control. Some experimental facts are indeed captured by theories based on the minimization of certain variables related to kinematics,[23] command,[17] torques,[51] or combinations thereof.[47] But these ideas shed little light on the underlying machinery. We have also not pursued an argument advanced by the present author[19] to the effect that the task of the neural motor system is not simply to produce the requisite torques but equally importantly, to preserve the integrity of the joints and bones during movement. The extensive literature of motor learning has also not been touched upon, and only passing mention has been made of the state-space approach.[35]

Despite these and other omissions, it is hoped that the uninitiated reader has now at least become familiar with the existence of the fertile grounds that have been opened up in the study of movement, though they may be in need of further seeding. There are challenging issues here for all comers, and all that is required is conceptual rigor, and some humility in the face of experimental facts.

ACKNOWLEDGMENTS

I am thankful to the colleagues who have discussed these ideas with me, and to the students at the Summer School who asked probing questions. The support of N.I.H. grant NS-19407 is also gratefully acknowledged.

REFERENCES

1. Alstermark, B., A. Lundberg, U. Norrsell, and E. Sybirska. "Integration in Descending Motor Pathways Controlling the Forelimb in the Cat. 9. Differential Behavioral Defects after Spinal Cord Lesions Interrupting Defined Pathways from Higher Centers to Motoneurones." *Experimental Brain Research* **42** (1981): 299–318.

2. Atkeson, C. G. "Learning Arm Kinematics and Dynamics." *Annual Review of Neuroscience* **12** (1989): 157–183.

3. Atkeson, C. G., C. H. An, and J. M. Hollerbach. "Estimation of Inertial Parameters of Manipulator Loads and Links." *Interna'l. J. Robotics Research* **5** (1986): 101–119.

4. Bullock, D., and S. Grossberg. "Neural Dynamics of Planned Arm Movements: Emergent Invariants and Speed-Accuracy Properties during Trajectory Formation." *Psychological Review* **95** (1988): 49–90.

5. Cole, K. J., and J. H. Abbs. "Coordination of Three-Joint Digit Movements for Rapid Finger-Thumb Grasp." *J. Neurophysiology* **55** (1986): 1407–1423.

6. Cruse, H., and M. Bruwer. "The Human Arm as a Redundant Manipulator: The Control of Path and Joint Angles." *Biological Cybernetics* **57** (1987): 137–144.

7. Denier van der Gon, J. J., A. C. C. Coolen, and H. J. J. Jonker. "Self-Organizing Neural Mechanisms Possibly Responsible for Muscle Coordination." In *Multiple Muscle Systems: Biomechanics and Movement Organization*, edited by J. M. Winters and S. L.-Y. Woo, 355–342. New York: Springer-Verlag, 1990.

8. Feldman, A. G., S. V. Adamovich, D. J. Ostry, and J. R. Flanagan. "The Origin of Electromyograms—Explanations Based on the Equilibrium Point Hypothesis." In *Multiple Muscle Systems: Biomechanics and Movement Organization*, edited by J. M. Winters and S. L.-Y. Woo, 195–213. New York: Springer-Verlag,1990.

9. Flash, T. "The Control of Hand Equilibrium Trajectories in Multi-Joint Arm Movements." *Biological Cybernetics* **57** (1987): 257–274.

10. Flash, T. "The Organization of Human Arm Trajectory Control." In *Multiple Muscle Systems: Biomechanics and Movement Organization,* edited by J. M. Winters and S. L.-Y. Woo, 282–301. New York: Springer-Verlag, 1990.

11. Georgopoulos, A. P. "Spatial Coding of Visually Guided Arm Movements in Primate Motor Cortex," *Canadian J. Physiology and Pharmacology* **66** (1988): 518–526.

12. Ghez, C., J. Gordon, M. F. Ghilardi, C. N. Christakos, and S. E. Cooper. "Roles of Proprioceptive Input in the Programming of Arm Trajectories." *Cold Spring Harbor Symposia in Quantitative Biology* **55**, in press.

13. Gielen, C. C. A. M., L. Ramaekers, and E. J. van Zuylen. "Long-Latency Stretch Reflexes as Co-ordinated Functional Responses in Man." *J. Physiology* **407** (1988): 275–292.

14. Gielen, C. C. A. M., and E. J. van Zuylen. "Coordination of Arm Muscles during Flexion and Supination: Application of the Tensor Analysis Approach." *Neuroscience* **17** (1986): 527–539.

15. Gielen, S., G.-J. van Ingen Schenau, T. Tax, and M. Theeuwen. "The Activation of Mono- and Bi-Articular Muscles in Multi-Joint Movements." In *Multiple Muscle Systems: Biomechanics and Movement Organization*, edited by J. M. Winters and S. L.-Y. Woo, 302–311. New York: Springer-Verlag, 1990.

16. Gurfinkel, V. S., Y. S. Levik, K. E. Popov, B. N. Smetanin, and V. Y. Shlikov. "Body Scheme in the Control of Postural Activity." In *Stance and Motion: Facts and Concepts*, edited V. S. Gurfinkel, M. E. Ioffe, J. Massion, and J. P. Roll, 185–193. New York: Plenum, 1988.

17. Hasan, Z. "Optimized Movement Trajectories and Joint Stiffness in Unperturbed, Inertially Loaded Movements." *Biological Cybernetics* **53** (1986): 373–382.

18. Hasan, Z. "Biomechanical Complexity and the Control of Movement." In *Lectures in the Sciences of Complexity*, edited by D. L. Stein. Santa Fe Institute Studies in the Sciences of Complexity, Lect. Vol. I, 841–850. Redwood City: Addison-Wesley, 1988.

19. Hasan, Z. "Biomechanics and the Study of Multijoint Movements." In *Motor Control: Concepts and Issues*, edited by D. R. Humphrey and H.-J. Freund. Chichester: Wiley, in press.

20. Hasan, Z., and G. M. Karst. "Muscle Activity for Initiation of Planar, Two-Joint Arm Movements in Different Directions." *Experimental Brain Research* **76** (1989): 651–655.

21. Hasan, Z. and D. G. Stuart. "Animal Solutions to Problems of Movement Control: The Role of Proprioceptors." *Annual Review of Neuroscience* **11**, (1988): 199–223.

22. Hinton, G.. "Parallel Computation for Controlling an Arm." *J. Motor Behavior* **16** (1984): 171–194.

23. Hogan, N. "An Organizing Principle for a Class of Voluntary Movements." *J. Neuroscience* **4** (1984): 2745–2754.

24. Hogan, N. "The Mechanics of Multi-Joint Posture and Movement Control." *Biological Cybernetics* **52** (1985): 315–331.

25. Hogan, N., and J. M. Winters. "Principles Underlying Movement Organization: Upper Limb." In *Multiple Muscle Systems: Biomechanics and Movement Organization*, edited by J. M. Winters and S. L.-Y. Woo, 182–194. New York: Springer-Verlag, 1990.

26. Hollerbach, J. M., and C. G. Atkeson. "Deducing Planning Variables from Experimental Arm Trajectories: Pitfalls and Possibilities." *Biological Cybernetics* **56** (1987): 279–292.

27. Houk, J. C., S. P. Singh, C. Fisher, S. Epstein, and A. G. Barto. "An Adaptive Sensorimotor Network Inspired by the Anatomy and Physiology of the Cerebellum." *Application of Neural Networks to Robotics and Control*, in press.

28. Jongen, H. A. H., J. J. Denier van der Gon, and C. C. A. M. Gielen. "Activation of Human Arm Muscles during Flexion/Extension and Supination/Pronation Tasks: A Theory on Muscle Coordination." *Biological Cybernetics* **61** (1989): 1–9.

29. Kaminski, T., and A. M. Gentile. "Joint Control Strategies and Hand Trajectories in Multijoint Pointing Movements." *J. Motor Behavior* **18** (1986): 261–278.

30. Karst, G. M., and Z. Hasan. "Direction-Dependent Strategy for Control of Multi-Joint Arm Movements." In *Multiple Muscle Systems: Biomechanics and Movement Organization*, edited by J. M. Winters and S. L.-Y. Woo, 268–281. New York: Springer-Verlag, 1990.

31. Karst, G. M., and Z. Hasan. "Initiation Rules for Planar, Two-Joint Arm Movements: Agonist Selection for Movements throughout the Workspace." Submitted.

32. Karst, G. M., and Z. Hasan. "Timing and Magnitude of Electromyographic Activity for Two-Joint Arm Movements in Different Directions." Submitted.

33. Kawato, M., K. Furukawa, and R. Suzuki. "A Hierarchical Neural-Network Model for Control and Learning of Voluntary Movement." *Biological Cybernetics* **57** (1987): 169–185.

34. Kawato, M., M. Isobe, Y. Maeda, and R. Suzuki. "Coordinates Transformation and Learning Control for Visually-Guided Voluntary Movement with Iteration: A Newton-Like Method in a Function Space." *Biological Cybernetics* **59** (1988): 161–177.

35. Kelso, J. A. S., G. Schoner, J. P. Scholz, and H. Haken. "Phase-Locked Modes, Phase Transitions and Component Oscillators in Biological Motion." *Physica Scripta* **37** (1987): 79–87.

36. Koshland, G. F., L. Gerilovsky, and Z. Hasan. "Activity of Wrist Muscles Elicited during Imposed or Voluntary Movements about the Elbow Joint." *J. Motor Behavior*, in press.

37. Kuperstein, M., and J. Rubinstein. "Implementation of an Adaptive Neural Controller for Sensory-Motor Coordination." *IEEE Control Systems* **9(3)** (1989): 25–30.

38. Lacquaniti, F. "Central Representations of Human Limb Movement as Revealed by Studies of Drawing and Handwriting." *Trends in Neuroscience* **12** (1989): 287–291.

39. Lacquaniti, F., and J. F. Soechting. "EMG Responses to Load Perturbations of the Upper Limb: Effect of Dynamic coupling between Shoulder and Elbow Motion." *Experimental Brain Research* **61** (1986): 482–496.

40. Macpherson, J. M. "How Flexible are Muscle Synergetics?" In *Motor Control: Concepts and Issues*, edited by D. R. Humphrey and H.-J. Freund. Chichester: Wiley, in press.

41. Miller, W. T., F. H. Glanz, and L. G. Kraft. "Application of a General Learning Algorithm to the Control of Robotic Manipulators." *Interna'l. J. Robotics Research* **6(2)** (1987): 84–98.

42. Nardone, A. and M. Schieppati. "Postural Adjustments with Voluntary Contraction of Leg Muscles in Standing Man." *Experimental Brain Research* **69** (1988): 469–480.

43. Nashner, L. M. "Fixed Patterns of Rapid Postural Responses among Leg Muscles during Stance." *Experimental Brain Research* **30** (1977): 13–24.

44. Nashner, L. M., and G. McCollum. "The Organization of Human Postural Movements: A Formal Basis and Experimental Synthesis." *Behavioral and Brain Sciences* **8** (1985): 135–172.

45. Nichols, T. R., and J. C. Houk. "Improvement in Linearity and Regulation of Stiffness that Results from Actions of Stretch Reflex." *J. Neurophysiology* **39** (1976): 119–142.

46. Pellionisz, A. J., and B. W. Peterson. "A Tensorial Model of Neck Motor Activation." In *Control of Head Movements*, edited by B. W. Peterson and F. Richmond, 178–186. Oxford: Oxford University Press, 1988.

47. Seif-Naraghi, A. H., and J. M. Winters. "Optimized Strategies for Scaling Goal-Directed Dynamic Limb Movements." In *Multiple Muscle Systems: Biomechanics and Movement Organization*, edited by J. M. Winters and S. L.-Y. Woo, 312–334. New York: Springer-Verlag, 1990.

48. Soechting, J. F. 'Elements of Coordinated Arm Movements in Three-Dimensional Space." In *Perspectives on the Coordination of Movement*, edited by S. A. Wallace, 47–83. Amsterdam: North-Holland, 47-83.

49. Soechting, J. F., and B. Ross. "Psychophysical Determination of Coordinate Representation of Human Arm Orientation." *Neuroscience* **13** (1984): 595–604.

50. Soechting, J. F., and C. A. Terzuolo. "An Algorithm for the Generation of Curvilinear Wrist Motion in an Arbitrary Plane in Three-Dimensional Space." *Neuroscience* **19** (1986): 1393–1405.

51. Uno, Y., M. Kawato, and R. Suzuki. "Formation and Control of Optimal Trajectory in Human Multijoint Arm Movement: Minimum Torque-Change Model." *Biological Cybernetics* **61** (1989): 89–101.

52. Zajac, F. E., and J. M. Winters. "Modeling Musculoskeletal Movement Systems: Joint and Body-Segmental Dynamics, Musculoskeletal Actuation, and Neuromuscular Control." In *Multiple Muscle Systems: Biomechanics and Movement Organization*, edited by J. M. Winters and S. L.-Y. Woo, 121–148. New York: Springer-Verlag, 1990.

Timothy R. Thomas† and Mychael Trong Vo‡
†Computing Division, Mail Stop B-258, Los Alamos National Laboratory, Los Alamos, NM 87545 and ‡Department of Electrical Engineering, California Polytechnical Institute, Pomona, CA 91768

Control of Tongue Movement Dynamics

INTRODUCTION

Speech perception is accomplished without any apparent effort and with considerable accuracy, despite the complexity and variability of the acoustic signal. Attempts to understand speech perception can be divided into three fundamental approaches. The first, called here a psychophysical approach, says that the ear and brain are so constructed that critical acoustic features of the speech signal are extracted and form an internal representation that serves as the basis of speech perception. Researchers following this approach might, for example, examine the ways in which cochlear mechanics alter the acoustic input, or try to determine the way in which the sensory nerve codes frequency information. Followers of the second, phonetic approach, argue that speech consists of a more or less linear sequence of phonemes, and that the problem of speech perception is to extract these sounds and to learn the complex set of rules that combine them into higher-level linguistic units. A third approach, based on the direct realism of James Gibson,[10] says that speech perception consists of extracting from the acoustic signal the movements of the speech articulators that the speaker intended to make, and then from that information constructing the linguistic elements necessary for determining meaning.

This last approach, usually described as the motor theory of speech perception,[13,14] says that listeners can detect what particular gestures (here defined as the movement of a speech articulator from a present location to a goal position) were made by a speaker, and that each word at its most basic level is described as a combination of these underlying gestures. The theory further specifies that it is not the actual movements of the speech articulators that are the invariant content that specifies a word; rather, it is the intention to make a particular gesture, such as the bilabial closure that begins the word "bad." With this approach the internal representation of a word is the same for both speech production and speech perception. Aside from the obvious efficiency of such an arrangement, there is substantial evidence that people can indeed find the gestures hidden in speech, and do use this information as the basis for speech perception.[5,7,8,17] However, it is not at present known how listeners extract this gestural content. If this theory has any validity, an artificial system designed to recognize speech in a human-like manner would need to find the intentions that underlie the movements that create speech. Unfortunately, before any such system could be designed, it is necessary to have in advance a good understanding of the properties of articulatory gestures and their underlying intentions—i.e., the parameters that describe them, their temporal extent, and how best to describe their relationship to actual movement.

Intentions are *prima facie* cognitive constructs that cannot be directly observed but must be inferred from the movements of the peripheral speech articulators. Thus, the problem of motor control is one of linking the cognitive domain of goals, intentions, plans, and commands to the real-world domain of objective events, anatomical movement, and physical effects. We, along with others[2,12] imagine that intentions take the form of qualitative mental commands that have a beginning, exist in an unchanging form for a period of time, and then end. Commands are taken to be goal oriented, in that they specify that some anatomical object, such as the tongue tip, be moved to a goal position, such as the alveolar ridge (the bony ridge behind the incisors where /t/, /d/ and /n/ phonemes are formed). Once a command has been initiated, execution is coordinated by a neuromotor effector system, probably located in and near Broca's area of the frontal lobe.[1] This system selects the muscles to be used and the degree of contraction required. A sequence of these commands form an *intention score* which, when executed, produces a coordinated sequence of motions. Different commands could be overlapped in time, especially if they involved the use of different groups of muscles, or they could be strictly sequential as in the case where the same anatomical object must reach one goal before moving on to the next. The relationship between commands and movements for the case of speech production has been extensively studied as a tool to explicate a phonological description of speech based on gestures.[4] The problem of harnessing the many degrees of freedom present in the speech musculature to the fewer degrees of freedom used to specify a goal-oriented command has also been studied in the task dynamic model of Kelso et al.[11] These efforts and others,[8] have at the least shown that the concept of a speech gesture command is a powerful construct that can greatly explicate the structure of speech both within and between languages.

From the perspective of the motor theory of speech perception, however, the problem is not how simple commands control complex musculature, nor how the phonological structure of speech is created by gestural movements, but rather what are the gestural intentions underlying speech, when do they begin and end, what are their parametric descriptions, and how can they be perceived. In an attempt to answer some of these questions, we will describe a model of speech motor control which uses a dynamical system to infer movements from a sequence of motor commands. We will then compare the model output to real movement observed during natural speech. Perhaps the most difficult articulator in this respect is the tongue. Since it is involved in the production of most consonants and all vowels, it is called upon to rapidly execute multiple gestures with a single constrained apparatus. The tongue also provides many opportunities for coarticulation effects (defined here as when one gesture is modified because of the influence of surrounding gestures) and so provides a challenging problem for understanding the gestural intention structure of speech.

The outcome of this exercise will be a set of model elements that were found to be necessary for reasonable agreement between actual and model-produced movements. Hopefully, these elements will correspond to the timing and descriptive properties of real intentions that existed in the brain of the speaker. Of course, this effort cannot prove that such elements actually exist, but it can serve a valuable heuristic function by generating testable hypotheses and by suggesting useful categories for a speech recognition system based on the motor theory. Furthermore, the insights gained here should permit one to discover the gestural intentions that underlie new utterances, not specifically modeled. It is the purpose of this lecture to report the results of our attempt to model the relationship between the gestural intentions of natural speech and the consequent movements of the tongue.

METHOD AND RESULTS
THE SPEECH SEGMENT SELECTED FOR MODELING

We gathered our data at the x-ray microbeam facility at the University of Wisconsin's Waisman Institute. This facility simultaneously records the acoustic signal and the movement during normal speech of 2.5 mm gold pellets attached to the tongue. This provides a two-dimensional record of the trajectory of those points on the tongue in synchrony with the acoustic wave form being produced. Data consisted of two speakers saying "tack tack tack" with pellets on the tongue tip, tongue body, and tongue dorsum. The basic tongue motions that form the word *tack* consist of raising the tongue tip to make an alveolar closure, followed by lowering the tongue for the vowel, and then raising the dorsum to produce a closure at the velar palate

TABLE 1 Stages of tongue movement for speaker GP4p for the middle "tack" of "tack, tack, tack."

Stage	Movement	Approximate Duration (ms)
1	Release of /k/ from preceding tack	103
2	Alveolar closure for initial /t/	108
3	Interword pause	71
4	From alveolar closure to /ae/ vowel position	135
5	Velar closure for final /k/	149
6	Velar release of /k/	124

(the rear portion of the hard palate, where the consonants /k/ and /g/ are formed). We choose to study the middle word to avoid end effects and to provide a known context. The stages of tongue motion used to produce this word are listed in Table 1.

THE TONGUE MOVEMENT MODEL

At the outset, our model is explicitly a model of the *movement* of the tongue as represented by the movement of the three pellets attached to it. It is not a model of the tongue itself like those of Coker,[6] Maeda,[15] or Kelso et al.[12] At the highest level of our model is an intention score or sequence of commands. The fundamental idea for such a score was taken directly from the gestural theory of Browman and Goldstein.[2,3,4] Intentions instruct the tongue to move from its present position to a target or goal location. Intentions, however, do not specify which particular muscle group to use, nor which anatomical part of the speech system to use. A mental intention, which is something like "close the vocal tract at the alveolar ridge," becomes, in the present model, "move the tongue tip pellet to the alveolar closure position." Note that we are not at all concerned about issues like damping which would be necessary to prevent a real tongue based-model from becoming unstable. We are only interested in producing a model-generated movement that corresponds to the actual movement observed. Whether that movement is produced by raising the jaw or by flexing the tongue is not at issue here. We assume that the neuromotor effector system is capable of rapidly and efficiently translating intentions into the goal-oriented, orderly motion of the pellets which we will model. Abbs[1] provides a convincing review of the literature that strongly supports the distinction made here between intentions and the specifics of motor execution.

In order to provide targets for the word *tack*, the speakers were asked to form the syllables *teh* and *keh* and the vowel /ae/ in isolation. The position of the first speaker's tongue during the vowel and at the moment of maximum height during each consonant is shown in Figure 1. These are assumed to be the target positions for the tongue when forming each of the phonemes for the word *tack*. It should be emphasized that we have assumed that all three pellets have a specific target position for each intention, rather than only the pellet actually making the stop has a target. This is different than other approaches[12] which assume the purpose of the movement is wholly specified by the place of maximum constriction, and the rest of the tongue is passively moved along because it is mechanically attached to the actively controlled portion. However, as Figure 1 shows, when making a velar stop, the speaker's tongue body pellet was higher than the velar pellet. In order to produce this position, a very explicit model of the mechanical properties of the tongue would be required, particularly since this position was sometimes reached from an initial position at the beginning of motion in which the tongue body was below the tongue dorsum. Of course, to introduce such a complex tongue model would be wholly antithetical to the purpose of the present study. In justification of our approach, it does not seem absurd to suppose that the brain is capable of controlling the position of the entire tongue. Rather, it seems more unlikely that the influence on acoustic resonance of the tongue body

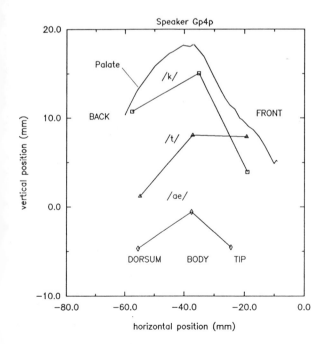

FIGURE 1 Vertical position of the tongue tip pellet as a function of time during the movement from the alveolar ridge to the vowel position of the word "tack."

position during motions to a consonantal stop would be left to the limitations of passive mechanics, rather than actively controlled so as to enhance clarity of communication in varying contexts.

We chose to model the movement by a second-order, undamped, linear differential equation $\ddot{x} + kx = 0$. This is equivalent to modeling the motion with an undamped spring of 0 rest length, stretched to half the distance between the initial position and the target, with the mass being moved set arbitrarily to 1. This spring is termed the primary spring. The reason for selecting this model can be seen in Figure 2, which shows the vertical position of the tongue tip as a function of time during the movement from the alveolar closure that begins the word *tack* to the position for the vowel /ae/. This roughly corresponds to the phoneme /t/ and is accompanied by a burst of energy in the acoustic signal. The distinct sinusoidal shape of this trajectory is exactly what is expected from the model we have selected. The duration of movement in this model is determined by the stiffness constant k, which was set so that $\pi/\sqrt{k}$ produced a duration equal to that which was actually observed during this movement. This stiffness was not changed thereafter. Since tongue pellet trajectories are measured in two dimensions, the model required an additional parameter of spring orientation, specified by a direction vector from the initial position to the target location.

Intentions are manifest within the model by the setting of the two spring parameters: direction and length. Execution of the commands automatically results

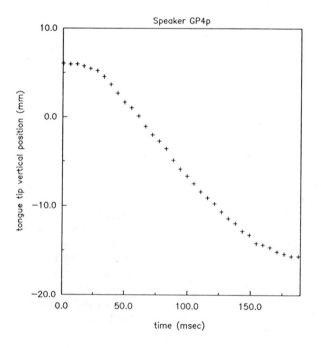

FIGURE 2 The target positions for the phonemes in the word "tack." Each phoneme is assumed to have a separate target position for each pellet.

from the dynamics of the system. Command parameters are not changed until the target is reached. At that point, that spring is eliminated and a new spring, with new parameters, is produced to move the pellet to the next target as required by the intention score.

After a target is reached, our subjects would sometimes pause before motion began to the next target. This was particularly evident during the period of time between reaching the alveolar closure for the initial /t/ of *tack* (which occurred immediately following the completion of the final part of the preceding word) and the release of that closure which produced the first acoustic signal of the word. This is labeled as an interword pause in Table 1, and is seen on the x-ray tracts as a time when the articulators do not move. Our model treats the pause as an intention with an onset and a duration. It does not have a target other than to remain in place for a specified period.

One major additional feature is needed to complete the motion model—a technique for handling coarticulation effects. Roughly, coarticulation means that the motions and targets used to produce a particular speech segment are different depending on the context of surrounding gestures.[16] In our example, one would expect the position that the tongue reaches during the vowel /ae/ in *tack* to be different

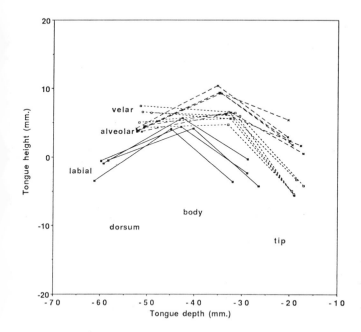

FIGURE 3 Tongue position for /u/ between labial, velar, and alveolar consonants.

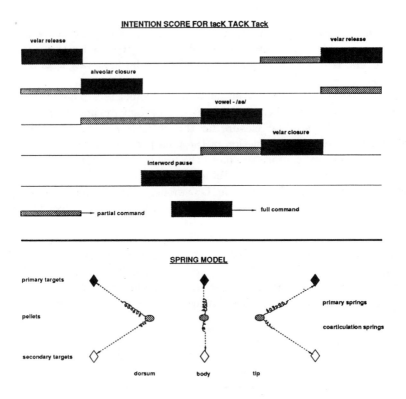

FIGURE 4 The intention score and spring movement model used to generate tongue tip, body, and dorsum pellet trajectories for the word "tack."

than that used to produce the vowel /ae/ in isolation. Since the gesture which follows the /ae/ in *tack* is a consonantal velar closure, the tongue would likely reach a position higher and further back than in the case of the isolated vowel. Figure 3, which was produced by our colleague Judith Hochberg, demonstrates this coarticulation effect for the vowel /u/.

We express this phenomenon in our model by adding a second spring, in addition to the primary one which propels the pellet toward the currently active target. This is configured as a coupled spring system with the second, coarticulation spring, set to move the pellet to the next target in the sequence, but with a lesser length so that the coarticulation spring has a smaller effect on movement than the primary spring. We found that stretching the spring to 1/10 of the distance to the subsequent target produced approximately the correct amount of coarticulation, and that value did not need to be altered for different combinations of targets. This idea is described as a partial command, in that it does not lead to reaching the target, only to moving somewhat in that direction. Stops, which need to seal the vocal tract, will only occur if the target is fully achieved. Thus they require a special consideration. Any attempt to coarticulate a stop with a following vowel would result in a failure to reach the palate. Therefore, we did not coarticulate the pellet which needed to make the stop. If we had been modeling more of the tongue than just the pellets, we could perhaps have achieved a stop with some other part of the tongue, and thus added a coarticulation that would have expressed itself as a change in the place at which the stop occurred. That refinement will need to be the result of future research based on more complete records of tongue movement—perhaps those that magnetic resonance systems will someday provide.

Figure 4 shows a schematic representation of the final model used in this study. The predicted motion of the pellets was obtained by simple superposition of the forces of the two springs in each coupled pair, iterated at 1 msec time steps. Figures 5 and 6 show a plot of both the actual and model-generated movement of the pellets during the motions from the alveolar release of the initial /t/ of *tack* to the release of the velar stop forming the /k/. The major difference between the two trajectories seems to be in the more linear motion of the model, whereas the actual tongue moved in a more curved, flowing pattern. This probably reflects some subtle technique to conserve momentum used by the speech system that is not included in our model. However, the general directions, velocities, and segmentation of the two trajectories show a substantial agreement. This was the outcome hoped for at the outset.

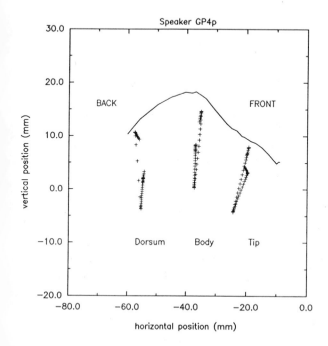

FIGURE 5 Model generated trajectories for the word "tack."

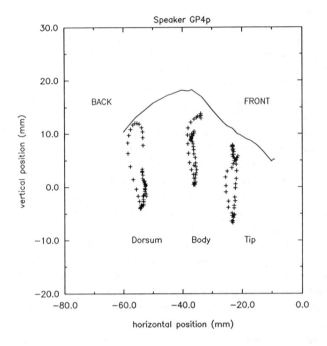

FIGURE 6   X-ray recorded trajectories for the word "tack."

DISCUSSION

The proposed value of the present effort was to elucidate the intention structure which is presumed to underlie the dynamics of speech articulation.[1] In this regard, several results should be noted: (1) separate intentions appear to be required for opening and closing movements; (2) pauses require intentions without targets; (3) intentions appear to involve a target position for the entire tongue, rather than only involving the portion of the tongue that will make the critical constriction; (4) since overshooting of targets did not occur in either the real or model-generated trajectories, intentions appear to be very efficiently realized—deviating only in the direction determined by the coarticulation with surrounding intentions; (5) intentions do not appear to involve the setting of a stiffness parameter to determine gestural duration, since all trajectories appear to be consistent with a single stiffness setting; and (6) coarticulation effects, which are definitely present, can be simply modeled by allowing the subsequent target to influence the movement toward the current target, and, within the model, the same degree of influence throughout the word is adequate to match the actual trajectories.

Clearly, caution would be advised at this point, since the above results are derived from the study of only two speakers saying a single word. While the proposed model does seem to reasonably describe the tongue movements in these conditions, it may well fail to do so with other speakers or other words. Nonetheless, the results obtained concerning the hypothesized nature of the intention structure of speech can well serve as hypotheses to be subjected to further, perhaps more imaginative testing in the future.

OTHER EVIDENCE

In addition to our modeling efforts, we have also collected a small amount of data from two other areas of speech science that provides further support for our approach to the control of tongue movement. The first area is that of speaking rate changes,[9] and the second is speech errors.

SPEAKING RATE CHANGES

Our model has proposed that the actual movements of the tongue are governed by simple dynamical equations whose parameters are set by intentions that determine the direction and the distance to be moved. The stiffness parameter which determined the rate of each movement was not changed throughout the utterance. This suggests that the velocity of the tongue movement is not a variable and will not change as a function of speaking rate. At first glance, this result seems unlikely, since we could have easily inserted a global stiffness parameter which would vary for speech at different rates. In order to test the hypothesis that such a parameter is not necessary, we asked our two speakers to say the nonsense word kooKOOK (stressed syllable in CAPS) three times each at careful, normal, and fast rates while we recorded by x-ray the position of the tongue pellets. The acoustic signal the second speaker produced, along with a plot of the tongue dorsum vertical position, is shown in Figure 7. Figure 8 shows the tongue dorsum velocity during that record. Clearly, the speaker accomplished the goal of talking faster since each succeeding group of three utterances took less time; however, a cursory glance at the velocity curve shows that velocity did not increase as a function of speaking rate. Table 2, which shows the mean peak velocities during each gesture, confirms this observation. This result is consistent with the outcome of the modeling effort, and confirms the value of that exercise. However, it leaves unanswered the question as to how the speakers were able to increase their speaking rate without increasing velocity.

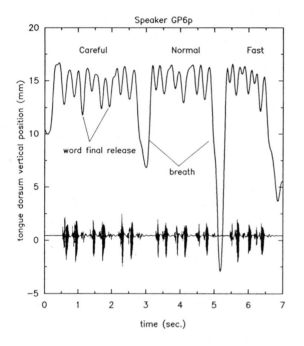

FIGURE 7 Tongue dorsum vertical movement for the nonsense word kooKOOK at three different speaking rates. The acoustic signal is superimposed.

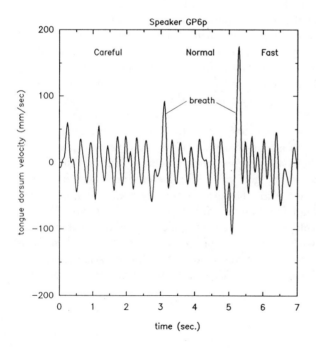

FIGURE 8 Tongue dorsum velocity for the nonsense word kooKOOK at three different speaking rates.

TABLE 2 Mean peak tongue dorsum vertical movement velocity (mm/sec) of speaker GP6p during each of the gestures in the nonsense word kooKOOK.

Speaking rate	velar release to 1st vowel	velar closure	Gesture velar release to 2nd vowel	velar closure	word final velar release
careful	36.9	33.5	35.2	34.9	48.4
normal	36.1	33.0	35.8	34.9	—
fast	21.6	23.3	45.2	41.2	—

Our modeling results do not offer any clear answers, but suggest that modifications in the intention score would be a likely place to adjust the speaking rate. The simplest such modification would be the total deletion of a gesture. An examination of Figure 7 shows that such deletions were in fact made. The release and subsequently required closure of the tongue dorsum for the final /k/ in kooKOOK occurred only during careful speech and were deleted during the first two tokens of normal and fast speech. These deletions primarily account for the shorter duration of normal when compared to careful speech. The perceptual cost of this deletion is that the final burst of acoustic energy that clearly signals the presence of the final /k/ is absent. Presumably, the closure for the final velar stop produces a sufficient acoustic effect to allow the listener to detect it, and the final release is added during careful speech only to enhance clarity. The fact that the release is present for the final word at each rate appears to reflect a subtle interaction with the breathing motion that ends each phrase. The tongue dorsum must be rapidly dropped away from the palate at the end of the phrase in order to permit an inhalation to begin the next phrase. The speakers seem to take advantage of this by using this motion to produce a final release sound, without any cost in terms of time. This emphasizes the subtlety with which the motor control of the tongue is planned.

Deletions, however, do not account for the differing duration of normal and fast phrases. This difference primarily results from a shortening of the duration of the interword and intersyllable pauses. This result fits quite nicely with the tongue motion model, in which pauses had durations that were arbitrarily set to match the actual data. Thus they provide a simple way to alter the overall rate of speech without changing the dynamics which control the motion of the tongue. However, during fast speech the speakers also used another technique to shorten overall duration—they reduced the size of the motion to the vowel for the unstressed syllable. This phenomenon does not fit well with our tongue motion model since it supposes an alteration in the target position for the unstressed, fast vowel. Such an alteration is difficult to incorporate into the model because there appears to be no compelling rationale for setting the exact location of new targets. Perhaps with careful study of many examples of these reductions, some underlying principle could be uncovered that would be consistent with the model, but at present this remains an area for further investigation.

SPEECH ERRORS

A second area that may provide insight into the fundamental validity of the model advocated here comes from an analysis of speech errors. If speech errors occur that could reasonably be traced to the supposed intention score, then the construct validity of that score would be strengthened.

What might be viewed as such an error occurred during the middle word of the three fast *kooKOOKs*. For that word, the voiced vowel sound for the first syllable (/koo/) was omitted, producing the word *k'KOOK* instead of *kooKOOK*. Figure 9 shows the acoustic signal, tracheal accelerometer record, and normalized vertical tongue positions during that phrase. The tracheal accelerometer record, shows that the glottal approximation gesture, which should have brought the vocal folds together and produced the voicing sound for the vowel, was deleted. Thus a voiceless vowel was produced. However, this deletion did not serve to shorten overall word duration, because the tongue body movement necessary for vowel production was still produced at a normal duration.

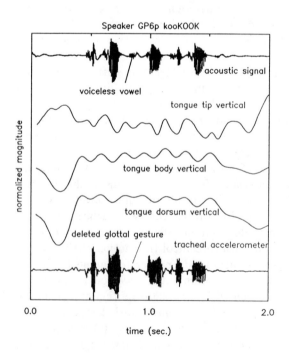

FIGURE 9
The acoustic signal, tongue vertical positions, and tracheal accelerometer record during the speech error kooKOOK→k'KOOK.

This error is consistent with the fundamental basis of our model, because it can reasonably be attributed to the simple deletion of one element of the intention score—the glottal approximation gesture. The tongue portion of the score was apparently unaltered because the tongue movement was normal for the unstressed vowel, as can be seen by comparing the tongue body position for the three words in the phrase. However, if an error is defined as an unplanned occurrence that reduces comprehension without serving another useful function, this deletion may not have been an error. Its function may actually have been to enhance clarity since it occurred during the most reduced motion of the tongue dorsum from the palate to the vowel position and, therefore, the resonance chamber in the mouth may have been incorrect for the intended vowel. Thus, rather than voicing the wrong vowel, a deletion may have beens elected in order to both simplify the motor task and avoid producing an incorrect sound. This final conjecture again illustrates that the present model is deficient in that it does not adequately account for reduced gestures.

SUMMARY

A model of the movement of the tongue was described which was based on a second-order dynamical system. Features of the model included the gestural intentions supposed by the motor theory of speech to be the distal objects of speech perception, and mechanisms for producing the simultaneous coarticulation of more than one gesture. A good agreement between model output and actual movement during natural speech was obtained. Additional experiments, taken from the manner in which speech rate changes are accomplished and from speech errors, provided support for the fundamental assumptions on which the model was based and showed that the model produced could provide a useful framework for the explanation of specific phenomena.

ACKNOWLEDGMENTS

We would like to thank both the Center for Nonlinear Studies at Los Alamos National Laboratory for providing generous facilities for this work and the Waisman Center on Mental Retardation and Human Development at the University of Wisconsin-Madison for essential help during the x-ray data collection phase of the research. The Department of Energy's Science and Engineering Research Semester Program directed by Ms. Sharon Dogruel provided financial support for Mr. Vo's participation.

REFERENCES

1. Abbs, J. H. "Invariance and Variability in Speech Production: A Distinction Between Linguistic Intent and Its Neuromotor Implementation." *Invariance & Variability in Speech Processes*, edited by J. S. Perkell and D. H. Klatt. Hillsdale, NJ: Erlbaum, 1986.

2. Browman, C. P., and L. Goldstein. "Articulatory Gestures as Phonological Units." *Phonology* **6** (1989): 201–251.

3. Browman, C. P., and L. Goldstein. "Tiers in Articulatory Phonology with Some Implications for Casual Speech." *Haskins Laboratories Status Report on Speech Research* **92** (1987): 1–30.

4. Browman, C. P., and L. Goldstein. "Toward an Articulatory Phonology." *Phonology Yearbook* **3** (1986): 219–252.

5. Best, C. T., M. Morrongiello, and R. Robson. "Perceptual Equivalence of Acoustic Cues in Speech and Nonspeech Perception." *Perception & Psychophysics* **29** (1981): 191–211.

6. Coker, C. H. "Synthesis by Rule from Articulatory Parameters." *IEEE Proc. Speech Commun. Process.* **Paper A9** (1967): 52–53.

7. Fitch, H. L., T. Halwes, D. M. Erickson, and A. M. Liberman. "Perceptual Equivalence of Two Acoustic Cues for Stop-Consonant Manner." *Perception & Psychophysics* **27** (1980): 343–350.

8. Fowler, C. A. "Segmentation of Coarticulated Speech in Perception." *Perception & Psychophysics* **36** (1984): 359–368.

9. Gay, T. M. "Mechanisms in the Control of Speech Rate." *Phonetica* **38** (1981): 148–158.

10. Gibson, J. *The Senses Considered as Perceptual Systems.* New York: Houghton Mifflin, 1966.

11. Kelso, J. A. S., E. Vatikiotis-Bateson, E. L. Saltzman, and B. Kay. "A Qualitative Dynamic Analysis of Reiterant Speech Production: Phase Portraits, Kinematics and Dynamic Modeling." *J. Acous. Soc. Am.* **77** (1985): 266-280.

12. Kelso, J. A. S., E. L. Saltzman, and B. Tuller. "The Dynamical Perspective on Speech Production: Data and Theory." *J. Phonetics* **14** (1986): 29-59.

13. Liberman, A. M., F. S. Cooper, D. P. Shankweiler, M. Studdert-Kennedy. "Perception of the Speech Code." *Psych. Rev.* **74** (1967): 431-461.

14. Liberman, A. M., and I. G. Mattingly. "The Motor Theory of Speech Perception Revised." *Cognition* **21** (1985): 1-36.

15. Maeda, S. "An Articulatory Model of the Tongue Based on a Statistical Analysis." *J. Acoust. Soc. Am.* **65** (1979): S22.

16. Perkins, W. H., and R. D. Kent. *The Functional Anatomy of Speech and Hearing.* Boston, MA: Little Brown, 1987.

17. Spoehr, K. T., and W. J. Corin. "The Stimulus Suffix Effect as a Memory Coding Phenomenon." *Memory & Cognition* **6** (1978): 538-539.

Leif H. Finkel† and Gerald M. Edelman‡
†Department of Bioengineering and Institute of Neurological Sciences, University of Pennsylvania, Philadelphia, PA 19104-6392 and ‡The Neurosciences Institute and The Rockefeller University, 1230 York Avenue, New York, NY 10021

Models of Somatotopic Map Organization

Five recent theoretical models of somatotopic map organization are reviewed. The models are evaluated with regard to their ability to account for a range of experimental findings on receptive field plasticity and dynamic map organization in somatosensory cortex. The models of Kohonen and of Schulten and Ritter, which are analytic in nature, are contrasted with those of Finkel, Edelman, and Pearson, of Miller and Stryker, and of Grajski and Merzenich, which all feature extensive computer simulations.

INTRODUCTION

Topographic neural maps have long been of interest to anatomists and physiologists as organized input/output structures. In such maps, the activation of each neuron can be thought of as representing a particular point in a feature space or stimulus dimension (e.g., auditory frequency) and there is a regular, ordered, and locally continuous mapping between the feature space and its neural representation. The map need not be, and indeed is usually not, globally continuous. Such maps usually display transformations, magnifications, and distortions in addition to redundant or truncated representations. Topographic maps have been characterized at many levels of the nervous system, from the spinal cord to the frontal cortex.

Recently, two observations have attracted the attention of neural modelers to topographic maps. First, the neurocomputational power of maps has come to be appreciated, and it is now generally thought that maps provide the basis for large-scale neural processing. Second, a series of experimental observations have shown that, in a number of brain regions, the functional organization of maps is *dynamic* and undergoes adaptive changes as a function of ongoing input stimulation. It thus becomes important to understand the rules underlying topographic map organization and to try to relate these rules to known biophysical, synaptic, anatomical, and physiological processes.

We will review several recent models of map organization. These models consider only one aspect of the mapping problem, namely, what determines the local organization of the map. This, in turn, implies discussing the basis for three mapping properties: (1) local continuity, (2) size of receptive fields (how much of feature space does each neuron survey), and (3) local magnification factor (how much map surface is devoted to a given area of feature space). Furthermore, our survey will be limited to models concerned with the somatosensory cortex. Recent experimental observations on maps in areas 3b and 1 of the monkey somatosensory cortex have revealed a host of dynamic phenomena. Although similar map changes have been observed in other somatosensory cortical areas, in subcortical stations of the somatosensory system, and in motor cortex, the work of Merzenich, Kaas, and their colleagues provides the richest source of modeling data.[13,15]

We will not review the experimental results here as they have been extensively and thoughtfully reviewed elsewhere.[1,15] Suffice it to say that areas 3b and 1 of monkey somatosensory cortex contain topographic maps of the body surface. In normal animals, there is a great deal of variability between the maps found in different individuals (for example, in the area devoted to representing the thumb), and there are also significant changes in any one individual's map over time. These temporal changes involve shifts in the locations of map borders with resulting changes of "what" is represented "where" in the cortex. A series of perturbation experiments[13,15] explored changes in the maps resulting from stimulation of small regions of the hand, cutting the median nerve, amputation of one or more digits, joining two fingers to produce a syndactyly, and various other procedures which change the probability or correlation of input from the skin. The general result of all such experiments was that the map readjusted, often within hours but usually over days and weeks, to reflect the new input conditions. In the readjustment there are several major rules which appear to be followed: (1) except at representational discontinuities, the receptive fields of nearby cells overlap considerably and the amount of overlap falls off monotonically with distance between the cells; (2) there is a distance limit of roughly 800 microns (probably determined by anatomical constraints) beyond which a de-innervated representation cannot spread; (3) there is an *inverse* relationship between the size of the representation and the size of the receptive fields of cells in that region; and (4) as map borders shift, they remain sharp.

It is worth pointing out that a large body of experiments and models have been devoted to the related problem of how organized maps arise developmentally.

Many of the mapping principles that we will discuss have been observed in the developmental situation, particularly in the well-studied projection of the retina onto the optic tectum.[2] There may even, indeed, be common mechanisms governing map organization in the developmental, regenerative, and normal adult adaptive state. In fact, map plasticity observed in the rodent whisker barrel system[23] may well be such a case. Another is the ocular dominance columns studied by many authors.[16]

The models which we review differ markedly in their biological realism and in their mathematical tractability. Despite these differences, several common themes will emerge that appear to be general properties of any biologically based mapping model. The critical components appear to be the network anatomy, the choice of synaptic modification rule, and the role of local groups of neurons in determining map function. Each of the models, however, makes use of these components in a unique fashion.

THE KOHONEN MODEL

A number of early neurocomputational models were devoted to the question of how cortical cells develop selectivities, such as the selectivity of visual cortical cells for orientation or ocular dominance.[20,22] However, the related but more complicated question of how organized maps of such selective cells arise was not addressed until Kohonen's pioneering paper.[14] In such a map, one neural surface (often a peripheral receptor sheet such as the skin or the retina) is mapped onto another neural surface (occasionally via several intermediate maps) with the preservation of topographic relationships—i.e., activation of nearby units in the peripheral sheet leads to activation of nearby units in the map. Kohonen presented his arguments in the general framework of maps of abstract features, maps of "patterns of activity" rather than just of points or local areas of input. And in later papers he developed his algorithm to handle more cognitive tasks such as speech recognition. But for our purposes, the most important aspects of the paper have to do with the organization of an initially unstructured neural map.

Kohonen proposes a simple algorithm to carry out map organization. It consists of four steps:

1. An array of units (usually 2-D) receives a set of inputs from a receptor sheet and each unit computes an output based on simple nonlinear summation processes.
2. There is a mechanism, whose exact implementation is left unstated, which compares the outputs of all units in the array, and selects the unit with the maximum output.
3. Certain kinds of local interactions lead to the concurrent activation of the selected unit *and* its nearest neighbors in the array.
4. An adaptive process (involving synaptic plasticity) then operates to increase the output of the selected units to the present input.

More formally, the inputs are given by a vector

$$\mathbf{x} = [x_1, x_2, \ldots, x_n]^T \tag{1}$$

where T represents the matrix transpose, and the synaptic weights of unit i are given by

$$\mathbf{m_i} = [m_{i1}, m_{i2}, \ldots, m_{in}]^T. \tag{2}$$

The output of unit i is then given by

$$\mathbf{s_i} = \mathbf{m_i^T x}. \tag{3}$$

And the adaptive process (step 4, above) of increasing the response of a selected unit, say unit i, is achieved by rotating the weighting vector (2) towards the input vector (1)

$$\mathbf{m_i}(t+1) = \frac{\mathbf{m_i}(t) + a\mathbf{x}(t)}{\|\, \mathbf{m_i}(t) + a\mathbf{x}(t)\, \|_E} \tag{4}$$

where E denotes the Euclidean norm. Kohonen points out that this process is a form of *unsupervised* "learning" since the direction of the rotation is always the same (towards the input) regardless of the consequences of the process on the behavior of the system.

It is shown in several simulations of two-dimensional maps, that starting from a random initial condition (i.e., the weighting vectors $\mathbf{m_i}$ randomly assigned), a global organization emerges after several thousand iterations of the adaptive process. Of course, the adaptive process specified by Eq. (4) is an intrinsically local process, affecting only a unit and its eight nearest neighbors. However, mutually incompatible changes produced at neighboring locations will tend to cancel each other out; thus, only sets of changes with long-distance coherence will survive. Kohonen also observes that global organization usually appears to propagate inward from the boundaries of the network. This follows from the fact that the boundaries have less neighbors to satisfy, and thus attain an appropriate global organization before the interior of a map can do so.

In an attempt to embed a degree of neural structure in the model, Kohonen considers a network composed of local excitatory interactions and long-distance inhibitory interactions. In such a network, as others before him also found, there is a natural tendency of the network to develop clusters of small groups of activity of roughly the same dimensions as the spread of local excitatory interactions. In a series of simulations and analytical studies, he argues that once these groups have formed (by whatever means), they then undergo the same adaptive process as was discussed above for selected units.

Most interestingly, Kohonen investigates the case of a synaptic sum rule in which the total synaptic strength of a cell is maintained constant (through normalization). One version of this rule is given by

$$m_{ij}(t+1) = \frac{m_{ij}(t) + as_i(t)(x_j - x_b)}{\left[\sum_{j=1}^n m_{ij}(t) + as_i(t)(x_j - x_b)\right]^{1/2}} \tag{5}$$

where x is the presynaptic input, x_b is an effective background value of the input, and s is the postsynaptic triggering frequency. Equation 5 is, essentially, a normalized modified Hebb rule.[8]

Given this synaptic sum rule embedded in a network with activity clusters already formed, Kohonen makes the important discovery that the network obeys a magnification rule. Namely, as the frequency of activation of a particular input region is increased, the fraction of the map devoted to the stimulated region is proportionately increased. He does not comment on the related physiological property, that as map representation increases, receptive field size decreases. We will consider this subject in detail below.

Finally, Kohonen makes a series of observations on various types of pathological mapping phenomena that occasionally occur, for example, the collapse of a two-dimensional map into one or more one-dimensional maps. These aberrations usually follow from inconsistencies in the range of the local excitatory and inhibitory interactions.

Kohonen's model does not incorporate much anatomical or physiological detail. He does make some speculations regarding the possible roles of various cortical cell types (chandelier cells, bipolar cells, etc.) which he proposes may play roles analogous to basis functions in determining the proper range of synaptic interactions. Overall, however, the argument is a mathematical one. Nonetheless, Kohonen clearly sets up the problem of how map order can be imposed by structured inputs, he anticipates the importance of local groups of cells, and he makes a series of remarkable observations about the major properties of topographically organized maps.

SCHULTEN'S VERSION OF THE KOHONEN MODEL

Although we will not review their work in detail here, it is important to point out that Schulten, Ritter, and coworkers have taken the basic Kohonen model and explicitly applied it to the case of the somatosensory system. They consider the phenomenology of map plasticity in monkey somatosensory cortex as elucidated by the work of Merzenich and others. Schulten's results show that an organized map of the hand of a monkey can be formed using a formalized version of Kohonen's rules. They also show a simulation in which one of the "fingers" is amputated. In this case, as is found experimentally, borders in the simulated map move so as to occupy the zone of representation of the missing digit.[17]

Schulten and Ritter use analytic techniques to examine the stationary state of the Kohonen mapping with regard to the problem of the magnification factor. In contrast to Kohonen's arguments, they find that in the general case of the two-dimensional map, the magnification factor (i.e., the amount of map devoted to a given peripheral region) *cannot* be expressed as any simple function of the frequency of stimulus input. Furthermore, in the case of one-dimensional maps, the magnification factor appears to depend on the stimulus frequency raised to the 2/3 power, rather than on the linear relationship found by Kohonen.

These differences point out the importance of the exact assumptions made by each modeler. Moreover, they illustrate that analytical techniques, useful and powerful as they are, very quickly reach the limit of their applicability in dealing with even such simplified versions of complex systems.

THE SM (SOMATOSENSORY MAP) MODEL OF FINKEL, EDELMAN, AND PEARSON

Finkel, Edelman, and Pearson have developed a model of topographic map organization in somatosensory cortex which attempts to account for the major experimental results obtained over the last few years by Merzenich, Kaas, and their colleagues. This SM (somatosensory map) model is biologically based, i.e., it incorporates the basic elements of cortical anatomy and physiology, and it features a realistic synaptic modification rule that formally resembles the mechanism of operation of the NMDA (N-methyl-D-aspartate) receptor. In addition to the question of the origin of topography, considerable attention is paid to receptive field properties, and these properties are directly measured in the model. Unlike previous models, the emphasis of the treatment is on simulation, rather than analytical techniques.

The SM model[24,6,19] is based on the theory of neuronal group selection, a general theory of neural function developed by Edelman.[25,4] Neuronal group selection proposes that the nervous system operates as a selective system, in a manner somewhat analogous to the operation of natural selection during evolution. However, in the nervous system, the units of selection are small groups of strongly interconnected neurons, and the mechanism of selection involves changes in synaptic efficacies (rather than differential changes in gene frequencies). In the SM model, neuronal groups play the pivotal role, acting as the basic units of organization of the map. The model stands or falls on the central premise that topography and receptive field organization arise from collective and cooperative interactions among groups of neurons.

The SM network consists of 1512 cells, both excitatory and inhibitory, which make extensive local interconnections. Each cell also receives a set of extrinsic inputs from a sheet of 1024 receptors that represents the hand of the monkey. There is a mirror-symmetric projection to the network from receptors on the front (glabrous surface) and back (dorsal surface) of the hand, arranged such that, the front and back of a given digit both project to the same region of the network. This double projection, together with the fact that each receptor contacts units over roughly 10% of the network leads to the *anatomical* possibility of a given point on the hand being represented over a wide range of the network. In other words, many alternative *physiological* maps (those obtained by microelectrode recording of receptive fields) are possible given the anatomical substructure. Note that the extrinsic projections are *assumed* to be topographically ordered at the start of the simulation (i.e., the developmental origins of topography are *not* studied). The problem of concern, then, is to determine the rules governing the organization and reorganization of the observed physiological map.

While the extrinsic projections are *anatomically* ordered, their initial synaptic weightings are random, as are those of the intrinsic connections. The simulations commence with a protocol for stimulating the receptor sheet. Small (3 × 3 receptor) regions of the hand are stimulated in random sequences for several hundred to several thousand iterations (each location on the hand is stimulated approximately 10 times). What inevitably follows, regardless of the size, shape, frequency, or pattern of stimulation (within rather large limits) is the emergence of neuronal groups. These are local collections of cells which have greatly increased the strengths of their connections to each other and have weakened their connections to cells not in the group. The analysis of how groups emerge and what accounts for their stability properties has been given elsewhere.[19] The main point, however, is that group formation arises from an interplay between the underlying anatomy, the physiological properties of the modeled cells, and the voltage-dependent properties of the synaptic modification rule. Once groups form they are stable, although they undergo a continual process of modification in which cells on the borders of adjacent groups can be traded back and forth.

The critical property of groups for the SM model is their cooperativity—the strong connections between the cells of a group lead to similar firing patterns among them, which in turn, leads to the development of highly overlapped receptive fields. Thus, in a sense, the receptive field becomes a collective property of local group of cells, rather than the intrinsic property of individual units. (In this sense, it resembles the concept of the segregate as proposed by Whitsel and colleagues[21]). Groups constantly modify the location and size of their receptive fields as a result of inter-group competition. The source of this competition is the amount and character of activation received by the group. The main aim of the SM model, then, is to examine how receptive field structure and map organization depend upon the characteristics of input stimulation to a network organized into neuronal groups.

The simulations show that if the receptor sheet is stimulated in different patterns, the resulting maps differ in detail; however, one always obtains a topographic representation of the hand which qualitatively resembles that found in cortical areas 3b and 1 of the monkey. The map is always topographic with discrete areas of representation of the front and back of the hand—it turns out that the borders between these regions correspond to borders between neuronal groups.

Perturbations in the input received by the network result in shifts in map borders that closely correspond to those observed *in vivo*. For example, when the probability of stimulation of a particular hand region was increased, the size of the representation of that region increased (by an order of magnitude). This expansion was at the expense of those regions that were formerly represented in adjacent areas of the network. Conversely, when stimulation of a hand region was decreased by cutting the median nerve (which subserves innervation of the median half of the front of the hand), the representation of that hand region disappeared. In its place there emerged an intact, topographic representation of the corresponding back of the hand (subserved by the ulnar nerve), as would be expected from the underlying anatomy.

Neuronal groups play a critical role in these shifts in map borders, and in fact, there is a curious characteristic of the shifts which argues most strongly for the necessity of groups. It has been observed experimentally, that as map borders shift, they always remain sharp. In other words, one does not see a "defocusing" of a representation followed by a resharpening; rather, the borders shift, sometimes by prodigious amounts, but at all times there is a relatively crisp border between, for example, the front and back of digit 1. This "shift with sharpness" is difficult to explain with most models of map organization because they must pass through a state in which connection strengths are partially strengthened and weakened (with concomitant changes in the receptive fields). In the SM model, the sharpness of the physiologically observed map borders is due to the fact that they occur at the borders between neuronal groups. Since group borders are one cell thick, the transitions between receptive fields will always be crisp. Shifts in map borders will occur as groups change the size and location of their receptive field locations on the skin, but at all times, the transitions will be sharp. It is the cooperative nature of the intra-group dynamics coupled with the competitive processes operating between groups that then accounts for the observed map properties.

The existence of groups also provides a natural explanation for both the magnification rule (size of representation is directly proportional to input stimulation) and the inverse rule (size of representation is inversely proportional to receptive field size). The argument depends upon two assumptions[24]: (1) that the anatomical size of a group changes slowly relative to changes in the size of its receptive field, and (2) that the relative overlap of receptive fields of adjacent groups remains fairly constant, regardless of actual receptive field size, due to the nature of group-group competition. Given these assumptions, which have been repeatedly borne out in simulations, the size of a representation in cortex will depend upon the *number* of groups with receptive fields located in the area of interest. Since the overlap between receptive fields of adjacent groups is constant (e.g. 30%), the smaller the receptive fields, the more groups required to span the space, and the larger the corresponding representation.

In a number of simulations,[18] we, in fact, observed the correct workings of both the magnification and inverse rules. However, for most of our mapping simulations, receptive field sizes were driven to their minimal limits, i.e., they were only 1-receptor wide. This was done to allow the finest scale map possible with limited resources. The consequence of this action was that receptive fields could not decrease further in size, making it impossible fully to test the above rules. Grajsky and Merzenich have subsequently investigated the operations of these rules in a similar set of simulations, and we will discuss their findings below.

The SM model represents an attempt to simulate the emergence of a number of rules of map organization in a realistic, biologically based simulation. At its core is the idea that interactions among neuronal groups control the ongoing organization of the observed physiological map from the manifold of possible anatomical maps. The model also makes a number of experimental predictions, some of which have already been confirmed.[19]

THE MILLER-STRYKER MODEL

Miller and Stryker have recently proposed a model whose main subject is the formation of ocular dominance columns in the visual cortex. As discussed above, this places their model in the domain of models of the *development* of cell selectivity. However, we consider their work here because they claim that their model also accounts for the emergence of organized somatotopic maps, and because they present a criticism of the SM model which merits discussion.

The model proposes that three anatomically-physiologically based functions together determine the segregation of the developing cortex into discrete regions (ocular dominance columns): (1) An *arbor function*, which defines the anatomical spread and synaptic strengths of the connections made by a thalamic afferent onto cortical cells, (2) an *afferent correlation function*, which defines the averaged correlation between sets of afferents different distances apart (this function is determined by the types of inputs received), and (3) a *cortical interaction function*, which defines the effect of one active synapse (on one particular cell) upon another active synapse (on another cell) as a function of cortical distance. The cortical interaction function lumps together the net effects of all direct and polysynaptic interactions between two given cortical cells.

The basic equation of the model then describes the change in the synaptic strength of thalamo-cortical afferents. The change in synaptic strength, S, of left eye-dominated thalamic afferents originating at thalamic location x' in the thalamus onto a cortical cell at location x in the cortex is given by

$$
\begin{aligned}
\frac{dS^L(x, x', t)}{dt} = & \lambda A(x - x') \sum_{y,y'} I(x - y) \big[C^{LL}(x' - y') S^L(y, y', t) \\
& + C^{LR}(x' - y') S^R(y, y', t) \big] - \text{DECAY}^L(x, x', t) ,
\end{aligned}
\tag{6}
$$

where x, y are cortical locations, and x', y' are topographically related thalamic locations. In this equation, $A(x - x')$ is the arbor function, $I(x - y)$ is the cortical interaction function, and $C(x' - y')$ are afferent correlation functions for Left-Left and Left-Right afferent combinations. (λ is a constant and the DECAY term is a local decay in synaptic strength). There is a symmetrical equation for S^R [We have slightly modified Miller and Stryker's notation to aid consistency with the other equations in this paper].

The majority of effort is then expended in analyzing, with both analytical techniques and simulations, the properties of this equation for various functional forms of A, I, and C. The LGN, and two-cortical layers (Left and Right) are modeled as three 25×25 arrays of units with periodic boundary conditions. The arbor functions chosen are nearly identical to those defined in the SM model above. Cortical interaction functions are generally assumed to be difference-of-gaussian-type functions (although several functional forms are investigated). The correlation functions used reflect the various stimulus perturbation protocols developed by Hubel and Wiesel and others in cat and monkey visual cortex (monocular deprivation, binocular deprivation, etc.).

The model is very successful in producing ocular dominance columns that match those found experimentally. Of course, there have been a host of other successful models in this domain-periodic solutions are a well-known property of the type of first-order differential equation given above. Moreover, while great attention is paid to the eigenfunction-based origin of the periodicity of such columns, there is almost no discussion of why the columns are "columns," i.e., why one obtains elongated regions of similar eye dominance as opposed to blobs or other possible shapes.

The ultimate usefulness of the model, then, must rest upon the claim that the three defined functions, A, I, and C, are the critical determinants of cortical organization. We consider this issue below in the section where the comparison to a SM model is made.

THE GRAJSKI-MERZENICH MODEL

Grajski and Merzenich have developed a simulation of somatotopic map organization which addresses the problem of the inverse rule.[10] Their network is very similar to the SM model, with excitatory and inhibitory cells, diverging afferents from the skin, and extensive local cortico-cortical connections. Altogether, 900 cells and 57,600 synapses are simulated, and the skin is represented by a 15×15 array of units divided into three "fingers." An important innovation is the inclusion of a subcortical layer of relay cells which both project to the cortical network and receive connections back from the cortical network. Such multi-layer networks with reentrant connections[9] offer the possibility of richer network dynamics. In fact, however, Grajski and Merzenich found that they could obtain similar results with a reduced two-layer network.

The model makes use of two types of normalization. First, the effect of each synaptic input on a cell is normalized by the total number of inputs of that class (i.e., from the same type of cell in the same network). Second, and more importantly, the total synaptic strength of each cell is normalized to a constant value. Synaptic normalization has been invoked in a number of previous models (for example, by Kohonen), and has some biological plausibility in that a cell must have a finite amount of resources that may be allocated among its synapses. However, in the present context, synaptic normalization effectively assumes the answer in that strengthening of one set of connections to a cell obligately weakens all other sets. To see this argument, we must consider the actual simulations carried out.

Grajski and Merzenich start with a network which has been topographically refined through several cycles of stimulation (interestingly, they require 10–12 passes over the skin to reach a steady state which is identical to that required by the SM model). The resulting map shows the required properties that receptive fields overlap considerably (up to 70%), receptive field overlap falls off monotonically with distance, and cortical magnification is proportional to the frequency of stimulation. A simulation was then performed in which the probability of stimulation to a small region of the skin was increased (by a factor of 5). As in the SM model, this results in an expanded cortical representation. In addition, Grajski and Merzenich found a smaller, but noticeable expansion in the subcortical representation. Most importantly, the cortical receptive fields located on the stimulated region were seen, on average, to decrease in size. Interestingly, subcortical receptive fields remained unchanged in size.

The converse experiment was also simulated. A region of cortex is "lesioned," i.e., the cells representing a given skin region are removed. The authors found that, under these conditions, with the Hebbian synaptic rule used, the intact representation is firmly entrenched. Thus, all cortico-cortical and cortical afferent connections are randomized and cortical excitation is artificially enhanced. Then, a new representation is seen to emerge in which receptive field sizes are markedly larger in the cortex and smaller in the subcortex.

Thus, the simulation appears generally to obey the inverse rule—although there are distinct differences in the behavior of the cortical layer and the subcortical layer. Moreover, there is no quantitative or qualitative relationship shown here; rather, just the sign of the changes are in the appropriate directions. Finally, the issue of synaptic normalization can be considered by the following argument. Under the action of a Hebbian rule, increasing the stimulation to a region of skin will cause strengthening of the afferent connections from that region. Cells which previously received only subthreshold inputs from the stimulated region will, after synaptic strengthening, include this region in their receptive fields. This accounts for the expansion of the representation (note that Kohonen achieved such a result using normalization). However, cells which previously had their receptive fields in the stimulated region will also undergo synaptic strengthening. If these cells have their synaptic strengths renormalized after the stimulation, then all connections from non-stimulated regions will be automatically weakened. This will cause certain previously effective connections to become subthreshold, and will result in a *decrease* in receptive field area. Thus, synaptic renormalization, *by itself*, implements the inverse magnification rule under a Hebbian rule.

The Grajski-Merzenich model makes an important contribution in considering what processes are sufficient to account for the inverse rule. In addition, they make a number of experimental predictions about subcortical changes which are non-intuitive. As we discuss below, such predictions are perhaps the most important results of any theoretical model.

A COMPARISON OF THE MODELS

The five models considered here differ in the emphasis placed on analysis versus simulation and in the amount of biological realism they capture. There is also an important historical path, with later models clearly building upon their predecessors. For our purposes, which are related to the biological details, the most instructive comparisons can be drawn between the three most recent and most biological models, the SM model, and those of Miller & Stryker and Grajski & Merzenich.

As Miller and Stryker point out, their model is nearly identical to the SM model for the case of "same-eye" correlations (the usual case in which correlations of inputs in the same eye are higher than those between eyes) and in which the correlations fall off at about 1-arbor radius. Their criticisms of the SM model, then, devolve around the role of groups in organizing cortical maps.

Miller and Stryker claim that the SM model predicts topographic disorder within maps on the scale of the size of a group. As mentioned above, in cortical areas 3b and 1, there are discrete patches of representation of the dorsal and glabrous skin of the hand. The SM model explicitly suggests that these patches are composed of *multiple* groups, not single groups as Miller and Stryker argue. In fact, one prediction of the SM model is that the width of such patches might be observed to fall in discrete steps as additional groups are added. It has been clearly stated[24,6,19] that groups are on the order of 50 to 100 microns in diameter, whereas dorsal/glabrous patches typically have dimensions of hundreds to thousands of microns. The types of cooperative phenomena which cause adjacent groups to develop similar receptive field domains (dorsal versus glabrous) were not explicitly investigated in the SM simulations—this was because incorporation of the long distance cortical connections needed to effect such interactions would have made the simulations computationally prohibitive. However, from a theoretical point of view, it seems clear that such local clustering effects would occur on the group level just as they do at every other level of such systems.

The question of group scale also arises in Miller and Stryker's claim that receptive field location should shift in an orderly and uniform manner across the cortex, not in the discrete fashion predicted by the SM model. Again, they are correct on the macroscopic scale—however, there are no experimental results on the scatter in receptive field locations of cells located within 50 microns of each other. Furthermore, receptive field clustering was deliberately maximized in the SM model,[19] in order to investigate the effects of such shifts. It is trivial to arrange for any desired degree of receptive field shifts within a group by varying the synaptic parameters.

In this regard, there is a critical difference in the synaptic modification rules used by the two models. As with all correlation models of synaptic modification since Hebb,[12] both models increase synaptic strength when presynaptic inputs are correlated with postsynaptic activity. The conundrum is always how to *decrease* synaptic strength. To oversimplify somewhat, we may say that in the SM model synaptic weakening occurs when there is postsynaptic activity but no presynaptic activity. In Miller and Stryker's model, synaptic weakening occurs when there is presynaptic activity but no postsynaptic activity. Thus, each model has chosen a different case of non-correlation between pre- and post-synaptic activity. Miller and Stryker make their choice based on reported experimental results using muscimol, a GABA agonist which inhibits cortical cells from firing, and leads to ocular dominance columns in which the less active (occluded) eye is dominant. The choice in the SM model has other sources of experimental support.[7] In fact, in a comprehensive study of synaptic modification mechanisms, Finkel and Edelman[8] point out that there are large families of related synaptic rules, and most likely, a number of different rules operate concurrently in various brain regions. Although many of these rules will generate similar topographic structures, as indeed both of the models discussed here demonstrate, particular rules show strengths and weaknesses in the context of the anatomy of the network in which they operate.

This raises perhaps the most important difference between the models. Whereas the SM model simulates individual excitatory and inhibitory cells, in the Miller-Stryker model there is only one type of unit, and all cortical interactions depend upon the cortical interaction function—i.e., the anatomy is replaced by an averaged, effective function. This simplification prevents the model from considering the effects of intracortical plasticity—for that would require changes in the function for each individual cell. However, without intracortical plasticity, one is only looking at half of the problem (actually far less than half since the overwhelming majority of connections received by most cortical cells come from other cortical cells). In addition, such structures as groups could never emerge, by definition, without considering changes in cortical connection strengths. The introduction of cortical plasticity would greatly complicate the Miller-Stryker model, and it remains to be shown whether, under such circumstances, their choice of synaptic rule would still lead to column segregation.

The Grajski-Merzenich model shares with the Miller-Stryker model the implicit goal of identifying a few, simple neural processes that can account for a range of experimental results. One pitfall in this approach is that the range of simulations is usually so narrow (one synaptic rule, one network architecture, a small number of stimulation protocols, etc.) that alternative explanations cannot be ruled out. Like the SM model, however, the Grajski-Merzenich model actually simulates a network, and allows the investigator to examine such things as receptive fields and patterns of excitation. The simplifications and abstractions made in the Miller-Stryker model allow more analysis but at the expense of such simulation detail.

Whether or not neuronal groups exist is ultimately an empirical question which must be answered experimentally. Recent findings[11] suggest that they do exist. It is of independent interest that such structures tend to spontaneously emerge in networks. In the models considered here, Kohonen and the SM model used groups (or group-like structures), while Miller-Stryker and Grajski-Merzenich did not. It is clear from the results that topographic organization can emerge without groups, requiring only a structured (correlated) input, a reasonable anatomy, and a Hebbian-type synaptic rule. Groups appear to be necessary for generating map shifts that retain sharpness. Finally, while the inverse rule has been shown to operate independently of groups under conditions of synaptic normalization, it remains to be demonstrated whether groups are necessary for the inverse rule if one does not assume synaptic normalization.

CONCLUSION

The utility of any model rests in its ability to account for the greatest range of observations with the smallest number of assumptions, and in its ability to generate new, non-intuitive predictions which can be tested experimentally and used to improve (or falsify) the model. It must be frankly admitted that, to date, the number and quality of experimental predictions generated by neural models has not been overwhelming. This may be due to the way in which such models are used. Usually, investigators abandon their simulations after accounting for some portion of presently known results. However, this should be the beginning of the modeling, rather than its end. Detailed simulations can serve as experimental laboratories, and should be used to carry out investigative forays into new procedures and the possibilities of designing experimental protocols. In part, this has not been done because of the special purpose nature of many of these simulations—carrying out a new experiment requires extensive amounts of re-programming. The emergence of more general purpose simulators will undoubtedly aid in this ongoing effort.

ACKNOWLEDGEMENTS

Leif Finkel would like to thank the following agencies for support of his research: The Whitaker Foundation, the Office of Naval Research, the McDonnell-Pew Program in Cognitive Neuroscience, and the University of Pennsylvania Research Foundation.

REFERENCES

1. Allard, T., and M. M. Merzenich. "Some Basic Organizational Features of the Somatosensory Nervous System." In *Connectionist Modeling and Brain Function: The Developing Interface*, edited by S. J. Hanson and C. R. Olsen. Cambridge: MIT Press, 1990.
2. Cowan, W. M., and R. K. Hunt. "The Development of the Retinotectal Projection: An Overview." In *Molecular Basis of Neural Development*, edited by G. M. Edelman, W. E. Gall, and W. M. Cowan, 389–428. New York: Wiley, 1985.

3. Edelman, G. M. "Group Selection and Phasic Reentrant Signalling: A Theory of Higher Brain Function." In *The Mindful Brain*, edited by G. M. Edelman and V. B. Mountcastle, 51–100. Cambridge: MIT Press, 1978.

4. Edelman, G. M. *Neural Darwinism*. New York: Basic Books, 1987.

5. Edelman, G. M., and L. H. Finkel. "Neuronal Group Selection in the Cerebral Cortex." In *Dynamic Aspects of Neocortical Function*, edited by G. M. Edelman, W. E. Gall, and W. M. Cowan, 653–695. New York: Wiley, 1984.

6. Finkel, L. H. "A Model of Receptive Field Plasticity and Topographic Map Reorganization in the Somatosensory Cortex." In *Connectionist Modeling and Brain Function: The Developing Interface*, edited by S. J. Hanson and C. R. Olsen, 164–192. Cambridge: MIT Press, 1990.

7. Finkel, L. H., and G. M. Edelman. "Interaction of Synaptic Modification Rules Within Populations of Neurons." *Proc. Natl. Acad. Sci. USA* **82** (1985): 1291–1295.

8. Finkel L. H., and G. M. Edelman. "Population Rules for Synapses in Networks." In *Synaptic Function*, edited by G. M. Edelman, W. E. Gall, and W. M. Cowan, 711–757. New York: Wiley, 1987.

9. Finkel, L. H., and G. M. Edelman. "Integration of Distributed Cortical Systems by Reentry: A Computer Simulation of Interactive Functionally Segregated Visual Areas." *J. Neuroscience* **9** (1989): 3188–3208.

10. Grajski, K. A., and M. M. Merzenich. "Hebb-Type Dynamics is Sufficient to Account for the Inverse Magnification Rule in Cortical Somatotopy." *Neural Computation* **2** (1990): 71–84.

11. Gray, C. M., and W. Singer. "Neuronal Oscillations in Orientation Columns of Cat Visual Cortex." *Proc. Natl. Acad. Sci. USA* **86** (1989): 1698–1702.

12. Hebb, D. O. *The Organization of Behavior*. New York: Wiley, 1949.

13. Kaas, J. H., M. M. Merzenich, and H. P. Killackey. "The Reorganization of Somatosensory Cortex Following Peripheral Nerve Damage in Adult and Developing Mammals." *Annual Review Neuroscience* **6** (1983): 325–356.

14. Kohonen, T. "Self-Organized Formation of Topologically Correct Feature Maps." *Biological Cybernetics* **43** (1982): 59-69.

15. Merzenich, M. M., G. Recanzone, W. M. Jenkins, T. T. Allard, and R. J. Nudo. "Cortical Representational Plasticity." In *Neurobiology of Neocortex*, edited by P. Rakic and W. Singer, 41–67. New York: Wiley, 1988.

16. Miller, K. D., and M. P. Stryker. "The Development of Ocular Dominance Columns: Mechanisms and Models." In *Connectionism in Neuroscience*, edited by C. Olsen and S. Hansen, 255–350. New York: Springer, 1990.

17. Ritter, H., and K. Schulten. "On the Stationary State of Kohonen's Self-Organizing Sensory Mapping." *Biological Cybernetics* **54** (1986): 99–106.

18. Pearson, J. C., L. H. Finkel, and G. M. Edelman. Unpublished observations.

19. Pearson, J. C., L. H. Finkel, and G. M. Edelman. "Plasticity in the Organization of Adult Cerebral Cortical Maps: A Computer Simulation Based on Neuronal Group Selection." *J. Neuroscience* **7** (1987): 4209–4223.

20. Takeuchi, A., and S. Amari. "Formation of Topographic Maps and Columnar Microstructures in Nerve Fields." *Biological Cybernetics* **35** (1979): 63–72.

21. Whitsel, B. L., O. Favorov, M. Tommerdahl, M. Diamond, S. Juliano, and D. Kelly. "Dynamic Processes Govern the Somatosensory Cortical Response to Natural Stimulation." In *Organization of Sensory Processing*, edited by J. S. Lund, 84–116. New York: Oxford University Press, 1988.

22. Willshaw, D. J., and C. von der Malsberg. "How Patterned Neural Connections Can Be Set Up by Self-Organization." *Proc. Roy. Soc. (Lond.)* **B194** (1976): 431–445.

23. Woolsey, T. A., and H. van der Loos. "The Structural Organization of Layer IV in the Somatosensory Region (SI) of Mouse Cerebral Cortex." *Brain Research* **17** (1970): 205–242.

Richard E. Michod
Department of Ecology and Evolutionary Biology, University of Arizona, Tucson, AZ 85721;
(602) 621-7517; Bitnet: michod@arizrvax

Sex and Evolution

CONTENTS

PARADIGMS FOR NATURAL SELECTION

INTRODUCTION

Darwin[16] deduced the process of natural selection from three generalizations about nature: (i) that organisms undergo a "struggle" to survive and reproduce and that (ii) variation and (iii) heritability exist in traits which affect an organism's ability to successfully compete in this "struggle." Darwin argued that biological systems which satisfy these conditions continually organize themselves by adapting to their environment.

There are two aspects to these conditions that are not generally realized. First, in addition to applying to the emergence of order (design) in biological systems, these conditions apply to the emergence of ordered phenomena in nonliving systems such as lasers, the dynamics of which obey Darwin's propositions.[5] Second, and perhaps more important for the biologist, is that these conditions provide for a rich spectrum of dynamical behavior which the process of natural selection can exhibit. In this section, I use this dynamical behavior as a means of classifying natural selection. I argue that different dynamics of natural selection give rise to different characterizations of the process of evolution, which in turn give rise to different paradigms for the evolutionary process. The adaptationist program will be seen as one of the several possible paradigms.

This dynamical approach should not be seen as a substitute for the usual method of classification of natural selection based on the hierarchy of living systems.[31] The hierarchical approach is based on the levels at which the natural selection process acts, for example, gene, organism, kin group, group, deme, and species. Rather, the dynamical approach distinguishes qualitative features of the evolutionary process associated with the different dynamics of natural selection. This approach was first used by Eigen[18] and later by Eigen and Schuster.[19] It was subsequently developed and applied to a variety of problems in evolutionary biology, including the origin of adaptation[5] and protein-mediated replication[38] in simple systems of molecular replicators, the origin and evolution of sex,[6] and the origin and evolution of distinct species.[7,8]

DARWINIAN DYNAMICS

Dynamical systems of differential or difference equations for the rate of change of the abundance of the ith entity with time $\partial N_i/\partial t$ satisfy Darwin's propositions above if the following conditions are met[5]:

i.   autocatalytic growth: $\partial N_i/\partial t$ depends on N_i, in such a way that

$$\underset{N_i \to 0}{\mathcal{L}} \frac{\partial N_i/\partial t}{N_i} = 0 \qquad (1)$$

so that spontaneous creation is forbidden (mutations may occur but must not dominate the dynamics);

ii. variation subscripted by i;

iii. continuity of the parameters of the equations through time; and

iv. the realized birth rate is decreased (alternatively, the death rate is increased) by "resource limitation," which is expressed by a function $f(R)$ in which R is a decreasing function of the abundances of the various entities.

There are a variety of different equations which satisfy these conditions and the examples in latter sections should help clarify matters. The requirement of resource limitation in (iv) above embodies Darwin's emphasis on competition. However, it must be realized that competition at the level of individuals is only one of the ways in which selection can occur at the level of gene frequencies. Cooperation, even altruism, at the level of individuals can lead to selection of genes and change in gene frequencies (e.g., kin selection, for review see Michod[37]; reciprocal altruism, see Brown et al.[12]).

FITNESS AND ADAPTEDNESS

The term fitness has a variety of senses in evolutionary biology (for discussion, see Byerly and Michod[13]). First, there is fitness in the sense used by Fisher[22] as a rate of increase of a type. Usually, it is most appropriate to express the rate of increase on a per-capita basis. Consequently, the Fisherian fitness of an entity i is simply $(\partial N_i/\partial t)(1/N_i)$.

Second, there is fitness in the sense of the expected number of offspring produced by a type. Fitness in this sense is usually termed the "Darwinian" or "Wrightian" fitness, and is sometimes expressed relative to other types in the population. This is the usual sense of fitness in population genetics.

Finally, there is fitness in the sense of adaptedness which was Darwin's usage in the phrase "survival of the fittest." This usage of the term fitness is especially confusing in evolutionary discussions. In the phrase "survival of the fittest," the term survival refers not only to life history survival but to survival over evolutionary time or evolutionary success. In this chapter, I use the per-capita rate of increase,

Fisherian fitness, as the ultimate measure of evolutionary success. Consequently, the phrase "survival of the fittest" predicts that there is a direct relationship between the adaptedness of a type (measured below by its adaptive capacities) and the type's per-capita rate of increase. In any Darwinian model of evolution, the Fisherian fitness of a type is a function of its adaptive capacities. However, in addition to the adaptive capacities, Fisherian fitness is a function of the genetic system, mating system, reproductive system, and the environment. As we will see, in many cases these other factors interfere and the adaptive capacities are not sufficient for predicting the outcome of selection. Consequently, it seems worthwhile examining the functional relationship between adaptedness and Fisherian fitness in more generality.

MODEL FRAMEWORK

For the moment, let us limit the adaptations under consideration to those which affect the coefficients b_i and d_i. Consider two autogenic types with densities N_1 and N_2. The realized birth rate is assumed to depend on the level of available resources, R. We could also include resource effects on the death rate, but for simplicity we ignore this. Some density dependence is already implicit in R, since R is a decreasing function of the density of both types. In what follows, R refers to those density-dependent elements which are shared by both competitors. The R term is, after all, the term that mediates the competition. There are, however, elements in density dependence that are not shared by both competitors, such as the chances of mating in a sexual species. In general, such effects can be included by multiplying the birth and death capacities for a type by the type's density raised to the u and v powers, respectively. This yields the following pair of equations (after Eigen and Schuster,[19] pp. 28–32; note that Eigen and Schuster consider only the case $v = 0$, which is realistic for the problems they deal with):

$$\frac{1}{N_1}\frac{\partial N_1}{\partial t} = b_1 R N_1^u - d_1 N_1^v$$
$$\frac{1}{N_2}\frac{\partial N_2}{\partial t} = b_2 R N_2^u - d_2 N_2^v \tag{2}$$

with,

$$R = K - N_1 - N_2.$$

I first give examples to illustrate some factors which affect u and v. The case $v < 0$ describes a decrease in mortality rate as density increases. Such a case is expected whenever social interactions are important. Social defense against predators is effective when there are many individuals in the population, but is less effective as density decreases. The case $v > 0$ describes an increase in mortality as density increases. Such a case is expected whenever type-specific disease is important (this is the mechanism that concerned Darwin as discussed below). For example, parasites or predators that fixate on common types decrease mortality for rare types.

The case $u > 1$ describes a decrease in reproductive success when rare. A large number of mechanisms can influence this case: difficulty of encountering a suitable mate and consequences thereof, inbreeding depression, outbreeding depression,[50] etc. Most mechanisms expected to produce $u > 1$ are either directly or indirectly related to sexual reproduction. The case $u < 0$ corresponds to mechanisms that increase reproductive success when rare but which are not included in the limiting resource R. In the case of an interbreeding population, a rare male mating advantage would fall into this category.

CONDITION FOR INCREASE WHEN RARE

I now assume that the b's and d's are such that either species reaches a positive equilibrium density, say, K_1 or K_2, when alone in the environment. Now assume the type 1 is at its equilibrium and type 2 is very rare, in which case it can be shown that type 2 can invade if

$$N_2^{u-v}\frac{b_2}{d_2} > N_1^{u-v}\frac{b_1}{d_1}.\tag{3}$$

There are three cases that can be distinguished based on Eq. (3): $u = v$, $u > v$, and $u < v$. These cases provide the basis for our discussion of natural selection paradigms.

SURVIVAL OF THE FITTEST ($u = v$)

If $u = v$, competition leads to replacement if (using Eq. (3))

$$\frac{b_2}{d_2} > \frac{b_1}{d_1}.\tag{4}$$

In this case, the ratio b_i/d_i is related to the carrying capacity K_i, and the replacement condition (4) can be shown to be formally equivalent to having $K_2 > K_1$. This is the simplest form of the adaptationist paradigm. In essence, there is no limit to how much larger K_2 must be compared to K_1; any infinitesimal increase in adaptedness is expected (statistically) to cause replacement. In other words, there is no limit to how rare the invader can be, so long as it is more adapted than the original type, it will replace it.

This is the adaptationist paradigm. It can also be termed "Malthusian" evolution, after the law of Malthus upon which it depends ($u = 0$; assuming for the moment that $v = 0$). A version of the adaptationist paradigm was set forth by Darwin[16]:

Under nature, the slightest difference of structure or constitution may well turn the nicely balanced scale in the struggle for life, and so be preserved. (p. 83)

Each new form will tend in a fully-stocked country to take the place of, and finally to exterminate, its own less improved parent or other less-favored forms with which it comes into competition. (p. 172)

There are several concepts implied in these quotes: the adaptive advantages involved can be infinitesimally small; they begin to increase as extreme rarities, insofar as they are a property of a single individual; the preservation of the new type is a statistical phenomenon; it is expected that ultimately the less well-adapted form will be replaced, at least locally, by the more adapted. In the language of dynamical systems, Darwin was describing an instability. He proposed that a population or community in nicely balanced equilibrium is always unstable to a perturbation in the form of either a more adapted individual or a more adapted species. As a result, the population or community changes as a new more-adapted type replaces the old one.

Equation (4) gives the condition for the replacement of one reproductively isolated population by another. In this case, the adaptationist paradigm is fragile, insofar as it is true only for the special case $u = v$. In some nonliving systems, such as lasers and primitive RNA (see Eq. (6) below), this condition is met. However, in the real world, the question of whether such a precise requirement can be strictly true is problematical.

SURVIVAL OF THE FIRST ($u \; \textit{¿} \; v$)

If $u > v$, the second species cannot replace the first if it invades when rare, independently of whether it is more adapted (as expressed in terms of b's and d's). If a more adapted type and a less well adapted form are equally common, then replacement can take place and the less well adapted form is driven to extinction. The case of $u > v$ is described by the phrase "survival of the common" or more accurately "survival of the first." We have used the term "cost of rarity"[7,8] to summarize the net effect of all mechanisms that tend to make $u > v$, yielding survival of the first. These involve, for example, the difficulties of finding a mate and social interactions. This paradigm also applies to the origin of proteins,[18,19,38] as discused below. Eigen[18] and Eigen and Schuster[19] in their discussion of hypercycles term this paradigm "once for ever" evolution.

The basic correlates of such a paradigm are[7,8,28]: a tendency to inhibit evolution leading to stasis and biological conservatism, and distinctness of species and coarse-grained adaptation to environmental grades. The details of how some of these consequences arise are discussed in the final section (see also Bernstein, Byerly, Hopf and Michod[7,8]) and are usually associated with sexual reproduction.

COST OF COMMONNESS ($u \ ; v$)

If $u < v$, the rare type can always invade the common type independently of whether it is more or less adapted (recall that N_2 is to be evaluated at zero in Eq. (3)). However, neither type can drive the other to extinction. Consequently, this case leads to a diversity of types. If I were to continue the terminology used for the other paradigms, I would label this case "survival of anybody." This suggests biological nonsense and obscures the interesting implications this case has for common populations. As discussed in more detail below, a version of this paradigm concerns competition in the presence of parasites that fixate on the common type. This cost of commonness was discussed by Darwin[16]:

> When a species, owing to highly favorable circumstances, increases inordinately in numbers in a small tract, epidemics—at least, this seems generally to occur with our game animals—often ensue: and here we have a limiting check independent of the struggle for life. (p. 70)

The case of $u < v$ is perhaps the most difficult to interpret biologically, since, in the case of extreme rarity, it leads to infinite Fisherian fitnesses. Nevertheless, this case does provide biologically useful insights into common populations. Let us consider disease (parasites, funguses, insects on plants, etc.) as a mechanism causing v to be large because parasites fixate on common populations and keep them from expanding even when resources are abundant. If we suppose that there are species-specific diseases (parasites, fungi, etc.) as well, then if any one species gets too abundant, epidemics occur. As a result, there are resources not utilized that are available for other species. Hence there is the potential for increased diversity at the species level as new species can invade by utilizing the unused resources. However, the density of the new species will at some point be checked by parasites leaving, perhaps, some resources still unused. This mechanism will continue until all available resources are used leading to a high degree of diversity along an environmental grade.

Up until now we have assumed that u and v are the same for different species. Consider, what happens when a species, either with biology corresponding to $u > v$ or $u = v$, arises in an established population with biology $u < v$. Assume that evolution in the $u < v$ species had proceeded in the manner discussed in the last paragraph. The two cases, $u > v$ or $u = v$, make different predictions. The only difference now for the case of $u > v$ is that there are a number of common types in the population which themselves expanded under the cost of commonness paradigm and which now inhibit the expansion of the new type. In the case of $u = v$, the concept of survival of the fittest still applies. The adaptationist paradigm is now in the form used by Darwin in the quote above, in which new improvements, no matter how slight, are expected to expand into the population. However, the resulting extinction of current types is a diffuse process because they are all so similar. One of the current types is expected to go extinct, but it may not be easy to determine which it is. It should be realized that this kind of competitive extinction need not occur among competing species which have expanded under the cost of rarity

$(u > v)$. In this case, the competing species will be coarsely packed and a species with biology $u = v$ could invade without driving others to extinction (rare species with $u > v$ continue to be prevented from invading by the cost of rarity; see Figure 3 below).

INTERBREEDING POPULATIONS

These paradigms also exist when one considers the replacement of one type with another within a randomly mating Mendelian population. If one considers a single locus with two alleles a and A, then whether one allele replaces the other depends on the adaptedness of the pairwise combinations of alleles (W_{AA}, W_{Aa}, and W_{aa}). If the heterozygote Aa is less adapted than the homozygotes aa and $AA(W_{AA} > W_{Aa} < W_{aa})$, then neither allele, when rare, can expand into a population dominated by the other. Successful invasion, when common, is allowed, and replacement takes place. If the heterozygote is more adapted that the homozygotes ($W_{AA} < W_{Aa} > W_{aa}$), then either allele can expand when rare, but neither is eliminated. Only in the case when the heterozygote is intermediate in adaptedness (say $W_{AA} > W_{Aa} > W_{aa}$) does the adaptationist paradigm hold so that there is a direct relationship between adaptedness and Fisherian fitness of genotypes (see Michod[39] for more discussion).

In the case where the types were reproductively isolated, the adaptationist paradigm was fragile since it depended on strict equality of u and v. However, in the case of free mixing, all three cases are equally robust in that they all depend upon inequalities. Consequently, no one of the three cases can be taken to be inferior to the others on the basis of the theoretical issues discussed here (but see Michod[36,39,40] and Byerly and Michod[13]).

THE EARLY REPLICATORS
THE FIRST REPLICATOR

A molecular replicator is a molecule that makes copies of itself. DNA and RNA are the best known examples of such molecules. They are made up of long strings of nucleotide bases, which behave like characters in an alphabet. Information in early replicators was probably likewise contained in a linear sequence of characters.

But where did the characters come from and how were they strung together? We can only guess. One view is that the characters—the nucleotides—were available in a kind of chemical alphabet soup present in small pools or water droplets, created from physical and chemical reactions that went on on the primitive earth. At some point, these characters linked with one another to form short strings.

Let N_i be the density of a particular nucleotide string i. Under conditions of prebiotic evolution, short-length oligonucleotides form spontaneously from more primitive precursor molecules (see, for example, Schuster[49]). The dynamic associated with the change in density of such an entity is

$$\frac{\partial N_i}{\partial t} = \beta_i - \delta_i N_i,\tag{5}$$

where β_i and δ_i are the rates of the forward and backward reactions, respectively. Such a dynamic is not a Darwinian dynamic, since it describes growth from zero density (Eq. (1) is not met). The parameters β_i and δ_i are intrinsically related, since both stem from the activation energy between the precursor molecules and the oligonucleotide product (Figure 1). There is no independent birth and death process for such an entity; consequently, there is no life history.

At some point a molecule arose with the capacity of self-replication. Replication in RNA and DNA results from the fact that individual nucleotides are attracted to one another. This attraction is ultimately based on electrostatic interactions, which are by their very nature complementary. In the case of DNA, A and T link up, as do G and C. In the primordial pool, likewise, the elements of a string such as $AATG$

SPONTANEOUS – CREATION ("birth" and "death" process coupled)

REPLICATOR (birth and death process decoupled)

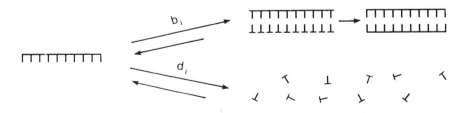

FIGURE 1 Diagram of chemical reactions associated with spontaneous creation and replication. These single reactions are probably a simplification of many intermediate steps. See text for further discussion. Figure taken from Michod[38]; courtesy of the American Society of Zoologist ©1983.

(the "parent" strand) could attract the other characters available in the alphabet soup, and thus form a complementary string *TTAC* (the "daughter" strand). Such a pair of strands held together by hydrogen bonds is often stable. However, under certain conditions (high temperature, for example) the bonds would weaken and the strands separate. The two single strands could then drift apart, and when conditions change to favor attraction again, they would be free to attract new complementary strands. This cycle of strand building and separation could continue indefinitely, given an ample supply of chemical building blocks, and thus the number of *AATG* and *TTAC* strings in the primordial pool could multiply.

Since the parent strand serves as a template for the creation of the daughter strand, this process is known as *template-mediated* replication. Enzyme-free, template-mediated polymerization of nucleotides of length 30–50 has been reported.[33,32,54] However, to my knowledge, a full cycle of replication—polymerization, separation of the strands, and polymerization of the daughter strands—has not yet been described in an experimental system. Although there are considerable practical problems, one can easily imagine how, in principle, template-mediated replication could continue indefinitely on the primitive earth.

A molecule capable of this kind of template-mediated replication may enter into two very different kinds of reactions (Figure 1). It may serve as a template for polymerization of a complementary strand or it may be degraded into its mononucleotide components, with respective rates b_i and d_i. (I assume that the spontaneous rate of formation of the replicator, β_i, is negligible in comparison to b_i. Consequently for the replicator, d_i, represents the same process as did β_i and δ_i of spontaneous creation (Figure 1, Eq. (5)) and is numerically equal to δ_i, since $\beta_i \sim 0$.) With the origin of template-mediated replication, there existed a simple life history described by the adaptive capacities b_i and d_i. These capacities measure, respectively, the rate of replication under unlimited resources, which are free nucleotides, and the rate of death, due probably to hydrolysis of the molecule.

The capacity to replicate is a basic feature of life. A second basic feature is metabolism. Metabolism simply refers to the conversion of high energy compounds (food) to low energy compounds (waste). Physical laws require metabolism to maintain a highly ordered structure such as an organism or a molecular replicator. For the replicator, the high energy compounds were simply the nucleotide building blocks, which must be in an energized state to link up with each other in the string. Once a bond is formed between the nucleotides in the daughter strand, the nucleotides are in a lower energy state. Energy sources present in the environment, such as sunlight, may have energized the nucleotide building blocks.

What kinds of properties would these strings of nucleotide characters have? In other words, what would they look like and what would they do, besides copy themselves? Probably, each string folded and twisted itself into a unique shape, determined by its particular linear sequence of characters, and the physical properties of the environment, such as the acidity, salinity, and temperature of the pool.

For these early molecules, shape was probably very important. Death of the molecule would result from breakage of the bonds that held the string together or changes to the characters themselves. If a molecule were especially twisted and compact, it might stand less chance of being broken apart. On the other hand, compact shape probably would not be conducive to replication. Because replication requires that nucleotides be able to move into positions next to their complementary characters in the replicating molecule, a more open shape probably would facilitate the copying process while a more compact shape might retard it. In any event, the molecule's three-dimensional shape would strongly affect the molecules capacity to survive and reproduce.

For natural selection to occur, variation is needed. As the molecules copied themselves, mistakes probably arose quite often. The physical forces that attract A to T or G to C are not completely faithful. Laboratory experiments indicate that one out of ten such nucleotide attractions would be erroneous, occurring as they must have without the aid of enzymes or other cellular machinery. (In comparison, the error rate for DNA copying in modern organisms is 1 out of 10 to 100 billion.) So, for every ten characters in our early replicator, one "wrong" nucleotide would be placed in the daughter strand. (There might be a C across from an A, instead of the thermodynamically preferred T.) What happened then?

In complementary nucleotide pairing, mistakes can easily be copied. So the error—technically called a *mutation*—could reproduce along with the rest of the "correct" nucleotides so long as it did not interfere with the new molecule's capacity (based on its shape) to replicate and persist in its environment. Mutability is also a basic feature of living systems, for without it they could not adapt to their environment by natural selection.

Once such a mistake copied itself, there would have been two different types of molecules in the population (say, type 1 and type 2), one defined by the sequence $AATG$ (or $TTAC$) and the other by the new sequence $TCAC$ (or $AGTG$). To the extent that these two molecules had different structures and properties, they may also have had different abilities to replicate or survive. The new molecule may have been more stable or able to replicate faster than the original one because of its particular shape. If so, it would come to predominate in the population, while the old one might go extinct. Of course, it could just as easily have been the case that the old type would have been better able to replicate or persist. Either way, the molecules would go through the process of natural selection, and the types with greater ability to survive and reproduce would come to dominate the population.

Let us be a bit more formal. Bernstein, Byerly, Hopf and Michod[5] have argued that a replicator, like single-stranded RNA, which undergoes template-mediated replication, obeys the following equation

$$\frac{\partial N_i}{\partial t} = \frac{b_i R}{r_i + R} - d_i. \tag{6}$$

The variable R is the concentration of available mononucleotide building blocks and is assumed to be a monotonic decreasing function of $\sum_{i=1}^{n} N_i$. The capacities b_i and d_i have already been described. There is, in addition to these capacities, a third capacity, r_i, which involves the interaction of the replicator with its resources, which are primarily mononucleotide building blocks. The constant r_i is a combination of rate coefficients of the formation and dissociation of the template-resource complex and is inversely related to the capacity of the molecule to utilize its resources. Different nucleotide sequences would have different shapes and supramolecular structures, described by different b_i, d_i, and r_i values, which would ultimately be determined by the nucleotide sequence.

The condition for increase of one type of replicator, say 2, in a population dominated by another type, say 1, is[5]

$$\frac{b_2 - d_2}{d_2 r_2} > \frac{b_1 - d_1}{d_1 r_1} . \tag{7}$$

If condition (7) is satisfied, type 2 will increase to fixation. Condition (7) is similar to condition (4) above in that it depends only on the adaptive capacities of the types and does so in such a way that the type of replicator which is most adapted, in the sense of having larger b_i, or smaller d_i or r_i, is selected. Consequently, in such a system the more adapted replicator, as determined by its b_i, r_i, and d_i, will come to predominate in the population from arbitrary initial density.

SEX AMONG THE EARLY REPLICATORS

Individual replicating molecules were probably not well isolated from one another. Parts of one molecule could easily break off and attach to other molecules. As discussed in more detail below, this "unintentional" breakage and rejoining of nucleotide strings is directly analogous to the recombination of DNA molecules that goes on during sex. In other words the living world was born sexual. There was little individuality and much promiscuity at this early stage of life. It wasn't that evolution "wanted" it that way; it just so happened that nothing could be done about it, until the replicators became more advanced and improved their individual integrity.

The free exchange of parts (nucleotide strings) would have meant the creation of new sequences of nucleotides with new properties, some of which might have improved a molecule's capacity to survive and reproduce. What's more, the exchange of parts would have allowed for the repair of certain errors that would often occur under the conditions of intense sunlight and heat prevailing on the primordial earth.

Ultraviolet light (UV) found in sunlight is harmful to nucleotide strings even when these strings occur in a modern cell. So for naked molecules in the primordial soup, UV light would have been extremely harmful, especially when one considers the high intensity of UV light on the primitive earth (there was no ozone layer on the primitive earth to help absorb the UV light as there is today). UV light can break nucleotide strings, link one complementary strand to the other thereby preventing it from separating and replicating, or cause spontaneous changes in the chemical structure of the nucleotide bases so that they lose their attraction for their complementary nucleotide. Under such conditions, some molecules might take on some very odd members—not characters in the alphabet (A, T, G, or C) but things altogether different, say a □ or a ¥. This kind of error—technically called *damage*—might make it impossible for the molecule to reproduce at all.

Here is how the free exchange of replicator parts could help in coping with such damages. (This same process also helps get rid of mutations. If two damaged molecules have their errors at different positions in the nucleotide sequence, they could both break apart and exchange halves, so that there could be two new molecules—one fully repaired and the other with both damages. By such a process, the replicators would have been able to maintain a relatively healthy population. Free exchange would continually purge the population of errors, by collecting them in wastebasket molecules, and at the same time generate error-free molecules that could continue to replicate, thereby giving the replicators a kind of immortality. We will return to the role of sex in maintaining the immortality of DNA shortly.

PROTEIN-AIDED REPLICATION

However, template-mediated replication had its limits. Although such replicators may persist indefinitely, they could not go on and diversify into ever more complex structures—such as organisms—for they could not make anything, other than more strings. The assembly of any structure, including strings of nucleotides, is easier if certain tools are available. For example, a model airplane is easier to put together on a level table with good lighting and clamps and vises to hold the parts in place while the glue dries. Likewise, one can imagine how template-mediated replication might proceed faster and more accurately if there were tools to hold the strands in place, position the nucleotides, keep out unwanted dirt or harmful pollutants, and so on. Initially, physical objects available in the molecule's environment, perhaps clumps of clay or the surfaces of water droplets, might have been used to aid in the assembly of the daughter strand by stabilizing the intermediate structures involved in replication. However, it would be more advantageous for the molecule to be able to create its own tools—especially if this ability could be passed on to offspring to ensure their survival and replication. This is where proteins come in.

Today, in living cells, DNA replication occurs within a factory of complicated tools that aid the template-mediated process described above. These tools are usually made of proteins. Like DNA, proteins are long sequences of characters, though in this case amino acids rather than nucleotides. Because there are twenty amino acids, instead of only four nucleotides, proteins come in a far greater number of varieties and so can exist in a variety of different shapes. How and why did evolution produce the first protein?

Proteins owe their existence to DNA; the sequence of a protein's amino acids is encoded in the sequence of nucleotides in a particular stretch of DNA. We can only guess at how this translation of DNA information into protein was first accomplished. Ever since the discovery in the mid-1950's of the code that translates genetic information into proteins, geneticists have speculated on the origin of the code and the first protein systems. It is possible that the primitive translation process involved direct physicochemical interactions between the polynucleotide molecule and amino acids.[57,26,25] However, I will not discuss this matter any further here.

More relevant to our concerns are the selective pressures that led to the evolution of proteins. Fortunately we have a better understanding of this issue.

Proteins, with their enhanced capacity to produce different shapes, are well suited to make the tools that would help the replicators survive and reproduce. Once the molecular replicator developed the ability to make proteins, the replicator itself was free to evolve into almost any kind of shape, regardless of whether it was conducive to replication. (Remember, we noted that the optimal shape for replicating—a relatively open strand—left the molecule vulnerable to breakage.) Thus, the replicators with protein-making powers could be both strong and prolific by producing two different proteins, one that aided replication and one that promoted survival. Consequently, protein-catalyzed replication would have been far faster and accurate than replication without proteins. With this increased accuracy, the replicator could increase its length and the amount of information it contained.[19]

By farming out the replication and survival tasks to the proteins, the nucleotide sequence could specialize in a third task, storing (and acquiring through mutation and recombination) the information needed to make proteins. This division of responsibility between the executive function (the proteins) and the legislative or instructive function (the nucleotide sequence) is familiar to us, as just such a division exists in many social organizations and systems of government. In biology these two different responsibilities go by the technical names *genotype*, for the legislative/instructive responsibility, and *phenotype*, for the executive. Until the replicators evolved the capacity to make proteins, the genotype (genetic information) and the phenotype (structure or shape) were inextricably linked. With the evolution of a means to translate the information stored in the nucleotide sequence into proteins, the genotype and phenotype could specialize in their different functions.

Useful tools—such as proteins—are likely made only at some cost to the replicator who happens, either by mutation or recombination, to acquire the ability to make them. After all, it takes time and energy to make the proteins, time and energy that could be spent replicating. So long as the benefits of the protein outweighed these costs, protein-producing replicators should be favored by natural selection.

Following Eigen,[18] we consider a simpler form of density regulation than used above in Eqs. (6). We assume that the total number of replicators is maintained at some constant number N_T,

$$N_T = \sum_{i=1}^{n} N_i, \tag{8a}$$

by, at each time step, subtracting the total productivity of the system, Ψ, from the unlimited rates of growth, where

$$\Psi = \frac{\sum_{i=1}^{n} \partial N_i / \partial t}{N_T}. \tag{8b}$$

This gives the following differential equation in place of Eq. (6)

$$\frac{\partial N_i}{\partial t} = N_i(b_i - d_i - \Psi) \tag{9}$$

for template-mediated replication.[38] Although with this representation of competition, the Darwinian dynamic has a different form, selection in such a system still follows the adaptationist paradigm with b_i and d_i determining the outcome of selection.

For a replicator that makes a protein, the following equation applies using Eqs. (8a) and (8b),

$$\frac{\partial N_i}{\partial t} = N_i(b_i - d_i + BN_i - C - \Psi), \tag{10}$$

in which B is the protein-catalyzed replication rate ($B \gg b_i$) and C is the cost of making the protein. The realized rate of protein-mediated replication, BN_i^2, stems from the assumption that the concentration of the protein is proportional to the concentration of the replicator that makes it, so that the probability per unit time of the replicator and the protein interacting is proportional to N_i^2. This kind of Darwinian Dynamic has been termed "hypercyclic" by Eigen[18] and corresponds to the survival of the first paradigm (since $u = 2 > v = 0$).

Now, consider competition between two replicators, one, say type 1, using template-mediate replication (Eq. (9)), and the other, say type 2, using protein-mediated replication (Eq. (10)). Mathematical analysis bears out what is easy to see intuitively—the protein-producing replicator cannot increase when rare, if the two competitors are otherwise identical ($b_1 = b_2 = b$, $d_1 = d_2 = d$). This is because the protein-producing replicator has trouble encountering the protein when it is rare. Nevertheless, it still must pay the cost, C, of producing the protein, so that its realized birth capacity is $b - C$ compared to b for the non-protein-producing replicator. Therefore, the hypercycle pays a cost of rarity and this leads to a kind of stasis in which it is difficult for a remarkable adaptation, such as protein-mediated replication, to get started.

The protein-producing hypercycle was probably worse off than this. The first protein was probably a general kind of catalyst and any replicator could benefit from it whether the replicator made it or not. Thus, the protein would be available to other competing replicators, such as the non-protein producing replicator, making it more difficult for the replicator that makes the protein to increase (since it pays the cost).

Clearly, some structure is needed to keep the proteins associated with the replicators that produced them. This may have come in the form of passive population structure[38] or self-structuring in an incompletely mixed medium.[11] In either event, these kinds of structures could have been intermediate steps on the way to the production of the first cell.

INDIVIDUALITY, ITS COSTS AND BENEFITS

Eventually, through a series of chance mutations and recombination events, some replicators acquired the ability to construct shelters around themselves, enabling them to horde their own proteins for their own exclusive use. Sooner or later this advantage would lead (probably around 2,000 or 3,000 million years ago) to the creation of the first cell. What a wonderful creation this was: protection and nutrition for the replicator all under the shelter of its own roof. No longer were its own proteins automatically shared by its neighbors. Cheating would no longer pay. With the advent of the cell, evolution would become a whole new ball game. Individuality had been born.

Individuality carried a certain cost, however. The replicator encased in a cell was forced to keep both its bad mutations (improperly sequenced nucleotides) and damages (non-nucleotide intruders on the sequence). Any errors the replicator acquired—by exposure to UV light, for instance—would now be trapped in the cell. The days of free exchange and easy access to spare nucleotides were over. Sex had come naturally and without effort to the naked molecules. Now imprisoned in a cell of their own invention, the replicators had to figure out an alternative way to make repairs. A new kind of sex had to be invented.

ORIGIN OF SEX

In this section, I summarize the results of mathematical models of the origin of sex[6,38] that motivate the experimental work on transformation discussed in the next section. First, we need to make some further assumptions about the primordial genome. I believe that these assumptions are generic, in the sense that they apply to any primordial genetic material, based on the properties that it must have, independent of the precise chemistry involved. We adopt the Eigen view that the hypercycle, a set of mutually independent interacting units, is an appropriate model for the primordial genome. The characteristics of hypercycles have been extensively studied by Eigen and coworkers.[19] The basic property of hypercycles used here is that they can stably maintain their component genes, so long as there is at least one copy of each gene present. The primordial genome is assumed to be segmented; that is, the different genes are assumed to be on separate fragments. No recombination enzymes are assumed to exist at this early stage of evolution of the genetic system.

As discussed in the previous section, at some point the primordial genome became encapsulated into a protocell. The precise details of how this came about are highly speculative[6,38]; however, once this occurred, the problems faced by the early protocell seem obvious, and again appear to be generic. The basic problem faced by the early protocells is gene deficiency. The causes of gene deficient cells are varied, but two basic causes are genetic damage and failure of assortment of genes from the parent cell into the daughter cell. Strategies for coping with gene deficient cells most probably involved *redundancy*, in one form or another.

Intra-cellular redundancy, or multiple copies of each gene in the protocell, allows for recovery from genetic damage, so long as all copies of a particular gene are not damaged at the same time. In addition to providing benefits as a result of recovery from genetic damage, high redundancy should help solve the assortment problem by increasing the chances that daughter cells have, at the least, one copy of each gene. So long as the protocell has at least one good copy of each gene, additional copies can be generated through the positive feedback inherent in the hypercyclic nature of the interactions between the genes. However, maintaining intra-cellular redundancy has costs, most obviously, the costs of replicating the additional copies of each gene. For simple systems, such as the protocell under consideration, these costs are likely to be a sizeable fraction of the resource, energy, and time budget.

Another strategy is for the protocell to remain at low redundancy (say, haploid) for most of the time, but periodically fuse with other cells (to become diploid) for genetic recovery, and then split into two daughter cells (haploid again), so as to reduce the costs of redundancy. This cycling of diploid and haploid states is the sexual cycle in its most basic form, whereas the strategy that maintains multiple copies of each gene without fusion and splitting is the asexual strategy.

Recombination involves the physical breakage and rejoining of DNA molecules. However, during the early stages in the sexual cycle, before genes became physically linked on the same chromosome, recombination in the modern molecular sense did not exist and was probably not necessary. Nevertheless, two of the main effects of recombination in modern systems, genetic variation and genetic recovery (repair), were present in the hypothesized sexual cycle of the protocell.

A number of explicit competition and population genetic models have been constructed embodying the ideas discussed above to determine the conditions under which a sexual strategy would evolve.[6,38] The basic assumptions of these models were as follows:

i.   Genetically deficient cells were produced both by damage of component genes and by assortment errors at defined rates.

ii. These "gene-dead" cells could be recovered by fusion, if fusion occurred in some specified window of time.

iii. Costs of redundancy were included by specifying that the rate of replication of a protocell was inversely related to its ploidy level (intra-cellular redundancy).

iv. Benefits of high ploidy (high intra-cellular redundancy) were included by assuming that a cell did not become gene-dead unless all of its copies of a gene were damaged. Furthermore, cells with high ploidy were more likely to successfully assort genes into daughter cells so that daughter cells had at least one copy of each gene.

v.   Costs of sex and fusion were also assumed.

The basic conclusion of these models was that the sexual strategy outcompeted the asexual strategy under most conditions except those of low levels of damage. Furthermore, conditions were specified under which a rare sexual mutation could increase in an asexual population as a result of its ability to mate with and to recover gene-dead cells, even though only one-half of its daughter cells contained the gene for sex. Consequently, these models and the ideas they embody provide an explanation for both aspects of the sexual cycle, fusion and splitting. Fusion evolved so as to use redundancy to recover genetic information that was lost through damage or assortment. Splitting evolved so as to reduce the costs of redundancy. These models show that it is possible for recovery from genetic damage to be the primary selective force molding the early stages of evolution of the sexual cycle. In the next section natural genetic transformation in bacteria is studied as a test case for this theory.

CONTINUED EVOLUTION OF SEX
BACKGROUND TO THE PROBLEM OF SEX

There are two basic aspects to all modern sexual systems. The first is *recombination*, defined as the breakage and rejoining of DNA (or RNA) molecules. The second is *outcrossing*, which refers to the fact that the two DNA molecules involved in recombination come from different individuals in the previous generation.

Sex has considerable short-term costs involving the cost of males,[35,55] high recombinational load,[50] lower genetic relatedness between parent and offspring,[52,53,56] and the cost of mating.[9,28] Such large costs imply that the adaptive benefit of sex must be comparably large. Yet the benefit of sex is now widely regarded by evolutionary biologists as one of the major unsolved problems of biology.

For most of the past fifty years, biologists had generally regarded the benefit of sex as a problem that had been solved. The traditional explanation is based on an obvious feature of sex. As a general rule, progeny resulting from sex are genetically different from their parents, whereas in asexual reproduction they are genetically the same as their sole parent. Sexually produced progeny not only differ from their parents, but also from each other (except for identical twins). Thus sex produces new genetic variants at each generation. By the traditional view, this genetic variation is the primary function of sex because it facilitates adaptation.

In the past fifteen years, much theoretical work has been done to define the genetic and ecological conditions under which sexually produced genetic variation can select for the sex trait (see, for example, Bell,[1] Williams,[55] Maynard Smith,[35] and Michod and Levin[44]). This work, rather than demonstrating the general applicability of the variation explanation, has shown that it operates only under rather limited conditions. Indeed, the general effect of sex on evolutionary advance may be conservative, just the opposite of what the variation explanation presupposes.[45] The realization that this longstanding explanation has serious problems in accounting for the ubiquity of sex in nature has led to the now widely expressed view, just mentioned, that the function of sex is a major unsolved problem in biology.

The hypothesis for the evolution of sex discussed here is based on another obvious feature of sex. Progeny are not only genetically different from their parents; they are also young. We consider this aspect of sex to be more fundamental than variation and it forms the basis of the repair hypothesis for the evolution of sex which we have developed elsewhere.[2,3,4,6,7,8,10,43] I now present an overview of this hypothesis along with the facts and principles upon which it depends. I have not tried to be complete here in my presentation or references to the literature and the papers just cited should be consulted for more complete discussion and documentation.

There are two aspects to youth.[7] The first aspect, which does not play a significant role in our explanation, relates to the process of development from fertilized egg to the mature organism. Development is thought to occur by a sequential switching on, or off, of genes. The developmental clock needs to be reset at each generation, and this presumably occurs when germ cells (egg and sperm) are formed. The second aspect of youth is that, whereas parents age, cells of the germ line do not, so that progeny do not reflect the aging of their parents. We argue that the lack of aging of the germ line results mainly from repair of the genetic material by meiotic recombination during formation of germ cells. Thus our basic hypothesis is that the primary function of sex is to repair the genetic material of the germ line. We refer to this as the repair hypothesis to contrast it to the traditional variation hypothesis.

When information of any kind is transmitted, it is subject to disruptions due to random influences which are collectively termed noise. This applies to the genetic material as well. In DNA and RNA genomes (RNA genomes occur in some viruses), noise occurs as a result of mutation and damage. These present a serious problem for survival, and early in evolution strategies evolved to reduce the deleterious effects of genetic damage and mutation.

We take care to distinguish between genetic damage and mutation, since the differences between them are basic to our hypothesis. Damages are abnormalities in the physical structure of DNA or RNA produced by reactive chemicals or radiation. Common examples are breaks in one or both strands of DNA, oxidatively altered nucleotides, and thymine dimers caused by UV. A damaged DNA molecule is no longer a regular sequence of standard base pairs. Damages frequently block replication or prevent gene expression, and a single unrepaired damage may be lethal to a cell. Damages are not replicated and thus are not inherited. Generally, damages can be recognized by enzymes which repair the genetic material.

Mutations, in contrast to damages, involve changes in the coding sequence of standard bases in DNA or RNA genomes. A mutated DNA molecule is still a regular sequence of standard base pairs. However, when expressed, mutations frequently lead to reduced or abnormal gene expression and thus often reduce fitness. Mutations are replicated and thus inherited. Since the regularity of the DNA molecule is not disrupted, mutations are not detectable by enzymes and thus are not repaired.

In summary, mutations are replicated but cannot be directly recognized by enzymes and cannot be repaired, whereas damages are not replicated but can be recognized by enzymes and repaired.

As mentioned above, sex has two fundamental aspects, recombination and out-crossing, which I now briefly discuss from the point of view of mutation and damage so as to give an overview of the repair hypothesis. Recombination refers to the process of breakage and exchange of segments of DNA between two homologous chromosomes present in the same cell. In multicellular organisms a key stage of meiosis, the process by which germ cells are formed, seems to be designed to promote recombination. We consider that the function of recombination is to repair genetic damage in the germ line. To illustrate this repair, we consider one type of drastic damage, the double-strand crosslink which connects the two strands of DNA. Recombinational repair of such double-strand damages could be accomplished by three stages. These stages also figure in current models of general homologous recombination: (i) physical removal of the double-strand damage resulting in a double-strand gap in the damaged chromosome, and (ii) physical transfer of good information from one strand of the undamaged homologue into this double-strand gapped region resulting in two single-strand gaps in each homologue which are then (iii) filled in by copying off the complementary strand.

The second fundamental aspect of sex is outcrossing which refers to the fact that the two homologous chromosomes which engage in recombination come from separate parents. This contrasts with asexual systems in which the two chromosomes present in any cell are from the same parent. For haploid organisms, like those discussed in the next section, recombinational repair of double-strand damages requires outcrossing since there is often no "backup" chromosome already present in the cell. In the case of more complicated diploid organisms we argue that outcrossing can be explained by considering the interplay of the expression of recessive and partially recessive deleterious mutations with recombinational repair.

In summary, we hypothesize that the two fundamental aspects of sex, recombination and outcrossing, have evolved in response to the two fundamental categories of error in genetic transmission, damage and mutation. Specifically, the need to repair damages selects for recombination and the need to mask the expression of deleterious recessive mutations in reproductive systems which have recombination selects for outcrossing (in diploids). In the next section I discuss an experimental system designed to test the role of sex in DNA repair in haploids. In haploids, both recombination and outcrossing are necessary for damage repair.

DNA DAMAGE AND THE EVOLUTION OF SEX

Natural genetic transformation provides an experimental system for testing these ideas about the role of DNA repair in the evolution of sex. Genetic transformation is a highly evolved attribute of cells and occurs naturally in a number of modern bacteria, including *Haemophilus* and *Bacillus*.[51] It results from the ability of cells to become "competent" to bind and take up DNA molecules from the environment. The subsequent incorporation of this DNA into the bacterial chromosome by the physical recombination of homologous strands is catalyzed by enzymes involved in both recombination and DNA repair. The acquisition of physiological competence for transformation in *B. subtilis*, and in other naturally competent bacteria, is a complex, energy-requiring, developmental process and not simply the consequence of passive entry of DNA into a cell.[17] (Nor is natural transformation in bacteria like *B. subtilis* similar to the artificially inducible uptake and transformation of *Escherichia coli* and other cells.) In nature the sources of transforming DNA have not been well defined but there is evidence that *B. subtilis* cells actively export DNA to the environment during growth. In addition, DNA is likely to be released from cells following cell death and lysis.[17,51]

There are similarities and differences between transformation and eukaryotic sex. Eukaryotic sex is coupled with reproduction, while in bacteria, sex and reproduction are decoupled. Both recombination and outcrossing are present in natural genetic transformation, but are not associated with either DNA replication or reproduction of the cell. Reproduction is strictly asexual or mitotic-like in all sexual prokaryotes, even though the cell may have previously had sex, that is recombined DNA from another individual.

The function of competence is similar to that of fusion or mating in that it promotes outcrossing. As with fusion, there are physiological costs to becoming competent and taking up exogenous DNA. For example, the development of the competent state is characterized by a decrease in nucleic acid synthesis, a latency in cell growth and multiplication, and the synthesis of a set of competence-specific proteins.[17,51] In addition, resident prophages are preferentially induced by UV in competent cells leading to a decrease in cell viability and the frequency of transformed cells. The prophages have evolved the capacity to get out of a threatened (damaged) cell. For this reason our experiments discussed shortly have been conducted on strains of *B. subtilis* that have been cured of these resident prophages.

Unlike sexual systems in more complicated organisms, transformation is not a reciprocal process in terms of the contribution of each "parent," there being a "donor" of genetic information (the transforming DNA) and a recipient (the competent cell).

Another difference between eukaryotic and prokaryotic sex, is that there are no costs of maintaining the extra genetic redundancy with transformation, since the additional template brought into the cell is discarded after it is used for repair. Transformation seems to have the best of both worlds; it provides DNA template for repair, while avoiding the cost of replicating extra chromosomes and the cost of lowered genetic relatedness between parent and offspring. A transformation-like process may have preceded cell fusion in the early evolution of the sexual cycle.

Although much is known about the mechanisms of DNA uptake, processing and recombination during transformation in bacteria such as *B. subtilis* (for review see Dubnau[17]), the primary function(s) of transformation and the evolutionary advantages it confers upon cells have received little attention and are not well understood. It has often been assumed that transformation evolved for the purpose of genetic variability, or simply as a means for nutrient and/or energy acquisition.[51] Recently, we have argued that the evolutionary function of transformation (and presumably competence development itself) in *B. subtilis* lies in its role in providing the cell with homologous DNA molecules for DNA repair.[42,58]

The repair hypothesis argues that genetic transformation in bacteria such as *B. subtilis*, like recombination in eukaryotes, evolved and confers a selective advantage as a system for DNA repair. To test this hypothesis our initial approach has been to study the relative densities of transformed cells (sexual cells) and total cells (primarily asexual cells) in cultures of *B. subtilis* as a function of the dose of a potent DNA damaging agent, such as ultraviolet radiation (UV) or hydrogen peroxide. We have done so under conditions in which cells are transformed before (DNA-UV) or after (UV-DNA) UV irradiation of the cultures.[42,58]

The results of experiments using a *B. subtilis* strain that is a wild type for recombination and repair show a qualitative difference in the relationship between the survival of transformed cells and total cells in the UV-DNA and DNA-UV treatments. In the UV-DNA treatment, cells transformed with homologous DNA (chromosomal markers) had a greater average survivorship than total cells. However, in DNA-UV treatments this relationship was reversed. Moreover, we find a consistent and qualitative difference between the UV-DNA and DNA-UV treatments in the relationship between the transformation frequency and UV dosage (see Figure 2 for a typical example of our results for transformation of wild-type cells using homologous DNA). The homologous DNA transformation frequency increases with UV dose in UV-DNA treatments but decreases with UV dose in the DNA-UV experiments, relative to the transformation frequency in undamaged cells. The transformation frequency is defined as the number of transformants divided by the number of total viable cells. Since the total cells are primarily non-competent cells, we view the transformation frequency as reflecting the ratio of sexual to asexual cells. In Figure 2 the change in transformation frequency as a function of UV dose is given by plotting the ratio of the transformation frequency at each UV dose to the transformation frequency at no UV dose. Points above (below) one indicate an increase (decrease) in the transformation frequency from what it was with no UV.

We have entertained two hypotheses as likely explanations for our results, both of which are consistent with the predictions of the repair hypothesis.[42,58] The survival of competent cells should increase relative to that of noncompetent cells in a UV-DNA experiment, either because (i) DNA damage directly increases the frequency or efficiency of transformation, or (ii) competent cells are more resistant to UV-induced damage as a result of an enhanced capacity for recombinational repair ("transformational repair"), and thus survive better in a damaging environment. Damages may increase transformation directly for several reasons. A damaged cell may increase its binding, uptake, and/or recombination with homologous transforming DNA. In addition, damages themselves may directly stimulate recombination thereby resulting in an increase in the transformation frequency without requiring a physiological response from the cell. Transformation using homologous DNA is completed rapidly, within a few minutes. Therefore, in DNA-UV experiments there can be no repair advantage derived from the uptake and integration of homologous DNA, since this occurs prior to the DNA-damaging treatment. Nor should there be any "induction" in the levels of competency or transformation in cells in DNA-UV experiments.

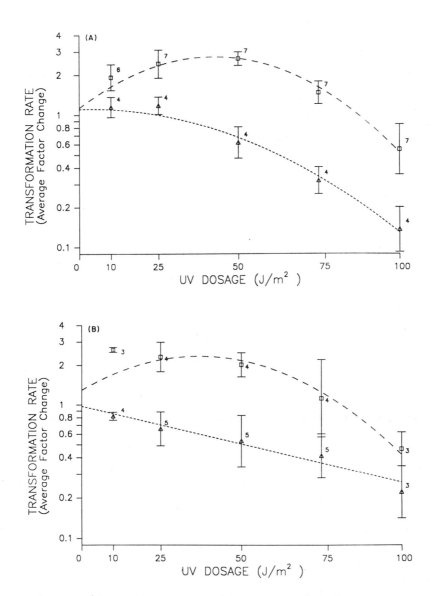

FIGURE 2 Change transformation in $B.\ subtilis$ as a function of UV dosage.
(A) Transformation using donor DNA purified from undamaged cells. (B) Transformation
using donor DNA purified from damaged (UV irradiated at $50\ J/m^2$) cells. Plots are
averages of the log factor change for each experiment. Error bars give average $\pm$
the standard error sample size is given near each mean. Regressions are as follows.
Figure 2A: DNA-UV regression: $y = 0.04377 + 0.00061x - 0.00010^{*}x^2$; UV-DNA
regression: $y = 0.05321 + 0.01794^{***}x - 0.00021^{****}x^2$. Figure 2B: DNA-UV
regression: $y = -0.00573 - 0.01015^{****}x$; UV-DNA regression: $y = 0.11315 +
0.01374^{*}x - 0.00019^{**}x^2$. $^{*}P < .05,\ ^{**}P < .01,\ ^{***}P < .001,\ ^{****}P < .0001$. Figure
taken from Hoelzer and Michod[27]; courtesy of the Genetics Society of America ©1991.

TABLE 1 Summary of Experiments on the Evolution of Transformation in *B. Subtilis*

Experiment	Factors Addressed	UV-DNA Result	DNA-UV Result
Donor Plasmid DNA	Induction of Recombination or Trans Repair	No Increase	No Increase
Donor Cloned *B. subtilis* Genes	Induction of Recombination or Trans Repair	Increase	Increase
recA1 Recipient	Induction of Excision Repair and SOS Repair	Increase	Smaller Increase
uvrA42 Recipient	Excision Repair	Increase	No Increase
Din Fusion Expression	Induction of SOS System	Same in Competent and Noncompetent Cells	
Donor Damaged DNA	Recombination Targeted to Damaged Sites	Increase	No Increase

There are other inducible factors not directly related to the transformation process, such as excision repair or SOS repair, which may influence the relative survival abilities of competent and noncompetent cells and these factors may contribute to the above-mentioned differences between transformed and nontransformed cells. For example, the SOS-like system for DNA repair and mutagenesis is elicited during competence development in *B. subtilis* cells, independent of any exposure to DNA damaging agents.[34]

To examine these alternative hypotheses and the contribution of the confounding factors just mentioned, we have performed experiments similar to those described above in which we have measured frequencies of homologous DNA-mediated transformation using as recipients two mutant strains that are deficient in excision and SOS repair. We have also conducted experiments using different kinds of donor DNA besides undamaged homologous chromosomal DNA. The additional kinds of donor DNA used were non-homologous plasmid DNA, cloned fragments containing *B. subtilis* genes, and damaged homologous chromosomal DNA. We have also employed the use of operon fusions to study damage inducible genes under the regulatory control of the SOS-like system as a means to assay for their expression in both competent and noncompetent cell fractions in response to DNA-damaging treatments.[58] The results of these experiments are summarized in Table 1.

Our experiments with the repair-deficient strains (recA1, Rec⁻, and uvrA42, Uvr⁻) demonstrate a similar qualitative difference in the transformation frequency between UV-DNA and DNA-UV treatments as seen with the wild-type strain (Figure 3–5 of Wojciechowski, Hoelzer and Michod[58]). These results indicate that our different results in the UV-DNA and DNA-UV experiments do not depend upon the induction of the SOS-like response or excision repair (Table 1). Moreover, the observed increase in the transformation frequency with UV cannot be explained just by a differential induction of the SOS-like system and its associated repair in competent cells, since SOS-like repair functions (din operons) are induced to a similar level in noncompetent cells (Table 5 of Wojciechowski, Hoelzer and Michod[58]).

The result, that transformation frequencies increase with UV in UV-DNA experiments, goes away if wild-type cells are transformed with non-homologous plasmid DNA, even though in DNA-UV experiments, plasmid transformation frequencies behave in a manner similar to that observed when cells were transformed with homologous chromosomal DNA (Figure 2 of Wojciechowski, Hoelzer and Michod[58]). This result is consistent with there being a repair advantage to homologous transforming DNA, but strongly suggests that the increase in homologous DNA transformation in UV-DNA experiments is not mediated through a rec-independent process such as an increase in the binding or uptake of DNA. However, the increase in both DNA-UV and UV-DNA experiments using cloned *B. subtilis* genes suggests that damages directly stimulate transformation.

In all of our experiments described above, DNA purified from undamaged cells is used to transform recipients that have been treated with UV radiation. Since under natural conditions the DNA available to transforming cells probably comes from cells, living or dead, that have been exposed to the same damaging environment as the recipients, an important question is whether the results change if cells are transformed with damaged DNA. To address this, transforming DNA was isolated from a competent culture that had either been exposed to UV radiation ($50\,J/m^2$) or not ($0\,J/m^2$), and these two kinds of DNA were used as donor in subsequent UV-DNA and DNA-UV experiments with a wild-type strain.[27] Like our previous results,[42,58] and using donor DNA isolated from undamaged competent cells as control (Figure 2(a)), we[27] find a similar qualitative difference in the frequency of transformation (about two fold at a fluence of 25 to $50\,J/m^2$ to the recipients) between the two treatments when competent cells are transformed with DNA isolated from damaged competent cells (Figure 2(b)). There is no statistical difference between experiments in which the donor DNA was either damaged or undamaged. The difference between the UV-DNA and DNA-UV regressions is highly significant ($P = .0001$) for both damaged and undamaged donor DNA.

The result that damaged donor DNA yields the same result as undamaged donor DNA suggests that transformation is targeted to damaged sites. If this were not the case and donor DNA was recombined at random (no targeting), then there should be no net advantage to transformation in the experiments in which donor and recipient DNA were equally damaged. This would be the case since, some of the time, damaged donor DNA would have been recombined for good recipient DNA and, some of the time, good donor DNA would have been recombined for bad recipient DNA. However, if recombination is targeted to damaged sites in the recipient's genome, damaged donor DNA should serve the purpose just as well as undamaged donor DNA. This is because it is highly unlikely that both the incoming DNA and recipient cell's DNA would be damaged at the same specific site. We are now seeking molecular confirmation of this interpretation that recombination is targeted to damaged sites.

There is abundant evidence that recombination functions in DNA repair both in haploids and diploids (for review see Bernstein, Hopf and Michod[10]). However, the problem of outcrossing is trickier. As already mentioned in haploids, outcrossing is necessary for efficient repair. However, in diploids this is no longer true. We now turn our attention to the problem of outcrossing in diploids. Specifically we wish to know why recombinational repair usually, but not universally, occurs in diploids in the context of outcrossing mating systems instead of selfing mating systems.

MUTATION AND THE EVOLUTION OF SEX

Haldane[24] argued that, in equilibrium populations, the effect of deleterious mutation on average fitness depends primarily on the mutation rate and is independent of the severity of the mutations, s. Specifically, the equilibrium average fitness is e^{μ_H}, where μ_H is the haploid genomic mutation rate.

We have shown that Haldane's result extends to a variety of reproductive systems, assuming multiplicative fitness effects of the mutations.[29] Specifically, we have shown that Haldane's principle holds exactly for haploid sex, haploid apomixis, and facultative haploid sex. In the cases of diploid automixis with terminal fusion, diploid automixis with central fusion, and diploid selfing, Haldane's principle holds exactly for recessive mutations, and approximately for mutations with some heterozygous effect. In the cases of apomixis with higher than diploid ploidy levels, diploid endomitosis, and haplodiploidy, we have shown that Haldane's principle holds exactly for recessive lethal mutations. In addition, we have extended Haldane's result to various mixtures of the above-mentioned reproductive systems. In the case of diploid outcrossing sexuals, we were unable to obtain an exact analytic result, but presented arguments and computer simulations which showed that Haldane's result extends to this case as well in the limit as the number of loci gets large.

Although diverse reproductive systems are equally fit at equilibrium, different reproductive systems harbor vastly different numbers of recessive genes at equilibrium. These different numbers of mutations may create transient selective pressures on individuals with reproductive systems different from that of the equilibrium population.

For example, theoretical estimates for the case of recessive lethals were that selfing maintained $2s$ and outcrossing maintained $\sqrt{N\mu_H}$ mutations, where N is the number of functional genes per genome. The number of functional genes is, in higher organisms, a number in between 40,000 and 100,000, and the haploid genomic mutation rate, μ_H, is a number between 0.5 and 1.0.

Bernstein, Byerly, Hopf, and Michod[7] argued verbally that these different numbers of mutations could create transient selection favoring outcrossing, even though all reproductive systems are equally fit at equilibrium (as far as mutational load at multiplicative loci is concerned). For example, in a completely selfing population, an outcrosser would mask all recessive lethals and have a fitness of one relative to "the fitness of selfers," which equals 0.60 if $\mu_H = 0.5$. Likewise in an outcrossing population, a rare selfer would most likely express one of the many hidden recessive lethals, whereas outcrossers would have a fitness of e^{μ_H}. Without an explicit model taking into account, among other factors, the more effective transmission of selfing alleles than outcrossing alleles,[21] it is difficult to evaluate this verbal argument. For this reason the masking model has been developed.[46]

Consider a single Mendelian locus with two alleles, M and m, that modify the probability of selfing, α, and probability of outcrossing, $1 - \alpha$. There are three genotypes at this locus, MM, Mm and m, with probabilities of selfing being α_1, α_2, and α_3, respectively. Now assume that there are three fitness phenotypes which relate to the state of the genome regarding deleterious mutations: Het, Hom-wt, and Hom-mut. The three phenotypes considered represent extreme states of the genome as far as the masking hypothesis is concerned. Hom-wt individuals do not contain any mutations; Het individuals contain mutations but don't effectively express them because the mutations are heterozygous; and Hom-mut individuals both contain mutations and express them because the mutations are homozygous. More specifically it is assumed that Het are heterozygous for at least one locus at which rare recessive or partially recessive deleterious mutations are segregating (the others are the wild type); Hom-wt are homozygous for wild-type alleles at all of these loci; and Hom-mut are homozygous for deleterious mutations at one or more loci. The fitnesses of these three types are assumed to be 1, $1 - hs$, and $1 - s$ for Hom-wt, Het, and Hom-mut, respective of the exact number of loci which are heterozygous and homozygous. The parameter h measures dominance and the parameter s is the selection coefficient against the mutations. In a real population there is a distribution of different types, each with different numbers of homozygous and heterozygous loci and this distribution evolves as the mating system evolves. Here, I consider just three types but allow their frequencies to change as the mating system evolves.

The underlying genetical control of the three phenotypes Hom-wt, Het, and Hom-mut is not explicitly represented in the model, although assumptions are made concerning the transmission of these phenotypes which reflect genetical issues (Table 2). In Table 2 the phenotypes of the parents are given across the top and the phenotypes of offspring down the left side. Outcrossing parents are assumed to produce all Het offspring, regardless of whether the parent is Hom-wt, Het, or Hom-mut. This is because mates will likely mask the expression of each other's recessive, or nearly recessive, deleterious mutations, since they likely carry mutations at $different$ loci.

Although the offspring of outcrossing parents are assumed to be the same, selfed offspring differ depending upon which of the three parental types produce them. In the masking model, the fitness of an offspring produced by selfing depends on which parent (*Het, Hom-mut,* or *Hom-wt*) produced it. Consequently, the average fitness of selfed offspring changes as the population frequencies of the parental types change. This is a basic difference that distinguishes the masking model from many inbreeding depression models that assume a constant fitness of selfed offspring (e.g., Feldman and Christiansen[20] among many).

Consider, first, *Hom-mut* parents. Since they are homozygous for deleterious mutations at one or me loci, all their offspring must also be homozygous for these mutations. Additional mutations that may occur during the parents' lifetime will only enhance the fact that all their selfed offspring must be *Hom-mut*. Second, consider *Hom-wt* parents. Although *Hom-wt* parents begin life with no mutations, they may accumulate an additional mutation with probability $1 - \mu_D$, the genomic diploid mutation rate. With probability μ_D, no new mutations occur and all their selfed offspring will be *Hom-wt*. If a new mutation occurs (again with probability μ_D), their selfed offspring will be *Hom-wt, Het,* or *Hom-mut* with probabilities 1/4, 1/2, and 1/4, respectively. (The occurrence of multiple mutations is ignored.)

Finally, consider *Het* parents. As a result of recombination, *Het* parents may produce offspring of any of the three types. In reality, the frequencies of these types among selfed offspring of *Het* parents is determined by the multi-locus segregation and recombination probabilities at loci that are polymorphic for deleterious mutations. However, the masking model sidesteps this genetical issue by assuming constant probabilities, q and k, that a *Het* parent produces by selfing *Hom-wt* and *Het* offspring, respectively, with $1 - k - q$ being the probability that a *Het* parent produces *Hom-mut* offspring by selfing. The parameter k measures the masking ability of selfers relative to a masking ability of 1 for outcrossers. The parameter q measures the probability that selfing purges deleterious mutations from the genome (by recombination and segregation), producing an offspring that is free of deleterious mutations. In a real population, k and q will change as the frequency of deleterious mutations evolves in response to the mating system.

TABLE 2 Transmission of Phenotypes in Masking Model (out = outcrossed)

	Hom-wt		Het		Hom-mut	
	selfed	out	selfed	out	selfed	out
Hom-wt	$1 - \mu_D + (\mu_D/4)$	0	q	0	0	0
Het	$\mu_D/2$	1	k	1	0	1
Hom-mut	$\mu_D/4$	0	$1 - k - q$	0	1	0

Nine variables are needed to describe this system corresponding to the frequencies of the nine types (three genotypes at the mating system locus combined with three fitness phenotypes). The system of nine nonlinear difference equations is presented elsewhere.[46] Using this system of equations, we have studied the stability of either complete outcrossing or complete selfing to invasion by mutant alleles at the mating-type locus that promote the alternative reproductive system. In most cases, outcrossing is stable when common to the increase of rare alleles promoting selfing so long as s is near 1 (mutations are lethal or sublethal), h is near 0 (mutations are nearly recessive), $k < 1/2$ (outcrossers produce twice as many heterozygotes as do selfers), and $q < 1/4$ (the probability that a heterozygous selfer purges its genome of mutations is less than one-fourth). The conditions for outcrossing to increase when rare vary with the different equilibria considered. If the selfing population is fixed for mutant homozygotes, outcrossing often increases. If the selfing population has wild-type individuals (whose genomes are purged of deleterious mutations), it is more difficult for outcrossing to increase and it can only do so if the mutation rate is high and the probability of purging is close to zero.

The parameters k and q will usually be less than $1/2$, with q being less than k and close to zero. For any particular locus, the frequency of heterozygotes among selfed offspring must be $1/2$ by Mendel's Laws. However, the probability that a selfer produces either heterozygotes or wild-type homozygotes at all such fitness determining loci will be less than $1/2$, so long as the selfer is heterozygous at more than two such loci. For example, if an individual carried n mutations in heterozygous condition, all on separate chromosomes, the probability that a selfer produces an offspring that is heterozygous at all loci is $(1/2)^n$. However, we also include in Het the individuals that are heterozygous at $n - i$ loci and homozygous wild-type at i loci $(0 < i < n)$. The overall probability of Het offspring is then

$$\sum T_{i=0}^{n-1} \binom{n}{i}(\frac{1}{2})^{n-i}(\frac{1}{4})^i$$

which is equal to $1/2$, if $n = 1, 2$ and less than $1/2$ if $n > 2$. The probability that a selfer produces an offspring which is homozygous for the wild-type allele at all loci (in other words, a Hom-wt offspring) is $q = (1/4)^n$. For these reasons, $1/2$ and $1/4$ are taken to be upper limits for k and q, respectively. Most offspring, one minus the probabilities of being Het and Hom-wt, will be Hom-mut, that is homozygous for deleterios alleles for at least one of the loci. With five such loci, more than 75% of the offspring will be Hom-mut.

Available data indicates that the stability conditions discussed above for outcrossing are likely to be met. For recessive lethal alleles $(s = 1.0, h = 0)$, any k and q such that $h + q < 1/2$ is sufficient for outcrossing to be stable when common. If there are more than two lethal alleles in the genome, this condition is expected to be satisfied. For human populations "almost everyone carries the equivalent of more than one lethal recessive gene in the heterozygous condition."[14] After their review of the data, Lande and Schemske[30] conclude that "typical individuals in large outbreeding populations are heterozygous for one or more nearly recessive lethal (or sublethal) genes." The level of lethal mutations will be much lower in predominantly selfing populations,[28,30] and once selfers are able to purge the genome of mutation, outcrossers have more difficulty getting started. The conclusion of the masking model, that outcrossing is favored when common if there is more than two semi-lethal mutations, is probably widely met in Nature.

SEX AND DARWIN'S DILEMMAS

The presence of sexual reproduction forces us to abandon the rigidly adaptationist view of evolution held by Darwin and by many subsequent thinkers. The reason for this is simply that the coming together of two individuals to produce offspring creates difficulties in reproduction when the species (or new coadapted type) is rare, with a resulting decrease in fitness, and a failure of the law of geometric increase. Ironically, it is in this limit of rarity that the law of Malthus, which is based on a density-independent fitness, is supposed to be valid. The Law of Malthus requires $u = 0$ (assuming $v = 0$), while with sexual reproduction $u > 0$.

The differences between the adaptationist paradigm (Malthusian evolution) and non-Malthusian evolution (survival of the first paradigm) are most striking when one considers the problem of the distinctness of species.[7,8,28] In a continuous environment, the adaptationist position leads to an indefinite proliferation of types, each specializing to different combinations of resource (see Figure 3(a)). Based on such reasoning Darwin posed a dilemma that has not been adequately addressed by the adaptationist paradigm[16]:

> Why, if species have descended from other species by insensibly fine gradations, do we not see innumerable transitional forms? Why is not all nature in confusion instead of the species being, as we see them, well defined? (p. 171)

> As according to the theory of natural selection, an interminable number of intermediate forms must have existed, linking together all the species in each group by gradations as fine as are our existing varieties. It may be asked: Why do we not see these linking forms all around us? Why are not all organic beings blended together in an inextricable chaos? (p. 462)

One adaptationist position taken by Darwin[16] argues that species intermediate to current species are not observed because (i) the niches which are intermediate between those occupied by any two species are small, consequently (ii) an intermediate (or linking) species, if it existed, would have a small population size, and (iii) small populations evolve slower than large populations and therefore will be eliminated by competition from the more common species to either side.

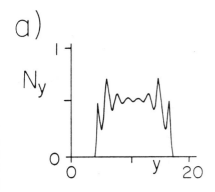

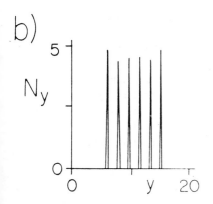

FIGURE 3 Density N_y of type of organism labeled y along a resource grade for (a) the case where the types are clones with no cost of rarity and (b) the case where the types are sexually reproducing species with a cost of rarity. Both distributions are the final, stable result reached after long times. In both calculations, all types were assumed present initially. In both cases there is a continuous resource grade over the interval 0 to 20. Resource abundance is, in the absence of consumption by the types, taken to be trapezoidal in shape; the trapezoid is constant over the interval 3 to 17, and falls to zero at the points 0 and 20. Figure taken from Bernstein et al.[9]; courtesy of the Journal of Theoretical Biology ©1985.

Darwin realized that this resolution to his dilemma of missing links in habitat space gave rise to a second dilemma of missing links in time because all these intermediate species which go extinct by competition should exist in the fossil record. As is well known, he resolved this second dilemma by appealing to "the extreme imperfection of the geological record."[16]

As discussed elsewhere,[5] if most living systems obeyed the adaptationist paradigm, nature *would* exist in "confusion" as an, apparently, senseless proliferation of types, each adapted to some slight difference in habitat. The fact that nature exists as a collection of, more or less, well-defined species suggests that the adaptationist position is incorrect. As shown in Figure 3(b) the cost of rarity destabilizes a continuous distribution of species and produces distinct species even on a continuous environmental grade.[7,8,28] Thus we expect asexual types, which have no intrinsic cost of rarity, to exhibit the proliferation expected by Darwin and to be more finely adapted than sexual species to environmental variation. Consequently, we feel that the above dilemmas posed by Darwin depend on his adaptationist perspective and can be resolved by the cost of rarity which is the basis of the survival of the first paradigm.

SUMMARY

We have viewed natural selection as a dynamical process depending on heritable adaptive capacities, genetic and reproductive systems, and the density and frequency of types. Doing so led us to consider a simple classification of these dynamics. From this classification we arrived at three paradigms: the survival of the fittest or adaptationist paradigm (which is based on the Law of Malthus), and two non-Malthusian paradigms, the survival of the first (which is based on a cost of rarity), and the cost of commonness. We noted that the adaptationist position is fragile, depending as it does on a strict equality of parameters ($u = v$). With these paradigms in mind, we investigated the origin of basic features of the biota: the presence of life-history, the separation of the genotype and phenotype, individuality, sex, and the apparent immortality of the DNA. For simple molecules replicating by template-mediated polymerization of complementary strands, the phenotype (shape and length) was initially coupled with the genotype (sequence of nucleotides). The more adapted sequences prospered in such a system, but diversification was severely limited without the production of proteins. For the phenotype to diverge from the genotype required the replicator to make proteins. The problem of the origin of proteins was the first encounter with non-Malthusian (cost of rarity) evolution. Later on, with the invention of sex between individual cells, evolution would become non-Malthusian for good. Population structure in the form of passive or self structure may have helped in the evolution of the first protein. The evolution of the individual cell was the outcome of the need for structure in the distribution of proteins. However, with the invention of individuality, evolution had to solve a new problem. Genetic errors became trapped inside the cell creating gene-deficient cells. Sex had come easily to the naked replicators, but now imprisoned in a cell of their own creation, the replicator had to figure a way to deal with genetic error. Sex had to be reinvented.

Sex has two basic components, recombination and outcrossing. In haploids both components are necessary for efficient damage repair. An experimental system, natural transformation in bacteria, provides a test of these ideas. Once diploidy emerged as the dominant stage of the life cycle, outcrossing was no longer needed for damage repair, although recombination still maintained this function. In diploids, recessive deleterious mutations build up and may prevent the abandonment of outcrossing in favor of closed systems of damage repair such as selfing. Mathematical work bearing on this idea supports the view that deleterious mutation can generate a barrier to the evolution of selfing. However, this barrier is by no means absolute.

With the evolution of sex between cells and organisms, the evolutionary process became non-Malthusian. The non-Malthusian character of evolution in sexual populations explain the stability of species as distinct entities and can resolve Darwin's two fundamental dilemmas: missing links in habitat space and missing links in time.

ACKNOWLEDGEMENTS

I thank Harris Bernstein and the late Fred Hopf for many stimulating discussions on the material covered in these lectures and Sally Otto and Todd Gayley for discussion and comments on the manuscript. This work was supported by NIH Grants HD-19949 and GM-36410.

REFERENCES

1. Bell, G. *The Masterpiece of Nature: The Evolution and Genetics of Sexuality.* Berkeley, CA: University of California Press, 1982.
2. Bernstein, H. "Germ Line Recombination may be Primarily a Manifestation of DNA Repair Processes." *J. Theor. Biol.* **69** (1977): 371–380.
3. Bernstein, H. H. "Recombinational Repair may be an Important Function of Sexual Reproduction." *BioScience* **33** (1983): 326–331.
4. Bernstein, H. H., Byers, and R. E. Michod. "Evolution of Sexual Reproduction: Importance of DNA Repair, Complementation and Variation." *Am. Nat.* **117** (1981): 537–549.
5. Bernstein, H. H., C. Byerly, F. A. Hopf, R. E. Michod, and G. K. Vemulapalli. "The Darwinian Dynamic." *Quart. Rev. Biol.* **58** (1983): 185–207.
6. Bernstein, H. H., C. Byerly, F. A. Hopf, and R. E. Michod. "Origin of Sex." *J. Theor. Biol.* **110** (1984): 323–351.
7. Bernstein, H. H., C. Byerly, F. A. Hopf, and R. E. Michod. "Genetic Damage, Mutation, and the Evolution of Sex." *Science* **229** (1985): 1277–1281.
8. Bernstein, H., H. C. Byerly, F. A. Hopf, and R. E. Michod. "Sex and the Emergence of Species." *J. Theor. Biol.* **117** (1985): 665–690.
9. Bernstein, H., H. F. A. Hopf, and R. E. Michod. "The Role of DNA Repair in Sexual Reproduction." In *Sociobiology of Reproductive Strategies in Animals and Man*, edited by Rosa, C. Vogel, and E. Voland. London: Chapman and Hall, 1989.
10. Bernstein, H., F. A. Hopf, and R. E. Michod. "The Molecular Basis of the Evolution of Sex." *Advances in Genetics* **24** (1987): 323–370.
11. Boerlijst, M. C., and P. Hogeweg. "Spiral Wave Structure in Pre-biotic Evolution: Hypercycles Stable Against Parasites." Submitted to *Physica D.*
12. Brown, J. S., M. Sanderson, and R. E. Michod. "Evolution of Social Behavior by Reciprocation." *J. Theor. Biol.* **99** (1982): 319–339.
13. Byerly, H., and R. E. Michod. "Fitness and Evolutionary Explanation." *J. Biology and Philosophy*, in press.
14. Cavalli-Sforza, L. L., and W. F. Bodmer. *The Genetics of Human Populations.* San Francisco: W. H. Freeman, 1983.

15. Crow, J. F. and M. Kimura. *An Introduction to Population Genetics Theory.* Minneapolis, MN: Burgress, 1983.

16. Darwin, C. *On the Origin of Species by Means of Natural Selection.* London: John Murray, 1859.

17. Dubnau, D. A. "Genetic Transformation in Bacillus subtilis." In *The Molecular Biology of the Bacilli,* edited by D. A. Dubnau, Vol. 1, 148–175. New York: Academic Press, 1983.

18. Eigen, M. "Self-Organization of Matter and the Evolution of Biological Macromolecules." *Naturwissenschaften* **58** (1983): 465–523.

19. Eigen, M., and P. Schuster. *The Hypercycle: A Principle of Natural Self-Organization.* Berlin: Springer-Verlag, 1983.

20. Feldman, M. W., and F. B. Christiansen. "Population Genetic Theory of the Cost of Inbreeding." *Am. Nat.* **123** (1983): 642–653.

21. Fisher, R. A. "Average Excess and Average Effect of a Gene Substitution." *Ann. Eugen.* **11** (1941): 53–63.

22. Fisher, R. A. *The Genetical Theory of Natural Selection,* 2nd ed. New York: Dover, 1958.

23. Haldane, J. B. S. *The Causes of Evolution.* London: Longmans, Green, 1932.

24. Haldane, J. B. S. "The Effect of Variation on Fitness." *Am. Nat.* **71** (1937): 337–349.

25. Hendry, L. B., E. D. Bransome, Jr., M. S. Hutson, and L. K. Campbell. "First Approximation of a Stereochemical Rationale for the Genetic Code Based on the Topography and Physicochemical Properties of 'Cavities' Constructed from Models of DNA." *Proc. Natl. Acad. Sci. U.S.A.* **78** (1981): 7440–7444.

26. Hendry, L. B., E. D. Bransome, Jr., and M. Petersheim. "Are There Structural Analogies Between Amino Acids and Nucleic Acids?" *Origins of Life* **11** (1981): 203–221.

27. Hoelzer, M. and R. E. Michod. "DNA Repair and the Evolution of Transformation in the Bacterium Bacillus Subtilis. III. Sex with Damaged DNA." *Genetics,* in press.

28. Hopf, F. A., and F. W. Hopf. "The Role of the Allee Effect in Species Packing." *Theor. Popul. Biol.* **27** (1985): 27–50.

29. Hopf, R., R. E. Michod, and M. Sanderson. "The Effect of Reproductive System on Mutation Load." *Theor. Popul. Biol.* **33** (1988): 243–265.

30. Lande, R., and D. W. Schemske. "The Evolution of Self-Fertilization and Inbreeding Depression in Plants. I. Genetic Models." *Evolution* **39** (1985): 24–40.

31. Lewontin, R. C. "The Units of Selection." *Ann. Rev. Ecol. System.* **1** (1970): 1–18.

32. Lohrmann, R., P. K. Bridson, and L. E. Orgel. "Efficient Metal-Ion Catalyzed Template-Directed Oligonucleotide Synthesis." *Science* **208** (1980): 1464–1465.

33. Lohrmann, R., and L. E. Orgel. "Efficient Catalysis of Polycytitylic Acid-Directed Oligoguanylate Formation by PB2+." *J. Mol. Biol.* **142** (1980): 555–567.

34. Love, P. E., M. J. Lyle, and R. E. Yasbin. "DNA-Damage-Inducible (din) Loci Are Transcriptionally Activated in Competent Bacillus subtilis." *Proc. Natl. Acad. Sci. USA* **82** (1985): 6201–6205.

35. Maynard Smith, J. *The Evolution of Sex.* Cambridge: Cambridge University Press, 1978.

36. Michod, R. E. "Positive Heuristics in Evolutionary Biology." *Br. J. Philos. Sci.* **32** (1981): 1–36.

37. Michod, R. E. "The Theory of Kin Selection." *Ann. Rev. Ecol. Syst.* **13** (1982): 23–55.

38. Michod, R. E. "Population Biology of the First Replicators: On the Origin of the Genotype, Phenotype and Organism." *Amer. Zool.* **23** (1983): 5–14.

39. Michod, R. E. "Constraints on Adaptation with Special Reference to Social Behavior." In *The New Ecology: Novel Approaches to Interactive Systems*, edited by P. W. Price, C. N. Slobodchikoff, and W. S. Gaud, 253–279. New York: Wiley, 1984.

40. Michod, R. E. "On Adaptedness and Fitness and Their Role in Evolutionary Explanation." *J. History of Biology* **19** (1985): 289–302.

41. Michod, R. E., and B. R. Levin. *The Evolution of Sex: An Examination of Current Ideas.* Sunderland, MA: Sinauer Associates, 1988.

42. Michod, R. E., M. F. Wojciechowski, and M. A. Hoelzer. "DNA Repair and the Evolution of Transformation in the Bacterium Bacillus Subtilis." *Genetics* **118** (1989): 31–39.

43. Michod, R. E. "What's Love Got To Do With It? The Solution to One of Evolution's Greatest Riddles." *The Sciences* (May/June 1989).

44. Michod, R. E., M. F. Wojciechowski, and M. A. Hoelzer. "Origin of Sex in Prokaryotes." In *Molecular Evolution, UCLA Symposia on Molecular and Cellular Biology*, edited by M. Clegg and S. Clark. New Series, Volume 122, 1989

45. Michod, R. E. "Sex as a Conservative Force in Evolution." In *Organizational Constraints on the Dynamics of Evolution*, edited by G. Vida and E. Szathmary. Manchester: Manchester University Press, 1990.

46. Michod, R. E., and T. W. Gayley. "On Evolution of Outcrossing and Sex." *Am. Nat.*, in press.

47. Muller, H. J. "Some Genetic Aspects of Sex." *Am. Nat.* **66** (1932): 118–138.

48. Roughgarden, J. *Theory of Population Genetics and Evolutionary Ecology: An Introduction.* New York: Macmillan, 1979.

49. Schuster, P. "Prebiotic Evolution." In *Biochemical Evolution*, edited by H. Gutfreund, 15–87. Cambridge: Cambridge University Press, 1980.

50. Shields, W. M. "Philopatry, Inbreeding, and the Adaptive Advantages of Sex." State University of New York, Albany, New York, 1979.

51. Stewart, G. J., and C. A. Carlson. "The Biology of Natural Transformation." *Ann. Rev. Microbiol.* **40** (1986): 211–235.

52. Uyenoyama, M. K. "On the Evolution of Parthenogenesis: A Genetics Representation of the 'Cost of Meiosis.'" *Evolution* **38** (1984): 87–102.

53. Uyenoyama, M. K. "On the Evolution of Parthenogenesis. II. Inbreeding and the Cost of Meiosis." *Evolution* **39** (1985): 1194–1206.

54. Van Roode, J. H. G., and L. E. Orgel. "Template-Directed Synthesis of Oligoguanylates in the Presence of Metal Ions." *J. Mol. Biol.* **144** (1980): 579–585.

55. Williams, G. C. *Sex and Evolution.* Princeton: Princeton University Press, 1975.

56. Williams, G. C. "Kin Selection and the Paradox of Sexuality." In *Sociobiology: Beyond Nature/Nurture? Reports, Definitions and Debate,* edited by G. W. Barlow and J. Silverman, 371–384. Boulder, CO: Westview Press, 1980.

57. Woese, C. R. *The Genetic Code: The Molecular Basis for Genetic Expression.* New York: Harper Row, 1967.

58. Wojciechowski, M. F., M. A. Hoelzer, and R. E. Michod. "DNA Repair and the Evolution of Transformation in Bacillus subtilis. II. Role of Inducible Repair." *Genetics* **121** (1989): 411–422.

II Seminars

Mark E. Nelson and James M. Bower
Computation and Neural Systems Program, Division of·Biology 216-76,
California Institute of Technology, Pasadena, CA 91125

Dynamics of Neural Excitability

An understanding of the dynamic properties of single neurons is essential for analyzing the behavior of networks of neurons and for developing theoretical models of neural information processing. In this chapter we introduce the generalized Hodgkin-Huxley formalism for modeling voltage-dependent and time-dependent changes in membrane conductance that underlie neural excitability. As an example, we examine the Morris-Lecar model, a simple two-conductance system that exhibits a variety of interesting dynamic properties including oscillation, bistability, and threshold behavior. We will illustrate the use of phase space techniques for analyzing the dynamic properties of this simple model of electrical excitability in a single neuron.

THE GENERALIZED HODGKIN-HUXLEY MODEL

In 1952, Hodgkin and Huxley published a mathematical model for the flow of electric current across the nerve membrane of the squid giant axon.[5] At the time it was proposed, the Hodgkin-Huxley (HH) model represented a breakthrough in the understanding of the biophysical events associated with neuronal signal processing and it continues to serve as a foundation for most of the biophysical modeling of

1990 Lectures in Complex Systems, SFI Studies in the Sciences of
Complexity, Lect. Vol. III, Eds. L. Nadel and D. Stein, Addison-Wesley, 1991

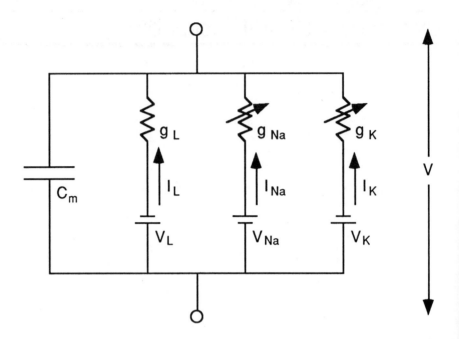

FIGURE 1   Electrical equivalent circuit proposed by the Hodgkin-Huxley model for a short segment of squid giant axon. The variable resistances represent voltage-dependent conductances.

single neurons that is done today.[7] Hodgkin and Huxley proposed that a segment of nerve membrane could be modeled by an electrical equivalent circuit of the type shown in Figure 1. For their model of the squid giant axon, they postulated three independent ionic currents, a sodium current I_{Na}, a potassium current I_K, and a non-specific "leakage" current I_L, as well as a capacitive current associated with the thin membrane that surrounds the axon.

The behavior of an electrical circuit of the type shown Figure 1, can be described by the equation:

$$C_m \frac{dV}{dt} + I_{ion} = I_{ext} \tag{1.1}$$

where C_m is the membrane capacitance, V is the potential difference across the membrane, I_{ion} is the net ionic current flowing through the membrane channels, and I_{ext} is an externally applied current. Each component of the ionic current I_j has an associated conductance g_j and an equilibrium potential V_j (the potential at which no current flows). The net ionic current flowing across the membrane is given by:

$$I_{ion} = \sum_j I_j = \sum_j g_j(V - V_j).$$
(1.2)

In the case of the squid giant axon model, we have:

$$I_{ion} = I_{Na} + I_K + I_L = g_{Na}(V - V_{Na}) + g_K(V - V_K) + g_L(V - V_L).$$
(1.3)

NONLINEAR CONDUCTANCES

In general, the ionic conductances g_j are not constant, but can depend on other biophysical parameters like membrane voltage or ionic concentrations. In the HH model for the squid giant axon, for example, the leakage conductance g_L is assumed to be constant, but the ionic conductances g_{Na} and g_K are voltage-dependent. The biophysical basis for this voltage dependence can be traced to the molecular interactions that occur when ions attempt to pass through microscopic pores or channels in the neuron membrane. The macroscopic conductances in the HH model arise from the combined effects of large numbers of these microscopic channels. For the purpose of understanding how the HH model characterizes these voltage-dependent conductances, it will be convenient to use the following metaphor for describing the behavior of membrane channels.[4]

Each membrane channel can be thought of as containing a number of gates. Each gate can be in one of two states, "permissive" or "non-permissive." When all of the gates for a particular channel are in the permissive state, ions can pass through the channel and the channel is considered "open." If any of the gates are in the non-permissive state, ions cannot flow and the channel is "closed." Each type of gate i is characterized by a state-variable w_i that represents the fraction of gates of that type that are in the permissive state. The fraction of gates that are in the non-permissive state is thus $(1 - w_i)$. Transitions between the permissive and non-permissive states are assumed to obey first-order kinetics:

$$\frac{dw_i}{dt} = \alpha_i(V)(1 - w_i) - \beta_i(V)w_i$$
(1.4)

where $\alpha_i(V)$ and $\beta_i(V)$ are voltage-dependent rate constants that describe the non-permissive→permissive and permissive→non-permissive transition rates as a function of membrane voltage. If the membrane voltage V is held at a fixed value, the state variable w_i eventually reaches a steady state value as $t \to \infty$. This equilibrium value is given by:

$$w_{i,\infty}(V) = \frac{\alpha_i(V)}{\alpha_i(V) + \beta_i(V)}.$$
(1.5)

The time course for approaching this equilibrium value is described by a simple exponential with time constant $\tau_i(V)$ given by:

$$\tau_i(V) = \frac{1}{\alpha_i(V) + \beta_i(V)} \ . \tag{1.6}$$

Using Eqs. (1.5) and (1.6), we can rewrite Eq. (1.4) as:

$$\frac{dw_i}{dt} = \frac{w_{i,\infty}(V) - w_i}{\tau_i(V)} \tag{1.7}$$

The macroscopic conductance for a particular channel type is proportional to the number of channels in the open state which is, in turn, proportional to the probability that all of the constituent gates are in their permissive state. Thus, the macroscopic conductance due to channels of type j is proportional to a product of the state variables w_i of the constituent gates:

$$g_j = \bar{g}_j \prod_i w_i \tag{1.8}$$

where $\bar{g}_j$ is the peak conductance when all of the channels are open (all of the constituent gates in the permissive state). For example, in the HH model for the squid giant axon, the Na channel consists of four gates, three identical gates of a type labeled "m" and one gate of type "h." Each gate type has a corresponding state variable, m or h, that obeys Eqs. (1.4)-(1.7). In this model, the macroscopic Na conductance is given by:

$$g_{Na} = \bar{g}_{Na} m^3 h \tag{1.9}$$

where $\bar{g}_{Na}$ is the peak conductance. Similarly, the K channel is modeled as consisting of four identical "n" gates and the macroscopic K conductance is given by:

$$g_K = \bar{g}_K n^4 \tag{1.10}$$

where $\bar{g}_K$ is the peak K conductance.

THE MORRIS-LECAR MODEL

To illustrate the dynamic properties that arise in HH-type models, we are going to consider a model system that is somewhat simpler than that of the squid giant axon, namely the Morris-Lecar model of membrane excitability in the barnacle muscle fiber.[8] The electrical equivalent circuit for this model is identical to that shown in Figure 1 for the squid giant axon, except that in the barnacle muscle, the Na current is replaced by a Ca current. The behavior of this system is described by a similar set of equations:

$$C_m \frac{dV}{dt} + I_{ion} = I_{ext} \tag{1.11}$$

$$I_{ion} = g_L(V - V_L) + g_{Ca}(V - V_{Ca}) + g_K(V - V_K) \qquad (1.12)$$

In this model, the Ca and K channels are each modeled as containing a single gate.

$$g_{Ca} = \bar{g}_{Ca} m \qquad (1.13)$$

$$g_K = \bar{g}_K n \qquad (1.14)$$

Thus, not only do the state variables m and n represent the fraction of gates in the permissive state, they also represent the fraction of Ca and K channels in the open state. The state variables m and n obey equations of the form given in Eq. (1.7):

$$\frac{dm}{dt} = \frac{m_\infty(V) - m}{\tau_m(V)} \qquad (1.15)$$

$$\frac{dn}{dt} = \frac{n_\infty(V) - n}{\tau_n(V)} \qquad (1.16)$$

The equilibrium values are modeled by an expression with a sigmoidal dependence on membrane voltage:

$$m_\infty(V) = \frac{1}{2}\left[1 + \tanh\left(\frac{V - V_1}{V_2}\right)\right] \qquad (1.17)$$

$$n_\infty(V) = \frac{1}{2}\left[1 + \tanh\left(\frac{V - V_3}{V_4}\right)\right] \qquad (1.18)$$

where V_1, V_2, V_3, and V_4 are parameters that determine the offset and slope of the sigmoidal function. The time constants with which these variables approach the equilibrium values are modeled as bell-shaped functions of the voltage:

$$\tau_m = \frac{\bar{\tau}_m}{\cosh[(V - V_1)/2V_2]} \qquad (1.19)$$

$$\tau_n = \frac{\bar{\tau}_n}{\cosh[(V - V_3)/2V_4]} \qquad (1.20)$$

For subsequent analyses in this chapter, it will be convenient to consider a simplification of the Morris-Lecar model in which the time constant for the Ca system is assumed to be much shorter than that for the K system. In this case, the Ca state variable m can be considered to achieve its equilibrium value instantaneously:

$$m(V, t) = m_\infty(V). \qquad (1.21)$$

Furthermore, it will be convenient to recast the Morris-Lecar equations in a dimensionless form in which the coefficients are of order unity. This is accomplished by rescaling voltages by V_{Ca} and conductances by a reference value[9] $\bar{G}_{ref}$. All subsequent analyses in this chapter will make use of the following equations for the reduced Morris-Lecar model in dimensionless form:

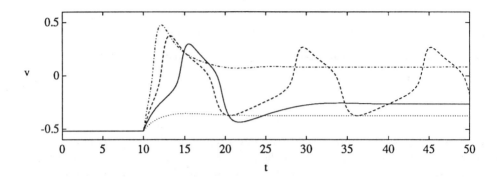

FIGURE 2 Responses of the reduced Morris-Lecar model to a series of external current injections. The parameters for this simulation are $v_1 = -0.01, v_2 = 0.15, v_3 = 0.0, v_4 = 0.30, \bar{g}_{Ca} = 1.1, \bar{g}_K = 2.0, \bar{g}_L = 0.5, v_K = -0.7, v_L = -0.5, \bar{\tau}_n = 5.0$. Dotted line: subthreshold response ($i_{ext} = 0.1$); Solid line: single action potential ($i_{ext} = 0.2$); Dashed line: train of action potentials ($i_{ext} = 0.3$); Dot-dashed line: depolarization block ($i_{ext} = 0.5$).

$$\frac{dv}{dt} + i_{ion} = i_{ext} \tag{1.22}$$

$$i_{ion} = \bar{g}_{Ca} m_\infty(v)(v - 1) + \bar{g}_K n(v - v_K) + \bar{g}_L(v - v_L) \tag{1.23}$$

$$\frac{dn}{dt} = \frac{[n_\infty(v) - n]}{\tau_n(v)} \tag{1.24}$$

$$m_\infty(v) = \frac{1}{2}\left[1 + \tanh\left(\frac{v - v_1}{v_2}\right)\right] \tag{1.25}$$

$$n_\infty(v) = \frac{1}{2}\left[1 + \tanh\left(\frac{v - v_3}{v_4}\right)\right] \tag{1.26}$$

$$\tau_n(v) = \frac{\bar{\tau}_n}{\cosh[(v - v_3)/2v_4]} \tag{1.27}$$

Figure 2 illustrates some of the different types of behaviors that can be generated by this model. In each simulation trial, an externally applied current i_{ext} is switched on at time $t = 10$ and held at a constant value thereafter. Depending on the magnitude i_{ext}, the system is observed to respond in qualitatively different manners. When i_{ext} is relatively low, the membrane voltage v increases smoothly to a new resting level somewhat above the original level, as illustrated by the dotted line. For somewhat larger values of i_{ext}, a single event known as an action potential occurs immediately following the current step as illustrated by the solid line. For even larger values of i_{ext}, a continuous train of action potentials is observed as shown by the dashed line. Finally, for large enough i_{ext}, the trains of action potentials disappear and the system is seen to stabilize at a new resting level as shown by the dot-dashed line in Figure 2.

PHASE SPACE ANALYSIS

is a powerful technique for developing insights and intuition into the behavior of nonlinear systems. It has been used to analyze the dynamic properties of neural systems like the Morris-Lecar model just described[8,9] and the Hodgkin-Huxley model of the squid giant axon.[1,2,3] The fundamental approach taken by phase space analysis techniques is to characterize the behavior of a dynamical system geometrically rather than numerically. In order to do this, the system behavior is represented in an abstract phase space where the coordinates of the space are the dependent variables of the equations of motion. The state of the system at any particular time can thus be represented by a single point in this space and particular solutions are represented by the paths traced out by this phase point as the independent variable (time) evolves. Phase space analysis characterizes the behavior of possible solutions by identifying certain points and boundaries in the phase space that determine the general behavior of all paths in the space.

PHASE PLANE REPRESENTATION OF THE MORRIS-LECAR MODEL

As an example of applying phase space techniques to an analysis of neural excitability, we will explore the behavior of the reduced presented above (Eqs. (1.22)-(1.27)). This model has two dependent variables v (proportional to membrane voltage) and n (proportional to K conductance) that evolve as a function of the independent variable t (time). Figure 2 illustrated several qualitatively different behaviors exhibited by this system in response to current injections. We will use phase-space techniques to understand how these different types of solutions arise in this model.

Figure 3 shows the phase plane trajectories for the four cases shown in Figure 2. It is instructive to follow one of the trajectories in detail to gain an understanding of how to interpret phase plane trajectories. For this purpose, we will follow the trajectory corresponding to the generation of a single action potential which is plotted as a solid line in Figures 2 and 3. Figure 2 represents time explicitly along the horizontal axis, while in Figure 3 time is represented implicitly. In Figure 3, one should imagine the system starting out at the point labeled "R" in the lower-left corner of the plot and following the solid trajectory around in a counter-clockwise direction as time evolves. As the phase point moves in the phase plane, it traces out a trajectory that can be correlated with the time-course of an action potential. The action potential trajectory can be broken up into four phases: (a) a rightward movement of increasing v and slowly increasing n, corresponding to the upstroke of the action potential; (b) an upward movement of increasing n and slowly decreasing v, corresponding to the "plateau" of the action potential; (c) a leftward movement of decreasing v and nearly constant n, corresponding to the downstroke of the action potential; and (d) a final downward movement of decreasing n returning the system to a new rest point.

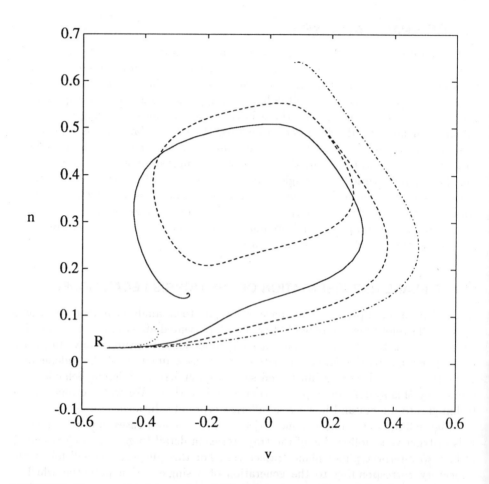

FIGURE 3   Responses of the reduced Morris-Lecar model to a series of external current injections viewed in the (v,n) phase plane. Same parameters and conditions as in Figure 2. Dotted line: subthreshold response; Solid line: single action potential; Dashed line: train of action potentials; Dot-dashed line: depolarization block.

The slope of the trajectory at any point in the (v,n) phase plane is given by the ratio of dn/dt to dv/dt at that point. Figure 4 shows the slope and direction of the phase trajectory in the (v,n) phase plane for the parameters used in Figures 2 and 3. It will be useful to identify points in the phase plane where $dn/dt = 0$ and $dv/dt = 0$. These sets of points, referred to as nullclines, correspond to points where the trajectories are horizontal or vertical. For the reduced Morris-Lecar model, the v nullcline can be found from Eqs. (1.22)-(1.23) by setting $\dot{v} = 0$:

$$n = \frac{-\bar{g}_{Ca} m_\infty(v)(v-1) - \bar{g}_L(v - v_L) + i(t)}{\bar{g}_K(v - v_K)} \qquad (\dot{v} = 0) \qquad (2.1)$$

The n nullcline occurs where $\dot{n} = 0$, which can be found from Eq. (1.24)

$$n = n_\infty(v) = \frac{1}{2}\left[1 + tanh\left(\frac{v - v_3}{v_4}\right)\right] \qquad (\dot{n} = 0) \qquad (2.2)$$

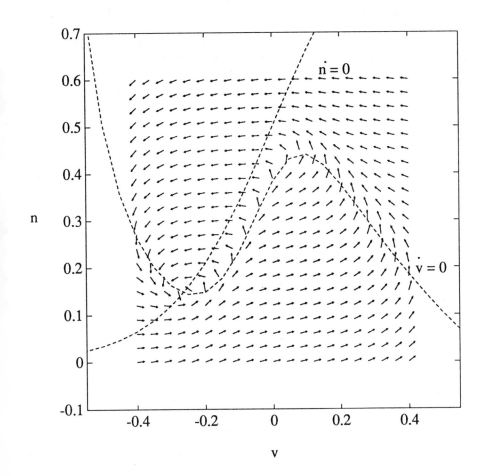

FIGURE 4 (v, n) phase plane for the reduced Morris-Lecar model. Same parameters as Figures 2 and 3 with $i_{ext} = 0.20$, corresponding to the case which generates a single action potential. Arrows indicate the slope and direction of the phase trajectory. Dashed lines: $\dot{v} = 0$ and $\dot{n} = 0$ nullclines.

The v and n nullclines are plotted as dashed lines in Figure 4. For the particular set of parameters selected, the nullclines intersect at a single point for which both $\dot{v} = 0$ and $\dot{n} = 0$. The behavior of the system near such singular points is an important determinant of the qualitative behavior of solutions to the equations of motion. In this case, the unique singular point is asymptotically stable, which is to say, that for nearby starting points the system will tend to the singular point as t approaches infinity. In general, however, it is also possible for such points to be unstable, in which case the system would tend to move away from the singular point. This brings us to the subject of stability analysis which is a technique used to characterize the behavior of the system near singular points.

STABILITY ANALYSIS

In order to describe the behavior near singular points, the first step is to linearize the system of equations about that point. In the case of the reduced Morris-Lecar model (Eqs. (1.22)-(1.27)), the linearized equations for small disturbances (v',n') about a singular point are:

$$\frac{dv'}{dt} = -\frac{\partial i_{ion}}{\partial v}v' - \frac{\partial i_{ion}}{\partial n}n' \tag{2.3}$$

$$\frac{dn'}{dt} = \frac{1}{\tau_n}\frac{dn_\infty}{dv}v' - \frac{1}{\tau_n}n' \tag{2.4}$$

Solutions to this set of equations are exponentials of the form: $e^{\lambda_1 t}, e^{\lambda_2 t}$ where λ_1 and λ_2 are roots of the equation:

$$\lambda^2 + \left(\frac{\partial i_{ion}}{\partial v} + \frac{1}{\tau_n}\right)\lambda + \frac{1}{\tau_n}\left(\frac{\partial i_{ion}}{\partial v} + \frac{\partial i_{ion}}{\partial n}\frac{dn_\infty}{dv}\right) = 0. \tag{2.5}$$

We can simplify this expression by defining two new conductances. First, consider the steady-state current defined by:

$$i_{ss}(v) \equiv i_{ion}(v, n_\infty(v)). \tag{2.6}$$

We define the steady-state conductance g_{ss} to be:

$$g_{ss}(v) \equiv \frac{di_{ss}}{dv} = \frac{\partial i_{ion}}{\partial v} + \frac{\partial i_{ion}}{\partial n}\frac{dn_\infty}{dv}. $$

The first term can be thought of as the instantaneous conductance g_{inst}, which characterizes how the current changes with voltage before n has a chance to change:

$$g_{inst} \equiv \frac{\partial i_{ion}}{\partial v}. \tag{2.8}$$

Using these two definitions, we can rewrite Eq. (2.5) as:

$$\lambda^2 + (g_{inst} + \frac{1}{\tau_n})\lambda + \frac{1}{\tau_n}g_{ss} = 0. \tag{2.9}$$

The roots of the equation are given by:

$$\lambda = \frac{-(g_{inst} + \frac{1}{\tau_n}) \pm \sqrt{(g_{inst} + \frac{1}{\tau_n})^2 - 4\frac{g_{ss}}{\tau_n}}}{2}. \tag{2.10}$$

If $g_{ss} > 0$, there are five cases to consider[6]:

1. $(g_{inst} + 1/\tau_n)$ positive; $(g_{inst} + 1/\tau_n)^2 > 4g_{ss}/\tau_n$. The roots are negative and real; the solution is a pair of decaying exponentials; the system is stable.
2. $(g_{inst} + 1/\tau_n)$ positive; $(g_{inst} + 1/\tau_n)^2 < 4g_{ss}/\tau_n$. The roots are complex with negative real parts; exponentially decaying oscillations, stable.
3. $(g_{inst} + 1/\tau_n) = 0$. This is a special case in which Eq. (2.9) reduces to the equation for a simple harmonic oscillator; this condition is unlikely to hold exactly in biological systems.
4. $(g_{inst} + 1/\tau_n)$ negative; $(g_{inst} + 1/\tau_n)^2 < 4g_{ss}/\tau_n$. The roots are complex with positive real parts; exponentially increasing oscillations; unstable.
5. $(g_{inst} + 1/\tau_n)$ negative; $(g_{inst} + 1/\tau_n)^2 > 4g_{ss}/\tau_n$. The roots are real and positive; increasing exponentials; unstable.

In summary, the singular point is unstable when $(g_{inst} + 1/\tau_n)$ is negative, i.e.:

$$-g_{inst} > \frac{1}{\tau_n}. \tag{2.11}$$

Thus instability only occurs when g_{inst} (the slope of the instantaneous $i - v$ relationship) is negative. Furthermore, g_{inst} must be sufficiently large with respect to the normalized rate constant of the accommodative process. In general, the faster the accommodative process, the more stable the system will be.

PHASE PLANE ANALYSIS OF THE MORRIS-LECAR MODEL

With these stability analysis techniques in hand, we now return to our phase plane analysis of the reduced Morris-Lecar model. Figures 5-8 show v vs. t and (v, n) phase plane plots for the four cases originally shown in Figures 2 and 3. In three of the cases the singular point has complex roots with negative real parts, implying that the singular point is stable. In these cases, the system eventually comes to rest at the intersection of the nullclines. In Figure 7, the singular point has complex roots with positive real parts, implying that the singular point is unstable. In this case, the system spirals around the singular point in a continuous stable oscillation.

The cases shown in Figures 5-8 don't encompass all of the possible behaviors that can be exhibited by the reduced Morris-Lecar model. As a final example, we

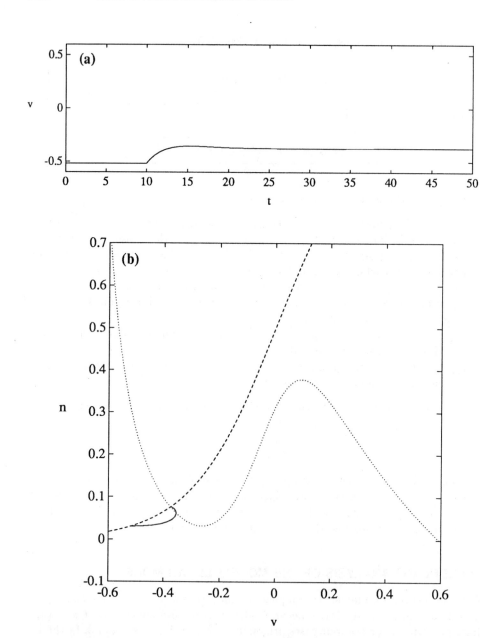

Figure 5 Subthreshold response—reduced Morris-Lecar model, $i_{ext} = 0.1$. Same parameters as Figure 2. Singular point at (-0.375, 0.076) is stable. (a) v vs. t (b) (v, n) phase plane. Dashed line: $\dot{n} = 0$ nullcline. Dotted line: $\dot{v} = 0$ nullcline.

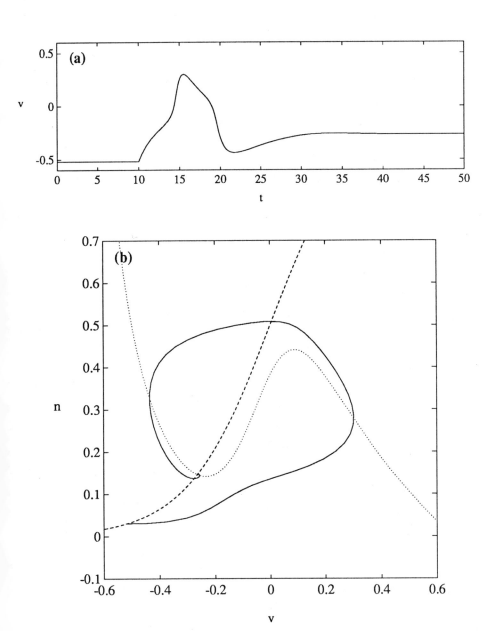

Figure 6 Single action potential—reduced Morris-Lecar model, $i_{ext} = 0.2$. Same parameters as Figure 2. Singular point at (-0.264, 0.146) is stable. (a) v vs. t (b) (v, n) phase plane. Dashed line: $\dot{n} = 0$ nullcline. Dotted line: $\dot{v} = 0$ nullcline.

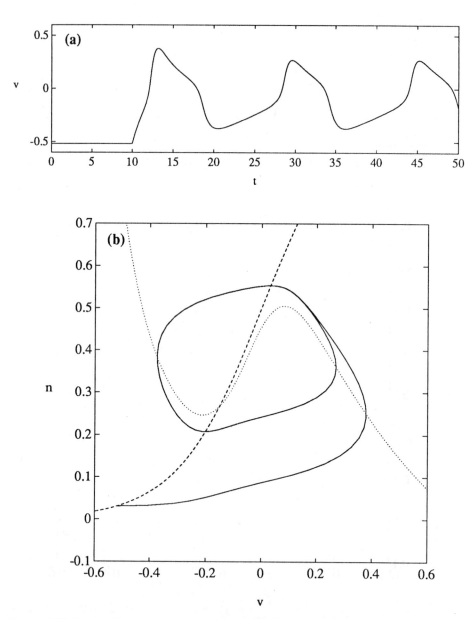

Figure 7 Train of action potentials—reduced Morris-Lecar model, $i_{ext} = 0.3$. Same parameters as Figure 2. Singular point at (-0.152, 0.266) is unstable. (a) v vs. t (b) (v, n) phase plane. Dashed line: $\dot{n} = 0$ nullcline. Dotted line: $\dot{v} = 0$ nullcline.

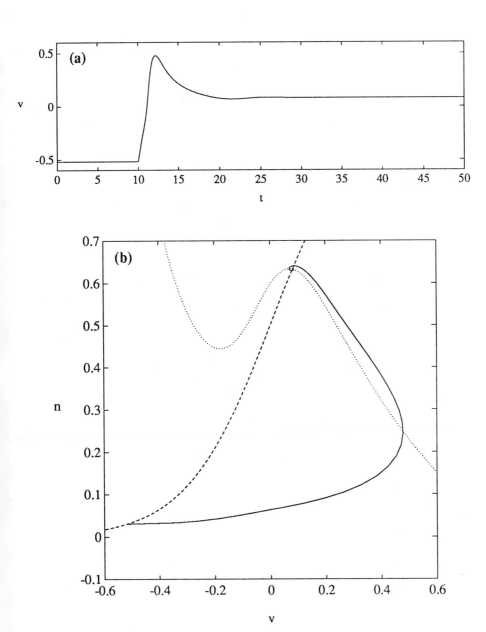

Figure 8 Depolarization block—reduced Morris-Lecar model, $i_{ext} = 0.5$. Same parameters as Figure 2. Singular point at (0.082, 0.633) is stable. (a) v vs. t (b) (v, n) phase plane. Dashed line: $\dot{n} = 0$ nullcline. Dotted line: $\dot{v} = 0$ nullcline.

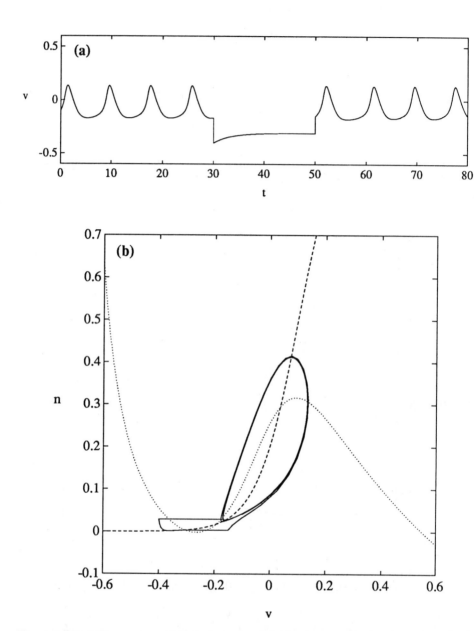

Figure 9 Bistability—reduced Morris-Lecar model. The parameters for this simulation are $v_1 = -0.01, v_2 = 0.15, v_3 = 0.1, v_4 = 0.145, \bar{g}_{Ca} = 1.0, \bar{g}_K = 2.0, \bar{g}_L = 0.5, v_K = -0.7, v_L = -0.5, \bar{\tau}_n = 0.87, i_{ext} = 0.075$. The singular point at (-0.307, 0.004) is stable, the singular points at (-0.192, 0.018) and (0.036, 0.294) are unstable. Brief current impulses are applied at $t = 30$ and $t = 50$ to switch between the oscillatory mode and the stable rest state mode. (a) v vs. t (b) (v, n) phase plane. Dashed line: $\dot{n} = 0$ nullcline. Dotted line: $\dot{v} = 0$ nullcline.

will consider an interesting case in which there are multiple singular points. As shown in Figure 9, slight modifications to the model parameters can result in a situation in which the nullclines intersect at three points. Stability analysis shows that the leftmost singular point is stable, while the other two singular points are unstable. Depending on the initial conditions, the system will either move toward the low-v singular point and come to rest there or it will oscillate about the high-v unstable points. The result is a bistable system with a stable resting state coexisting with a stable oscillation. The system can be switched between these two modes, as illustrated in Figure 9, by applying brief current impulses.

In summary, we have seen that even a simple two-variable single-neuron model is capable of exhibiting a surprisingly wide variety of dynamic responses. In actuality, the complexity of most neurons is much greater than that of the barnacle muscle fiber. Neurons typically have numerous voltage-dependent and chemical-dependent membrane channels and their repetoire of responses is much greater than that exhibited by the reduced Morris-Lecar model. It is important to keep in mind the potential complexity of single-neuron dynamics when trying to understand the behavior of neural networks and when trying to develop models of neural information processing.

REFERENCES

1. FitzHugh, R. "Thresholds and Plateaus in the Hodgkin-Huxley Nerve Equations." *J. Gen. Physiol.* **43** (1960): 867–896.
2. FitzHugh, R. "Impulses and Physiological States in Theoretical Models of Nerve Membrane." *Biophys. J.* **1** (1961): 445–466.
3. FitzHugh, R. "Mathematical Models for Excitation and Propagation in Nerve." In *Biological Engineering*, edited by H. P. Schwan. New York: McGraw Hill, 1969.
4. Hille, B. *Ionic Channels of Excitable Membranes.* Sunderland, MA: Sinauer Associates, 1984.
5. Hodgkin, A. L., and A. F. Huxley. "A Quantitative Description of Membrane Current and its Application to Conduction and Excitation in Nerve." *J. Physiol.* **117** (1952): 500–544.
6. Jack, J. J. B., D. Noble and R. W. Tsien. *Electric Current Flow in Excitable Cells.* London: Oxford University Press, 1983.
7. Koch, C., and I. Segev. *Methods in Neuronal Modeling: From Synapses to Networks,* Cambridge, MA: MIT Press, 1989.
8. Morris, C., and H. Lecar. "Voltage Oscillations in the Barnacle Giant Muscle Fiber." *Biophys. J.* **35** (1981): 193–213.
9. Rinzel, J., and G. B. Ermentrout. "Analysis of Neural Excitability and Oscillations." In *Methods in Neuronal Modeling: From Synapses to Networks,* edited by Koch, C. and I. Segev. Cambridge, MA: MIT Press, 1989.

M. A. V. Gremillion
Center for Nonlinear Studies, Los Alamos National Laboratory, Los Alamos, NM 87545, and Department of Neurosciences, University of California at San Diego, La Jolla, CA 92093

Linking Structure and Function: Information Processing in the Brain

INTRODUCTION

The field of complex systems is a new research specialization which investigates the complicated dynamics of systems from physics, math, computer science, and more recently, biology. It examines entire systems, not simply their components, attempting to understand the properties which arise from system interactions and to establish a framework in which system behavior of all kinds can be explored and compared. Using primarily modeling and mathematical techniques, it has tackled the economy, adaptive and pattern recognition systems, and systems characterized by chaotic dynamics (see this volume).

Biology presents the greatest challenge to complex systems research. Large-scale biological problems—biochemical networks, the immune system, evolutionary processes, the brain—are quite different in both type and degree of complexity than problems previously solved by science. But the wealth of data in these areas means that they are rapidly becoming amenable to a complex systems approach—the analysis of a system *as* a system, as more than the sum of its parts.

Arguably the most complex system in existence is the mammalian brain. Huge amounts of anatomical, physiological, chemical, and behavioral data have been gathered on the brain. These data, however, have not been well integrated. Increasing specialization has led to a fracturing of the field in which the physiologists who study cell response and function are not aware of details of the system's structure, and vice versa. Each subdiscipline submits its own explanation of system behavior, virtually without considering information from the others.

This is a serious problem, because, since the inception of the field of neuroscience, the relationship between structure and function in the nervous system has proved not only revealing but a fundamental principle of system organization.[8] The *structure-function relationship* posits that structure and function are well correlated in the brain—a simple statement, but one with tremendous power if applicable throughout the nervous system. Under this postulate, the structure of molecules, neurons, circuitry, and topology all provide keys to their function and to the resulting system behaviors. Further, it implies system function is comprehensible only in the framework of system structure; to ignore the structure is to ignore valuable constraints which can lead to totally new theories of function.

Traditionally, theories of function in neuroscience have emerged from physiology. Physiologists have suggested a number of means by which information in the brain can be processed—through receptive fields, through feature extraction—yet the principles underlying the generation of these phenomena are not well understood. One source of confusion is the unit of information processing that the brain uses. Is it the single cell? The module? Distinct areas? Some unknown?

A complex systems approach would be to examine the overall structure and function of the system and to attempt to establish a common framework for information-processing interactions. New hypotheses and experimental directions might arise from the new framework, as well as implications for information processing in the brain and perhaps other complex systems.

This paper will use the structure-function relationship as a basis for exploring units of information processing. It will examine the brain as a whole, first providing the non-specialist with an short overview of the structure and some of the functions or outputs of the brain. It then very briefly reviews three of the prominent theoretical concepts that have emerged in the last few decades: receptive fields, feature extraction, and parallel processing.

Next, it addresses the question of information processing and outlines the structures which have traditionally been proposed to be the basic unit of information processing. An alternative unit on which information processing in the brain might be based is then proposed, and data outlined to support it. Finally, the implications of this different mode of processing are discussed, both for the brain and for other complex systems.

BRAIN STRUCTURE

The brain is a spatially hierarchical structure—that is, structure exists and is meaningful on many scales. Anatomically, the brain can be divided into two sets of spatial scales: single cells and their smaller components, and cell groups, combinations of which comprise increasingly larger and more complicated architectures. (See Figure 1.)

A single neuron is composed of many smaller structures, including dendritic and axonal arborizations, membrane specializations such as spines and synapses, and vast numbers of molecular complexes serving as ion channels, pumps, and receptors as well as the usual cell organelles.[16] Virtually all of these structures help determine a cell's membrane dynamics and thus its action potential, or output, patterns.[7,15,27]

Single cells can then be grouped: into classes with distinctive morphology, chemistry, and position in a layer and structure; into layers comprised of many classes of neurons.[16,24,33] A range over a many-layered surface, whose boundaries are often determined through physiological response patterns, makes up a distinct area; areas together constitute an anatomical structure.(For evolutionarily older structures, this hierarchy is not as well defined.) Structures—like the occipital cortex, hippocampus, and cerebellum—are generally distinguishable from the surrounding tissue by the naked eye. Two roughly matching sets of structures form each of the hemispheres of the brain.[16]

Connecting within and between layers, areas, and structures is the microcircuitry of the system—the very specific set of connections of each neuron type to others. Connectivity includes the placement and density of connecting synapses onto the cell, and the neurotransmitters and receptors involved. Feedback within the circuitry, both positive and negative as well as more indirect forms, abounds. From the microcircuitry arise the anatomical columns seen in structures such as visual cortex, in which inputs from the two eyes alternate.[24,33] (See Figure 2.)

Circuit interactions in which spatial relationships are preserved, termed topographical mapping, are among the most common form of connection in mammalian brain (see Figure 2). This connectivity pattern is maintained for most sensory and motor systems and many limbic connections. While they maintain spatial relationships, virtually all topographical mappings are *by neighborhood*. That is, the point-to-point mappings often envisioned, in which cell A connects to cell B only, are extremely rare if not nonexistent in the brain; instead, by virtue of the (often large) width of dendritic and axonal arbors, a single cell projects to a neighborhood (or many neighborhoods in different layers) of cells, and a cell receives from a neighborhood (or many neighborhoods) of cells.[19,24,32]

Convergence and divergence give estimates of how many cells of a type project to a single receiving cell, and how many cells of a type a single cell projects to.[23] *Coverage factor* gives the number of cells which, by virtue of their overlapping arbors, cover a single point in the layer in which they arborize.[38] While quantitative

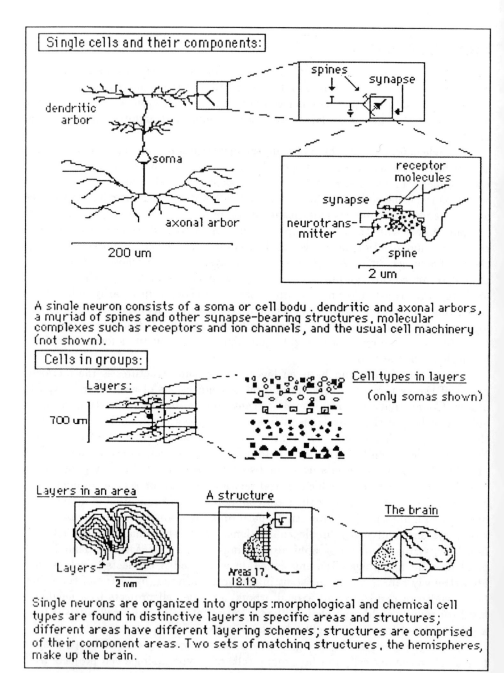

Single cells and their components:

spines synapse

dendritic arbor

soma

axonal arbor

200 um

receptor molecules

synapse

neurotrans-mitter

spine

2 um

A single neuron consists of a soma or cell body, dendritic and axonal arbors, a myriad of spines and other synapse-bearing structures, molecular complexes such as receptors and ion channels, and the usual cell machinery (not shown).

Cells in groups:

Layers:

700 um

Cell types in layers
(only somas shown)

Layers in an area

A structure

The brain

Layers

2 mm

Areas 17, 18,19

Single neurons are organized into groups: morphological and chemical cell types are found in distinctive layers in specific areas and structures; different areas have different layering schemes; structures are comprised of their component areas. Two sets of matching structures, the hemispheres, make up the brain.

FIGURE 1   Hierarchies of spatial scales in the brain.

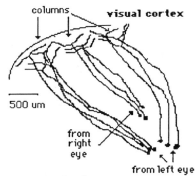

500 um

from
right
eye

from left eye

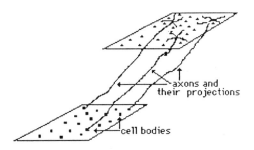

axons and
their projections

cell bodies

Anatomical columns

A schematic of the ocular dominance columns in visual cortex. Columns perpendicular to the surface of the cortex are formed through the inter-leaving of inputs from the two eyes.

Topographical mapping

A major percentage of the connections in the brain are topographically mapped: when a set of cells projects to a layer, spatial relationships are preserved even though each cell projects to a neighbor-hood rather than a point.

FIGURE 2   Circuitry Interactions.

estimates of coverage factors are not widely available, one reliable estimate of the coverage of LGN cell axons to layer 4 of visual cortex was between 400 and 800—that is, the axons of between 400 and 800 different cells covered each point in the layer.[19] A basket cell also in cortex was found to contact roughly 300 pyramidal cells,[19] indicating that high convergence and divergence from one cell type to another is therefore not unusual. This fact has strong implications for how information is processed by the nervous system.

CHEMICAL ANATOMY

The hierarchy of spatial scales is to some degree correlated with the brain's chemical anatomy—the distribution of neurochemicals over a single cell, over groups of cells, and over layers, areas, and structures. Neurochemicals alter membrane potential through the activation of channels and receptors in the receiving cell, thus changing the firing pattern of individual cells as well as groups of neurons.

Neurochemicals fall into a variety of classes[6,13,16,21] (see Table 1). "Local" neurotransmitters are generally short-acting amino acids such as glycine, glutamate, or gamma-aminobutyric acid whose actions are synaptic, and the number of cells any one cell carrying these neurotransmitters contacts is usually quite small, probably only thousands.[6] The "classical" neurotransmitters—dopamine, serotonin, acetylcholine, and noradrenaline—have a longer timecourse of action, and often act diffusely rather than synaptically. Since very small clusters of each of these types

TABLE 1 Chemical Anatomy

Neurochemical type	Examples	Time constant of effect	Type of release
Fast neurotransmitters	aspartate glutamate glycine GABA	5-20 msec	synaptic
Classical neurotransmitters	noradrenaline acetylcholine serotonin dopamine	50 msec to seconds	diffuse (extra-synaptic)
Neurohormones	melatonin estrogen prolactine vasopressin	minutes to days	distributed: receptor-based
Neuropeptides	substance P, neuropeptide Y, somatostatin	variable	variable

of neurons project throughout virtually the entire brain, a cell whose neurotransmitter is one of these probably contacts many millions of cells.[6] Neurohormones such as the sex, pituitary, and corticosteroid hormones, act through receptors spread throughout the brain; their timecourse is often hours or days, and includes effects on genomic expression.[6] About another class of chemicals, the neuropeptides, not much is known of their timecourse or actions, but they are found in virtually every region of the brain.[6,16]

Since any given neuron has hundreds to thousands of synapses as well as additional diffuse and extrasynaptic stimulation, it is generally receiving all manner of neurochemical activation spanning the spectrum of chemical effect, timecourse, and distribution. Interactions of neurochemicals suggest a more complicated picture than simple spatiotemporal correlations of structural and chemical anatomy, but initially this framework may be a useful one for categorizing populations of cells.

BRAIN FUNCTION

Brain function is neither as easily quantifiable, nor as investigable, as the structure of the nervous system. In the same way that it is easier for two people to agree on the physical components of a flute than on the meaning or even the format of its musical sounds, analyses of neural systems structure are more straightforward than analyses of system output. A functional interpretation is more clearly an *interpretation*.

Many assessments of function exist, from the molecular to the entirety of the brain. Markers of molecular and synaptic behavior are now available; chemical and pharmacological experiments can be used to assess the effects of such manipulation on single cells, groups of cells, or the behavior of the entire organism. Physiological techniques can measure electrical activity of small patches of membrane, single cells, and the firing of small clusters of neurons. Metabolic activity can be visualized over areas of millimeters to centimeters.[16]

Large-scale patterns of electrical, magnetic, and metabolic activity in the brain can be visualized through imaging techniques such as the PET and CAT scans, nuclear magnetic resonance, and the electroencephalogram and magnetoencephalogram. Behavioral measures, which reflect the function or output of the entire organism, enable correlations of areas, structures, and hemispheres to particular capabilities.[16]

Through these methods, some generally accepted precepts of function have been developed. At small spatial scales, we have some insight into and mathematical models of the dynamics of ion channels and receptor proteins[7,15,18] and of the many processes that occur at individual synapses.[13] Governing equations have been formulated for the action potentials of single neurons,[7] and to a large degree it is understood how dendritic branching structure and spine position, density, and shape are integrated into the membrane potential fluctuations that generate an action potential.[27]

At larger spatial scales, areas and structures have been correlated to particular kinds of capacities or behaviors; left and right hemispheres are believed to have different strengths and capacities.[16]

But for decades now, the primary focus of neurophysiologists—and of much of neuroscience—has been on the behavior of single cells in response to a stimulus. It is thus from single-cell neurophysiology that many of the major paradigms have arisen. Through it, the concepts of *receptive field, feature extraction*, and *parallel processing streams* have influenced virtually all areas of neurobiology.

CURRENT FUNCTIONAL INTERPRETATIONS

The single cell receptive field has played a major role in theoretical neuroscience, generating the idea of feature extraction by particular cells and producing a means by which the dimensions of sensory systems could be mapped onto streams, areas, and structures. The following outlines some of the conclusions, issues, and problems associated with these concepts.

Receptive fields of single cells are the region of sensory space within which, and the stimuli to which, the neuron "best" responds. The stimuli that produces the highest rate of firing is believed to be the preferred stimulus, and cells that respond maximally to a stimulus are believed to be "extracting" the features of the stimulus. A visual cell, for instance, can prefer stimuli moving at a certain velocity and/or direction, a particular orientation or a certain spatial frequency. Successively higher areas—those receiving information that has been processed by increasingly large numbers of previous neurons—are considered to be extracting progressively more specific features.

This perspective has been an experimentally fruitful one, but fundamental difficulties exist. First, the idea of increasingly finer extraction of features by successively "higher" cells leads to the "grandmother cell"—that cell which would extract the features of, and recognize, "grandmother," or whatever other specific entities and objects exist in the visual field. Numerous logical as well as physiological arguments have asserted the improbability of this, but why and how this does *not* occur has rarely been addressed. Next, most ideas of receptive field and "best" behavior neglect the temporal dimension of neuronal output, looking only at a cell's average rate of firing over fairly long time scales (e.g., a second). It is unlikely that temporal information is completely irrelevant: recent work[9] has demonstrated that there are often temporal patterns in the spike trains of neurons, and the simple fact that neurons are continuously processing temporal information implies that temporal properties of its output are important.

Further, not all of a cell's responsiveness lies within the classically defined receptive field. Allman[1] and colleagues have determined that cells throughout the visual pathway are influenced by stimuli far outside their classical receptive fields. This counters the idea that individual neurons process a discrete, rigidly defined portion of the sensory space.

Feature extraction is thought to occur when single cells segregate particular dimensions of the stimulus—for instance, color, contrast, motion, and spatial frequency in the visual system. Individual neurons responding to a small set of stimulus characteristics then project to other classes of cells higher in the system, establishing a pathway through which only certain types of information flow .

Pathways specific to only particular kinds of information are the parallel processing streams. In this scenario, information is extracted, differentially manipulated through the separate pathways, and sent on to different "higher" areas for processing segregated by stimulus qualities.[33] Correlations of brain areas and structures to function seem to support feature extraction; for example, single cells in area V4 of the visual cortex show strong physiological responses to color stimuli and not to motion, while those in area MT respond strongly to motion and not to color.[31]

However, not all of the evidence is conducive to this hypothesis. Significant anatomical evidence suggests that visual "parallel streams" interact in the first stages of cortical processing,[40] making unlikely truly distinct channels. Recent studies revealed that, functionally, visual information is not, in fact, segregated very stringently.[11,31] Data on segregation of function by area, as in V4 and MT, is generally taken as supporting area/function relationships. However, in virtually all cases what has been demonstrated is the relationship of that brain component to the behavior, not the necessity of its presence to manifest it.[31] It appears that many functions ostensibly performed by a specific area can also be performed without it. This calls into question both truly separate parallel channels, and the strict hierarchy supposedly necessary to achieve parallelism.

Despite an understanding of the behavior of single cells, and of how areas and structures apparently correlate to different functions, there is a critical lack of knowledge about how hierarchical levels—from single neurons to behavior—are united. A theory that could unify these levels would be useful in its ability to explain single cell as well as areal data.

UNITS OF INFORMATION PROCESSING

The hierarchy of spatial scales, and the functional perspectives presented above, provide a preliminary framework for an understanding of information processing in the brain. Taken together, functional measures and the anatomy of the system should provide strong constraints on proposed mechanisms of information processing as well as the predicted behavior of the relevant units of the system. Information processing units have been proposed at all scales. However, rarely have anatomical and functional data been interpreted in the light of each other; certainly the anatomical evidence has often been ignored by schemes to simply generate the behavior without worrying how mechanism maps onto structure.

TRADITIONAL UNITS OF INFORMATION PROCESSING

Previously proposed units of information processing have focused primarily on two main themes: the single cell,[3] and some form of the module/column.[22,34,35] However, many other units have either been suggested, or are at the core of certain assumptions about the mechanisms of information processing. The following list details some of these, as well as some of the problems associated with each.

THE SINGLE CELL. Ever since the neuron doctrine of Ramon y Cajal,[28] the fundamental bias, even dogma, of neuroscience has been that the single cell is the unit of information processing in the brain.[3,16] Though many other investigators have focused on areas and structures, pathways and streams, it has been assumed that the single cell—so easily realized as an entity through our microscopes and with our electrodes—is the relevant unit. Yet this most obvious structural level may not be the most critical level for understanding the brain's information processing. A mosaic is composed of a myriad of tiny tiles, but looking at a single tile tells you very little about the picture which many tiles together compose. Similarly, it may be important to look at cells in relation to each other, and even more important to look at a large group of cells, in order to understand how information is represented.

Moreover, a single cell receives from, and projects to, on the order of hundreds or thousands of cells. Measurements of single synapses indicating that individual synapses are usually worth less than 0.1 mV, also support that single neurons of themselves have very little influence, and that a small number of neurons are unlikely to be responsible for the generation of another cell's receptive field.[2] Given these facts, and the high degree of convergence and divergence that has been anatomically demonstrated, it seems that groups of cells, however defined, are more likely the functional units of the system.

THE MORPHOLOGICAL CELL TYPE. Anatomists have long drawn some major distinctions between neuron types. The simplest classifications, spiny vs. nonspiny, or pyramidal vs. nonpyramidal, unfortunately do not have counterparts physiologically: often cells of two different types behave in very similar fashions, or cells of the same type have different receptive field properties.[19] Nor have distinctions based more on morphology—divisions such as basket cells, chandeliers, spiny stellates, etc.—been capable of differentiating the physiological responses patterns of these cells.[19] Without a functional correlate, such a scheme is clearly inadequate.

LAYERS. The distinctness with which cells are separated into anatomical layers and the fact that inputs and outputs can be said to project to or from a layer provides a strong argument for the layer as a processing unit. However, it is difficult to imagine a truly specific role being assigned to a particular layer in, for instance, the cortex, since usually the different cell types in a layer receive from, and project to, different cell types, layers, areas, and structures.[32] The oft-repeated "input" and "output" layers thus clarify the *nature* of neither the encoding or processing of information through a given system. That layers are important is clear; how or why they are important is not.

COLUMNS AND MODULES. The proposal of the column as an information-processing unit has been made by many people at different times.[8,22,35] While it has been suggested for a few other structures, the module/column has been noted or theorized primarily in the cortex. Thus, it has limited value as the unit of information processing throughout the entire brain. However, because of the importance of the cortex, the cortical module or column idea has had a significant effect on how information processing is viewed in neuroscience.

Columns and modules fall into two categories: anatomical and functional. Anatomical units such as ocular dominance columns in primary visual cortex are generated by the known columnar arrangements of afferents, in this case the alternation of inputs from left and right eyes. Functional columns, in contrast, were found through single cell recordings in which cells below and above each other tend to have the same preference for a stimulus quality such as orientation. The anatomical analogues of functional columns, however, have been difficult to find.

Columns as information-processing units, rather than structural entities, have numerous problems.[34] Columns are not information-processing entities in that they are not themselves a means of segregating information. Recordings from many cells in a column will demonstrate that within the column are many different kinds and degrees of response, and for many dimensions of the stimulus (e.g., direction, color, velocity), not even an ordered set of responses.[24,35] There is no unitary columnar response.

Neither are columns or modules a unit for transferring information. Proponents of columns generally require that connections within a column be stronger than between them; it is not clear that this is so, as massive horizontal projections as well as the normal overlap of convergence and divergence insure that individual columns are highly connected to each other. Suggestions that entire columns could be "turned on" as separate units have neither been proven nor are, perhaps, provable.[25] While vertical organization clearly exists, the usefulness of columns in suggesting hypotheses for testing has yet to be proved.[34]

STREAMS OR CHANNELS; AREAS. A large portion of the brain is responsive to sensory information. One theory of information flow is derived from the visual system: the idea that information is partitioned into streams or channels that flow in parallel, which are kept distinct through their circuitry. Distinct classes of cells in the retina, for instance, are thought to project to entirely different populations of cells, first in the thalamus, then in primary visual cortex, eventually leading to distinct areas whose connections are segregated by function.

Since they rely on anatomical data and are supported by physiological data, streams and areas capture some of the the essential features of information flow. Different kinds of information (at least in the visual system) are indeed mostly segregated; however, it is likely that at many points within the stream or area, significant amounts of information transfer occur between supposedly separate "channels."[11,31,40] Since streams or areas both consist of large and complex circuitries however, it seems more likely that they are composed of information-processing units rather than each serving as a single integrated unit.

ANATOMICAL STRUCTURES. Structures such as the cortex, hippocampus, cerebellum, etc., in that they have been noted as distinct physical entities for centuries, have often been considered as having distinct "roles" in the brain. While early scientists often made a one-to-one correspondence of a given structure to a function, recent indications are that while a certain structure may have a certain function as its primary job, many other structures interact with it to produce the function. It is also likely that each structure plays a role in many behaviors.

GROUPS. Edelman[12] has proposed the "group" as a unit of information processing. The group is a functionally defined entity: that set of cells which responds to a certain set of inputs. Groups are defined (1) in contrast to each other at a given time and (2) in contrast to the previous group membership at a different time over long time scales in which the driving inputs have been altered.

Edelman's theory captures an essential point: that populations are processing information. But its virtual disconnection from the actual anatomy we know to exist—which is in large part stereotyped across animals of a species and across species—ignores fundamental architectural constraints on information processing, and renders specific predictions difficult at best. For one, the lack of correspondence or even acknowledgement of existing anatomical structure makes it difficult to understand how interactions of "groups" take place through the existing circuitry. The very arbitrariness of the definition would lead one to think that the system is a homogeneous network where connectivity is primarily dependant on the driving force of external stimuli—a hypothesis that is not supported by anatomical data indicating strong similarities in circuitry across animals before and after exposure to the environment.

While each of the hypothesized units above may not be "wrong," each has significant flaws. Further, none has the power of explanation or prediction we would expect from an actual information-processing unit. In the same way acceptance of the neuron as a morphological unit dramatically altered the 19th century perspective of the brain, an information-processing unit should help restructure our concepts of its information flow.

AN ALTERNATIVE INFORMATION-PROCESSING UNIT

It seems clear, then, that any serious discussion of information processing in the brain must consider the wealth of anatomical data and its implications. It is also critical that a unit of information processing be capable of communicating across the vast numbers and organization types of brain structures: a unit of information-processing must be a "currency" which can be utilized easily across all or most of the brain.

These constraints are strong ones. How can we correlate experimental findings with a functional anatomical group that would be a candidate for an information-processing unit?

The brain is composed of neurons. Each individual neuron has a number of qualities, including specific morphology, placement, connectivity to other neurons, and a set of intrinsic properties related to its membrane and chemistry (see Table 2). But each of these properties of individual neurons is actually a *characteristic* property of a class of neurons to which that cell belongs. The properties of a class spring from the developmental timecourse through which neural cells evolve, a process which will determine not only a cell's position in the layers of the cortex or another structure, but the genetic and chemical environment that will engender the growth and structure of its arbors and its connectivity pattern.

Many categorizations of cell types based on some of these factors have been proposed: morphological cell types, spiny and non-spiny, excitatory and inhibitory. But each of these captures only a small fraction of the differences we see between cells, and does not serve to fully distinguish them either structurally or functionally.

A more precise classification based on all of these factors is the cell type-in-a-layer, or *celltial*. The members of a celltial share all of the qualities outlined in Table 2: morphological, connective, intrinsic, and chemical. Further, a celltial *acts* as an unit—high convergence and divergence insures that members of a celltial cannot be considered meaningfully in isolation. Each celltial has its own properties: a characteristic density and spacing, and the coverage factors of its dendritic and axonal arbors to its projection layers. It connects to other celltials with a particular density and weighting of synaptic connections to each type. Because each celltial, by virtue of its connectivity, has a different position in the information flow, each celltial plays a different role in the system circuitry.

Classifications of this kind have already been made in well-documented systems. The retina, the cerebellum, and the hippocampus are all systems with small enough numbers of cell types that they have been differentiated by layer, morphology, chemistry, and connectivity.[2,16,38,39] The large numbers of cell types and layers in the cortex have so far prevented the same degree of comparison, but available data[10,40] supports similar classification.

TABLE 2 Properties of Single Cells

Morphology	Membrane properties
• some size and shape • dendritic arbor size, shape, density, and projection pattern • axonal arbor size, shape, density, and projection pattern • spiny or nonspiny	• axonal conductane time • refractory period duration • length (space) constant • resistance, conductance • density or ion channels, etc.
Chemistry	Connectivity
• releases specific neurochemical • expresses receptors for specific neurotransmitters/chemicals • time constants of various synapse types	• position in a layer/area/etc. • connections to specific cell types in specific layers • positioning of synapses from particular cell types • weighting or density of synapse types

Developmental, anatomical, and circuit data point to the brain's organization by celltial. During development, neurons do not migrate according to their morphology or transmitter type—these are expressed later. Instead, migrations of neurons during development are by layer. During migration, cells belonging to a certain layer are differentiated into cell types with specific morphology.[26,37] Cell types with the same morphology but in different layers are probably further differentiated in terms of chemistry, arborization size and placement, etc., due to the varying chemical and growth factor gradients across the layers.

Anatomical data offers wide support to the celltial categorization. Wassle et al.'s[39] determination that, even so morphologically close cell types as ON and OFF Beta ganglion cells are separately and regularly distributed is a strong indication that the cell type *within a layer* is an intrinsic grouping of the system. Every cell-type-in-a-layer so far examined in the retina has shown a rough mosaic distribution.[38] In other anatomical structures such as the cortex, cerebellum, or hippocampus, when distributions have been observed through immunocytochemistry or other precise staining, similar regularity has been observed.[10,40]

Investigation of the microcircuitry shows that the celltial is the fundamental component of microcircuitry throughout the brain. Topographical mapping, including feedback, is by *celltial*—not by individual neuron or by system. When one structure, area, or layer connects to another, it is through one or more celltials; when two neurons talk, they do so as members of celltials whose other members are doing the same. Celltials cut across boundaries of layer, area, and structure. Primary visual cortex does not feedback to the lateral geniculate: a celltial of pyramidals in layer 6 feeds back to specific celltials in the lateral geniculate nucleus. Though we simplify by speaking of layers and areas, the units which are truly interacting are the celltials.

Receptive field response patterns, which arise from circuitry interactions, are in effect generated by celltial properties and interactions. It makes sense for the members of a cell-type-in-a-layer to share the same receptive field properties, in that they each share the same *position* in the information flow—that is, roughly identical input-output relations. The same morphological class of cell (such as a pyramidal) in two different layers almost certainly has dendrites in different layers and hence each receives very different sets and sequences of information. That they are integrating information that is at least partially quite different should serve to differentiate their receptive fields, yielding the variety of physiological classes that we see in a layer. (This is not to say that different celltials in different layers may not share some response characteristics: similar patterns of circuitry—e.g., of feedback and feedforward—are probably capable of generating roughly similar response patterns. However, there are probably differences between similar response patterns in different layers that our somewhat crude analyses have failed to capture.) While this hypothesis remains to be proven, recent studies in the cortex[17,20] have found that pyramidal cells in the same layer of cortex with two different dendritic arborization patterns have distinctly different electrophysiological response patterns.

That the celltial should exist as a unit is not so striking: many of the requirements of celltial organization are actually assumptions of neuroanatomy. That cells can be categorized according to morphological type; that there are regularities in the distribution of the cell bodies; that cells in a layer with the same morphology share patterns of synapse placement, arbor projections, and physiological response pattern[17,21,24,38,39,40]—all of these reflect an order in the nervous system that anatomists have been documenting for a hundred years.

THE CELLTIAL AS THE UNIT OF INFORMATION PROCESSING

"We have difficulties trying to imagine a kind of computation which makes equally good sense in sensory, motor, language-related and other areas of the brain, in the man, in the cow, moose or alligator."

—Valentino Braitenberg[5]

From work of the last few years on systems such as neural networks, cellular automata, and the like, we know that distributed processing is not only a viable, but a formidable means of information encoding and manipulation. We also know that information in the brain is distributed, in many complex ways; it appears that this distributed aspect may be necessary to generate the patterns that result in, or create, speech, thought, memory, and perception. But we do not understand the codes and mechanisms which the brain uses.

Celltials as information-processing units are a powerful way to analyze the brain. Single cells become members of a class, and presumably are predictable as such; areas and functions become defined by the celltials that comprise them and the role celltials play in their circuitry. A celltial organizational framework has implications for how receptive fields are generated, for how circuitries interact, for how the system manipulates information. It provides testable hypotheses and suggests new directions for study and analysis.

It is important to understand exactly how a celltial operates. The interaction between two celltials can be understood as a process of spatial and temporal smoothing. First, each point, or neuron member of a celltial, is averaging over a set of points which are the inputs to its dendritic field. If we take two celltials A and B and project A onto B, then to a first approximation, it doesn't matter exactly which of A's cells onto a single B cell within a given time frame are activated, simply the number of them . The B cell will have virtually the same response since A projects to a characteristic position on any B cell's arbor, and the length constant and time constant of its responses to A are roughly the same. (See Figure 3.)

Information is smoothed over the entire celltial in the transmission of information from one celltial to another in two ways: convergence and divergence. The output from A is smoothed, because each cell in A projects (diverges) to many cells in B; and B, too, is smoothing over its input since each cell in B is receiving

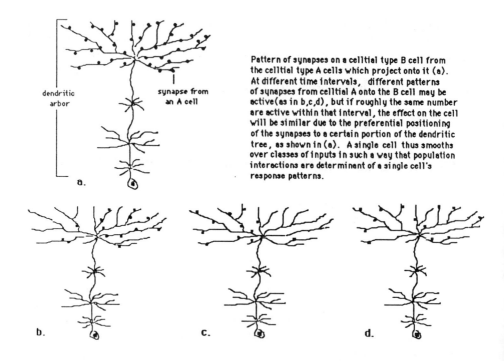

Pattern of synapses on a celltial type B cell from the celltial type A cells which project onto it (a). At different time intervals, different patterns of synapses from celltial A onto the B cell may be active(as in b,c,d), but if roughly the same number are active within that interval, the effect on the cell will be similar due to the preferential positioning of the synapses to a certain portion of the dendritic tree, as shown in (a). A single cell thus smooths over classes of inputs in such a way that population interactions are determinant of a single cell's response patterns.

dendritic arbor

synapse from an A cell

a.

b.   c.   d.

FIGURE 3   Smoothing over celltials.

from many cells in A(convergence). Cells in the same neighborhood are receiving spatially and temporally correlated input. The degree of overlap, of convergence and divergence, gives the fineness of grain of the smoothing.

Smoothing serves multiple purposes as has been pointed out in other contexts. First, the degree of *redundancy* one would expect from this arrangement would quite likely be effective in reducing the effects of both spatial and temporal noise within the system, since averaging of signals increases the signal-to-noise ratio. This is true both for single cells—whose output depends on the behavior of its input populations, and not on any single driving input—and for the celltial as a whole. The inherent disadvantage of using noisy elements like neurons is countered by the advantages of population encoding.

Secondly, the fact that the representation is a distributed entity rather than a point mapping enables the smooth representation of objects rather than a representation which must be "cut," as with computer pixels, only at certain spatial positions dictated by the density of cells. Counter to paradigms in which single cells operate as discrete elements, representation by a celltial is continuous rather than discrete.

This is clearly useful for representation of the real world, which *is* continuous. Beyond this, a continuous representation may be required by encoding of the inputs and outputs of the brain. Recent data indicates that both sensory (input) and motor (output) systems use continuous representation of information. In the retina, neuroscientists have long puzzled over the purpose of the gap junctions between light-sensitive photoreceptors. A lattice of discrete photoreceptors, the cones, serves as an intermediary between the external world and its internal visual representation; but the gap junctions(through which currents are shared between cells) between cones effectively *merge* point information as the first step of visual information processing.[36] Furthermore, recent work on the motor output system indicates that it must use some sort of population encoding of muscle movement rather than the previously assumed point-to-point control.[30] Why both input and output would utilize a non-discrete form of encoding is puzzling unless the representation of information throughout the system is a continuous one rather than comprised of discrete points.

Representation, including feature extraction, has traditionally been presumed to operate within single cells of the nervous system. Particular cells in the visual cortex respond extremely well to moving light edges, firing most frequently when presented with them compared to other stimuli. Thus they could be, and have been, considered edge detectors. The usual viewpoint has been that other cells not firing at their peak frequency are *not* responding to the signal. However, given that many other cells of a celltial also respond to that stimulus, a different interpretation is that the stimulus representation is the spatiotemporal pattern across each of the celltials capable of responding to the stimulus.

In other words, it is not a few cells encoding the stimulus—it is the entire population of a given celltial. A given spatiotemporal pattern across a cell type in a layer will signal both stimulus qualities and, since spatial relationship is inherent within a celltial, position. The so-called random bursting or firing of a single cell, when seen in the context of the firing patterns of the other members of the celltial, may no longer seem random but appear as an integral part of the spatiotemporal pattern of the celltial of which it is a member. Temporal phenomena such as oscillations or patterns in spike trains may be meaningful only in the context of the celltial spatiotemporal pattern to which they belong.

Given that celltials are the components of circuitry, what can we say about the nature of their interaction? The interaction between any two celltials is not simply a function of one or the other of the celltials, but of both—of the intersection of the set of axonal arbors with the set of dendritic arbors. That intersection will determine how many A cells project onto each B cell. The projection of A onto C, where C has a very different dendritic radius in that layer than B, may lead to a considerably different effect of A onto C than A onto B (see Figure 4). To be precise, it is the density and arborization size of both projecting and receiving celltials that will determine the effect of one on the other. From this we can derive that *information is relative*, depending on which celltial is reading the information. In this case, energy in the form of action potentials only becomes "information" when it is integrated/interpreted by another celltial.

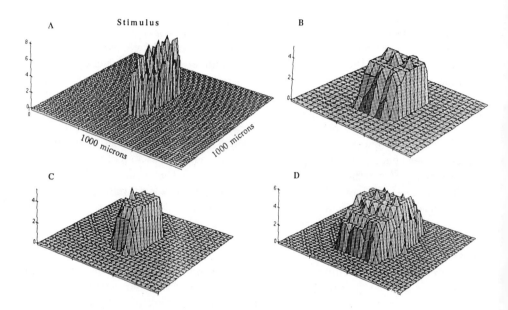

FIGURE 4 Effect of changes in celltial parameters on celltial response patterns.
Each plot represents the summed spike output of a layer of cells. A is the excitatory
stimulus for each of the other three celltials, whose output is shown in B, C and D.
The only differences between the celltials in B, C, and D are the density of the cells
that comprise them and/or the diameter and shape of their dendritic trees. A. Stimulus:
3249 cells, axonal arbor diameter 150 um. B. Density is 529, elliptical dendritic arbor
is 50 um × 150 um. C. Density is 900, circular dendritic arbor is 30 um. D. Density
is 1089, circular arbor is 100 um. X and Y axis in each of the drawings are the same
1000 microns × 1000 microns: Z axis, the number of spikes that occurred over 35
msec, are labeled. Some of the differences displayed by B, C, and D are in spatial
extent, peak number of spikes (120 vs. 200 Hz for C vs. D), accurate representation of
stimulus shape, and speed of fall-off with distance. This is the simplest possible circuit
(excitation to the cell type); more complicated circuits could amplify or otherwise modify
the differences in celltial response shown here.

    The relativity of information means that, of the ten or twenty different celltials
to which celltial A is sending information, each one may be seeing something different. This is because the interaction of one celltial's output structure with another
celltial's input structure creates something different: the structure of the information *transfer*. A seminar speaker giving a lecture in a particular manner will be
received quite differently by the different members of the audience; so too will a
given celltial's output be differently perceived by those receiving input from it. Until information is received and interpreted it is meaningless, and the same "output"
can be interpreted in different ways by different receivers.

    Through this, individual celltials can serve multiple "functions"—that is, a
given celltial can encode multiple aspects of a stimulus, various aspects of which can

be separated into components through the interpretations of its receiving celltials. In terms of information processing, combining a number of feature responses into a single celltial and sending that signal out to another system is far more efficient than having many cell types in a layer or system do so.

The representation of a stimulus, then, is neither static, nor necessarily a single entity. If relativity of information is true of the nervous system, then we can begin to understand how such a compact system is capable of the miraculous complexities of perception, cognition, language, and motor skills.

CONCLUSIONS

The major strength of a theory based on celltial coding of information is that it is derived directly from the brain's structure, and not from some arbitrary conception of how certain functions *might* be accomplished. Since the fundamental code of the nervous system must not only be mapped onto structure, but generated by structural interactions, it is critical that we acknowledge the anatomical constraints in generating hypotheses of information processing.

This paper proposes some fundamental re-evaluations of structural and functional data, and their relationship. First, a functional anatomical group, the celltial, is defined. A celltial is a class of cells in a layer who share the same morphological type, intrinsic membrane properties, connectivity, and chemistry. A celltial is a distinct anatomical entity; it has a characteristic density, regularity of distribution, and position in the information flow.

A celltials is a functional entity as well: it is fundamentally a component of circuitry. Celltials unify the hierarchy of spatial scales through their communication across layers, areas, structures, and hemispheres.

Next, since all of the members of a celltial share the same position in the information flow, a specific prediction of this theory is that they should also share receptive field attributes. A single cell's receptive field, derived from celltial interactions, is thus best noted with respect to the behavior of its neighbors in the celltial. It follows that feature extraction is a property of celltials rather than of single cells.

Further, celltials are units of information processing in the brain. Information is distributed across these lattice-like structures; given the high convergence and divergence of one celltial to another, the representation of information is effectively continuous rather than discrete, despite the use of discrete anatomical units.

Finally, the interaction of any two celltials results in a structure of information transfer. This structure can vary widely between any two celltials, and the information thus received by two celltials from the same input celltial can be quite different. The kinds of computations generated by the interactions of these two-dimensional lattices of cells are likely to be quite different from those we have imagined for single cell interactions. Such computations may have global or emergent qualities which make them equally capable of being applied to processing in sensory, motor, and other areas of the brain.[14]

How can information about this kind of system be applicable to other complex systems, natural or unnatural? What principles can we derive from the brain that make sense in the context of other information-processing systems, or other biological systems?

One possibility is that using structure-function relationships, we can propose units of structure or function previously unknown. Further, that in a given hierarchical system, there may be a *central* level of function/structure which may provide the best window on how the system transfers or otherwise manipulates matter, energy, or information. This level will probably differ depending on the property one is interested in; for the brain, the hypothesis here is that the population level of the celltial offers a "best description" of information flow. Understanding of other processes, such as genetic expression, plasticity, or global cognitive properties such as memory, or attention, may have different "central levels."

Next, in natural information-processing entities such as nervous systems, population interactions rather than the interactions of single elements may hold the key to system properties and behavior. For artificial information-processing systems like neural networks and cellular automata, this implies that differentiating the system into populations with differing temporal and spatial ranges will increase the order of complexity which it can handle, and the information it can store. The success of layered networks is one instance of the success of this idea.[4]

Spatiotemporal patterns—rather than spatial or temporal averages of system behavior—are clearly important. Interactions of these patterns through complex circuitries form the basis for information-processing capacities, or for other emergent behaviors of the organism.

The discussion of information in the context of complex architectures suggests that we reconsider the meaning of "information" and its transfer in complex systems. Information as a stable, quantifiable, and objective entity may not exist in such systems, where the "meaning" or value of the information is dependent on the interaction of recipient and sender. To stretch a bit further, many transfers of energy or matter may also be, in the context of a given system, the transfer of information as well.

Representation of information is a fundamental issue, both in natural neural-systems and in artificially constructed systems. The linear/sequential representations we are accustomed to in language, motor capabilities, and computers have biased us away from examining other kinds of representation far more suited to, and developed by, the brain's architecture. Assuming the brain and other systems work through webs of complicated distributed processing, it is time to explore new mechanisms of information encoding, transfer, and transformation.

ACKNOWLEDGMENTS

I would like to thank André Longtin, David Brown, Susan Coghlan, Bryan Travis, and Marty Sereno for useful discussions; and the Center for Nonlinear Studies, Los Alamos National Laboratory for its support.

REFERENCES

1. Allman, J., F. Miezin, and E. McGuinness. "Stimulus-Specific Responses from Beyond the Classical Receptive Field." *Ann. Rev. Neurosci.* **8** (1985): 407-430.
2. Andersen, P. "Properties of Hippocampal Synapses of Importance for Integration and Memory." In *Synaptic Function*, edited by G. Edelman, W. Gall, and W. Cowan, 403–429. New York: John Wiley, 1987.
3. Barlow, H. "Single Units and Sensation: A Neuron Doctrine for Perceptual Psychology." *Perception* **1** (1972): 371–394.
4. Barto, A. "From Chemotaxis to Cooperativity: Abstract Exercises in Neuronal Learning Strategies." In *The Computing Neuron*, edited by R. Durbin, C. Miall and G. Mitchison, 73–98. Reading, MA: Addison-Wesley, 1989.
5. Braitenberg, V. "Cortical Architectonics: General and Areal." In *Architectonics of the Cerebral Cortex*, edited by M. A. B. Brazier and H. Petsche. New York: Raven Press, 1978.
6. Cooper, J., F. Bloom, and R. Roth. *The Biochemical Basis of Neuropharmacology*. New York: Oxford University Press, 1989.
7. Cronin, J. *Mathematical Aspects of Hodgkin-Huxley Neural Theory*. Cambridge: Cambridge University Press, 1987.
8. Creutzfeldt, O. D. "The Neocortical Link: Thoughts on the Generality of Structure and Function of the Neocortex." In *Architectonics of the Cerebral Cortex*, edited by M. A. B. Brazier and H. Petsche, 357–383. New York: Raven Press, 1978.
9. Dayhoff, J. "Detection of Favored Patterns in Spike Trains." Ph.D. Dissertation, University of Pennsylvania, 1980. Available from University Microfilms International, Ann Arbor, Ml.
10. DeFelipe, J., S. Hendry, and E. Jones. "A Microcolumnar Structure of Monkey Cerebral Cortex Revealed by Immunocytochemical Studies of Double Bouquet Axons." *Society for Neuroscience Abstracts* **106.2** (1990): 240.
11. Dobkins, K., and T. Albright. "Color Facilitates Motion in Visual Area MT." *Society for Neurosciences Abstracts* **502.12** (1990): 1220.
12. Edelman, G. *Neural Darwinism*. New York: Basic Books, 1987.
13. Edelman, G., W. Gall, and W. Cowan. *Synaptic Function*. New York: John Wiley, 1987.

14. Gremillion, M. A. V. "An Anatomical Substrate for Information Processing." In preparation.
15. Jack, J., D. Noble, and R. Tsien. *Electric Current Flow in Excitable Cells.* New York: Oxford University Press, 1983.
16. Kandel, E., and J. Schwartz. *Principles of Neural Science.* New York: Elsevier, 1985.
17. Larkman, A., and A. Mason. "Correlations Between Morphology and Electrophysiology of Pyramidal Neurons in Slices of Rat Visual Cortex. I. Establishment of Cell Classes." *J. Neurosci.* **10** (1990): 1407–1414.
18. Mackey, Michael. *Ion Transport Through Biological Membranes.* Lecture Notes in Biomathematics, Vol. 7. New York: Springer-Verlag, 1975.
19. Martin, K. A. C. "From Single Cells to Simple Circuits in the Cerebral Cortex." *Qtr. J. Exp. Phys.* **73** (1988): 637–702.
20. Mason, A., and A. Larkman. "Correlations Between Morphology and Electrophysiology of Pyramidal Neurons in Slices of Rat Visual Cortex. II. Electrophysiology." *J. Neurosci.* **10** (1990): 1415–1428.
21. McGeer, L., J. Eccles, and E. McGeer. *Molecular Neurobiology of the Mammalian Brain.* New York: Plenum Press, 1988.
22. Mountcastle, V. "An Organizing Principle for Cerebral Function: The Unit Module and the Distributed System." In *The Mindful Brain*, edited by G. Edelman and V. Mountcastle, 7–50. Cambridge, MA: MIT Press, 1978.
23. Orban, G. *Neuronal Operations in the Visual Cortex.* New York: Springer-Verlag, 1984.
24. Peters, A., and G. Jones. *Cerebral Cortex, Volume 1: Cellular Components of the Cerebral Cortex.* New York: Plenum Press. 1984.
25. Popper, K., and J. Eccles. *The Self and Its Brain.* Berlin: Springer-Verlag, 1981.
26. Rakic, P. "Specification of Cerebral Cortical Areas." *Science* **241** (1988): 170–176.
27. Rall, W. "Cable Properties of Dendrites and Effects of Synaptic Location." In *Excitatory Synaptic Mechanisms*, edited by P. Andersen and J. Jansen, 175–187. Oslo: Universitet forlarget, 1970.
28. Ramon y Cajal, S. *Histologie du Systeme Nerveux de l'homme et des Vertebres II*, translated by L. Azoulay. Paris: Maloine, 1911.
29. Rose, D., and V. Dobson. *Models of the Visual Cortex.* New York: Wiley and Sons, 1985.
30. Schieber, M. "How Might the Motor Cortex Individuate Movements?" *TINS* **13** (1990): 440–445.
31. Schiller, P., and N. Logothetis. "The Color-Opponent and Broad-Band Channels of the Primate Visual System." *TINS* **13** (1990): 392–398.
32. Shepherd, G. *The Synaptic Organization of the Brain.* Oxford: Oxford University Press, 1990.
33. Stone, J. *Parallel Processing in the Visual System.* New York: Plenum Press, 1983.
34. Swindale, N. "Is the Cerebral Cortex Modular?" *TINS* **13** (1990): 487–492.

35. Szentagothai, J. "Specificity Versus (Quasi-) Randomness in Cortical Connectivity." In *Architectonics of the Cerebral Cortex*, edited by M. A. B. Brazier and H. Petsche, 77–97. New York: Raven Press, 1978.

36. Vardi, N., E. Herzberg, and P. Sterling. "Gap Junction Distribution in Cat and Monkey Retina Visualized with Monoclonal Antibody to Connexin32." *Society for Neurosciences Abstracts* **441.9** (1990): 1076.

37. Walsh, C., and C. L. Cepko. "Clonally Related Cortical Cells Show Several Migration Patterns." *Science* **241** (1988): 1342–1345.

38. Wassle, H., and Riemann, H. "The Mosaic of Nerve Cells in the Mammalian Retina." *Proc. R. Soc. Lond B* **200** (1978): 441–461.

39. Wassle, H., B. Boycott, and R.-B. Illing. "Morphology and Mosaic of On- and Off-Beta Cells in the Cat Retina and Some Functional Considerations." *Proc. R. Soc Lond. B* **212** (1981): 177–195.

40. Yoshioka, T., and Lund, J. S. "Substrates for Interaction of Visual Channels Within Area V1 of Monkey Visual Cortex." *Society for Neurosciences Abstracts* **295.1** (1990): 707.

Hans B. Sieburg
HIV-Neurobehavioral Research Center, Department of Psychiatry, M-003-H, University of
California, San Diego, CA 92093

Physiological Studies *in Silico*

This lecture describes a software think tank and its applications to the
study of cytokine networks in the immune and nervous systems. Special consideration is given to the neuro-immunological effects of Human
Immune-deficiency Virus (HIV) infection.

INTRODUCTION

A typical course on neurophysiology will describe many of the physiological systems
of the brain starting with (1) the physical and chemical phenomena involved in the
functions and activities of the individual cell; then (2) focusing on the physiology of
particular cell populations, e.g., the monoamine system, the dopamine system, the
auditory system, and the visual system; and (3) ending with considerations at the
system level that include the integration of individual subsystems and behavior.
From a wider perspective, this topic organization aims at the understanding of
how the physical and/or chemical interaction, or communication, between objects
at different levels of organization (cells, cell populations, subsystems) can lead to

the creation of micro-environments of self-organized behavior that influence the organization, function and behavior of objects on other levels.

Starting in the late 1960s a rapidly growing variety of neuro- and immuno-modulators have been discovered which are responsible for regulating the maturation, growth, and responsiveness of particular cell populations. It was commonly observed that these modulators, which we refer to as cytokines, are effective locally and in very low quantities. When administered in large quantities, these substances are cytotoxic. As a consequence, immune-system-derived cytokines (immunokines) have been used to battle cancers. Although cytokines are not specifically directed at a stimulus, but regulate its processing in an unspecific manner, they have been implicated in the generation and recall of long-term memory, the focusing of attention, and more recently, in degenerative effects of aging due to the progressive loss of regulatory capacities. The mechanisms by which specific responses are calibrated by non-specific regulators remains an important unsolved question in physiology.

In the first part of this lecture, I will present a brief survey of the complex physiologies of the immune system and the brain in order to settle on a common biological language which is applied throughout the lecture. The second part covers my own modeling system, called the cellular device machine, which my coworkers and I have used to simulate normal immune and nervous system behaviors, as well as the pathogenesis of infectious diseases, in particular infection by the Human Immune-deficiency Virus (HIV). To describe this approach as experimental at an equal level with *in vitro* and *in vivo* techniques of biological research, I chose the term *in silico*. The final section of the lecture covers the retrospects[26] and prospects of modeling the neuro-immunology of HIV-infection on the CDM development system.

COMPLEX PHYSIOLOGIES: THE IMMUNE SYSTEM AND THE BRAIN

The immune system is composed of large and heterogeneous populations of cells capable of recognizing, processing, and deactivating harmful substances termed antigens. The cell populations are functionally interrelated by intricate networks of cooperative processes. Many of these processes are evoked by antigen and subsequently evolve into complex patterns of cellular signaling through interdependent secreting and binding of cytokines. Cytokines are regulatory molecules that are released by cells and affect the movement or metabolism of other cells. During infections or due to neurological stresses, normal cytokine secretion patterns can change. Experimental and clinical research has indicated that cytokines are effective in tumor therapy, but, in higher than physiologic concentrations, can also cause toxic effects on surrounding healthy cells and tissues. Important characteristics of the immune system are the capacity to distinguish self from non-self and the long-term preservation of antigen-specific events due to a process defined as immunological memory. All cells of the immune system derive from the pluripotent stem cells which reside in the bone marrow. Coming from there, immune system cells undergo a series of differentiation steps in specific micro-environments, e.g., T cells in the thymus, before they enter the blood circulation as mature cells. From there, via the lymphatic system, they migrate into the secondary lymphoid tissues such as the lymph nodes. Here they remain, organized in distinctive patterns, before they recirculate into the system. Therefore, other than of cells and proteins, the immune system consists of fluids for transport, i.e., blood and lymph, and of organs and tissues supporting cell development and/or antigen trapping. Antigen trapping in these highly organized cellular clusters provides the threshold concentrations and tight cell packing favorable to a successfully launched immune response.

Important immune system cell types are the mononuclear phagocytes, e.g., monocytes and macrophages, and the lymphocytes, e.g., B cells and T cells. Macrophages can nonspecifically recognize, process, and present antigens to activate specific T lymphocyte clones. B cells also process and present antigen to T lymphocytes, but the preceding recognition process is specific. T lymphocytes activated after the recognition of antigen presented in the context of self-molecules on the surface of macrophages or B cells can secrete cytokines that affect the differentiation and proliferation behavior of B cells, T cells, and macrophages. T cells are therefore regarded as pivotal transducer elements that recognize antigen-specific signals and convert them into antigen-nonspecific cytokine signals, which ultimately enhance antigen-specific responses such as the secretion of specific antibody by plasma B cells or the lysis of infected cells by cytotoxic T cells. Large subsets of both the macrophages and the T lymphocytes carry a surface marker denoted as the cluster designation 4 (CD4) molecule. This receptor is involved in important immunological functions, and is also known to be the target receptor of the Human Immuno-deficiency Virus (HIV). To obtain more detailed information about the immune system and its physiology, I recommend the reading of Roitt et al.[20] and Oppenheim and Shevach.[18]

Since the brain as a system is already described in another lecture of this summer school (also see Gazzaniga[6]), I will focus on a summary of the cell types of the nervous system as far as they are relevant for the later discussion of immune and nervous system interactions. Much of this summary was taken from Kandel and Schwartz[11] and Gilma and Newman.[7] There are two classes of cells in the nervous system: the nerve cells (or neurons) and the neuroglial cells (or glia). The neuron is thought to be the primary functional and anatomic unit of the nervous system. The typical neuron has four morphological regions: cell body, dendrites, axon, and presynaptic terminal of the axon. Each region has a distinctive function. The cell body contains a nucleus and the apparatus needed for the synthesis of macromolecules. Dendrites and the axon are processes protruding from the cell body. They can be distinguished by the direction of information flow with respect to the cell body. Dendrites are branching processes that receive stimuli and conduct impulses generated by those stimuli toward the nerve cell body. They are therefore called afferent processes. The axon of a nerve cell is a single fiber extending to other parts of the nervous system or to a muscle or a gland (), sometimes over a considerable distance (1 meter or more). The term axon, in a physiologic sense, applies to a fiber that conducts away from a nerve cell body, and thus is an efferent process. Any long fiber, however, may have the anatomic properties of an axon regardless of the direction of conduction. Near its end the axon divides into many fine branches, each of which has an ending called the presynaptic terminal. The terminals of a presynaptic cell contact the receptive surface of other cells, called postsynaptic cells, and transmit, by chemical or electrical means, information about the activity of the neuron to other neurons or to effector cells (e.g., muscle cells). The point of contact is known as the synapse. The receptive surfaces consist of the membrane of the cell body, the dendrites, or the synaptic terminal of the axon.

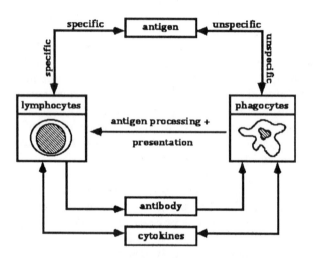

FIGURE 1 Although this schematic does little justice to the actual means by which an immune system controls the vast expanse of a host's body, it nevertheless depicts in a good approximation the way in which the flow of information is organized between the various subsystems involved in launching a debilitating response against the intrusion by foreign pathogens, generically termed antigens ("*antibody generators*"). In a real human host, several trillion variations of this diagram are implemented to protect against a universe of challenging antigens. A response is initiated once the intruder has breached a sophisticated line of exterior defenses such as the skin, stomach acid, mucus, etc., and encounters the host's interior defenses consisting of the cells and proteins of the innate and adaptive immune system. The innate immune system recognizes antigen unspecifically, and the adaptive immune system recognizes antigen specifically by virtue of a large variety of anticipatory molecules, termed antigen receptors. The most commonly known such antigen receptors are the antibodies. Should the agent persist or challenge the host a second time, the adaptive subsystem is able to mount a largely (continued)

FIGURE 1 (continued) enhanced secondary response due to a process termed immunological memory. In comparison, the secondary response mounted by the innate subsystem, which is without the ability to generate memory, is the same as the primary response. Both subsystems make a crucial distinction prior to launching a response. Namely, they can distinguish between host-characteristic, or "self," antigens, against which responses are usually suppressed, and "non-self" antigens. The innate and adaptive subsystems are not independent. Rather, they are functionally integrated by a large number of cooperative processes between their cellular components. Many of these processes are initiated by antigen and subsequently evolve into complex patterns of cellular signaling through interdependent secreting and binding of immunomodulators, called cytokines. In the schematic shown here, phagocytes which are part of the innate immune system, bind, process and present antigen to the lymphocytes which are part of the adaptive immune system. Upon stimulation and subsequent exchange of cytokines, lymphocytes undergo a number of differentiation and proliferation steps which lead to large clones of antibody-producing B lymphocytes and/or cytolytically active T killer lymphocytes. The process of antigen presentation marks therefore a critical conversion of unspecific information into specific response patterns which, when disturbed during disease, can lead to severe deficiencies in a host's immune reaction.

Communication between neurons can occur through electrical or chemical synaptic connections. Electrical synapses consist of low-resistance connections between the membranes of two neurons where a change in potential in a presynaptic neuron can be transmitted to a postsynaptic neuron with little attenuation. Electrical synapses are found extensively in the invertebrate and lower-vertebrate nervous system and at some sites in the mammalian nervous system. Chemical synapses are the predominant type of interneuronal communication in the mammalian brain. Action potentials in the presynaptic neuron cause the release of neurotransmitters from synaptic vesicles. The transmitter substance traverses the synaptic cleft and either excites (depolarizes) or inhibits (hyperpolarizes) the postsynaptic neuronal membrane. The synaptic potentials produced in the postsynaptic membrane are graded responses, in contrast to the all-or-none behavior of the action potential transmitted along axons. Depolarizing responses lead to impulse generation and/or transmitter release, whereas hyperpolarization opposes these.

As I stated before, there is a second class of cells in the nervous system, called glial cells, which in the central nervous system (CNS) outnumber the nerve cells by nine to one. Glial cells serve, in part, as supporting elements, a role played by connective tissue cells in other parts of the body. They also segregate groups of neurons from each other and may have additional, perhaps nutritive, functions. Lastly, certain glial cells make up the fatty insulating myelin sheaths around axons which are thought to be essential for achieving high-speed conduction. Although the membrane potential of glial cells can be altered by changes in external potassium concentration produced by impulses in nerve cells, glia are not directly involved in signaling information. Glial cells, which lack axons, are generally divided into five major classes: astrocytes (Figure 2), oligodendrocytes, microglia, ependymal cells, and Schwann cells. Oligodendrocytes form and maintain the myelin sheaths of the central nervous system. The Schwann cells form the myelin sheaths around the axons of peripheral nerves. Microglia are phagocytic cells that form part of the nervous system's defense against infection and injury.

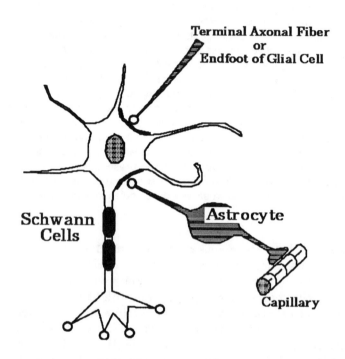

FIGURE 2 This picture shows some essential features of the interaction between glial cells—shown here are an astrocyte and a Schwann cell—and nerve cells. As stated in the main narrative, Schwann cells and oligodendrocytes form the fatty insulating myelin sheaths around axons of the peripheral (the former) and the central nervous system (the latter). The astrocytes are commonly subdivided into two subclasses: protoplasmic and fibrous. Each class is characteristically associated with a different part of the neuron. Fibrous astrocytes contain many filaments and are found extensively in areas of the central nervous system (CNS) containing myelinated axons (these regions are called white matter because of the whitish appearance of myelinated axons in unstained, freshly cut brain sections). The protoplasmic astrocytes contain few filaments and are associated with the cell bodies, dentrites, and synapses of neurons. These components of neurons are clustered together in regions called gray matter because large collections of nerve cell bodies and dendrites are gray in appearance in brain sections. Fibrous and protoplasmic astrocytes establish bridges between blood capillaries and neurons and are therefore thought to have a nutritive function. In the picture, I indicated points of contact between secretory and receptive surfaces by black circles and black curved lines. Astrocytes cannot develop action potentials, but they are highly permeable to potassium and become depolarized if the extracellular (continued)

FIGURE 2 (continued) concentration of potassium increases. They are thought to take up extracellular potassium during intense neuronal activity and to buffer potassium concentration in the extracellular space. They also take up and store neurotransmitters. Astrocytes are sensitive to a wide variety of insults to CNS tissue. Depending on the noxious agent, they may respond with cytoplasmic swelling, accumulation of glycogen, fibrillar proliferation within the cytoplasm, cell multiplication, or a combination of these reactions. They are frequently the cells that form a permanent scar or plaque after destruction of neuronal elements.

There is currently little to no doubt that the immune and nervous system communicate directly. The evidence for this ranges from direct invasion of the other system's domain, e.g., lymph nodes are enervated and macrophages cross into brain tissue, over the adaptation of one system's cell types to the others micro-environments, e.g., astrocytes and microglia are phagocytic cells closely related to the phagocytic cells of the immune system, to the shared usage of modulators, e.g., the cytokines interleukin 1 (IL-1[7] IL-3, IL-6[8]); Tumor Necrosis Factor (TNF)[6] granulocyte-macrophage colony stimulating factor GM-CSF[9] and G-CSF[9] are used both in the brain and the immune system. Also there are feedback loops between modulators such as the adrenocorticotropic hormone (ACTH) and IL-1.[11] Some of these immune/nervous system interactions are portrayed in Figure 3.

In conclusion of this introductory part of the lecture, I will summarize some shared structural and functional properties which make the immune and nervous systems a fascinating and, at the same time, dreadful subject for simulation. First, both are anticipatory systems. This terminology describes that all recognition elements (e.g., receptors on the surface of cells) are expressed prior to the encounter with "things." Therefore, in a situation where a "thing" encounters a population of cells expressing a mixed repertoire of recognition elements, the "thing" will select for a subset of recognition elements (in the case of receptors, this means the ligand binds to a receptor), and thus becomes a "stimulus." Only after a "stimulus" is established, the anticipatory system will be required to select, i.e., whether to focus attention on (respond to) the "stimulus" or not. In the immune system this process is called "Self–Non-self Discrimination," and its proper functioning is an important component of our health comfort. In the brain, this wider concept of "attention focusing" can help to explain cognitive processes such as learning, which involves diverse subsystems and strongly depends on the degree of recall of previously established "stimuli." Therefore, I would name the capacity to establish memory as the second important characteristic of immune and nervous systems. Memory is likely to be the result of morphologic and metabolic changes which establish activated cells as long-lived products of recognition/response events. In other words, the life expectancy of a cell is dependent on a history of interaction opportunities which, when taken, prolong its life span. Third, environments ("where in space?") are critical for achieving a rapid response to a "stimulus." This is evidenced by

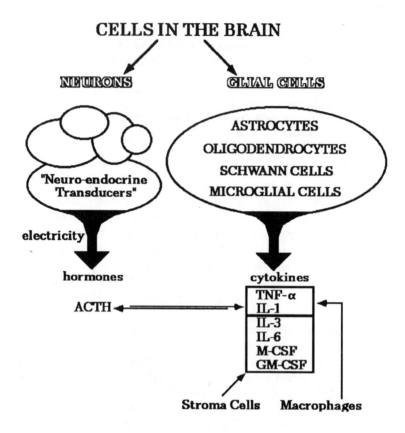

CELLS IN THE BRAIN

FIGURE 3 This slide depicts a process of activation/inhibition feedback between so-called neuroendocrine transducers, and glial cells. The term "neuroendocrine transducers" is used to describe nerve cells which can be electrically or chemically stimulated to secrete hormones. Such hormones, e.g., the adrenocorticotropic hormone (ACTH) secreted from the anterior pituitary during stress, can stimulate certain glial cells, such as the astrocytes, but also brain macrophages, to secrete interleukin 1 (IL-1), IL-6, and Tumor Necrosis Factor (TNF). Recent research has shown that IL-1 down-regulates the production of ACTH. This result supports the general belief that hormones are key mediators of many of the interactions between the central nervous system and the immune system.

the existence of tightly packed cellular regions, such as the lymph nodes and the hippocampus, where short-range communication between cells facilitates stimulus processing, and most likely also reduces the probability of chaotic regulatory interactions. These areas are potentially important for the generation and recall of distributed long-term cellular memory. Fourth, the coordination of biological clocks ("when in time?") is critical for proper recognition and response. More specifically, the "right" receptor populations must be expressed in a sufficiently large subpopulation at just the "right" point in time (the "no receptor–no see" axiom). Given the diversity of "things" in the universe, anticipatory systems are therefore forced to entertain a large and diverse repertoire of recognition elements, thus a large number of cell types and clones. Whether this helps to explain why, in the immune system, the number of lymphocytes (10^6 for humming birds, 10^9 for mice, 10^{12} for humans, 10^{18} for whales) seems to be correlated linearly with a species' body volume, I do not know yet.

THE CDM DEVELOPMENT SYSTEM FOR MODELING COMPLEX PHYSIOLOGIES ON THE COMPUTER

In this section, I want to discuss a software think tank which we described previously.[23,24,25,26] We have shown that this package allows us to simulate large and heterogeneous physiologic systems; in particular, complex processes of cellular communication that involve a large number of excitatory and inhibitory chemical signals which influence the maturation, proliferation, and migratory behavior of cells. Biological cells are represented in this system as cell devices, which is why the think tank was named a cellular device machine (CDM).

To cope with the tremendous amount and variety of present-day biomedical research information in the most flexible way, we designed the CDM as an interactive multi-media system[2] (Figure 4). It consists of three major components: The "Scriptor," the "Simulator," and the "Evaluator." In the "Sketchbook" mode of the "CDM Scriptor," ideas can be mapped out in different ways: as drawing, as text, or as software fragments. Parts, or all, of a "Sketchbook" page then can be transferred into the "Object" mode of the "CDM Scriptor," to become models of a simulation experiment on the "CDM Simulator." All "Scriptor" objects are mathematical graphs, i.e., combinations of circles, arrows, and rectangles. The idea behind this approach is to let the experimenter use a high level of visual abstraction in representing biological entities to enhance his or her focus on the main objectives of the problem to be solved. Furthermore, this all-graphical method does not require additional training of the non-specialist user. Once a "Scriptor" object is designated as a model, its graphical representation is interpreted as code written in our simple programming language called "SLANG." During a "CDM Simulator" experiment, a large number of such "scripts" interact in a cellular automaton environment. Their interaction results in population dynamics which are recorded for comprehensive symbolic analysis in the "CDM Evaluator."[1]

I am presently undecided about the possibility or usefulness of comparing object-oriented multi-media software and expert systems. Rather than scolding the latter in favor of the former, we designed the CDM development system to amalgamate the best of both approaches by creating a dynamical setting, similar to a brain, where large amounts of stored and categorized data can interact and newly associate. Under the dynamical paradigm we are not only able to decide static [problem ⇒ solution] pathways, a task for which expert database systems are traditionally optimized but, moreover, are able to use all available information to show us the outcome, over time, of hypothetical relationships between data fragments when we know a big-picture solution (e.g., "Make a vaccine!") yet are in higher need of clearly phrasing a problem (e.g., "What causes AIDS?"). The capacity for surprise by trailing effects back to causes, i.e., detailing inverse [solution ⇒ problem(s)] pathways, distinguishes the CDM as an easy-to-use development system for modeling biology on the computer.

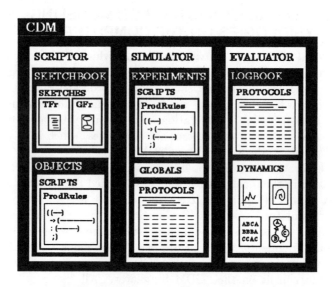

FIGURE 4   This figure depicts an overview of the CDM Development System (CDM-DS). This system consists of three software packages which are interrelated by file-transfer capabilities. The major data-structures in the CDM scriptor are the "SKETCHBOOK" and the "OBJECTS." The "SKETCHBOOK" consists of pages which contain "sketches" composed of "text fragments" (TFr) and "graphic fragments" (GFr). The "OBJECTS" data structure consists of cell devices, i.e., representations by mathematical rules, called scripts, which model the differentiation behavior of biological cells or idealized biochemical properties of factors modulating such behavior. The smallest unit of a script is the "production rule" whose make-up is explained in more detail in the mathematical narrative. In the CDM scriptor, scripts are generated from GFr's, which greatly facilitates the usage of the CDM-DS by a non-specialist user untrained in the simulator language ("SLANG"). The main data-structure in the CDM simulator is the "EXPERIMENT," which in turn consists (continued)

FIGURE 4 (continued) of scripts, a series of globals which pertain to general initial conditions such as cell device population sizes, the initial distribution of populations in their simulation environment, and parameters describing their mobility. Before launching an actual simulation it is possible to edit, or add new, scripts by directly modifying the SLANG production rules. The outcome of a simulation executed under the CDM-DS is a protocol which is the third component of an "EXPERIMENT" data structure. This protocol contains information regarding the time course and purpose of a simulation, observations made while the simulation was running, as well as numerical data which, for example, describe the evolution of population sizes over time. To interpret the outcome of an experiment, protocol files are accessed by the third package, named the CDM EVALUATOR. The EVALUATOR separates descriptive experiment information from numerical experiment information, and places the former at the beginning of a document termed "REPORT." Following this narrative are the numerical data plotted and examined by a variety of mathematical methods ranging from classical time series analysis (left upper icon in the "DYNAMICS" box), over phase-space analysis (right upper icon), to symbolic dynamic analysis (lower icons). The latter evaluation technique is the most intriguing among the three, since we can rewrite the symbolic strings (lower left-hand icon) as mathematical graphs (lower right-hand icon), which describe the global experiment dynamics in a form similar to the cell device graphs initially used to represent local dynamical relationships. The advantage of this approach is that the transition from local to global dynamics can be anchored into a single mathematical theory, namely the theory of graphs. Having this single mathematical framework greatly facilitates accessing the logical correctness of experiments, guarding against artifacts, and deriving important numerical quantifiers of dynamical behavior, such as the topological entropy.

A more mathematical description of the CDM is this. Consider a locally finite graph G, i.e., for every vertex $v \in VG$, the vertex set, we have

$$\#N(v) := \{w \in V : \exists\, e \in E_G : v = t(e) \text{ and } w = o(e)\} < \infty.$$

Here, $N(v)$ denotes the "neighborhood" of the vertex v, E_G denotes the edge set of G, $t(e)$ the origin of edge e, and $t(e)$ the terminal of e. Now let M be a set of Mealy automata $A = (Q, \Sigma, \Delta, \delta, \lambda, q_0)$,[9] where Q, Σ, Δ denote the state set, input set, and output set, respectively, and $\delta : Q \times \Sigma \to Q$, $\lambda : Q \times \Sigma \to \Delta$ are the input and output functions, and q_0 denotes the initial state. Finally, let

$$\forall\, v \in V_G : \exists!\, S \in P(M) : v = S,$$

i.e., the vertices of G represent finite collections of Mealy automata. A *Cellular Device Machine* (CDM) then is a pair (G, M), where G and M are described above. According to this definition, we can think of a CDM as a locally finitely connected network of smaller machines. The information flow in this network is event-driven since outputs derive from inputs associated with a state transition.

To implement a CDM (G, M) on a general-purpose workstation such as an Apple Macintosh computer, we used a cellular automaton model with the following specifications. A schematic overview of this model is presented in Figures 4 and 5. Let $B(x, y, z) = 1$ for $x - y > z$, $= 0$ otherwise, where x, y, z are non-negative integers. For a fixed positive integer L let $P(L) := \{p = p_1 p_2 \ldots p_L : p_i \in GF(2)\}$ be the set of all pattern elements (PELT's) of length L, where $GF(2)$ denotes the

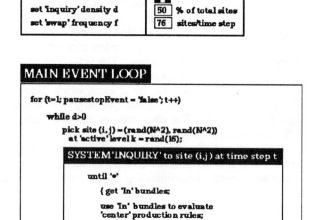

FIGURE 5   Up to 256 parameters can be incorporated into one single simulation experiment. Of these, 16 are declared as "active" and 240 declared as "passive." This terminology is applied to indicate which of the parameter objects owns a full script composed of one or more production rules ("active"), and which are only present as bundles ("passive"). For each "active" object, the simulator provides a cellular automaton layer, i.e., a rectangular array of sites with periodic boundary conditions. Therefore each layer is isomorphic to a torus. Sites (i, j) are vertically connected across all layers. During the initialization of a simulation experiment, particular "GLOBALS" are chosen by the user, such as the shapes of the input and output neighborhoods, the number of system "inquiries" per time step, and the degree of movement allowed before a new time step is entered. Examples of such settings are given in the right half of the "GLOBALS" box shown in the upper row of this figure. It should be noted that the "GLOBALS" expand the degrees of freedom traditionally granted in (continued)

FIGURE 5 (continued) cellular automaton simulations. For example, in order to be able to appropriately account for the secretion of modulators by cells, we require that not only a center site is updated but that also changes take place in an output neighborhood, whose shape usually differs from that of the input neighborhood from where signals are taken up. Furthermore, to account for the amount of movement among cells (or the lack of such), we use the "swapping frequency" f. Lastly, the amount of explicit update in the parallel cellular automaton environments is regulated by the "inquiry" density d. While traditionally d is 100%, we use a space filling, random covering algorithm to emulate the update of a sub-automaton, chosen randomly per time step, which may, as shown in the sample setting, occupy only 50% of the area of the original cellular automaton. This strategy has two advantages. First, if the space filling algorithm is chosen correctly, the sub-automaton processes the same amount of information per time step than the full automaton, which carries out a certain amount of redundant processing. Second, being able to adjust the update density d to the amount and complexity of the production rules involved in an experiment, helps to effectively reduce processing time. To illustrate this point I added the C language pseudo-code for the simulator's main event loop in the lower box of this figure.

alphabet $\{0,1\}$ of two elements. One of the simplest pattern recognition methods on $P(L)$ derives from the Hamming distance

$$\forall\, p_1, p_2 \in P(L) \colon H(p_1, p_2) := \sum (p_{1i} + p_{2i}) \bmod 2,$$

which calculates the number of 1's in the sum modulo 2 of p_1 and p_2. Here, the sum ranges from 1 to L. Using the Hamming distance, we can define the recognition function $R(p_1, p_2; \Theta) := B(H(p_1, p_2), 0, \Theta)$, which is 1 iff the number of 1's in the sum modulo 2 of p_1 and p_2 is greater than the non-negative integer threshold Θ, and 0 otherwise. It is, of course, possible to use other recognition functions or larger alphabets, e.g., we have tested the value of a variety of one-dimensional cellular automaton rules to this effect, but empirically the Hamming distance has turned out to be very effective in the biological context we are working in.

Now let $A \in M$ be a Mealy machine with state set $Q(A) = \{q_0, q_1, \ldots, q_n\}$, input set $\Sigma(A) = \{a_1, \ldots, a_m\}$ and output set $\Delta(A) := \wp(\{b_1, \ldots, b_k\})$. Here, $\wp(D)$ denotes the power set of a set D. Any $x \in Q(A), \Sigma(A), \Delta(A)$ is of the general form $x = (c, A, x, n)$, where $c \in P(L)$ and n is a non-negative integer, called the eigenlife, whose purpose we will discuss below. Given states $q, q^* \in Q(A)$, an input $a \in \Sigma(A)$, and outputs $b_{i(1)}, \ldots, b_{i(s)} \in \Delta(A)$ such that $q = o(e)$ and $q^* = t(e)$ with edge $e = (a)$ in the transition graph $G(A)$ of A, we can use R to define functions $\lambda : Q(A) \times \Sigma(A) \rightarrow \Delta(A)$ and $\delta : Q(A) \times \Sigma(A) \rightarrow Q(A)$ as follows: $\delta(q, a) = q^*$, iff $R(q, a; \Theta) = 1$, and $= q$, otherwise; $\lambda(q, a) = \{b_{i(1)}, \ldots, b_{i(s)}\}$, iff $R(q, a; \Theta) = 1$, and $= \{\varepsilon\}$, otherwise. Here, ε denotes the empty symbol. Both functions can be combined to transfer graphically programmed Mealy automata and additional quantitative information into machine-interpretable production rules of the general form:

$$(q.[K, M]a) \rightarrow (q^*; > N_1 b_{i(1)}, \ldots, > N_s b_{i(s)}, < a; \bullet) : (q; \bullet); .$$

Here, $K, M, N_1, \ldots, N_s$ denote positive integers, where $[K, M]$ represents the number $J, K \leq J \leq M$, of Mealy machines $A' \in N(A)$ such that $a \in \Delta(A')$, and $N_1, \ldots, N_s$, respectively, describe the numbers of vertices in $N(A)$ which are to receive outputs $b_{i(1)}, \ldots, b_{i(s)}$. The production rules can be read as "if ... then ... else" conditions, where "$\rightarrow$" indicates "then" and "$:$" defines "else." The symbol "$>$" marks the $N_j b_{i(j)}$ as outputs, and "$\bullet$" indicates a jump condition which returns control from a production rule to (G, M). The instruction "$< a$" indicates that the signal "a" is to be taken up by "q" during recognition and therefore to be removed from the simulation environment. If all neighborhoods $N(v)$ are chosen to be of the same type, then a CDM (G, M) behaves like a cellular automaton dynamical system (on the graph G). The transition rules $F : C(G) \rightarrow C(G)$ are given by the restriction maps

$$F(g)(v) : N(V) \rightarrow \prod \left(Q(A) \times \Sigma(A) \times \Delta(A) \right),$$

where $C(G) := \{g : g : V_G \rightarrow \prod(Q(A) \times \Sigma(A) \times \Delta(A))\}$ denotes the set of all configurations.

A practical usage for the "$\bullet$" terminator symbol arises in the context of feedback loops. For example, suppose we want to model the interaction between brain

TABLE 1 Scripts for ACTH/IL-1 Exchange

$(AP0.S) \rightarrow (AP1; > ACTH):(AP0; \bullet);$
$(AP1.IL-1 \rightarrow (AP0; \bullet):(AP1; > ACTH);$

$(MP0.ACTH) \rightarrow (MP1; > IL-1):(MP0; \bullet);$
$(MP1.ACTH) \rightarrow (MP1; > LM-1):(MP0; \bullet); .$

TABLE 2 Alternative Description for the IL-1-Sensitive Anterior Pituitary Cells

$(AP0.IL-1) \rightarrow (AP1; > ACTH):(AP0; \bullet);$
$(AP1.IL-1) \rightarrow (AP0; \bullet):(AP1; > ACTH); .$

macrophages and anterior pituitary cells as they may occur during stress reactions of a host.[22,8] In this case, a stress signal S turns on a process B, the release of adrenocorticotrophic hormone (ACTH), which stays on and in turn induces a process C, the release of interleukin 1 (IL-1) by brain macrophages. As has been shown in many clinical and *in vitro* experiments, IL-1 (process C) can down-regulate the production of ACTH (process B). Therefore, the ACTH/IL-1 exchange establishes a feedback loop where down-regulation is represented by the placement of a terminator symbol into the relevant production rule. More precisely, we would use

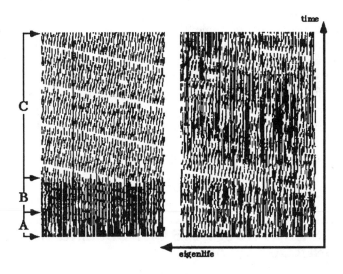

FIGURE 6 Sections from the eigenlife profiles of two simulation experiments of immune responses to a normal (left) and a persisting (right) antigen are shown. The eigenlives of individual cell devices are plotted horizontally, where a black pixel indicates a non-zero and a white pixel indicates a zero eigenlife value, versus time on the vertical axis. The "normal" antigen causes an initial immune response (area A) which dissipates slowly as the agent is overcome (area B), after which the system returns to a normal cell death and cell regeneration (from bone marrow stem cells) (area C). The eigenlife monitors the survival time of cells participating in immune responses. One of the cell devices shows non-zero eigenlife for the whole time period covered. In the second profile, i.e., the immune response against a persisting agent, the system never returns to a fully normal state. Such immune responses are expected to generate large numbers of memory cells, which have "seen" antigen and whose lifespan is much larger than that of normal cells and which therefore have been established as long-lived products of the encounter with antigen. The profiles shown were generated by linearizing two-dimensional cellular automata.

the scripts in Table 1 to describe this mechanism. Here, AP and MP denote an anterior pituitary cell and a macrophage; respectively, AP0 and MP0 are the inducible states capable of recognizing the signals S and ACTH. AP1 and MP1 are the releasing states where ACTH or IL-1 are secreted (indicated by the symbol ">") upon induction. If S or ACTH are not available, AP and MP will remain in their inducible states as indicated by the "else" portions of the first two production rules. An anterior pituitary cell in the induced state AP1 will stop to produce ACTH (sic the terminator symbol!) upon recognition of IL-1, and return to its inducible state. The induced macrophage, which depending upon availability, will continue to respond to ACTH by producing more IL-1, will cease production if the signal concentration falls below recognition level (placement of the terminator following the "else" (":") symbol). We experimented with an alternative description for the IL-1 sensitive anterior pituitary (AP) cells, where the stress-signal S is IL-1. This view results in the script shown in Table 2.

The simplest possible adaptation strategy in our Macintosh model of a machine (G, M) is the event-driven increase of the life span of $A \in M$, called *eigenlife*. The eigenlife is defined as follows. For $Q(A) = \{q_0, q_1, \ldots, q_n\}$, let $\{m(q_0), m(q_1), \ldots, m(q_n)\}$ be a set of fixed positive integers called the masses of the states. Let $q, q^* \in Q(A)$ with edge $e = (a)$, $a \in \Sigma(A)$, $q = o(e)$, $q^* = t(e)$. Then $E(A)(q^*) = E(A)(q) + m(q^*) \bullet R(q, a; \Theta) - 1$. Thus, for each automaton A, this defines an integer-valued map $E(A) : C(A) \rightarrow Z$. Figure 6 features the eigenlife profiles of two simulation experiments and explains how they can be applied to interpret system performance.

RETROSPECTS AND PROSPECTS OF MODELING THE NEURO-IMMUNOLOGY OF AIDS

Infection by the Human Immuno-deficiency Virus (HIV) has been shown to induce a large variety of pathological defects in the nervous system[10,19] and the immune system.[5,14] The AIDS dementia complex, a progressive neurological syndrome characterized by white and gray matter depletion and abnormalities in cognition, motor performance, and behavior, has been indicated as a common and important cause of morbidity in patients in advanced stages of infection.[17] The critical basis for the immuno-pathogenesis of HIV infection is the progressive depletion of CD4+ T lymphocytes, a condition eventually resulting in a profound immunosuppression. The understanding of HIV infection is greatly complicated by the complexity of HIV's target systems, but there seem to be common grounds on which neurologic and immunologic disease develop which are amenable to computer-aided studies.

An important example for this are what I call "closed loop dynamics." I use this nomenclature to describe in an intuitive way the biological equivalent of stationary attractors in dynamical systems theory. "What are immunological or neurological attractors?" and "Why do I emphasize stationarity?" The answers to these questions can be found in the understanding that, in order to fight disease properly, the immune system must maintain a very broad responsiveness at all times. As I pointed out in the first part of this lecture, immune responsiveness is reflected in the number of cell types, their capacity to communicate, the diversity of the antigen repertoire, the amount of hematopoetic control, and cytokine regulation. All of these mechanisms interact to prevent the over- or under-representation of any particular system feature. In a similar way, one can argue that, for a brain to function properly, a broad neurologic responsiveness must be maintained which, for example, allows one to maintain the capacity to rapidly shift attention (this capacity is strongly impaired in autistic individuals). Using this viewpoint, a dynamicist understands that immunological or neurological attractors are the result of the respective system focusing on a challenge. In a healthy state, each system maintains these foci for the duration of a response but afterwards rebounds back into diversity. This process of focusing and unfocusing describes a non-stationary phenomenon very similar to the emergence and decline of a hurricane. As we can easily imagine, stationary hurricane or permanently installed immune responses, such as in allergic reactions or against latent viral infections, can cause great discomfort. To summarize the idea, a dynamical view of the diseased immune system entails the search for events which can cause stationary response foci that aid in maintaining an infectious agent rather than serve to obliviate it.

Indications for the existence of "closed loop dynamics" have been found in a number of clinical and experimental studies reviewed in a recent article by Zeda Rosenberg and Anthony Fauci.[21] This paper stresses the central role of cytokines in establishing the dynamic process inevitably leading to immune disease: "Thus, it can be postulated that the natural history of HIV infection is dependent to a large extent on the aggregate immune system response to a wide variety of stimuli. Over time, the production of cytokines that can either activate or suppress immune-cell function as well as HIV expression results in a cumulative increase in the levels of HIV replication that are sufficient to cause a self-sustaining cycle [a "closed loop dynamic"; added by HBS] of HIV expression and multiple cytokine production. A gradual increase in HIV expression can result in the moderate decline in the number of CD4$^+$ T cells that is observed during the clinically latent phase of HIV infection. Once a sufficient virus load is present in the body, a cycle of virus-induced cytokine production and cytokine-induced virus expression is initiated. Such a cycle of escalating HIV production may be responsible for the sharp decrease in CD4$^+$ cells that is often seen years after initial virus infection." In the brain, the predominantly infected cell types are the macrophage and the microglia. It is, therefore, clear that since HIV causes dysfunction without apparently infecting nerve cells, secondary mechanisms must be responsible for a large variety of the neurological disorders observed in AIDS patients. The two hypotheses currently receiving the most attention are that a protein made (by/because of) HIV is toxic to neurons and that cytokines released from HIV-infected macrophages or microglia cells, or astrocytes, are neurotoxic. The precise meaning of neurotoxicity—leading to cell death, or demyelination, or altered signal transmission or something else—is presently debated. I have decided to focus my own work on cytokine, neuro-transmitter, neuropeptide, and hormone secretion patterns, since I am more likely to find "closed loop dynamics" with long-range effects on CNS function in this area than in any other. Another reason for this decision is that we have already extensively investigated the cytokine secretion patterns in the immune response against HIV and other infectious agents, and are therefore in a position to compare immune- and neuro-pathogeneses. The different aspects of our findings in a B-cell, T-cell, macrophage artificial immune system are summarized in the model of the global dynamics of HIV infection in an individual shown in Figure 7.[26]

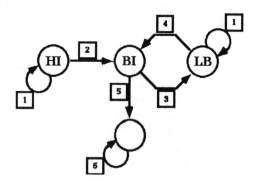

FIGURE 7 Phase graph of the global HIV/AIS dynamical system. By exposing the system to different experimental conditions involving the single HIV antigen in free or cell-associated form, a combination of HIV and another antigen, or exogenous applications of IL-2, we were able to identify a preliminary four-stage model of HIV pathogenesis which has one entry state (HI), a loop (BI) → (LB) → (BI), and a terminal (ID). The vertices define the following global dynamical states: (HI) the *h*ost's *i*nitial immune system status by the time of virus introduction, (BI) a *bi*furcation state from where the system can reach one of two other states, namely the *l*atency *b*asin (LB) and the *i*mmune *d*eficiency state (ID). The edges, as depicted by the transitions between the dynamical states, derive as follows. We showed that a host will remain, as indicated by the self-transition (1), in the normal, broadly responsive, and functionally diverse immune system status (HI) as long as HIV exists only in the cell-associated form, and the environment is kept antigen free. A transition, marked (2), to the next following state (BI) will occur however, if HIV manifests in free form or if other infectious agents materialize. When (BI) is entered for the first time, we find increased levels of virus titer. Our data show that a rapid transition, marked (3), into (LB) will occur due to the binding of HIV to CD4+ cells or due to the uptake of virus particles by antigen-presenting cells. When (BI) is entered a second or more times, we find, in addition to the elevated virus titer, increased cytokine concentrations (except for IL-2), detectable anti-HIV antibody titer, reservoirs of latently infected macrophages and T cells that the immune system is unable to clear, and an increase in the population of cytotoxic T cells. (LB) is characterized (continued)

FIGURE 7 (continued) by well-established reservoirs of latently infected macrophages and CD4$^+$ T cells and low virus titer. Since the latent form is favorable to HIV's natural life cycle, one may think of (LB) as a dynamically attracting state, hence the term basin. If (LB) is entered a second or more times, we are likely to find newly acquired patterns, e.g., diminished IL-2 responsiveness in CD4$^+$ T cells, sero-conversion, and increased IL-2 responsiveness in the CD8$^+$ cell population. Our simulations showed that the host remains in the latency basin unless other infections materialize, at which point a transition, labeled (4), into the bifurcation state (BI) occurs. Arriving again at (BI), the immune system status is marked with uncertainty, since a variety of immunological forces (e.g., cytokines, cytotoxic T cells, or abundant, but unspecific, antibodies) may interact chaotically, and therefore make it difficult to predict the future destination state: The system may progress towards immune deficiency, i.e., enter (ID), or go into remission, i.e., enter (LB). The actual causes for the transition from (BI) to (ID), labeled (5), are presently unknown. Empirical studies have shown that the immune-deficiency state can be maintained for varying periods of time, a possibility that is indicated by the self-transition (6). The global dynamical model discussed here can be used to calculate the rate of disease progression in terms of the number of times the system traverses the loop (BI) → (LB) → (BI) and the speed at which individual transitions occur. This indicates the possibility of using phase graphs to predict epidemiology by systematically modifying initial immune system statuses and the state-transition conditions. The CDM development system can again be used to implement this approach.

TABLE 3 Influences of Stress Factors and Secondary Infections (Figure 8)

$(AP0.S) \rightarrow (AP1; >ACTH):(AP0;^\bullet);$
$(AP1.IL-1) \rightarrow (AP0;^\bullet):(AP1; >ACTH);$

$(MP0.ACTH) \rightarrow (MP1; >IL-1,IL6, >TNFa;^\bullet):(MP0;^\bullet);$
$(MP1.ACTH) \rightarrow (MP1; >IL-1,IL6, >TNFa):(MP0;^\bullet); .$

$(MP0.HIV) \rightarrow (MP0^*;^\bullet);$
$(MP0^*.ACTH) \rightarrow (MP1^*; >IL-1,IL6, >TNFa, >HIV;^\bullet):(MP0^*;^\bullet);$
$(MP1^*.ACTH) \rightarrow (MP1^*; >IL-1,IL6, >TNFa, >HIV):(MP0^*;^\bullet); .$

$(AS0.IL6) \rightarrow (AS1; >AS1;^\bullet):(AS0;^\bullet);$
$(AS1.gIFN) \rightarrow (AS2; >IL-1, >TNFa):(AS1;^\bullet);$
$(AS1.IL6) \rightarrow (AS1; >AS1;^\bullet): (AS1;^\bullet);$
$(AS2) \rightarrow (AS0;^\bullet);$

Our approaches to modeling the effects of HIV infection on the central nervous system have so far focused on the effects of HIV infection on known communication pathways between astrocytes, microglials, brain macrophages, and neurons. Figure 8 shows, for example, how stress factors and secondary infections can influence adrenocorticotropin secretion, HIV expression and secretion from latently infected macrophages, and the changes in size of the reservoir of infected macrophages. In this example, we used the scripts in Table 3. The production rule (AS2) → (AS0; •) in line 4 of the astrocyte script is special since the lack of a specified input signal and an alternative "else" condition indicates automatic return upon next system "inquiry" to the state shown on the right. The star "*" alongside the macrophage state labels "MPx," where x can be "0" or "1," sets the infected cells apart from the non-infected cells. These experiments can be expanded or altered in various ways. As I indicated in the section about the CDM development system, one could chose IL-1 as the stress factor. Also, the pain pathway can be accounted for to some extent, since it was reported in the literature that macrophages can be stimulated by Substance P or Substance K into secreting IL-1, IL-6 and TNF-α.[15] The outcome of these experiments and of others will be reported in a forthcoming publication.

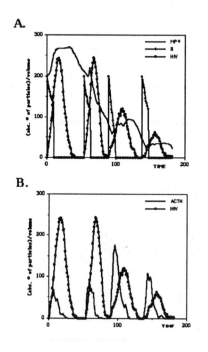
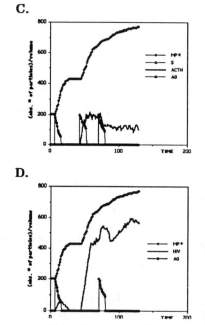

FIGURE 8   (continued)

FIGURE 8   (continued) These time-series plots depict the population size of latent HIV-infected macrophages, and the concentrations of a stress factor S, adrenocorticotropic hormone (ACTH), and a secondary antigen AG. Population sizes and concentrations are all expressed as absolute number of particles per volume of 32 x 32 sites. A volume of 32 x 32 (or 16 x 16, or 64 x 64) sites is automatically reserved for each of the parameters—12 in the case shown here—of a simulation experiment. The plots shown derive from two experiments. Both scenarios contained populations of astrocytes, anterior pituitary cells, and uninfected and latent HIV-infected macrophages. In the first experiment, I injected, at four different points in time, equal amounts of a stress factor S to study its effect on adrenocorticotropin secretion and HIV expression. I limited the half-life of S to 10 time steps. Part A of this figure shows that the spikes indicating S-injection degrade at different velocities due to the uptake of S particles by stress-factor-sensitive anterior pituitary cells, before the remaining free S particles decompose upon reaching their half-life. The differing amounts of S uptake are correlated with differing amounts of ACTH secretion (depicted in part B). The ACTH curve in part B also reveals a small increase of the hormone following each of the larger humps. These increases derive as a consequence of macrophage stimulation leading to IL-1 secretion and high HIV titers. The concentration of HIV particles follows a surprisingly smooth pattern which begins in phase, and ends out of phase, with the ACTH concentration. Because I arranged the experiment in such a way that there is an overall decline in the population of latently infected macrophages—because many become productively infected and die—it is clear from the curve in part A that stress responses lead to increases in the size of this population. In the second experiment, productive HIV infection is not considered cytotoxic, which is why we see an overall increase in the population of latently infected cells. In order to characterize the difference in effect between neurological and immunological stressors, I first challenged the pituitary-astrocyte-macrophage system with a second antigen (AG), then presented the system with the neurological stressor S, and finally challenged again with AG. As shown in part D, the antigenic challenges have small effects compared to the increase in HIV secretion caused by the stressor S. Interestingly, when after the second challenge with AG the system was left to itself, the cycle of ACTH secretion (the ACTH curve is shown in part C) and HIV secretion never stopped. This experiment therefore presents us with an example of a "closed loop dynamic" where latent HIV infection has become sufficiently established to mimic a neurological stress response (with interleukin 1 as a mediator). In a clinical or wet-laboratory setting, these findings would indicate a correlation between persistently elevated levels of adrenocorticotropin, and a large reservoir of latently infected macrophages; the former is more easily measured than the latter.

In conclusion, let me mention two open questions which can be tackled by simulations on the CDM development system. First: "What are the long-term effects of HIV strain diversity on the immune status of an infected individual, i.e., the natural balance between immuno-potentiation and immuno-suppression?" Second: "How come that HIV turns off hematopoesis?" Intuitively, I think that both questions address a common theme from two different angles, namely, HIV immune evasion due to virus adaptability. Here, we must understand the term "virus adapt-ability" in the widest possible sense to include not only species diversity due to random mutations, but also the more refined and subtle processes leading to ad-justments of HIV's reproductive mechanisms to the life cycle of a specific cell type. If, for example, HIV were to use IL-2-driven T-cell proliferation to propagate its genetic information without secretion of virus particles, the reservoir of latently infected T cells could grow without immune recognition. By the same token, but on a much larger scale, HIV might be able to adapt to the intricate workings of the cytokine communication network. Redirection of the resources in this network for the benefit of virus spread may at the same time suppress hematopoesis. One important reason why we need to look for such subtle effects rather than blatant and immediate cytotoxicity, is that they take time; and time is, in more than one sense, the very essence of HIV pathogenesis. Answers therefore to the questions I mentioned above will show us how to arrest post-infection illness and will guide us in developing plans for preventive strategies, such as vaccines.

CONCLUDING REMARKS

This two-hour lecture can be seen, understood, or criticized in different ways. Be-cause this lecture was given in the context of a summer school on complex systems, I intended to present to you a personality profile of the science that my group and I pursue, as opposed to a linear parade of polished results. Especially, I wanted you to see some of the why's, the how's and the where-to's of our work which uses findings, thoughts, and techniques from four sciences all too often regarded as independent: mathematics, immunology, computer science, and neurology. The introductory section about immuno- and neuro-physiology intended to establish a common biological language, an understanding of the complexities involved, and to explain why the nature of the biological questions we try to answer requires us to work with discrete dynamical systems rather than with the otherwise more frequently used differentiable ones. The second section, where I introduced and dis-cussed our model development system, also contains a fair amount of mathematics. Apart from its entertaining value, the purpose of this mathematical interlude is two-fold: First, I wanted to clarify where one can find the mathematical foundations of our discrete modeling approach, namely in the theory of graphs. Secondly, I wanted to make the point that using an object-oriented software system for modeling bi-ology means using a complex dynamical system to simulate complex dynamical systems. Therefore, in order to successfully guard against artifacts, it is necessary to take great care in filtering out the system's baseline dynamics mathematically, a task that is as boring as "Definition, Theorem, Proof," but certainly worthwhile when it comes to applying computer simulation to medicine. The third section then picks up the biomedical theme to illustrate in some special cases how the CDM de-velopment system can be used to carve out the dynamics of an infectious disease affecting both the immune and nervous systems. A full discussion of the immunological component can be found in Sieburg et al.[26] More results on the neurological complications of HIV infection will be presented in a forthcoming publication.

ACKNOWLEDGMENTS

The work reported in this paper is supported by NIMH grant R29 MH45688-02 to the author.

REFERENCES

1. Alekseev, V. M., and M. V. Yakobson. "Symbolic Dynamics and Hyperbolic Dynamic Systems." *Physics Reports* **75(5)** (1981): 287–325.
2. Ambron, Sueann, and Kristina Hooper, eds. *Interactive Multimedia*. Redmond: Microsoft Press, 1988.
3. Beutler, B., and A. Cerami. "The Biology of Cachectin/TNF—A Primary Mediator of the Host Response." *Ann. Rev. Immunol.* **7** (1989): 625–55.
4. Dinarello, Ch.A. "Biology of Interleukin 1." *FASEB J.* **2** (1988): 108–115.
5. Fauci, A. S. "The Human Immunodeficiency Virus: Infectivity and Mechanisms of Pathogenesis." *Science* **239** (1988): 617.
6. Gazzaniga, M. S. "Organization of the Human Brain." *Science* **245** (1989): 947–952.
7. Gilma, S., and S. W. Newman. *Essentials of Clinical Neuroanatomy and Neurophysiology*, 7th edition. Essentials of Medical Education Series. Philadelphia: Davis, 1989.
8. Gorman, J. M., and R. Kertzner. "Psychoneuroimmunology and HIV Infection." *J. Neuropsych. & Clin. Neuroscience* **2(3)** (1990): 241–252.
9. Hopcroft, J. E., and J. D. Ullman. *Introduction to Automata Theory, Languages and Computation*. Reading, MA: Addison-Wesley, 1979.
10. Johnson, R. T., J. C. McArthur, and O. Narayan. "The Neurobiology of Human Immunedeficiency Virus Infections." *FASEB J.* **2** (1988): 2970.
11. Kandel, E. R., and J. H. Schwartz. *Principles of Neural Science*, 2nd edition. Amsterdam: Elsevier, 1985.
12. Khansari, D. N., A. J. Murgo, and R. E. Faith. "Effects of Stress on the Immune System." *Immunology Today* **11** (1990): 170–175.
13. Kishimoto, T. "The Biology of Interleukin 6." *Blood* **74** (1989): 1–10.
14. Levy, J. A. "Mysteries of HIV: Challenges for Therapy and Prevention." *Nature* **333** (1988): 519.

15. Lotz, M., J. H. Vaughan, and D. A. Carson. "Effect of Neuropeptides on Production of Inflammatory Cytokines by Human Monocytes." *Science* **241** (1988): 1218–1221.
16. Malipiero, U. V., K. Frei, and A. Fontana. "Production of Hemopoietic Colony-Stimulating Factors by Astrocytes." *J. Immunol.* **144** (1990): 3816–3821.
17. Navia, B. A., B. D. Jordan, and R. W. Price. "The AIDS Dementia Complex: I. Clinical Features." *Ann. Neurol.* **19** (1988): 517.
18. Oppenheim, J. J., and E. M. Shevach. *Immunophysiology.* Oxford: Oxford University Press, 1990.
19. Price, R. W., B. Brew, J. Sidtis, M. Rosenblum, A. C. Scheck, and P. Cleary. "The Brain in AIDS: Central Nervous System HIV-1 Infection and AIDS Dementia Complex." *Science* **239** (1988): 586.
20. Roitt, I. M., et al. *Immunology,* 2nd edition. London: Gower Medical Publishing, 1989.
21. Rosenberg, Z. F., and A. S. Fauci. "Immunopathogenic Mechanisms of HIV Infection: Cytokine Induction of HIV Expression." *Immunol. Today* **11** (1990): 176–180.
22. Rubinow, D. R. "Brain, Behaviour, and Immunity: An Interactive System." *J. National Cancer Institute* Monographs **10** (1990): 79–82.
23. Sieburg, H. B., and C. E. Müller-Sieburg. *Nuclear Physics B* (Proc. Suppl.) **2** (1987): 615.
24. Sieburg, H. B. "A Logical Dynamic Systems Approach to the Regulation of Antigen-Driven Lymphocyte Stimulation." In *Theoretical Immunology, Part One,* edited by A. S. Perelson, 273–294. SFI Series in the Sciences of Complexity, Proceedings Volume II. Reading, MA: Addison-Wesley, 1988.
25. Sieburg, H. B. "The Cellular Device Machine: Point of Departure for Large-Scale Simulations of Complex Biological Systems." *Computers Math. Applic.* **121** (1990): 247–267.
26. Sieburg, H. B., J. A. McCutchan, O. K. Clay, L. Caballero, and J.J. Ostlund. "Simulation of HIV Infection in Artificial Immune Systems." *Physica D* **45** (1990): 208–228.

André Longtin
Complex Systems Group and Center for Nonlinear Studies, Theoretical Division B213, Los Alamos National Laboratory, Los Alamos, NM 87545

Nonlinear Dynamics of Neural Delayed Feedback

Neural delayed feedback is a property shared by many circuits in the central and peripheral nervous systems. The evolution of the neural activity in these circuits depends on their present state as well as on their past states, due to finite propagation time of neural activity along the feedback loop. These systems are often seen to undergo a change from a quiescent state characterized by low-level fluctuations to a oscillatory state. We discuss the problem of analyzing this transition using techniques from nonlinear dynamics and stochastic processes. Our main goal is to characterize the nonlinearities which enable autonomous oscillations to occur and to uncover the properties of the noise sources these circuits interact with. The concepts are illustrated on the human pupil light reflex (PLR) which has been studied both theoretically and experimentally using this approach.

---

1. INTRODUCTION
NEURAL DELAYED FEEDBACK

By neural delayed feedback system (NDFS), we mean an assembly of one or several populations of neurons forming a feedback loop, i.e., a path from the input of a neural system to its output and back to its input. In each of these populations, the evolution of neural activity at time t is a function of the input to *and* output of this population from the past all the way up to time t. This delayed action arises because of the finite velocity at which signals (graded potentials or action potentials) are transmitted between these neural populations. That mathematical models of the phenomena we will be discussing should take these delays into account is a consequence of the fact that the delays are of the same order of magnitude or larger than the response time of these systems (see Section 3).

In general, there is a certain arbitrariness in designating a given neural activity as the input or output of a system. These are often chosen according to the experimentally measurable quantities. For example, the pupil light reflex controls the retinal light flux, equal to the product of light intensity and pupil area (the pupil is the hole in the middle of the colored part of the eyeball formed by the iris muscles). An input to this system is a variation in retinal light flux, due, for example, to a variation in the ambient light intensity; the output response is also a variation in retinal flux due to a change in pupil area. This example illustrates that the "neural" feedback loop need not comprise only neurons, but can include muscle cells and receptor cells.

TWO INTERESTING QUESTIONS ABOUT NEURAL SYSTEMS

One of the most interesting questions in the study of human neural systems in particular, as well as of physiological systems in general, is the origin of the ongoing fluctuations in the time course of measurable quantities. There has been much speculation, in the wake of recent studies (see, e.g., Glass and Mackey[6]) that this variability may be due in great part to deterministic chaos. The fact that certain low-dimensional deterministic systems behave in many respects like probabilistic systems (e.g., they generate invariant densities) has provided investigators with a new paradigm with which to reinterpret mechanisms theretofore attributed to noise. Most studies[2,6] involve either (1) modeling or (2) computation from time series of dynamical invariants such as the fractal dimension or Lyapunov spectrum. Determining whether aperiodic time series are the manifestation of a stochastic process (i.e., noise) or a deterministic process (chaos) is important theoretically, and each point of view suggests different kinds of experiments to further knowledge about the system under study.

Another equally important question is how oscillatory behavior either arises or is suppressed in neural systems. Oscillations in neural activity are ubiquitous within the central nervous system and between the peripheral and central nervous systems (e.g., in reflexes). Oscillatory neural activity can be associated either with health or pathology. An epileptic seizure is a dramatic transition to a pathological oscillatory state.[19] The respiratory rhythm,[6] as well as oscillations in the olfactory[4] and visual[7] cortex, seem to occur under normal healthy conditions. Much work has been invested in developing a theory for oscillation onset in neural systems.[1,9,11,13] Two aspects of these phenomena have received little attention: (1) the very noisy background on which the oscillatory activity is superposed and (2) propagation delays. While it is not necessarily useful to view all neural systems generating oscillations as delayed feedback systems, it is important to realize when delayed feedback is the key component leading to the oscillatory instability.

NONLINEAR DYNAMICS AND NEURAL MODELING

Nonlinear dynamics deals with the study of periodic and aperiodic oscillations. It has become clear in the last few decades that nonlinear dynamics is a useful tool to understand how neurophysiological systems undergo qualitative behavior changes as parameters are varied. Examples include single neuron dynamics[1,10,23] and physiolocal control systems.[6] These studies involve building sound physiological models and identifying the proper bifurcation parameter(s) leading to the oscillatory state. Essential to these studies has been the availability of data for guiding mathematical modeling and for validating model predictions. We will see in Section 4 that nonlinear dynamics has to be augmented to include the effects of noise in order to understand oscillation onset in NDFS's.

THE PARADIGM OF NEURAL DELAYED FEEDBACK SYSTEMS

In the case of human neural control systems, it is often very difficult to carry out experiments, and thus models are based on known anatomy, animal studies, and/or hypothetical mechanisms. There is one system which is amenable to experimental investigation and whose study holds hope for uncovering more functional principles of neural control: the human pupil light reflex (PLR). This was one of the systems to which concepts from cybernetics were first applied in the '50s, especially in the work of Stark and his colleagues. The essential feature of this reflex is that its feedback loop can be opened by an optical trick (see Section 2). This makes it possible to induce and control oscillations in this reflex by artificially increasing its feedback gain. Among its many interesting dynamical behaviors, the PLR exhibits "hippus" which is an ongoing pupil area fluctuation that occurs even under constant lighting conditions and seems to be due to neural noise injected into the reflex arc.[15,17,24,28] Of course, one of the main advantages of working with the human PLR is that the results are directly relevant for humans.

　　This paper deals with the properties of oscillatory NDFS's, and focuses in particular on the PLR. Section 2 describes experiments designed to induce oscillatory neural activity in the human PLR. Section 3 outlines a model used to study autonomous oscillations in the human PLR. We further describe the Hopf bifurcation in ordinary and delay-differential equations. Section 4 deals with the stochastic Hopf bifurcation in DDE's and the concept of physiological order parameter. Physiological irregularity is discussed in Section 5, and the paper ends with a conclusion and outlook on future investigations in Section 6.

2. THE HUMAN PUPIL LIGHT REFLEX

The human pupil light reflex is the involuntary response of the iris musculature to variations in the light intensity impinging on the eye. Light enters the eye through the pupil and falls on the retina. The reflex tries to control the retinal light flux, equal to the product of light intensity and pupil area. It is a negative feedback control system: an increase in retinal light flux (due, for example, to an increase in the ambient light intensity) is counteracted by a neural response which decreases the light flux (e.g., a decrease in pupil area).

This control loop can be opened using an optical trick first discussed by James Clerk Maxwell and popularized by Stark and Sherman.[27] and "Maxwellian view" involves shining a narrow beam of light down the center of the pupil, as in Figure 1(a). The beam is so narrow that iris movements cannot influence the retinal illumination, and thus the feedback loop is opened. It is possible to electronically close this loop with an analog signal proportional to pupil area provided, for example, by a pupillometer.[17,22,26,27] This signal can then be used to control the intensity of a light source in Maxwellian view. This procedure is termed "clamping" and allows the experimenter to substitute a different feedback in the place of the nonlinear feedback naturally present in the system. This involves designing an "area comparator" which electronically converts area values to light intensity values. The changes in retinal light flux are then due to changes in intensity, since the area of the light beam entering the pupil is fixed.

The data shown in Figure 2 was obtained by our collaborators at the Free University of Amsterdam using a reflectance-type pupillometer.[17,22] The area comparator is simply a linear amplifier relating pupil area to light intensity, as shown in Figure 1(b). The (positive) gain of this amplifier is adjustable, and sets the slope in Figure 1(b). From the point of view of pupil area A, the clamped reflex as a whole operates with the feedback of Figure 1(c), where the forcing on the pupil area is plotted versus the area itself. The slope here is negative because an increase in area causes an increase in light intensity which forces the area to decrease. Also, the saturation of the feedback at high and low area values is a consequence of the nonlinearities in the reflex. Hence, from the point of view of pupil area, the area comparator of Figure 1(b) actually produces a negative feedback configuration.

Figure 2 shows experimental recordings of pupil area versus time at four different gain settings. All the data is obtained from the same healthy subject within

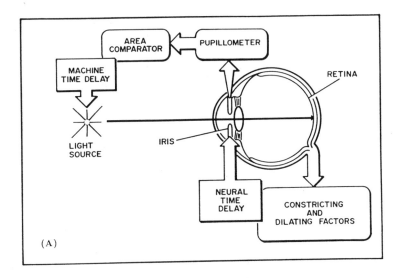

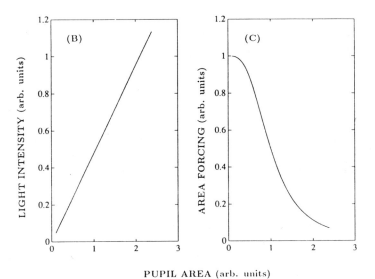

PUPIL AREA (arb. units)

FIGURE 1   (a) Schematic of clamped pupil light reflex. The narrow beam of light illuminates the retina in open-loop, i.e. the iris muscles cannot block the beam even under maximal constriction. The area comparator is the electronically synthesized external feedback which relates light intensity to measured pupil area. (b) Area comparator used here to produce oscillations. The gain is equal to the slope of this curve. (c) Forcing on the pupil area as a function of pupil area. Because an increase in light intensity causes a decrease in pupil area, this area comparator actually produces a negative feedback configuration as in the unclamped pupil light reflex. The nonlinearities of the reflex are responsible for saturation at high and low area values.

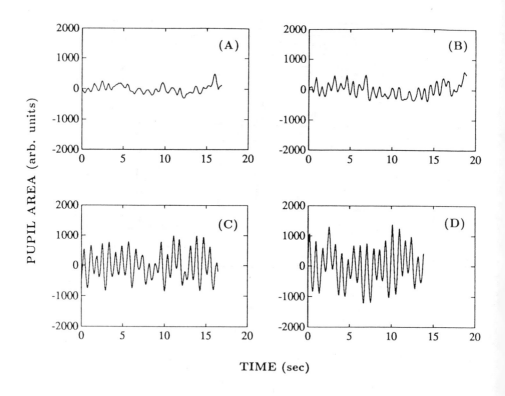

FIGURE 2 Pupil area as a function of time at four different gain settings: (a) 1.41; (b) 2.0; (c) 2.82; and (d) 4.0. Area is measured in arbitrary units (the same throughout the whole experiment) relative to the mean for a given record. The linear trend across one record has also been removed. The sampling rate is 50 Hz.

a short time span in order to maximize stationarity. The baselines have been corrected for drift. While the mean amplitude of the oscillation increases with the gain, the period varies only mildly. Further, the signals are strongly aperiodic despite the presence of a dominant frequency. From classical control systems theory based on linear transfer functions, one expects the system to begin oscillating as the gain is increased. While oscillations are definitely more prominent as the gain is increased, it is not clear at which point they appear (data at lower gain values than those shown substantiate this point). Pinpointing oscillation onset in this NDFS will be the focus of Section 4.

We must emphasize that the oscillations are not externally driven by some periodic modulation of the light intensity. Rather, the oscillations are autonomous in the sense that pupillary dynamics determine the retinal light flux. Once the system is set up in the high-gain configuration along with a mean light beam intensity, the oscillations begin spontaneously.

3. HOPF BIFURCATION IN NEURAL DELAYED FEEDBACK
THE HOPF BIFURCATION

This section focuses on the Hopf bifurcation, i.e., the transition from a stable equilibrium (fixed point) to a stable oscillation (limit cycle), as a parameter is varied. This bifurcation is characterized by the crossing of a pair of roots of the characteristic equation, obtained from the linearization of the nonlinear equations of motion, into the right-hand side of the complex plane. It is relatively straightforward to compute the condition for the roots to migrate across the imaginary axis and to verify the further condition that this crossing be non-tangential. However, this calculation only indicates how the fixed point goes unstable, but does not guarantee the existence of a stable limit cycle. This depends on whether the equations of motion satisfy certain non-degeneracy conditions.[8]

Hopf bifurcation analysis, although quite involved, can be used to compute the period and amplitude of the limit-cycle oscillation. In practice, one often resorts to numerical simulation of the equations of motion to obtain this information, as well as to determine whether the bifurcation is super- or subcritical. The basic difference here is that in the former case, the amplitude of the oscillation is zero at the bifurcation point. In the latter case, the amplitude is finite and the limit cycle can only be reached through some perturbation from the fixed point (so-called "hard excitation").[8]

The Hopf bifurcation theorem is essentially the same for a system of ordinary differential equations (ODE's) or delay differential equations (DDE's). However, the analysis is much more complicated for DDE's because they are infinite-dimensional systems evolving in a functional (Banach) space rather than the usual phase space spanned by the finite number of degrees of freedom of ODE's. This is a consequence of the fact that a family of solutions to a DDE is parametrized not by a finite vector of initial conditions but by a function on the interval $(-\tau, 0)$, where τ is the delay. Further, the characteristic equation for linear DDE's is a transcendental function. This implies that it cannot be solved analytically for its complex roots, and that it has an infinite number of roots.

MODEL FOR HIGH-GAIN OSCILLATIONS IN THE PLR

We now outline a mathematical model for nonlinear autonomous oscillations in the PLR which occur as the feedback gain is increased. Details can be found in Longtin and Milton.[16] Although the PLR is a spatially distributed system with thousands of parallel pathways, not much is known about the parameters characterizing the spatial features beyond the retina. What is known is where the neurons of one nucleus project to, and we have found that this is sufficient to explain the phenomena we are interested in. Thus our model is spatially homogeneous, and deals with one scalar variable, the pupil area $A(t)$. This area value is inversely proportional to the area of the iris muscles. The main iris muscle is a sphincter surrounding the pupillary margin. It reduces its diameter upon constriction, thereby stretching the iris tissues. Hence, iris area is proportional to the iris sphincter activity.

The retinal light flux is transduced into neural activity in the optic nerve (for simplicity, one can assume that stimulus strength is encoded in firing frequency of action potentials). This neural activity is processed by different brainstem nuclei and finally reaches the sphincter muscle. The result is that an increase in retinal light flux increases sphincter activity which decreases pupil area. The following model has been shown to predict the basic features of pupillary dynamics including oscillatory behavior[16]:

$$\frac{dg(A)}{dA}\frac{dA(t)}{dt} + \alpha g(A) = \gamma \ln\left[\frac{\phi(t-\tau)}{\bar{\phi}}\right] \tag{1}$$

where τ is the time delay between the falling of light on the retina and a change in pupil area. α is a rate constant for pupillary motion, $\phi(t)$ is the retinal light flux, and $\bar{\phi}$ is the threshold flux below which no pupil response occurs. The function $g(A)$ (shaped like the inverse of a sigmoid) relates area to iris sphincter activity. The logarithm accounts for the compression of the light flux scale in the transduction process at the retina. A rate constant of 3 sec^{-1} and a delay of 300 msec are typical.

To induce pupillary oscillations as in Figure 2, the retarded flux is given by

$$\phi(t-\tau) = I(t-\tau)A_{\text{beam}} \tag{2}$$
$$= \left[I_m + I_{\text{a.c.}}(A(t-\tau))\right]A_{\text{beam}} \tag{3}$$

where A_{beam} is the area of the light beam used in Maxwellian view, $I(t)$ is the light intensity, I_m is the mean light intensity, and $I_{\text{a.c.}}$ is the functional dependence of light intensity on pupil area synthesized by the area comparator (see Figure 1(b)). Under these conditions, the delay also comprises the external electronic delay. Equation (1) has a fixed point solution A^* given by the root of the transcendental equation obtained by setting the time derivative equal to zero. Equation (1) undergoes a supercritical Hopf bifurcation when the delay is increased past a critical value. This is also the case when the slope of the nonlinearity $g(A)$ evaluated at the fixed point decreases below a critical value. This corresponds to increasing the feedback gain. The amplitude of the limit cycle is constrained by the neuromuscular nonlinearity and the logarithmic compression at the retina. Near the bifurcation, the amplitude is predicted to grow as the square root of the deviation of the parameter from its value at the bifurcation point, while the period varies slowly across the bifurcation.

4. DEFINING OSCILLATION ONSET: PHYSIOLOGICAL ORDER PARAMETERS

The limit cycles in Figure 2 are irregular. Quantitative analysis of the data shows[17] that the amplitude does not grow as the square root of the distance from the bifurcation point, as the deterministic theory reviewed in Section 3 predicts. However, both the observed growth and the aperiodicity can be explained by hypothesizing that stochastic forces are also driving pupillary dynamics. Noise such as "hippus" can arise from the spontaneous activity of the neurons in the reflex arc, or from the activity of other neural pathways which impinge on those of the PLR but are not under experimental control. The power spectrum of hippus is approximately broad band up to a cutoff frequency of $\approx 1 Hz$, which indicates that it is a colored noise. Its correlation time can be roughly estimated as the inverse of the spectrum bandwidth, i.e., ≈ 1 sec.

In a noisy system, one is confronted with the very basic problem of defining what is meant by oscillation onset. Defining a "statistical bifurcation point" can be done with some help from nonequilibrium statistical mechanics. The idea is to find a quantity that exhibits a qualitative change as the bifurcation parameter is swept, just as in the deterministic Hopf bifurcation case. The obvious measure of oscillation, i.e., the height of the main peak in the power spectrum, does not undergo such a qualitative change.[5] Any amount of noise will reveal this peak, even when the parameter is well below the bifurcation point (the peak is a "noisy precursor" of the Hopf bifurcation; see Fronzoni et al.[5]). The peak simply grows as the parameter is swept across the deterministic bifurcation point. The autocorrelation function does not exhibit any qualitative change either, as it is obtained from the Fourier transform of the power spectrum.

However, the probability $\rho(x)$ that a signal $x(t)$ takes on a value between x and $x + dx$ does undergo a qualitative change from a unimodal to a bimodal shape. The distance between the peaks, which serves as an "order parameter," is a statistical measure of amplitude different from "the mean oscillation amplitude"; in fact, the order parameter can be zero even though the mean amplitude in the time series is not.

The probability densities for the area time series of Figure 2 are shown in Figure 3. Although one sees a widening of the density as the gain is increased, the order parameter is always zero. At the highest gain, however, one could argue that the order parameter is not zero, but it simply cannot be resolved from the short time series available. This is supported by numerical experiments. The behavior of this order parameter has been computed in Longtin et al.[17] by numerically integrating a simplified version of Eq. (1) having the same qualitative dynamics. The noise is modeled by an Ornstein-Uhlenbeck process which is an exponentially correlated colored noise (correlation time 1 sec). At each parameter in a set spanning across the deterministic bifurcation, the density $\rho(A)$ is computed from the simulated time series $A(t)$ and the order parameter is computed. We find that it is very important

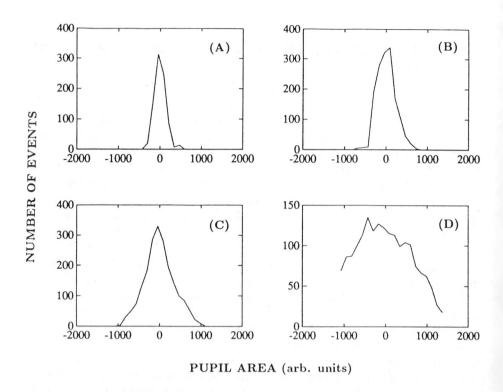

NUMBER OF EVENTS

PUPIL AREA (arb. units)

FIGURE 3   Area densities corresponding to the time series in Figure 2. The densities are 20 bin histograms of the area values spanning the range of values obtained with a 4.0 gain. The data in Figure 2 along with more data corresponding to total recording times of (a) 16.78, (b) 29.90, (c) 39.60, and (d) 35.96 sec has been used to construct the densities.

to allow for transients to die out, especially near the bifurcation point where the exchange of stability occurs. At this point "critical slowing down" occurs, which means that perturbations away from the attractor decay more and more slowly as the bifurcation point is approached. In practice, the solution has to be integrated over thousands of periods in order to resolve the growth of the order parameter. Hence, the poor resolution in Figure 3 is not surprising, given the length of the available time series.

Another interesting result is that the peaks of $\rho(A)$ do not correspond to those of the deterministic case (i.e., the extrema of the oscillation); in fact, the order parameter is smaller than the deterministic amplitude. The parameter value at which the order parameter becomes non-zero, i.e., the point which we define as "oscillation onset," is larger than the deterministic one. Noise postpones oscillation onset from a statistical point of view: on the limit-cycle side of the bifurcation, the system spends more time near the unstable fixed point than in the absence of noise. This effect has been reported in ODE's and is called a "noise-induced transition."[5] We have found this to occur with both additive and multiplicative noise.

Physiological order parameters have also been studied in the context of movement coordination in Haken et al.[9] Order parameters are very difficult to resolve on short experimental time series such as those shown in Figure 2. However, we believe that an understanding of their behavior is an essential backbone of any analysis. Furthermore, measurements of order parameters may still be used to validate a model despite their large standard deviation. But other measures may be more useful for short neurological time series, such as the behavior of period, amplitude and phase of oscillations. In fact, it was shown in Longtin et al.[17] that the noise intensity at which the predicted and measured curves of "mean amplitude versus gain" have similar shapes also yields similar shapes for the predicted and observed curves of "mean period" as well as "relative amplitude and period fluctuations versus gain." These last three curves further agree quantitatively with the data. While these quantities do not undergo qualitative changes, they are important characteristics of the oscillations and the proper prediction of their behavior serves as a strong test for the model.

5. ORIGIN OF PHYSIOLOGICAL IRREGULARITY

Although the mean behavior of NDFS's may be described by deterministic laws, upon closer scrutiny they are seen to generate aperiodic activities. For example, in the case of the PLR, hippus is always more or less manifest. This general observation raises the question of the origin and purpose of this irregularity. This has been a subject of great debate over the past few years (for a comprehensive review, see Pool[21]). The debate has centered on whether chaos, noise, or both underlie the irregularity. Thus the simpler view of simple dynamical motion with noise has received little attention. The reasons for this are varied, the major one being that fractal dimensions and other quantities can be cranked out of time series with relative ease, along with the immediate conclusion that the dynamics are principally chaotic. There are many problems with this view (for a review see Milton et al.[19]). It is safe to state in these cases that the steady-state dynamics involve only a few degrees of freedom. But the presence of chaos will remain uncertain until one can rule out the other kinds of (maybe noisy) dynamics which can yield, for example, similar dimension values. It is known,[14,20] for example, that the fractal dimension of attractors reconstructed from time series of colored-type noises such as $1/f$ (one-over-f) are also finite (although it diverges to infinity in the white-noise limit); thus results have to be interpreted with extreme care. Fractal-dimension algorithms are now being extensively used because (1) often there is no adequate theory for the system under study, or (2) system parameters cannot be varied to produce bifurcations leading to the chaotic state, or simply (3) they are readily available.

More stringent and self-consistent tests for chaos, based on prediction of time series, are being developed and hold hope for clarifying the source(s) of irregularity (for a review see Eubank and Farmer[3]).

Without further belaboring the merits and pitfalls of these methods, we wish to emphasize that simple dynamics with noise should also be considered as a candidate for irregularity. This is especially true in the vicinity of bifurcation points where the dynamics are dominated by the noise, due to critical slowing down. It is possible that even in simple NDFS's there are a few feedback loops that interact together and with different noise sources (there are at least three pathways in the human PLR; see Longtin[15]), and hence that critical behavior could occur around more bifurcation points.

Previous studies[24,25,28] have focused on the dynamics of hippus in the steady state and during transient responses to light; they have concluded that hippus is probably noise. Longtin et al.[17] have investigated hippus at oscillation onset by studying the critical behavior of a model DDE at a Hopf bifurcation; they have reached the same conclusion. This raises two questions. Is the noise really injected into the reflex arc with a one-second correlation time, or is it quasi-white but appears strongly colored because of the low-pass characteristics of this neuromuscular system, as suggested by Stanten and Stark?[24] This remains to be investigated. Also, what is the ultimate origin of hippus? Is it the trace of a chaotic process? Important clues may lie in the fact that the interaction of hippus with the dynamics of the reflex arc seems to be unidirectional, i.e., the dynamics of the arc are influenced by that of the noise, but not vice versa.

Assessing stochastic components of any dynamics requires proper characterization of the noise, namely its density, spectrum, and source. It is conceivable that characteristics of the noise itself may serve as bifurcation parameters in real NDFS's.

6. CONCLUSION AND OUTLOOK

NDFS's are inherently nonlinear, and under experimental control display a variety of dynamical behaviors. Stochastic components of NDFS's must be disentangled from deterministic ones in order to understand their behavior in the vicinity of bifurcation points. The special problem of oscillation onset leads one to define a physiological order parameter as the distance between the two peaks of the invariant density for the dynamics of the measured variable. While these peaks are extremely difficult to resolve on short time series, as is the case for the PLR experiments, it is nevertheless important to know this distance is less than the limit-cycle amplitude in the absence of noise. This means noise postpones the oscillation onset by stabilizing the fixed point. That noise is responsible for the observed aperiodicity can be proved by explaining the behavior of more easily measured quantities such as the mean and standard deviation of the period and amplitude of the area fluctuations. In this way, hypotheses about stochastic inputs to a system can be validated.

Numerous avenues remain to be explored. One is to understand the influence of distributed delays (which result from a distribution of nerve axon diameters) and of the spatial extent of NDFS's on oscillation onset. It may be that critical behavior and growth of order parameters different from what is expected from a spatially homogeneous system can be a signature of interactions between parallel pathways comprising the neural system. Tests that differentiate between additive and multiplicative noise would also be quite useful, as our analysis of pupil dynamics cannot resolve their effects. One needs a theory with which the mean and variance of the amplitude and period fluctuations can be analytically rather than numerically computed. Further, hardly anything is known about the behavior of stochastic DDE's (see, however, Kolmanovskii and Nosov,[12] Chapter 4). A Fokker-Planck-type theory for "generalized delayed Langevin equations" would be more than welcome as it would allow densities and order parameters to be calculable at least in principle. It is expected that this knowledge will become important as more neural systems involving delays are investigated.

ACKNOWLEDGMENTS

The author would like to thank Michael Mackey and John Milton with whom many of these ideas have been discussed, as well as Jelte Bos of the Medical Physics Department of the Free University of Amsterdam who has graciously provided the data. Financial support from NSERC (Canada), from the Complex Systems Group (T-13), and the Center for Nonlinear Studies (CNLS) at Los Alamos National Laboratory in the form of postdoctoral fellowships is greatly acknowledged.

REFERENCES

1. an der Heiden, U. *Analysis of Neural Networks.* Lecture Notes in Biomathematics, vol. 35. Berlin: Springer-Verlag, 1980.
2. Degn, H., A. V. Holden, and L. F. Olsen. *Chaos in Biological Systems.* New York: Plenum, 1987.
3. Eubank, S., and J. D. Farmer. "An Introduction to Chaos and Randomness." In *1989 Lectures in Complex Systems*, edited by E. Jen. SFI Studies in the Sciences of Complexity, Lect. Vol. II, 75–190. Redwood City, CA: Addison-Wesley, 1990.
4. Freeman, W. J. "Spatial Properties of an EEG Event in the Olfactory Bulb and Cortex." *Electroencephalogr. Clin. Neurophysiol.* **44** (1978): 586–605.

5. Fronzoni, L., R. Mannella,P. V. E. McClintock, and F. Moss. "Postponement of Hopf Bifurcations by Multiplicative Colored Noise." *Phys. Rev. A* **36** (1987): 834–841.

6. Glass, L., and M. C. Mackey. *From Clocks to Chaos: The Rhythms of Life.* Princeton: Princeton University Press, 1988.

7. Gray, C. M., and W. Singer. "Stimulus-Specific Neuronal Oscillations in Orientation Columns of Cat Visual Cortex." *Proc. Natl. Acad. Sci.* **86** (1989): 1698–1702.

8. Guckenheimer, J., and P. Holmes. *Nonlinear Oscillations, Dynamical Systems, and Bifurcations of Vector Fields.* New York: Springer, 1983.

9. Haken, H., J. A. S. Kelso, and H. Bunz. "A Theoretical Model Of Phase Transitions in Human Bimanual Coordination." *Biol. Cybern.* **51** (1985): 347–356.

10. Hayashi, H., and S. Ishizuka. "Chaos in Molluscan Neuron." In *Chaos in Biological Systems*, edited by H. Degn, A. V. Holden, and L. F. Olsen, 157-166. New York: Plenum, 1987.

11. Herz, A., B. Sulzer, R. Kühn, and J. L. van Hemmen. "Hebbian Learning Reconsidered: Representation of Static and Dynamic Objects in Associative Neural Nets." *Biol. Cybern.* **60** (1989): 457–467.

12. Kolmanovskii, V. B., and V. R. Nosov. "Stability of Functional Differential Equations." *Mathematics in Science and Engineering*, vol. 180. Orlando: Academic Press, 1986.

13. Li, Z., and J. J. Hopfield. "Modeling the Olfactory Bulb and its Neural Oscillatory Processings." *Biol. Cybern.* **61** (1989): 379–392.

14. Llosa, J., and J. Masoliver. "Fractal Dimension for Gaussian Colored Processes." *Phys. Rev. A Brief Report* (October 1990): in press.

15. Longtin, A. "Nonlinear Oscillations, Noise and Chaos in Neural Delayed Feedback." Ph.D. Thesis, McGill University, 1989.

16. Longtin, A., and J. G. Milton. "Modelling Autonomous Oscillations in the Human Pupil Light Reflex Using Nonlinear Delay-Differential Equations." *Bull. Math. Biol.* **51** (1989): 605–624.

17. Longtin, A., J. G. Milton, J. Bos, and M. C. Mackey. "Noise and Critical Behavior of the Pupil Light Reflex at Oscillation Onset." *Phys. Rev. A* **41** (1990): 6992–7005.

18. Milton, J. G., and A. Longtin. "Evaluation of Pupil Constriction and Dilation from Cycling Measurements." *Vision Res.* **30** (1990): 515–525.

19. Milton, J. G., A. Longtin, A. Beuter, M. C. Mackey, and L. Glass. "Complex Dynamics and Bifurcations in Neurology." *J. Theor. Biol.* **138** (1989): 129–147.

20. Osborne, A. R., and A. Provenzale. "Finite Correlation Dimension For Stochastic Systems with Power-Law Spectra." *Physica D* **35** (1989): 357–381.

21. Pool, R. "Is It Healthy To Be Chaotic?" *Science* **243** (1989): 604–607.

22. Reulen, J. P. H., J. T. Marcus, M. J. van Gilst, D. Koops, J. E. Bos, G. Tiesinga, F. R. de Vries, and K. Boshuizen. "Stimulation and Recording

of Dynamic Pupillary Reflex: The IRIS Technique. Part 2." *Med. Biol. Eng. Comp.* **26** (1988): 27–32.

23. Rinzel, J., and Y. S. Lee. "Dissection of a Model for Neuronal Parabolic Bursting." *J. Math. Biol.* **25** (1987): 653–675.

24. Stanten, S. F., and L. Stark. "A Statistical Analysis of Pupil Noise." *IEEE Trans. Bio-Med. Eng.* **13** (1966): 140–152.

25. Stark, L. "Stability, Oscillations and Noise in the Human Pupil Servomechanism." *Proc. IRE* **47** (1959): 1925–1939.

26. Stark, L. "Environmental Clamping of Biological Systems: Pupil Servomechanism." *J. Am. Opt. Soc.* **52** (1962): 925–930.

27. Stark, L., and P. M. Sherman. "A Servoanalytic Study of Consensual Pupil Reflex to Light." *J. Neurophysiol.* **20** (1957): 17–26.

28. Usui, S., and L. Stark. "A Model for Nonlinear Stochastic Behavior of the Pupil." *Biol. Cybern.* **45** (1982): 13–22.

Walter Fontana
Theoretical Division and Center for Nonlinear Studies, Los Alamos National Laboratory,
Los Alamos, NM 87545 USA and Santa Fe Institute, 1120 Canyon Road, Santa Fe, NM
87501 USA

Functional Self-Organization in Complex Systems

A novel approach to functional self-organization is presented. It consists of a
universe generated by a formal language that defines objects (=programs),
their meaning (=functions), and their interactions (=composition). Results
obtained so far are briefly discussed.

1. INTRODUCTION

Nonlinear dynamical systems give rise to many phenomena characterized by a
highly complex organization of phase space, e.g., turbulence, chaos, and pattern
formation. The structure of the interactions among the objects described by the
variables in these systems is usually fixed at the outset. Changes in the phase portrait occur as coefficients vary, but neither the basic qualitative relationships among
the variables, nor their number is subject to change.

In this contribution, I will be concerned with the class of systems that is,
in some sense, complementary. This class contains systems that are "inherently
constructive." By "constructive" I mean that the elementary interaction among
objects includes the possibility of building new objects. By "inherently constructive"
I want to emphasize that the generation of new objects from available ones is an

intrinsic, specific, non-random property of the objects under consideration. It is *not* primarily caused by noise.

A prime example of such a system is chemistry. Molecules undergo reactions that lead to the production of specific new molecules. The formation of the product object is *instructed* by the interacting reactants. This is to be distinguished from a situation in which new objects are generated by chance events, as is the case with copying errors during the replication process of a DNA string (mutations).

Examples for complex systems belonging to the constructive category are chemistry, biological systems like organisms or ecosystems, and economies. Clearly, nonlinearities and noise occur everywhere. In this note, I will be primarily concerned with the implications following from noiseless constructive properties. An example of a dynamical system built on top of these properties will be given, but the formulation of a general theory (if it exists) that combines nonlinear dynamical *and* constructive aspects of complex systems is a major problem for the future.

The "manipulation" of objects through other objects, as it occurs with molecules, might in principle be reduced to the behavior of the fundamental physical forces relevant for the particular objects. Quantum mechanics is an example with respect to chemistry. At the same time, however, the very phenomenon of manipulation introduces a new level of description: it generates the notion of "functionality." Objects can be put into *functional* relations with each other. Such relations express which object produces which object under which conditions.

The main assumption of the present work is: *The action of an object upon other objects can be viewed as the application of a computable function to arguments from the functions' domain of definition (which can be other functions). Functional relations can then be considered completely independently from their particular physical realization.*

There is no doubt that this is a very strong assumption, but such an abstraction is useful if we want to focus on a classification of functional relations in analogy to a classification of attractors in dynamical systems theory. It is also useful in defining toy models that capture the feedback loop between objects and the functions acting on them defined by these very objects. It is such a loop that identifies complex constructive systems.

"Function" is a concept that—in some mathematical sense—is irreducible. Mathematics provides since 1936, through the works of Church,[4,5] Kleene,[16] Gödel,[12] and Turing,[23] a formalization of the intuitive notion of "effective procedure" in terms of a complete theory of particular functions on the natural numbers: the partial recursive functions.

The following is a very brief attempt to explore the possibility of establishing a useful descriptive level of at least some aspects of constructive complex systems by viewing their objects as being machines = computers = algorithms = functions. For a more detailed exposition see Fontana.[10,11]

2. WHAT IS A FUNCTION?

A function, f, can be viewed (roughly) in two ways:

- A function as an *applicative rule* refers to the *process*—coded by a definition—of going from argument to value.
- A function as a *graph*, refers to a set of ordered pairs such that if $(x, y) \in f$ and if $(x, z) \in f$, then $y = z$. Such a function is essentially a look-up table.

The first view stresses the computational aspect, and is at the basis of Church's λ-calculus (see for example Barendregt[2]). The theory λ is a formalization of the notion of computability in precisely the same sense as the Turing machine and the theory of general recursiveness. However, λ is a very different and much more abstract approach.

Stated informally, λ defines inductively expressions consisting of variables. A variable is an expression, and every *combination* of expressions, wrapped in parentheses, is again an expression. Furthermore, a variable can be *substituted* by *any* other expression that follows the expression to which the variable belongs. An elegant notational (syntactic) structure provides a means for defining the scope of the variables in an expression.

In an informal notation this means that, if $f[x]$ and $g[x]$ are expressions, then $(f[x]g[x])$ is also an expression, and $(f[x]g[x])$ is equivalent to $(f[x := g[x]])$, where the latter denotes the expression that arises if every occurrence of x in $f[x]$ is replaced by the expression $g[x]$.

Intuitively, what is captured here is just the notion of "evaluating" a function by "applying" it to the argument. That is, consider $f[x]$ as denoting a function, then the *value* of that function when applied to the argument expression a is obtained by literally substituting a for x in $f[x]$ *and* performing all further substitutions that might become possible as a consequence of that action. If all substitutions have been executed, then an expression denoting the value of a under f has been obtained. In this way functions that can be identified with the natural numbers (numerals) or all computable operations on them, for example, addition and multiplication, can be defined.

Three features of functions in λ are important for the following:

- Functions are defined recursively in terms of other functions: imagine functions as being represented by trees with variables at the terminal nodes. This makes explicit that functions are "modular" objects, whose building blocks are again functions. This combinatorial representation, in which functions can be freely recombined to yield new functions, is crucial.
- Objects in λ can serve both as arguments or as functions to be applied to these arguments.
- There is no reference to any "machine" architecture.

Although the whole story is much more subtle, the preceding paragraphs should convey the idea of "function" that will be used in the framework described in the following sections. The point is that under suitable mathematical conditions λ-objects and composition give rise to a reflexive algebraic structure.

Turing's completely different, but equivalent, approach to computability worked with the machine concept. It was his "hardware" approach that succeeded in convincing people that a formalization of "effective procedure" had actually been achieved. The existence of a universal Turing machine implies the logical interchangeability of hardware and software. Church's world of λ is entirely "software" oriented. Indeed, in spite of its paradigmatic simplicity, it contains already many features of high-level programming languages, such as LISP. For a detailed account see Trakhtenbrot.[22]

3. THE MODEL

To set up a model that also provides a workbench for experimentation a representation of functions along the lines of the λ-calculus is needed. I have implemented a representation that is a somewhat modified and extremely stripped-down version of a toy-model of pure LISP as defined by Gregory Chaitin.[3] In pure LISP a couple of functions are pre-defined (six in the present case). They represent primitive operations on trees (expressions), for example, joining trees or deleting subtrees. This speeds up and simplifies matters as compared to the λ-calculus in which one starts from "absolute zero" using only application and substitution. Moreover, I consider for the sake of simplicity only functions in one variable. The way functions act on each other thereby producing new ones is essentially identical to the formalism sketched for λ in the previous section.

I completely dispense with a detailed presentation of the language (which I refer to as "AlChemy": a contraction of Algorithmic Chemistry). Figure 1 and its caption give a simple example for an evaluation that should depict what it is all about. The interested reader is referred to Fontana.[11]

The model is then built as follows.

1. *Universe.* A universe is defined through the λ-like language. The language specifies rules for building syntactically legal ("well-formed") objects and rules for interpreting these structures as functions. In this sense the language represents the "physics." Let the set of all objects be denoted by $\mathcal{F}$.

2. *Interaction.* Interaction among two objects, $f(x)$ and $g(x)$, is naturally induced by the language through function composition, $f(g(x))$. The evaluation of $f(g(x))$ results in a (possibly) new object $h(x)$. Interaction is clearly asymmetric. This could easily be repaired by symmetrizing. However, many objects like biological species or cell types (neurons, for example) interact in an asymmetric fashion. I chose to keep asymmetry.

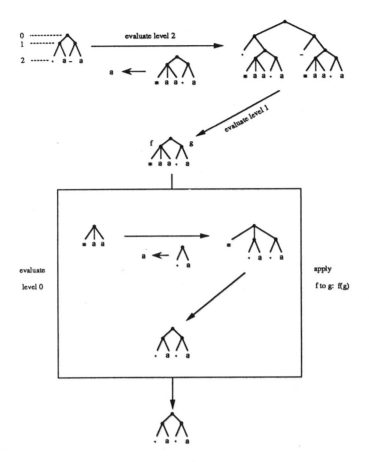

FIGURE 1   Evaluation example. The value of the expression ((+ a) (− a))
is computed when the variable a takes on the value ((∗ a a) (+ a)). The
interpretation process follows the tree structure until it reaches an atom (leaf). In this
case it happens at depth 2. The atoms are evaluated: the operators "+ " and "− " remain
unchanged, while the value of a is given by ((∗ a a) (+ a)). The interpreter backs
up to compute the values of the nodes at the next higher level using the values of
their children. The value of the left node at depth 1 is obtained by applying the unary
"+ "-operator to its sibling (which has been evaluated in the previous step). The "+ "
operation returns the first subtree of the argument, (∗ a a) in this case. Similarly,
the value at the right depth 1 node is obtained by applying the unary "− "-operator
to its argument ((∗ a a) (+ a)). The "− " operation deletes the first subtree of its
argument returning the remainder, (+ a). The interpreter has now to assign a value
to the top node. The left child's value is an expression representing again a function.
This function, (∗ a a), is labelled as f in the figure. Its argument is the right neighbor
sibling, (+ a), labelled as g. Evaluating the top node means applying f to g. This
is done by evaluating f while assigning to its variable a the value g. The procedure
then recurs along a similar path as above, shown in the box. The result of f(g) is
the expression ((+ a) (+ a)). This is the value of the root (level 0) of the original
expression tree, and therefore the value of the whole expression, given the initial
assignment. The example can be interpreted as "function ((+ a) (− a)) applied to
function ((∗ a a) (+ a))".

Note that "interaction" is just the name of a binary function $\phi(s,t)$ that sends any ordered pair of objects f and g into an object $h = \phi(f,g)$ representing the value of $f(g)$ (see Figure 2 for an example). More generally, $\phi(s,t) : \mathcal{F} \times \mathcal{F} \mapsto \mathcal{F}$ could be *any* computable function, not necessarily composition, although composition is the most natural choice. The point is that whatever the "interaction" function is chosen to be, it is itself evaluated according to the semantics of the language. Stated in terms of chemistry, it is the same chemistry that determines the properties of individual molecules and at the same time determines how two molecules interact.

3. *Collision rule.* While "interaction" is intrinsic to the universe as defined above, the collision rule is not. The collision rule specifies essentially three arbitrary aspects:

  a. What happens with f and g once they have interacted. These objects could be "used up," or they could be kept (information is not destroyed by its usage).

  b. What happens with the interaction product h. Some interactions produce objects that are bound to be inactive no matter with whom they collide. The so-called NIL-function is such an object: it consists of an empty expression. Several other constructs have the same effect, like function expressions that happen to lack any occurrence of the variable. In general such products are ignored, and the collision among f and g is then termed "elastic"; otherwise, it is termed "reactive."

  c. Computational limits. Function evaluation need not halt. The computation of a value could lead to infinite recursions. To avoid this, recursion limits, as well as memory and real-time limitations have to be imposed. A collision has to terminate within some pre-specified limits; otherwise, the "value" consists in whatever has been computed until the limits have been hit.

The collision rule is very useful for introducing boundary conditions. For example, every collision resulting in the copy of one of the collision partners might be ignored. The definition of the language is not changed at all, but identity functions would have now been prevented from appearing in the universe.

In the following I will imply that the interaction among two objects has been "filtered" by the collision rule. That is, the collision of f and g is represented by $\Phi(f,g)$ that returns $h = \phi(f,g)$ if the collision rule accepts h (see item (b) above); otherwise, the pair (f,g) is not in the domain of Φ.

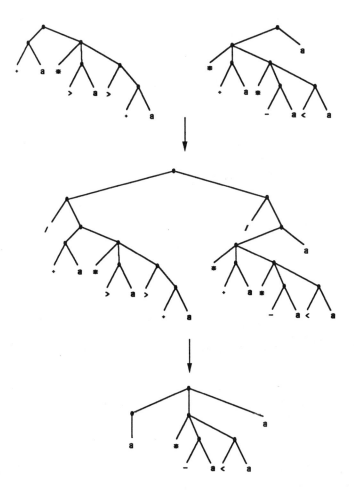

FIGURE 2 Interaction between functions. Two algorithmic strings (top) represented
as trees interact by forming a new algorithmic string (middle) that corresponds to
a function composition. The new root with its two branches and '-operators is the
algorithmic notation for composing the functions. The action of the unary '-operator
("quote"-operator) consists in preventing the evaluation of its argument. The interaction
expression is evaluated according to the semantics of the language and produces an
expression (bottom) that represents a new function.

4. *System.* To investigate what happens once an ensemble of interacting function
"particles" is generated, a "system" has to be defined. The remaining sections
will briefly consider two systems:

■ *An iterated map acting on sets of functions.* Let $\mathcal{P}$ be the power set, $2^{\mathcal{F}}$, of the set of all functions $\mathcal{F}$. Note that $\mathcal{F}$ is countably infinite, but $\mathcal{P}$ is uncountable. Let $\mathcal{A}_i$ denote subsets of $\mathcal{F}$, and let $\Phi[\mathcal{A}]$ denote the set of functions obtained by all $|\mathcal{A}|^2$ pair-interactions (i.e., pair-collisions) $\Phi(i, k)$ in $\mathcal{A}$, $\Phi[\mathcal{A}] = \{j : j = \Phi(i, k), (i, k) \in \mathcal{A} \times \mathcal{A}\}$. The map M is defined as

$$M : \mathcal{P} \mapsto \mathcal{P}, \ \mathcal{A}_{i+1} = \Phi[\mathcal{A}_i]. \tag{1}$$

Function composition induces a dynamics in the space of functions. This dynamics is captured by the above map M. An equivalent representation in terms of an interaction graph will be given in the next section.

■ *A Turing gas.* The Turing gas is a stochastic process that induces an additional dynamics on the nodes of an interaction graph. Stated informally, individual objects now acquire "concentrations" much like molecules in a test-tube mixture. However, the graph on which this process lives changes as reactive collisions occur. Section 5 will give a brief survey on experiments with the Turing gas.

4. AN ITERATED MAP AND INTERACTION GRAPHS

The interactions between functions in a set $\mathcal{A}$ can be represented as a directed graph G. A graph G is defined by a set $V(G)$ of vertices, a set $E(G)$ of edges, and a relation of incidence, which associates with each edge two vertices (i, j). A directed graph, or digraph, has a direction associated with each edge. A labelled graph has in addition a label k assigned to each edge (i, j). The labelled edge is denoted by (i, j, k).

The action of function $k \in \mathcal{A}$ on function $i \in \mathcal{A}$ resulting in function $j \in \mathcal{A}$ is represented by a directed labelled edge (i, j, k):

$$(i, j, k) : \ i \xrightarrow{k} j, \ \ i, j, k \in \mathcal{A}, . \tag{2}$$

Note that the labels k are in $\mathcal{A}$. The relationships among functions in a set are then described by a graph G with vertex set $V(G) = \mathcal{A}$ and edge set $E(G) = \{(i, j, k) : j = k(i)\}$.

A useful alternative representation of an interaction is in terms of a "double edge,"

$$(i, j, k) : \ i \xrightarrow{(i,k)} j \xleftarrow{(i,k)} k, \ \ i, j, k \in \mathcal{A}, \tag{3}$$

where the function k acting on i and producing j has now been connected to j by an additional directed edge. The edges are still labelled, but no longer with an element of the vertex set. The labels (i, k) are required to uniquely reconstruct the edge set from a drawing of the graph. The graph corresponding to a given edge set is obviously uniquely specified. Suppose, however, that a function j is produced by two different interactions. The corresponding vertex j in the graph then has four inward edges. Uniquely reconstructing the edge set, or modifying the graph, for example by deleting a vertex, requires information about which pair of edges results from the same interaction. Some properties of the interaction graph can be obtained while ignoring the information provided by the edge labels. The representation in terms of double-edges (i, j, k) has the advantage of being meaningful for any interaction function Φ mapping a pair of functions (i, k) to j, and not only for the particular Φ representing chaining. The double edge suggests that both i and k, are needed to produce j. In addition, the asymmetry of the interaction is relegated to the label: (i, k) implies an interaction $\Phi(i, k)$ as opposed to $\Phi(k, i)$. This representation is naturally extendable to n-ary interactions $\Phi(i_1, i_2, \ldots, i_n)$. In the binary case considered here, every node in G must therefore have zero or an even number of incoming edges.

The following gives a precise definition of an interaction graph G. As in Eq. (1) let $\Phi[\mathcal{A}]$ denote the set of functions obtained by all possible pair collisions $\Phi(i, k)$ in $\mathcal{A}$, $\Phi[\mathcal{A}] = \{j : j = \Phi(i, k), (i, k) \in \mathcal{A} \times \mathcal{A}\}$. The interaction graph G of set $\mathcal{A}$ is defined by the vertex set

$$V(G) = \mathcal{A} \cup \Phi[\mathcal{A}] \tag{4}$$

and the edge set

$$E(G) = \{(i, j, k) : i, k \in \mathcal{A}, j = \Phi(i, k)\}. \tag{5}$$

The graph G is a function of $\mathcal{A}$ and Φ, $G[\mathcal{A}, \Phi]$. The action of the map

$$M : \mathcal{A}_{i+1} = \Phi[\mathcal{A}_i] \tag{6}$$

on a vertex set $\mathcal{A}_i$ leads to a graph representation of M. Let

$$G^{(i)}[\mathcal{A}, \Phi] := G[\Phi^i[\mathcal{A}], \Phi] \tag{7}$$

denote the ith iteration of the graph G starting with vertex set $\mathcal{A}$; $G^{(0)} = G$.

A graph G and its vertex set $V(G)$ are closed with respect to interaction, when

$$\Phi[V(G)] \subseteq V(G); \tag{8}$$

otherwise, G and $V(G)$ are termed innovative.

Consider again the map M, Eq. (6). What are the fixed points of $\Phi[\cdot]$? $\mathcal{A} = \Phi[\mathcal{A}]$ is equivalent to (1) $\mathcal{A}$ is closed with respect to interaction and (2) the set $\mathcal{A}$ reproduces itself under interaction. That is,

$$\forall j \in \mathcal{A}, \ \exists \ i, k \in \mathcal{A} \text{ such that } j = \Phi(i, k). \tag{9}$$

Condition (9) states that all vertices of the interaction graph G have at least one inward edge (in fact, two or any even number). Such a self-maintaining set will also be termed "autocatalytic," following M. Eigen[7] and S. A. Kauffman[14,15] who recognized the relevance of such sets with respect to the self-organization of biological macromolecules.

Consider a set $\mathcal{F}_i$ for which Eq. (9) is still valid, but which is not closed with respect to interaction. $\mathcal{F}_{i+1}$ obviously contains $\mathcal{F}_i$, because of Eq. (9), and in addition it contains the set of new interaction products $\Phi[\mathcal{F}_i] \setminus \mathcal{F}_i$. These are obviously generated by interactions within $\mathcal{F}_i \subset \Phi[\mathcal{F}_i]$. Therefore, Eq. (9) also holds for the set $\Phi[\mathcal{F}_i]$, implying that the set $\mathcal{F}_{i+1}$ is autocatalytic. Therefore, if $\mathcal{A}$ is autocatalytic, it follows that

$$G[\mathcal{A}, \Phi] \supseteq G^{(1)}[\mathcal{A}, \Phi] \supseteq G^{(2)}[\mathcal{A}, \Phi] \supseteq \ldots \supseteq G^{(i)}[\mathcal{A}, \Phi] \supseteq \ldots. \tag{10}$$

In the case of strict inclusion, let such a set be termed "autocatalytically self-extending." Such a set is a special case of innovation, in which

$$\Phi[V(G)] \supseteq V(G) \tag{11}$$

holds, with equality applying only at closure of the set.

An interesting concept arises in the context of finite, closed graphs. Consider, for example, the autocatalytic graph G in Figure 3(b), and assume that G is closed. The autocatalytic subset of vertices $V_1 = \{A, B, D\}$ induces an interaction graph $G_1[V_1, \Phi]$. Clearly, $G[V, \Phi] = G_1^{(2)}[V_1, \Phi]$, which means that the autocatalytic set V_1 regenerates the set V in two iterations. This is not the case for the autocatalytic graph shown in Figure 3(a). More precisely, let G be a finite-interaction graph, and let $G_\alpha \subseteq G$ be termed a "seeding set" of G, if

$$\exists\, i, \text{ such that } G \subseteq G_\alpha^{(i)}, \tag{12}$$

where equality must hold if G is closed. Seeding sets turn out to be interesting for several reasons. For instance, in the next section a stochastic dynamics (Turing gas) will be induced over an interaction graph. If a system is described by a graph that contains a small seeding set, the system becomes less vulnerable to the accidental removal of functions. In particular cases a seeding set can even turn the set it seeds into a limit set of the process. Such a case arises when every individual function f_i in $\mathcal{A}$ is a seeding set of $\mathcal{A}$:

$$f_{i+1} = \Phi(f_i, f_i), \quad i = 1, 2, \ldots, n - 1$$
$$f_1 = \Phi(f_n, f_n). \tag{13}$$

Furthermore, suppose that G is finite, closed, and autocatalytic. It follows from the above that all seeding sets G_α must be autocatalytically self-extending, for example, as in Figure 3(b). If G is finite, closed, but not autocatalytic, there can be no seeding set. Being closed and not autocatalytic implies $V(G^{(2)}) \subset V(G)$. The vertices of G that have no inward edges are lost irreversibly at each iteration. Therefore, for some i either $G^{(i)} = \emptyset$, or $G^{(i)}$ becomes an autocatalytic subset of G.

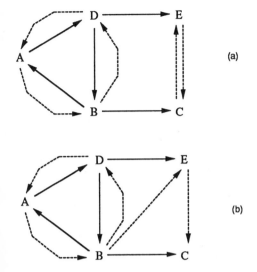

(a)

(b)

FIGURE 3   Interaction graph and seeding set. Two self-maintaining (autocatalytic) graphs. As outlined in section 4, a particular function, say D, is produced through interaction between A and B. Therefore, D has two incoming edges from A and B. The edge labels are omitted. The dotted line indicates that B is applied to the argument A (solid line). The graphs are self-maintaining because every vertex has incoming edges. The lower graph, (b), can be regenerated from the vertex subset $\{A, B, D\}$, in contrast to the upper graph, (a). Both contain parasitic subsets: $\{E, C\}$ in (a), and $\{C\}$, $\{E\}$ in (b).

In the case of innovative, not autocatalytic sets, i.e., sets for which

$$\Phi[\mathcal{A}] \not\subseteq \mathcal{A} \wedge \Phi[\mathcal{A}] \not\supseteq \mathcal{A} \tag{14}$$

holds, no precise statement can be made at present.

A digraph is called connected if, for every pair of vertices i and j, there exists at least one directed path from i to j and at least one from j to i. An interaction graph G that is connected not only implies an autocatalytic vertex set, but in addition depicts a situation in which there are no "parasitic" subsets. A parasitic subset is a collection of vertices that has only incoming edges, like the single vertices C and E in Figure 3(b), or the set $\{C, E\}$ in Figure 3(a). As the name suggests, a parasitic subset is not cooperative, in the sense that it does not contribute to generate any functions outside of itself.

All the properties discussed in this section are independent of the information provided by the edge labels (in the double-edge representation). Note, furthermore, that the above discussion is independent of any particular model of "function." It never refers to the implementation in the LISP-like AlChemy. The representation of function in terms of that particular language is used in the simulations reported briefly in the next section.

5. A TURING GAS

The interaction graph, and, equivalently, the iterated map, describe a dynamical system induced by the language on the power set of functions. This graph dynamics is now supplemented by a mass action kinetics leading to a density distribution on the set of functions in a graph. The kinetics is induced through a stochastic process termed "Turing gas."

A Turing gas consists of a fixed number of function particles that are randomly chosen for pairwise collisions. In the present scheme a reactive collision keeps the interaction partners in addition to the reaction product. When a collision is reactive, the total number of particles increases by one. To keep the number constant, one particle is chosen at random and erased from the system. This mimics a stochastic unspecific dilution flux. The whole system can be compared to a well-stirred chemical flow-reactor.

Three versions of the Turing gas have been studied. In one version the time evolution of the gas is observed after its initialization with N (typically $N = 1000$) randomly generated functions. In the second version the collision rule is changed to forbid reactions resulting in a copy of one of the collision partners. In the third version the gas is allowed to settle into a quasi-stationary state, where it is perturbed by injecting new random functions.

The following summarizes very briefly some of the results.

- *Plain Turing Gas.* Ensembles of initially random functions self-organize into ensembles of specific functions sustaining cooperative interaction pathways. The role of a function depends on what other functions are present in the system. A function, for example, that copies itself and some, but not all, others, acts as a pure self-replicator in the absence of those particular functions that it could copy. If some of them were present, the copy function would suddenly act "altruistically." The dynamics of the system is shaped by self-replicators (functions that copy themselves, but not the others present in the system), parasites (see section 4), general copy functions (identity functions), and partial copiers (functions that copy some, but not all functions they interact with). The "innovation rate," i.e., the frequency of collisions that result in functions not present in the system, decreases with time indicating a steady closure with respect to interactions (mainly due to the appearance of identity functions). If the stochastic process is left to itself after injecting the initial functions, fluctuations will eventually drive it into an absorbing barrier characterized by either a single replicator type, or by a possibly heterogeneous mixture of non-reactive functions ("dead system"), or by a self-maintaining set where each individual function species is a seeding set (section 4). The system typically exhibits extremely long transients characterized by mutually stabilizing interaction patterns. Figure 4 shows an interaction graph (in a slightly different representation than described in section 4; see caption) of a very stable self-maintaining set that evolved during the first 3×10^5 collisions starting from 1000 random functions. All functions present in the system differ from that initial set. The numbers refer to the function expressions (not shown) as they rank in lexicographic order. Function 17 is an identity function, although—for the sake of a less congested picture—only the self-copying interaction is displayed. Patterns like those in Figure 4 often include a multitude of interacting self-maintaining sets. In Figure 4, for example, deleting group I on the upper left still leaves a self-maintaining system (group II). Several other parts can be deleted while not destroying the cooperative structure. Sometimes these subsets are disconnected from each other with respect to interconversion pathways (solid arrows), but connected with respect to functional couplings (dotted lines). Figure 4 shows two groups of functions (indicated by I and II) that are not connected by transformation pathways (solid arrows). That is, no function of I is acted upon by any other function in the system such that it is converted into a function of group II—and *vice versa*. Group I, however, depends on group II for survival. The introduction of a "physical boundary" between I and II, cutting off all functional couplings among them, would destroy group I, but not group II.

■ *Turing Gas Without Copy Reactions.* Copy reactions, i.e., interactions of the type $f(g) = g$ or f, strongly influence the patterns that evolve in the Turing gas as described above. Forbidding copy reactions (by changing the collision rule, section 3) results in a rather different type of cooperative organization as compared to the case in which copy reactions and therefore self-replicators were allowed. The system switches to functions based on a "polymeric" architecture that entertain a closed web of mutual synthesis and degradation reactions. "Polymer" functions are recursively defined in terms of a particular functional "monomer" (Figure 5). The individual functions are usually organized into disjoint subsets of polymer families based on distinct monomers. As in the case of copy reactions, these subsets interact along specific functional pathways leading to a cooperativity at the set level. Figure 5(a) and 5(b) shows an example of two interacting polymer families. Neither family could survive without the other. Due to the polymeric structure of the functions the Turing gas remains highly innovative. A much higher degree of diversity and stability is achieved than in systems that are dominated by individual self-replicators. The high stability of the system is due to very small seeding sets. For example, everything in the system shown in Figure 5(a) and 5(b) follows from the presence of monomer 1 of type A (Figure 5(a)) and monomer 1 of type B (Figure 5(b)). Almost the whole system can be erased, but as long as there is one monomer A1 and one monomer B1 left, the system will be regenerated.

■ *Turing Gas With Perturbations.* The experiments described so far kept the system "closed" in the sense that at any instant, the system's population can be described by series of compositions expressed in terms of the initially present functions. An open system is modeled by introducing new random functions that perturb a well-established ecology. In the case without copy reactions, the system underwent transitions among several new quasi-stationary states (metastable transients), each characterized by an access to higher diversity. Systems with copy reactions were more vulnerable to perturbations and lost in the long run much of their structure.

A detailed analysis is found in Fontana.[11]

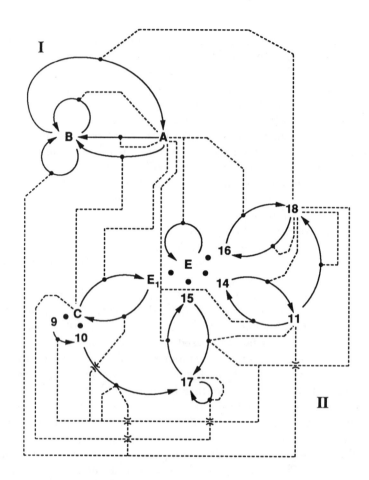

FIGURE 4 Interaction graph of a metastable Turing gas transient. The interaction
graph of the functions present in the system after 3×10^5 collisions is shown. The
system started with 1000 random functions, and conserves the total number of particles
(1000). The numbers denote the individual functions according to their lexicographic
ordering (not shown). Capital letters denote sets, where $A = \{1, 2, 3, 4\}$, $B =
\{5, 6, 7, 8\}$, $C = \{9, 10\}$, $E = \{12, 13, 14, 15, 16\}$, and $E_1 = \{12, 13\} \in E$.
Solid arrows indicate transformations and dotted lines functional couplings. A dotted
line originates in a function, say k, and connects (filled circle) to a solid arrow, whose
head is j and whose tail is i. This is to be interpreted as $j = k(i)$. Large filled circles
indicate membership in a particular set. Function 17 is an identity function. Note: all
dotted lines and solid arrows that result from 17 copying everything else in addition to
itself have been omitted. The function set is closed with respect to interaction.

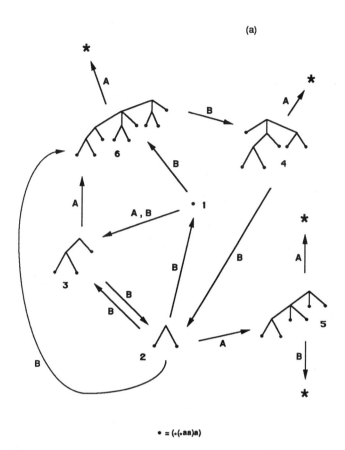

(a)

• = (•(•aa)a)

FIGURE 5   (a) Interaction graph of a metastable Turing gas state without copy reactions. The figures show the interaction pathways among two polymer families established after 5×10^5 collisions starting from random initial conditions. The tree structure of the functions is displayed. The leaves are "monomers" representing the functional group indicated at the bottom of each graph. Solid arrows indicate transformations operated by function(s) belonging to the family denoted by the arrow label(s). Stars in the transformation pathways represent functions that were not present in the system at the time of the snapshot ("innovative reactions"). Due to the polymeric architecture of the functions, the system remains highly innovative. There are always (at least) two functions in the system that a polymerizing function (like the monomer 1 in 5(a)) can combine in order to produce a third one not in the system. The "space limit" tag.

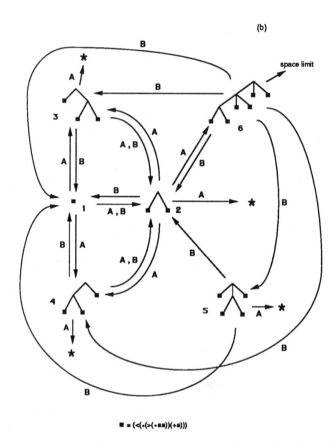

■ = (<(•(>(•aa))(+a)))

FIGURE 5 (continued.) (b) indicates that the corresponding reaction product would hit the length limitation imposed on each individual function expression (300 characters in the present case).

6. CONCLUSIONS

The main conclusions are:

1. A formal computational language captures basic qualitative features of complex adaptive systems. It does this because of:

    a. a powerful, abstract, and consistent description of a system at the "functional" level, due to an unambiguous mathematical notion of function;

b. a finite description of an infinite (countable) set of functions, therefore providing a potential for functional open-endedness; and

c. a natural way of enabling the construction of new functions through a consistent definition of interaction between functions.

2. Populations of individuals that are both an object at the syntactic level and a function at the semantic level, give rise to the spontaneous emergence of complex, stable, and adaptive interactions among their members.

7. QUESTIONS AND FUTURE WORK

The main questions and directions for the future can be summarized as follows.

1. Is there an equivalent of a "dynamical systems theory" for functional interactions? Can the dynamical behavior of the iterated map, Eq.(1), be characterized? Can examples be found that exhibit attractors other than fixed points? Can a classification of all finite self-maintaining sets of unary functions be made?

2. What is beyond replicator (Lotka-Volterra) equations? The standard replicator equation[8,13] on the simplex $S_n = \{x = (x_1, x_2, \ldots, x_n) \in \mathbb{R}^n : \sum x_i = 1, x_i > 0\}$,

$$\dot{x}_i = x_i \left(\sum_j a_{ij} x_j - \sum_{k,l} a_{kl} x_k x_l \right), \quad i = 1, \ldots, n, \tag{15}$$

considers objects i that are individual self-replicators. The Turing gas represents a stochastic version of its deterministic (infinite-population) counterpart[21]

$$\dot{x}_i = \sum_{j,k} a_{ijk} x_j x_k - x_i \sum_{r,s,t} a_{rst} x_s x_t, \quad i = 1, \ldots, n, \tag{16}$$

where in the present case the entry $a_{ijk} = 1$ iff function k acting on function j produces function i, and $a_{ijk} = 0$ otherwise, and x_i, $0 \le x_i \le 1$, represents the frequency of function i in the system.

What can be said about the behavior of Eq. (16)? What can be deduced from it for the finite population Turing gas? Note that the use of a formal language, like λ, allows a finite description of a particular instance of an infinite matrix $a_{ijk}, i, j, k = 1, 2, \ldots$.

3. An obvious extension of the present work is the multivariable case. The restriction to unary functions implied that only binary collisions could be considered. n-ary functions lead to $(n+1)$-body interactions. Suppose the interaction is still given by function composition. A two-variable function $f(x,y)$ then interacts with functions g and h (in this order) by producing $i = f(g,h)$. f acts with respect to any pair g and h precisely like a binary interaction law expression Φ.

However, f can now be modified through interactions with other components of the same system. This might have significant consequences for the architecture of organizational patterns that are likely to evolve. The extension to n variables is currently in preparation.

4. Future work includes a systematic investigation of the system's response to noise (section 5, item 3), either in the form of a supply of random functions, or in the form of a "noisy evaluation" of functions. What are the adaptive properties of the (noisy) Turing gas?

5. If some of the questions above are settled, then an extension to more sophisticated, typed languages (distinction among various object/data "types") that enable a compact codification of more complicated—even numerical—processes could be envisioned; always keeping in mind that processes should be able to construct new processes by way of interaction. The combination with a spatial extension or location of these processes then leads to Chris Langton's[17] vision of a "process gas." I leave it to the reader to further speculate where such intriguing tools might lead.

8. RELATED WORK

The coupling of a dynamics governing the topology of an interaction graph with a dynamics governing a frequency distribution on its vertices is a common situation in biological systems. The immune system,[6] development (ontogenesis),[19] and prebiotic molecular evolution[1] are but a few areas in which some modeling has been done.

The approach sketched here is related in particular to the pioneering work of S. A. Kauffman,[14,15] D. Farmer et al.,[9] R. Bagley et al.,[1] S. Rasmussen et al.,[20] and J. McCaskill.[18]

Kauffman, Farmer et al. and Bagley et al. consider a system of polymers, intended to be polynucleotides or polypeptides, each of which specifically instructs (and catalyzes) the condensation of two polymers into one or the splitting of one into two. This sets up the constructive part of the system and the related dynamics of interaction graphs. The production rate of individual polymers is then described by differential equations in terms of enzyme kinetics[1,9] providing a (nonlinear) dynamical system living on the interaction graphs. This approach represents one of the most advanced attempts to model a specific stage in prebiotic evolution.

Rasmussen and McCaskill made a first step towards abstraction, and the approach described here was prompted by their investigations. Rasmussen's system consists of generalized assembler code instructions that interact in parallel inside a controlled computer memory giving rise to cooperative phenomena. This intriguing system lacks, however, a clear cut and stable notion of functionality, except at the individual instruction level. McCaskill uses binary strings to encode transition-table machines of the Turing type that read and modify bit strings.

A model like the present Turing gas cannot provide much *detailed* information about a particular *real* complex system whose dynamics will highly depend on the physical realization of the objects as well as on the scheme by which the functions or interactions are encoded into these objects. Nevertheless, many phenomena that emerge, for example, in the polymer soup of Kauffman, Farmer et al., and Bagley et al. appear again within the approach described here. The hope then is that an abstraction cast purely in terms of functions might enable a quite general mathematical classification of cooperative organization. The question is if such an abstract level of description still can capture *principles* of complex *physical* systems. How much—in the case of complex systems—can we abstract from the "hardware" until a theory loses any explanatory power? I think the fair answer at present is: we don't know.

ACKNOWLEDGMENTS

This work is ultimately the result of many "reactive collisions" with John McCaskill, David Cai, Steen Rasmussen, Wojciech Zurek, Doyne Farmer, Norman Packard, Chris Langton, Jeff Davitz, Richard Bagley, David Lane, and Stuart Kauffman. Thanks to all of them!

REFERENCES

1. Bagley, R. J., J. D. Farmer, S. A. Kauffman, N. H. Packard, A. S. Perelson, and I. M. Stadnyk. "Modeling Adaptive Biological Systems." *BioSystems* 23 (1989): 113–138.
2. Barendregt, H. P. *The Lambda Calculus.* Studies in Logic and the Foundations of Mathematics, vol. 103. Amsterdam: North-Holland, 1984.
3. Chaitin, G. J. *Algorithmic Information Theory.* Cambridge: Cambridge University Press, 1987.
4. Church, A. "An Unsolvable Problem of Elementary Number Theory." *Am. J. Math* 58 (1936): 345–363.
5. Church, A. "The Calculi of Lambda-Conversion." "Princeton, NJ: Princeton University Press, 1941.
6. deBoer, R., and A. Perelson. "Size and Connectivity as Emergent Properties of a Developing Immune Network." *J. Theor. Biol.*, (1990): in press.
7. Eigen, M. "Self-Organization of Matter and the Evolution of Biological Macromolecules." *Naturwissenschaften* 58 (1971): 465–523.
8. Eigen, M, and P. Schuster. *The Hypercycle.* Berlin: Springer-Verlag, 1979.

9. Farmer, J. D., S. A. Kauffman, and N. H. Packard. "Autocatalytic Replication of Polymers." *Physica D* **22** (1986): 50–67.

10. Fontana, W. "Algorithmic Chemistry." Technical Report LA-UR 90-1959, Los Alamos National Laboratory, 1990. To appear in *Artificial Life II*, edited by D. Farmer et al. Santa Fe Institute Studies in the Sciences of Complexity, Proc. Vol. X. Redwood City, CA: Addison-Wesley, 1991.

11. Fontana, W. "Turing Gas: A New Approach to Functional Self-Organization." Technical Report LA-UR 90-3431, Los Alamos National Laboratory, 1990. Submitted to *Physica D*.

12. Gödel, K. Presented in his 1934 lectures at the Institute for Advanced Study. Quoted from: S. Kleene. "Turing's Analysis of Computability and Major Applications of it." In *The Universal Turing Machine: A Half-Century Survey*, edited by R. Herken, 17–54. Oxford: Oxford University Press, 1988.

13. Hofbauer, J., and K. Sigmund. *The Theory of Evolution and Dynamical Systems*. Cambridge: Cambridge University Press, 1988.

14. Kauffman, S. A. *J. Cybernetics* **1** (1971): 71–96.

15. Kauffman, S. A. "Autocatalytic Sets of Proteins." *J. Theor. Biol.* **119** (1986): 1–24.

16. Kleene, S. "Lambda-Definability and Recursiveness." *Duke Math. J.* **2** (1936): 340–353.

17. Langton, C. G. Personal communication, 1990.

18. McCaskill, J. S. In preparation.

19. Mjolsness, E., D. H. Sharp, and J. Reinitz. "A Connectionist Model of Development." Technical Report YALEU/DCS/RR-796, Yale University, 1990.

20. Rasmussen, S., C. Knudsen, R. Feldberg, and M. Hindsholm. "The Coreworld: Emergence and Evolution of Cooperative Structures in a Computational Chemistry." *Physica D* **42** (1990): 111–134.

21. Stadler, P., W. Fontana, P. Schuster, and J. H. Miller. "Towards a Mathematics of a Turing Gas." In preparation.

22. Trakhtenbrot, B. A. "Comparing the Church and Turing Approaches: Two Prophetical Messages." In *The Universal Turing Machine: A Half-Century Survey*, edited by R. Herken, 603–630. Oxford: Oxford University Press, 1988.

23. Turing, A. M. "On Computable Numbers with an Application to the Entscheidungs Problem." *P. Lond. Math. Soc. (2)* **42** (1936-1937): 230–265.

Stuart A. Kauffman

Santa Fe Institute, 1120 Canyon Road, Santa Fe, NM 87501 and Department of Biochemistry and Biophysics, School of Medicine, University of Pennsylvania, Philadelphia, PA 19104-6059

Random Grammars: A New Class of Models for Functional Integration and Transformation in the Biological, Neural, and Social Sciences

A novel class of models for functional integration and transformation in biological, neural, and social sciences based on random grammars is proposed. This class of models is a generalization of models of the origin of life in which catalytic polymers act on polymers and transform them to other polymers. Under certain conditions a phase transition occurs in which collectively self reproducing, or autocatalytic sets of polymers arise. Polymers can be viewed as symbol strings, and the transformations they mediate in acting upon one another can be characterized as algorithmic or as *implementing a grammar*. Autocatalytic sets are a certain kind of collective identity operator in such a grammar. Fontana's "Alchemy," which also yields autocatalytic sets of string processes, can be viewed as another specific grammar in which symbol strings act upon one another. This leads me to suggest studying parameterized families of grammars in a non-denumerably infinite *grammar space*. Strings instantiate a grammar via "enzymatic sites" which act on other strings and mediate the transformations mandated by the grammar.

In subsequent uses, symbol strings might be interpreted as chemical polymers in a prebiotic soup, as metabolites in an organism, as neurons in a

1990 Lectures in Complex Systems, SFI Studies in the Sciences of
Complexity, Lect. Vol. III, Eds. L. Nadel and D. Stein, Addison-Wesley, 1991   **427**

neural system, as components in the schemas of an ego structure, as concepts in a scientific theory, as goods and services in an economy, or as roles in a culture. The kinds of sets of strings which arise as strings act upon one another, then provide models of functional integration and transformation in the corresponding classes of systems. The hope is that analysis of grammar space will show that rather few broad "regimes" exist. Where those robust regimes map onto biological, neural, or social systems, it may be possible to account for functional integration and transformation in those systems due to membership in the corresponding "universality class."

New kinds of objects emerge, characterized by the sets of strings generated by strings acting upon strings in the potentially infinite space of symbol strings. These include "Jets," "Lightening Balls," "Mushrooms," "Eggs," "Filigreed Fogs," and "Pea Soup." Some of these objects can be either finite or infinite in string diversity, others are only finite or only infinite.

Injection of new strings into a system can cause transitions among these classes of objects. For example, an isolated self-reproducing "egg" can transform into an infinite system if a novel symbol string is injected. Or a collection of separate systems, each generating a small finite jet of strings, can jump to an infinite filigreed fog if the separate systems begin to exchange symbol strings. Here we find models of evolution without a genome in protoorganisms, of cultural transformation when an isolated culture come into contact with another culture, and of technological "take-off" at a critical complexity of interacting goods and services in an exchange economy. Models of this type may prove broadly useful.

INTRODUCTION

The general problem which arises in investigating the capacities of complex systems to adapt lies in understanding both the *functional* and the *dynamical* order which integrates these systems. *E. coli* "knows" its world. A wealth of molecular signals passes between a bacterium and its environment, which includes other microorganisms, and higher organisms. The signals entering the bacterium are harnessed to its metabolism and internal transformations such that, typically, the cell maintains itself, replicates, and passes its organized processes forward into history. Similarly, a colony of *E. coli* integrates its behavior. The organisms of a stable ecosystem form a functional whole. The niches occupied by each, jointly add up to a meshwork in which all fundamental requirements for joint persistence are met. Similar features are found in an economic system. The sets of goods and services comprising an economy form a linked meshwork of transformations. The economic niches occupied by each allow it to earn a living and add up to a web in which all mutually defined requirements are jointly met. Both biological and technological evolution consist in the invention of profoundly, or slightly novel organisms, goods or services which integrate into the ecological or economic mesh, and thereby transform it. Yet at almost all stages, the web retains a kind of functional coherence. Furthermore, the very structure and connections among the entities sets the stage for the transformation of the web. In an ecosystem or economic system, the very interactions and couplings among the organisms, or goods and services create the conditions and niches into which new organisms, or goods and services, can integrate. The "web" governs its own possibilities of transformation.[15]

Similar functional integration of roles, obligations, and institutions apply at societal levels. The revolution occurring in Eastern Europe and the U.S.S.R. in these *anni mirabili* is accompanied by a sense that the Soviet system is an integrated whole with the property that if one, or a few, features are removed or altered, the entire system must transform to something quite different—and whole. In June, 1989, the Communist leaders in China saw fit to kill their students in Tienamin Square. Why those leaders did so is clear: The students were demonstrating for increased democracy. Their government feared the consequences would transform Chinese Communism. In short, the puzzle is not to understand what China's leaders did, but rather to understand what they *knew*. In a real and deep sense the Chinese government knew that, were a few features of their system altered, the entire edifice stood in danger of dramatic transformation. What, indeed, did they know?

In the biological and social sciences we badly lack a body of theory, indeed even a means of addressing these issues: What is a functional whole and how does it transform when its components are altered? In this article, I shall develop an outline for a fresh approach to these important issues. The approach is based on the use of random grammars. The objects of the theory are strings of symbols which may stand for chemicals, goods and services, or roles in a cultural setting. Symbol strings act on one another, according to the grammar, to yield the same or other symbol strings. Thus, the grammar specifies indirectly the functional connections among the symbol strings. It defines which sets of strings, acting on other sets of strings, produce which sets of output strings. These mappings are the functional couplings among chemicals in a protoorganism, among a population of organisms in an ecosystem, and become the production technologies in an economy. Diverse grammars model diverse possible chemistries, or possible production technologies. By studying the robust features of functionally integrated systems which arise for many grammars that should fall into a few broad "grammar regimes," it should be possible to build towards a new theory of integration and transformation in biological and social sciences. Among the features we will find are phase transitions between finite and potentially infinite growth in the diversity of symbol strings in such systems. This phase transition may well underlie the origin of life as a phase transition in sufficiently complex sets of catalytic polymers, and a similar phase transition may underlie "take-off" in economic systems once they attain a critical complexity of goods and services which allows the set of new economic niches to explode supracritically.

My suggestion to study random grammars grows from work initiated by myself, in investigating *autocatalytic sets of polymers*,[14] thereafter carried on in collaboration with Doyne Farmer, Richard Bagley, Norman Packard, and Walter Fontana,[6]

In particular, I believe the recent extensions by Walter Fontana concerning autocatalytic sets are exciting and important.[7] All point towards a new way to investigate the emergence of functional integration and adaptation in complex systems.

INFINITE AUTOCATALYTIC SETS AND THE ORIGIN OF LIFE AS A PHASE TRANSITION

The invention of autocatalytic sets lay in an attempt to understand whether the origin of life necessarily required the self complementarity of RNA or DNA molecules, where the plus single strand is a template for its minus strand complement. It is just this feature which has commended DNA or RNA as the "first" living molecules, and has seemed, in principle, to rule out proteins, more readily formed in abiotic conditions, as the first living molecules. But what if a *set* of proteins might *collectively* catalyze their own formation from some simple building blocks, such as amino acids? In principle, such autocatalytic sets are nearly inevitable. The same principles apply to the emergence of self reproducing sets of single stranded RNA molecules. Here, in outline, is the model.

Consider a set of peptides, where a peptide is a short sequence of amino acids, each, a choice of one among several or twenty types. As the maximum length of such sequences, M, grows, the total number of polymers grows *exponentially*, and is nearly 20^M. This is an old idea. There are five new ideas. First, consider the most primitive reaction among peptides or single stranded RNA polymers. It consists in ligating two polymers into a longer polymer, or cleaving one into two shorter fragments. Such chemical reaction transformations can be represented by a triad of directed line segments leading from the two shorter fragments to a node, and from the node to the larger polymer constructed by uniting the two smaller polymers in a given left to right order. Now construct a *reaction graph*, showing all polymers as circular nodes, connected by reaction triads whose nodes are square. Use of arrows from the two smaller polymers to the square reaction node and from it to the larger polymer uniquely specifies the substrates and product of the ligation reaction. Recall that reactions are reversible. Thus, the same triad specifies the reverse cleavage reaction. Color the reaction line segments black to indicate that the reactions are *not catalyzed*. This collection of nodes and directed lines represents the uncatalyzed reaction graph.

The second new idea is to consider the ratio of reactions to polymers. Simple calculations show that the ratio of the number of reaction triads to the number of polymers is about M. This is a central observation, which, as we shall see, holds in very many contexts. The number of components, here peptides or RNA sequences, increases exponentially as M increases, but the number of transformations among the components, here reactions, increases even faster. Thus, the ratio of potential transformations among the components to the components increases as M increases.

The third new idea is to note that peptides or RNA polymers themselves can catalyze the ligation and cleavage of peptides or RNA polymers. Therefore, given a *model* showing which polymers catalyze which reactions, we can ask whether a set of polymers up to length M contains *an autocatalytic subset.* An autocatalytic subset of polymers is one having the property that each polymer has at least one formation reaction catalyzed by some other member of the set, and that connected sets of catalyzed transformations lead from some maintained "food set" to all the polymers in the set. As we see next, under a wide variety of models for which polymers catalyze which reactions, autocatalytic sets "crystallize." Note that since peptides or RNA sequences can be modeled as "strings" of letters, and because these strings "operate" on one another, that autocatalytic sets are just a kind of "algebra" or "algorithm" for mapping strings into strings. An autocatalytic set will be some kind of identity under that cluster of operations.

By hypothesis, none of these uncatalyzed transformations occur. Hence, a "soup" is unchanging. The simplest model of which polymer catalyzes which reaction is that each polymer has a *fixed probability, P,* of catalyzing each reaction. This is not a chemically realistic hypothesis, but is useful for the moment. Given this hypothesis, each peptide is asked, in turn, if it catalyzes each reaction, and answers "yes" with probability P, and "no" with probability $1 - P$. Note down, for each reaction, which polymers catalyze it. Color any reaction which is catalyzed red. Thus, red triads denote catalyzed reactions. Ask all polymers which reactions, if any, each catalyzes. When this process is complete, some fraction of the reaction triads may be colored red. This is the *catalyzed reaction subgraph* of the uncatalyzed reaction graph.

The fourth new idea is that when the maximum sized polymer, M, is large enough compared to any fixed probability of catalysis, P, then an autocatalytic set will crystallize. The intuition is simple. As M increases, the ratio of reactions to polymers increases. And the number of polymers increases. On average each polymer catalyzes a number of reactions equal to P times the number of reactions. As the ratio of reactions to polymers increases, by M increasing, there must come a critical size, M_c, at which almost every polymer has its formation catalyzed by some polymer. This is a phase transition in which a large connected cluster of catalyzed transformations is present in the catalyzed reaction subgraph. Careful analytic results show that, indeed, such a phase transition occurs, related to the giant components of random graphs.[4,5] After this transition, autocatalytic sets exist. Thus, in a critically complex mixture of catalytic polymers, autocatalytic sets will emerge. On this view, which I hold with fair conviction, life is an emergent "crystallization" of self reproductive metabolisms, based on polymer chemistry, chance, and number, rather than template magic.

The fifth new idea is that the catalyzed reaction graph can be "infinite" or finite. Modify the model as follows: Consider a set of polymers up to length L. Call this the "food set." Imagine that all polymers up to L are maintained by exogenous supply. Allow these to undergo ligation and cleavage reactions, if catalyzed by members of the food set, and form polymers of any length ranging up to $2L$. In turn, these new polymers are present as potential substrates, products and enzymes, together with all polymers in the food set. Iterate the procedure to assign polymers, at random, those reactions they may catalyze. Update the set of polymers up to length $4L$ which may now be formed. Over iterations, the set of polymers catalyzed "out of" the food set may continue to increase in diversity *without limit*. In this case, the growth of the catalyzed reaction graph, and set of polymers formed is *unbounded*, and the graph growth can be said to be *supracritical*. The graph of polymers engendered by this process of catalysis is infinite.

Conversely, the growth of the catalyzed reaction graph may stop at some point. No new polymers may be added to the system. Then, the growth of the graph is *subcritical*.

Fairly simple algebraic arguments suffice to define a phase transition in a parameter space whose axes are the probability of catalysis, P, and the size of the food set, a function of L. A line partitions the space into two regions. In the subcritical region, either the food set is too small or the probability of catalysis is too low, and graph growth is finite. Above the transition, graph growth is infinite. The critical scaling law is

$$B^{Lc*} = (1/2p)^{1/2} \quad \text{or} \quad P = B^{-2lc*}$$

where B is the number of kinds of amino acids considered. Because the same story obtains for single stranded RNA molecules, B is the number of kinds of nucleotides considered.

It is easy to unite the two pictures given and show that, in principle, *supracritical, hence, infinite, autocatalytic sets can exist.*

It is important to stress that in this analysis I have eschewed any discussion of the *thermodynamic requirements* driving synthesis of larger polymers from smaller ones. I shall continue to do so for the moment. That is, we are considering only the formally allowed transformations among hypothetical polymers engendered by catalysis of reactions, not yet the issue of the actual flow of real polymers over the space of catalyzed reactions under specific physical conditions. I stress however, that autocatalytic sets incorporating plausible kinetic and thermodynamic factors are feasible.[6]

RNA WORLD, CHEMICALLY PLAUSIBLE "TEMPLATE MATCH RULES" YIELD FINITE AND INFINITE AUTOCATALYTIC SETS

The existence of autocatalytic sets persists with more plausible models, based on template complementarity in single stranded RNA sequences, of which polymers catalyze which reactions. The work that follows has been carried out with Rick Bagley at the Santa Fe Institute.[16] Consider a population of single stranded RNA molecules. The hypothesis that life started with such RNA molecules is widespread, both because RNA molecules, called ribozymes, have now been found to catalyze reactions,[2] and because template replication of arbitrary single stranded RNA molecules seems so reasonable. Each strand is to line up free complementary nucleotides, then catalyze the joining of these to form a complementary strand which will melt off and repeat the cycle. Note that, were this to work, then *any* single stranded RNA molecule would be able to lead to synthesis of its *precise* complement, and via the latter, to its own synthesis. Such a ± pair would, of course, be an autocatalytic set. The + strand catalyzes the formation of the − strand, and vica versa.

In practice, unfortunately, this autocatalytic template replication of an arbitrary single stranded sequence has not yet succeeded. The only case which works uses a + strand comprised by C and G nucleotides with $C > G$ in composition. But the resulting − strand has $G > C$ and fails to work as a template.[10]

Suppose such reactions were, in general, to work, and note a further feature. Here we have a *finite* autocatalytic set, a + and − pair of template complement strands replicating themselves, but not necessarily reaching out into sequence space and replicating or catalyzing the formation of other sequences. Thus, we reach a simple, important conclusion. *In principle, autocatalytic sets can be either finite or infinite.*

Indeed, *finite RNA autocatalytic sets exist in practice.* The single stranded RNA hexamer $GGCGCC$ acts on two trimers, CCG and CGG, binds them via complementary base pairing, ligates them, and releases the resulting hexamer which, upon examination, is identical in $3' - 5'$ order.[10] Thus, a "soup" supplied with the hexamer and a population of trimers will form a *finite* autocatalytic set with three sequences as members, and convert the trimers largely to hexamers.

Note the critical point that the hexamer is acting as a specific *ligase*, recognizing specific trimers, not as a general polymerase. Thus, actual chemistry supports specific ligation by single stranded RNA molecules.

Strikingly, single stranded RNA molecules and specific ligation based on template recognition supports not only finite autocatalytic sets, but in principle, supports *supracritical, infinite* autocatalytic sets. Bagley and I have modified the rules for catalysis. We require that the potential catalytic strand "template match" the left and right, three or four nucleotides on the two potential substrate molecules, either perfectly or with perhaps one mismatch allowed. Matching here is base pair complementarity. Only if such sequences match, does the prospective "ribozyme" have a chance to be an actual catalyst for the reaction. The chance is proportional to the "match-strength," and polymer length. Again, leaving out thermodynamic issues, we find that with a critical complexity of polymers in the "soup" we obtain apparently infinite autocatalytic sets. The criterion for this judgement is that the number of new polymers being added to the system grows superexponentially over updates of the catalyzed reaction graph.[16]

Thus, the RNA world, with string matching, can give rise to both *finite and infinite* autocatalytic sets.

EVOLUTION AND CO-EVOLUTION IN AUTOCATALYTIC SETS.

At present Rick Bagley and Walter Fontana are implementing code to follow the capacity of autocatalytic sets to *evolve*. A "shadow" set of polymers derived by *spontaneous* reactions among the polymers in the set is assumed to form at low frequency. The "shadow set" is a suggestion of Doyne Farmer. If among the shadow set, one or a cluster of new polymers can catalyze their formation from themselves and the existing autocatalytic set, they are added to the set. Thus, the set evolves to a new autocatalytic set. In turn, old polymers may die out of the set, due to addition of real thermodynamic criteria which I will not discuss here. Thus, evolution can occur in a space of polymers and transformations among the polymers. In addition, autocatalytic sets are favorable objects to study *coevolution*. We need merely define interactions between such sets, whereby sequences made in one set can migrate to and enter another set. Such signal sequences may integrate into, or disrupt, the functional organization of each set in ways I will return to below. But, in short, such couplings and their coevolution affords the opportunity to study the emergence of shared functional integration in coevolving reproducing entities.

EMERGENCE OF A CONNECTED METABOLISM

Similar arguments, briefly presented here, apply to the emergence of a connected metabolism. Consider all organic molecules, counted in terms of the number of carbon atoms per molecule. As the number of atoms per molecule goes up, the number of organic molecules explodes very rapidly. But the number of legitimate reactions by which they transform rises even faster. Indeed, consider, in general, reactions with two substrates and two products. It might be thought of as a mapping of pairs of organic molecules into pairs of organic molecules. Occupied "cells" in such a matrix show the legitimate chemical reactions. It is easy to see that if each pair were able, on average, to undergo only one reaction, then the number of reactions is crudely the square of the number of organic molecules. Now consider two related hypotheses. First, cast a mixture of RNA or peptide sequences upon this space of organic reactions. If each sequence has some chance of catalyzing any reaction, then if enough sequences are cast upon the space, or if the space is complex enough, connected sequences of reactions will crystallize as a giant component in the reaction graph. A connected metabolism will emerge whole. The second related hypothesis notes that organic molecules can catalyze organic reactions. Under a variety of hypotheses about the distribution of catalytic activity among the organic molecules, autocatalytic sets should emerge.[16]

FUNCTIONAL INTEGRATION

Autocatalytic sets exhibit an emergent *functional integration*. Once the set of chemical reactions, ligation and cleavage, the set of polymer strings, and the assignment of which strings catalyze which reactions is made, the rest is emergent. The notion of function is contained in the idea of which strings catalyze the transformation of which other strings. The idea of an autocatalytic set is precisely that of *catalytic closure*. This is a first, root, central image of a *functional whole!* The members of an autocatalytic set get themselves "made" by members of the set. All niches necessarily are filled. It is just this sense of catalytic closure that is the start of an image of functional wholeness that I want to pursue. However, as we shall see in considering "catalytic jets," the requirement of closure is too strong.

AUTOCATALYSIS MUST BE VERY GENERAL

These ideas are very general. They rest upon the ideas that there is a set of objects and a set of transformations among the objects, and that the objects can themselves mediate the transformations. One, then notes that for a wide variety of such objects and transformations, the number of transformations grows more rapidly than the number of objects, hence with a wide variety of distributions of which object mediates which transformation, autocatalytic sets should emerge. One expects the ideas to have wide applicability.

FONTANA'S ALCHEMY

Walter Fontana has recently generalized the analysis of autocatalytic sets and catalyzed transformations in a useful way.[16] Note that, in the model of the origin of life, one is describing mappings of strings into strings, mediated by strings. Thus, as remarked above, this is, in general, some kind of algebraic or algorithmic transformation in which autocatalytic sets are certain kinds of identity operations of a cluster of the transformations and objects.

Fontana borrows a powerful algorithmic language (the progenitor of Lisp) derived from the λ-calculus, invented to be as powerful as universal Turing machines. Fontana's idea is to exploit the general idea of strings acting on strings as algorithms. In this hypothetical chemistry, Fontana does not require that mass be conserved. Two strings collide—the first is the "program" which acts on the second as an input. By construction, most strings are legal both as program and as input. Thus, most collisions between strings transform the second string into some single new string. Fontana defines a "Turing gas" in which a random collection of strings is placed in a "chemostat." After each productive collision between strings, the number of strings has increased by one. To supply a selective condition, Fontana randomly removes one string, hence holds the number of strings in the chemostat constant.

Fontana has carried out three kinds of numerical experiments. In the first, a set of 700 strings is allowed to interact by random collisions. At first only new strings are generated. But over time, more and more strings which are generated already have identical copies in the chemostat. Eventually a closed set of strings, an *autocatalytic set*, emerges. In this first set of experiments, the autocatalytic set is dominated by a general replicase, or a sequence that can copy itself and any other sequence. It is equivalent to a ribozyme which might copy itself and all others.

In the second set of experiments, Fontana disallows copying strings. Nevertheless, *closed collectively autocatalytic* sets of strings emerge. Thus, one set had some 45 kinds of strings present which mutually transformed into one another. These sets are the direct analogues of the autocatalytic polymer sets discussed above.

In the third numerical experiment, Fontana injected sets of 20 random strings into an evolving chemostat. He found that the terminal autocatalytic set differed from that which would have occurred without exogenous perturbation.

JETS AND AUTOCATALYTIC SETS: TOWARDS A NEW STRING THEORY

Whether we are considering the transformation of chemicals or goods and services in an economy, or a variety of other cases, it seems useful to consider the infinite set of binary strings as the "objects" under analysis. Then, in general, strings or sets of strings act on strings or sets of strings to yield strings. In general, such transformations are just mappings, or algorithms. The set of strings operated upon can be one or many. The set of strings carrying out the operations can be one or many. In general, the set of transformations will increase more rapidly than the set of strings. Thus, the general question is this:

For various kinds of random or nonrandom mappings of strings into strings, what kinds of "sets" of strings emerge? What we need is a way of generating families of algorithms, or grammars, or finite state automata which realize those algorithms (or grammars) and then to discover the kinds of functionally "generative" sets we obtain. I turn next to some intuitions about those sets, then return in the next section to consider ways of studying the space of possible grammars.

STRING SET "GEOMETRIES": JETS, LIGHTNING BALLS, MUSHROOMS, EGGS, (FIXED, TRAVELING, WOBBLY, ERGODIC AND HAIRY) FILIGREED FOGS, AND PEA SOUPS

Consider first a "Jet." Imagine a rule for polymer sets in which any string only catalyzed the ligation of strings both of which were larger than the catalyst string. Then, by construction, no *feedback loops* could form. All catalyzed transformations would lead to ever larger strings.

Let me define a Jet as a set of transformations among strings from some *maintained "founder" set of strings*, analogous to the food set, S_o, having the property that under the algorithmic transformations among the strings, each string is produced only by a unique set of "parent" strings, and is produced in a unique way. This is probably sufficient, but more than necessary. In any case, it leads to a Jet of string productions which never cycles back on itself.

Note that a Jet might be *finite*, or *infinite*.

A Lightning Ball is a Jet cut free from its founder set, free to propagate through string space until it dies out (if the jet is finite) or propagate forever (if the jet is infinite). Presumably a periodic or a quasi-periodic Lightning Ball which orbits back to the starting set of strings, or near the starting set of strings, is possible.

The "orbits," defined as the succession of sets of strings in the Lightning Ball, in string space might be chaotic or ergodic.

Let me next define a Mushroom. The first example is an autocatalytic set of polymers growing forth from a maintained "food set." Here a set of transformations jets up via a kind of stem free of feedback loops, then feedback loops begin to form creating the head of the Mushroom. In effect, a Mushroom is a Jet from a maintained founder set, with feedback loops.

Mushrooms are models of functional "bootstrapping." An immediate example is an autocatalytic peptide set with a sustained metabolism of coupled transformations from the food set. But perhaps another example is the technological evolution of *machine tools*. For example, the first tools were crude stones, then shaped stones which enabled formation of better tools, used to dig ore, then metal tools, then ultimately, the development of machine tools which themselves generate tools such as axes, chisels, and machine parts for other machine tools. Presumably the onset of agriculture is a similar example. Many more must exist in economic and cultural evolution, as well as organic evolution.

Like Jets, Mushrooms can be *finite or infinite*. Next, consider the Egg. The examples of hexamer and trimer RNA sequences which reproduce only themselves in RNA sequence space, and Fontana's two autocatalytic sets are eggs, whole in and of themselves. Eggs are self sufficient sets of algorithmic transformations with no need of a "stem" from a maintained founder set. In an egg, strings can produce arbitrary strings, hence can enter a closed set which finds only itself, free of all other strings. Let me reserve the term Egg for *finite closed autocatalytic sets*. An unchanging finite egg is a kind of *identity operator* in the process algebra or grammar by which strings act on strings such that the collection of processes produces precisely and only itself. Eggs may prove to be useful models of self confirming mythic or even scientific conceptual systems by which the outside world is parsed. They may also prove useful as models of cultural identity, integration and wholeness.

Presumably Eggs come in several types. We have already considered the Fixed Egg which maps into itself. But eggs might move through string space, creating Traveling Eggs. The former correspond to autocatalytic sets which are closed and hold to a fixed set of sequences and transformations. Traveling eggs, are rather like Lightning Balls, but contain feedback loops, and change composition in sequence space in various ways. Presumably, Wobbly Eggs, which orbit among a periodic set of sequences, or quasi-periodic set, might be possible; so might Chaotic eggs and Ergodic eggs. The set of strings in an ergodic egg would wander randomly over string space as the egg traveled. In addition, Hairy Eggs would be finite objects from which Jets or Mushrooms may extrude, perhaps stochastically if the production rules are activated probabilistically. Fontana may have found such structures with a stable core metabolism sending out a fluctuating flare of other strings.

The Filigreed Fog is an infinite supracritical autocatalytic set which may or may not have a stem to a sustained founder set, or, like an egg, might also float free. Unlike an egg, however, it is not bounded. Nevertheless, the Filigreed Fog is *limited* in that there are at least *some strings which can never be generated by the set*.

Finally, there is the Pea Soup, defined as an infinite set which, in principle, will eventually include all possible strings. It is intuitively clear that the autocatalytic set generated by the model in which each polymer has a fixed chance of catalyzing each reaction will form a Pea Soup if it is supracritical. Ultimately all strings could have their formation catalyzed by some string.

EVOLUTION AND STABILITY OF FUNCTIONAL SETS

Among the obvious questions about such sets are their stability and capacity to evolve. Consider an Egg. How many Eggs does the algorithmic set contain? A few? Many? Given a definition of one-mutant variants, is an Egg stable to all one-mutant variations in its composite strings? Two-mutant variants? Thinking of Eggs as attractors, how many are accessible from any Egg, and by how much of a mutation in the set of strings present? For example, Fontana began to study this by injection of exogenous strings. Can one jolt an Egg to another Egg? Similar questions apply to all the kinds of structures depicted. Such questions bear on the stability and capacity to evolve by "noise" of such functional sets.

Note that transformation of a Jet to a Jet, or Mushroom to another Mushroom, or Egg to Egg, or between such types of sets by a perturbation or mutation begins to get at our intuitions that the Soviet or Chinese political system is fragile, that a few minor changes in the coherent structure must lead to the replacement of many or most functional linkages.

DECIDABILITY PROBLEMS

A number of issues may be undecidable. For example, whether a given set of founder strings in a given algorithmic set of rules is subcritical or supracritical might be such an issue. It appears intuitively related to the halting problem. Similarly, in a Filigreed Fog, it may not be formally decidable that a given string cannot be produced from the initial set of strings by the grammar. I suggest below in an economic context where strings are goods, that such formal undecidability may map into the logical requirement for bounded rationality in economic agents, and an equal logical requirement for incomplete markets. Thus, such models may invite modification of neoclassical economic theory.

SIZE DISTRIBUTION OF AVALANCHES OF CHANGE

In autocatalytic polymer sets, addition of a new polymer may trigger the formation of many new strings, and elimination of old strings. In a technological web, addition of the automobile drove out the horse and many horse trappings. When Fontana injected random strings, a peripheral component of the autocatalytic metabolism tended to change. What is the size distribution of such avalanches? For example, Bak and colleagues have drawn attention to self-organized criticality in their Sand

Pile model.[1] At the critical state, a power law distribution is found with many small and few large avalanches. A similar distribution is found in Boolean networks at the edge of chaos, as well as in certain model ecosystems which have optimized joint fitness. Thus, we are led to ask what such avalanches of "damage," or changes, look like in the various objects discussed. For example, it might be the case for finite Jets that avalanches early in the Jet are large, and late in the Jet are small. Or avalanches might show a common distribution regardless of when unleashed in the lifetime of a Jet. Similar questions arise with respect to Fixed Eggs and Traveling Eggs. Perhaps a power law distribution obtains whenever sets go supracritical. Perhaps that is just when all damage is also infinite.

Note that these questions may allow us to begin to address such issues as the *sensitivity of history* to small perturbations. For history, too, is an unfolding of transformations among some indefinite, or infinite set of possibilities. Similarly, the evolution of autocatalytic sets in a world of polymers, with coevolution among the sets, captures both historical accidents and a kind of entropic exploration of the world of the possible.

SETS OF STRINGS ACTING ON SETS OF STRINGS: AGGREGATED TRANSFORMATIONS AS MACHINES TUNES THE RATIO OF TRANSFORMATIONS TO STRINGS

Consider Fontana's Turing-gas. A thousand strings interact with one another by random collisions. This parallels the studies on autocatalytic sets. Suppose that 100 different types of strings are present. Then, the chance that any specific string will undergo an ordered set of five of the transformations mediated by these strings is low. Consider instead a machine, or complex, made of a sequential aggregate of five kinds of strings, such that any string which encounters the aggregate undergoes sequentially all five transformations. Thus, the aggregate is a kind of "machine" made of simple transformations, which ensures a complex set of transformations. Now consider that if there were 100 kinds of strings in the gas, then there are $100^5 = 10^{10}$ of these combined fivefold transformations.

One implication of the use of an aggregate, or ordered set of five kinds of strings as a "machine" which acts on a single string or a set of strings, is that the number of "machines" is very much larger than the set of single strings. Since each machine carries out a compound transformation on an input string, this is equivalent to saying that construction of complex machines increases the ratio of potential transformations mediated by one machine to strings. Hence, achieving phase transitions to more complex supracritical sets becomes easier.

Another implication is that coordination of five strings in an ordered way into a machine alters effective time scales. If strings acted on one another in random pairwise interactions, a vast set of strings would have to be present in the system to assure that all of these complex transformations were sampled in reasonable time. Thus, we can think of the aggregation of primitive strings into aggregates, the *invention of machines*, as means to mediate specific compound transformations at high frequencies. Clearly this will alter the functional sets formed. This example makes it clear that *time scales matter*. A set of strings interact by some dynamics, as in Fontana's random collision dynamics. Altering the probabilities of string interactions profoundly alters which sets of composite transformations occur and which sets of strings arise.

The image is not a poor one. The machines in our economy form specific complex objects among a set of many other possible ones.

FROM ONE CHEMOSTAT TO MANY: COEVOLUTION AND PHASE TRANSITIONS

By introducing a multiplicity of chemostats which operate on strings internally, and may also exchange strings between the chemostats, we can explore models of coevolution, the emergence of competition or mutualism in biology, or economic trade between economic agents or units. In addition, phase transitions among the kinds of sets generated, jets, mushrooms, and so forth, can take place as a function of the number of chemostats which come to interact. As that number increases, the joint complexity of strings being operated upon can pass critical thresholds. Such transitions may model "take-off" in an economy or even intellectual community. I now discuss this in more detail.

The autocatalytic models, and Fontana's model, so far take place in a stirred reactor. All strings can interact with all strings. Consider instead a set of *chemostats or boxes*. Each box, to be concrete, begins with a sustained founder set, which are its sustainable natural resources. Each set of strings proliferates purely internally. So far this is nothing but the stirred reactor within one chemostat. Now let some of the strings be made for "export only." These exported strings may pass to other boxes. Those other boxes may be identified by spatial location, or some strings may "bind" to the box surface and serve as address strings. In either way, the invention of multiple chemostats or boxes serves to identify "individual" regions of "local processes" which may then *coevolve* with other such regions.

Among the first questions to consider are these:

Imagine that each box, granted its sustained founder set, yields only a finite Jet by itself. It may be the case when strings can be exported between boxes that some or all of them are lifted to a more complex level of activity. For example, the collective system might form an *infinite Jet, a finite or infinite Mushroom, a Filigreed Fog, or even a Pea Soup!* The clear point to stress is that collaborative interaction may transform a system from one to another of the types of functional sets. In particular, there may be a *critical level of complexity* for any given set of algorithmic transformations, leading with high expectation to each of these transitions. If so, what are these thresholds like? Might they, for example, bear on economies which are unable to expand in diversity of goods versus those which can explode? Does this concept bear on the consequences to isolated cultural systems when brought in contact with other isolated systems or a larger world culture? Do they bear on the scientific explosion following the Renaissance?

Consider the question of *functional integration between the boxes*. Each box can be thought of as a kind of country with natural resources, or a firm interacting with other firms, or perhaps an integrated functional organism. String inputs from other countries or organisms may perturb the internal dynamics of each box. In response, the box may "die," that is the Jet or other process might collapse to a sustained founder set or to nothing, the null set, or it might transform to some other more or less constant functional set. In the latter case, we have an image of entities which alter their internal structure in response to external couplings such that each is internally a stable sustained flux of collaborative processes in conjunction with the couplings to the other boxes. It is an image of stable signal relations among bacterial cells, or perhaps, as we see in more detail below, trade relations among nations endowed with different natural resources and histories of technological development. Are there many alternative attractors to such a system given the same founder sets to each box? How history dependent is it? How stable to perturbations?

Such coevolving "boxes" literally come to "know one another" and know their worlds. We must consider when and whether such systems are competitive or such systems coevolve mutualisms which optimize mutual growth rate or equivalently, utility. Indeed, I suspect that these processes must occur in biological and economic evolution.

DYNAMICAL STABILITY AS WELL AS COMPOSITIONAL STABILITY

The sets we considered above, jets, mushrooms, eggs, fogs, and so forth, deal with the *string composition* generated by different grammatical or algorithmic rules by which strings interact. But, in addition to the composition of such generated sets, it is also important to consider the dynamical aspects of such systems in terms of the "concentrations" of strings of each type over time. For example, an Egg might reproduce itself compositionally at a dynamical steady state or along a limit cycle in string space by which its constituent strings were successively produced. Presumably other orbits might suffice for an Egg to persist. Similar questions arise for jets, fogs, and other potential objects.

A critical difference between string systems and familiar dynamical systems is that the former operate in an indefinitely large state space, the latter in a fixed state space. For example, Boolean networks and other dynamical systems exhibit dynamical attractors *in a fixed state space*. The functional sets we are considering are, in a sense, evolving in an open state space of strings.

Boolean networks exhibit three main regimes of dynamical behavior: *chaotic, ordered, and complex.*[11,12,13,16,17] The transition between these is governed by a *phase transition*. In the ordered phase, *percolating frozen components*, where binary variables are fixed in active or inactive states, span the system leaving behind isolated islands of unfrozen elements free to fluctuate from 0 to 1 to 0 in complex patterns. In the chaotic regime, the unfrozen "spins" form a percolating component. In the complex region, which lies at the boundary between order and chaos, the frozen component is just melting and the unfrozen component is just percolating. Avalanches of "damage" due to perturbing the activity of single sites propagate on all length scales in a power law distribution in the complex regime. Damage only propagates a finite distance in the frozen ordered regime, typically within one unfrozen island. Damage propagates to a finite fraction of all spins in the infinite spin limit in the chaotic regime, exhibiting sensitivity to initial conditions in the chaotic regime.

A variety of tentative arguments suggest that systems in the complex regime on the edge of chaos can carry out the most complex computations and can "adapt" most readily.[16,17,19]

Our questions, with respect to functional sets, concern whether or not the analogues of frozen components form, and whether ordered, chaotic and complex behavior occurs. below, I return to show why I believe the answer is yes. I stress that these questions of *dynamical order* arise, in addition to and in conjunction with questions about the string compositional stability of string sets. One important new implication is that the *dynamical* behavior of a set of strings can control how it explore string space. For example, an infinite fog may not be populated because the system cannot pass bottlenecks in string space for dynamical reasons.

INFINITE BOOLEAN NETWORKS AND RANDOM GRAMMARS: APPROACHES TO STUDYING FAMILIES OF MAPPINGS OF STRINGS INTO STRINGS.

In order to study jets, eggs, fogs and functional interactions, we require mathematical models of the algorithmic interactions by which strings act on strings to produce strings. The autocatalytic polymer set with fixed probability of catalysis, P, the RNA string match rule Bagley and I have investigated, and Fontana's Alchemy are three choices of rules by which strings act on one another. The aim of this section is to consider alternative approaches to generate in some ordered way the set of "all possible" mappings of strings into strings. In fact, this cannot be done in an ordered way. The set of all such mappings involves the infinite power set of binary strings of infinite length acting on itself to produce the infinite power set of binary strings. This class of objects is not denumerably infinite. It maps to the reals. Consequently, any ordered approach to this problem requires simplifying at least to a denumerably infinite set of objects, categorized in terms of some parameters such that mappings of increasing complexity can be studied and such that these mappings fall into useful classes.

The aims of this endeavor should be stated clearly. I believe such mappings, grammars, or algorithms, sending strings or sets of strings, operated upon by strings or sets of strings, into strings or sets of strings, may provide useful models of molecular interactions or molecular machines in organisms, models of production technologies in economic systems, models of conceptual linkages in psychological, scientific ideational, or cultural systems. We surely do not, at this stage, have detailed understanding of such functional couplings among metabolites in organisms, of technological possibilities governing linked production technologies in economies, or among mythic or other elements in psychological or cultural systems. The hope is this: By exploring large tracts of "Grammar Space" we may find rather few "regimes" in each of which the same general behavior occurs in the sets of strings generated by the specific grammar. That is, just as exploration of random Boolean networks has revealed three broad regimes, ordered, complex, and chaotic, so too may exploration of grammar space reveal rather few broad regimes. We can then hope to map these broad regimes onto biological, economic, conceptual, or cultural systems. Thereby we may obtain models of functional couplings among biochemical, technological, or ideational elements without first requiring detailed understanding of the "physics" or true "laws" governing the couplings of functional elements of those diverse systems. We may find, in short, the proper Universality classes.

Next, I discuss two approaches to this task. The first, explores the representation of mapping of strings into strings via infinite Boolean networks. The second, considers the use of random grammars with definable parameters which allow grammar space to be explored.

A NATURAL INFINITE DIMENSIONAL STATE SPACE REPRESENTATION OF THE MAPPING OF STRINGS INTO STRINGS. INFINITE BOOLEAN NETWORKS

One natural representation for strings mapping into strings is an infinite dimensional state space of symbol strings which are finite but of arbitrary length. Consider binary strings of length L, where L can increase up to infinity. Order these in counting to infinity, beginning with the two "monomers" 0 and 1, then the four "dimers" $00, 01, 10, 11, \ldots$, followed by the eight trimers, and so on. At each string length, L, there are 2^L types of strings. This infinite list of string types whose lengths also increase to infinity can be ordered from a starting point, the monomer 0. Create two infinite matrices. The first "input matrix" is ordered such that each column denotes one specific binary symbol sequence, and the columns begin with the monomers at the right-most column of the matrix, the dimers to the left, the trimers to the left of the dimers, and so forth stretching to infinity in the left direction. Thus, each possible symbol string, from short to long, is assigned one column in the input matrix. The second "response" matrix is simply the mirror image of the input matrix. The response matrix lists the monomers in the *leftmost* two columns, the four dimers to their right, etc., stretching to infinity in the rightward direction. The "input" matrix has as its rows all possible combinations of the presence of absence of the possible types of symbol strings, starting with the row $(\ldots 00000)$ on top. There are an infinite number of rows in the input matrix. The cardinality of both the column and the row infinities is, of course, the same.

The positions of "1" values in each row of the input matrix represent which strings are present in that state of the world. The response matrix will show the *next state* of the world as strings act on strings to produce strings. By construction, the input and response matrices are mirror symmetric, hence to read the "next state" that is formed from each input state, the reading must be flipped from right to left for the input matrix and to left to right for the response matrix.

Alternative mappings from the input to the response matrix represent alternative mappings of the set of strings into itself. In order to proceed further, some further definitions are required. Let a machine, M^*, be an ordered collection of M strings. Let an input bundle, I^*, be an ordered set of I strings. The action of M^* operating on I^* will yield an ordered output set of strings, O^*.

Any row of the input matrix has a finite number of "cells" with the value 1, representing the fact that each row represents a unique combination of presence or absence in the "world" of strings up to some length. The possible machines built of these strings might be limited to a specific maximum number of strings, for example, five, or might range up to the finite total number of strings present in that state of the world. Call the latter maximum size machine "unbounded" in the sense that as rows further down the infinite input matrix which represent the presence of yet more symbol sequences, are considered, still more complex machines can be built. Due to the ordered way that the input matrix is constructed, it is possible, given any constraint on which ordered sets of strings count as legitimate machines M^* or input bundles I^*, to uniquely number each machine, M^*, and input bundle I^*. In a moment I shall use such unique numberings to produce a deterministic mapping from the current state of the "world" to the next state.

The choice to include all possible "unbounded" machines as legitimate machines or unbounded input bundles as legitimate bundles specifies the the power set of strings operating on itself as the mathematical entity of interest. This is clearly the widest interpretation. It allows the generation of the maximal number of strings in the presence of a fixed set of strings that is possible under *any* interpretation of the kinetics in which strings collide with one another and act on one another. Other choices are more limited. For example, we might wish to assume that the largest machine any string might enter into was the only machine which formed and acted. In that case, transformations mediated by smaller machines would be inhibited by the capacity to form the larger machine. Here, instead, we assume that all legitimate machines able to be formed are present and carry out their transformations on all possible legitimate input sets. Thus, for the moment I, therefore, choose the widest full power set interpretation of machines and input sets.

If a string is acted upon and transformed, we need to choose whether the initial string remains in the system or not. The natural interpretation is that the string is "used up" or *disappears* in the transformation. (Note in chemistry, the back transformation always occurs. This is not, in general, the case, however.)

With these assumptions, the *next state* of the world is a mapping from the present state, given by some *Boolean functions in the response column*. Such a system can be thought of as a discrete time, autonomous synchronous infinite automaton. The dynamics of this infinite automaton gives the way strings engender strings in the potentially infinite space of strings.

Three further assumptions lead us to a canonical and ordered way to generate a *sensible series of families of transformations.*

1. Note that we can parameterize such transformations by *the largest size machines allowed, M^*, and the largest size input sets to a machine allowed, I^** Thus, the system might at the moment have 1000 strings, but only machines with five or fewer strings, or input bundles I^* with five or fewer strings, might be allowed. In any *row* in the input matrix there are a finite number of sites with "1" values, corresponding to strings present in the system, say N^*. The the maximum number of machines is (N^{*5}). The maximum number of input bundles is similarly (N^{*5}). The product of these is the maximum number of pairs of "machines" and "input bundles." Determinism demands that for each, there is a unique outcome.

2. In order to retain denumerablity, we need a rule which limits the number of output strings, given the number of input strings. Call this a limit on the "output spray," S^*. Given that as well, then given N^* in the input row we have a maximum limit on the number of strings present in the *next* state of the world.

3. Next, and again to retain denumerability, note that we can choose to delimit the *maximum length* of a new string produced by machines whose maximum member length is M, and whose maximum input bundle string length is L, to some finite bound which increases with M, L, or $M + L$. Thus, in the origin of life model from a food set, at each iteration, the maximum string length doubles because one imagines ligating two strings present in the system. Any such bounding choice is a *third parameter*, which in effect creates an *expanding cone* down the rows of the response matrix. The cone asserts that maximum string length can only grow so fast as a function of lengths of strings already present in the system.

Given these bounds, then, we have, for each row of the input matrix, a bound on the maximum number of strings which can be present in the next state of the world, and a bound on the maximum length of those strings. (Note, parenthetically, that a constraint exists between the maximum rate of cone expansion of lengths of strings, and the total number of output strings from any input state of the world. There must be enough possible strings in the "space" allotted to accept the new strings).

THE QUENCHED DETERMINISTIC VERSION: MAPPING THE INFINITE POWER SET OF BINARY STRINGS INTO ITSELF

The mathematical object we are considering in the Boolean idealization where all allowed transformations occurs is really a mapping of the now denumerably infinite power set of N strings into itself. That is, consider a row of the input matrix. It contains a set of N strings. But the power set of ordered pairs of strings, ordered triads of strings, ordered Nads of strings, is just the set of all possible machines,

M^*, constructible from those strings. Similarly the sets of single strings, ordered pairs of strings, etc., are the set I^* of possible input bundles. I^* and M^* are the same power set in the limit when machines and bundles with N strings in them are allowed. As remarked above, identify each unique ordered set of strings which is a machine, M_i, with a unique *number*. Similarly, identify each unique ordered set of strings which is an input bundle, I_i with a unique *number*. Then the pair of numbers, i, j, specify a unique machine input pair, hence must always have a fixed output bundle, $O_{i,j}$. The output, of course, is bounded by the output spray S.

Given this, then it is possible to define for each machine input pair, regardless of *which state of the world it occurs in*, hence which row of the input matrix, a unique output bundle. This assures both determinism, and constraints within the family M, I, and S. Since the number of strings and their lengths comprising machines, input bundles, and output sets are all finite and bounded for any unique machine input pair, it follows that we can generate *all possible finite number and length legitimate output sets* which might be generated by the unique machine and input pair.

Thus, it follows that, in terms of the parameters giving maximum machine, bundle and "output spray" sizes, and cone angle or string length amplification factor, as well as the *deterministic machine input pairs mapping into unique output bundles*, we can consider *all possible* mappings of the infinite set of strings into itself. Hence, within these parameterizations *we can explore all possible dynamical behaviors of this family of systems*

I note that this construction is also effectively for large finite systems beginning with small symbol strings. It is feasible to use the identifying numbers of each machine and input bundle, i, j, as seeds to a random number generator which specifies uniquely the output bundle, $O_{i,j}$. Thus, it is not necessary to hold massive memory files for the mapping between each input state and its successor output state of the world. Such simulation does not seek to study all possible mappings, but a large random sample of mappings for different values of the parameters of the model.

THE ANNEALED MODEL

Given the bounds on the response matrix above, we may consider a simpler "annealed" model which may prove useful. Consider, for each row of the input matrix and the bounds on amplification, output spray and so forth, *all possible ways of filling rows in the response matrix with 1 and 0 values, consistent with those constraints*. Each way corresponds to a well defined transformation of the infinite set of strings into the infinite set of strings, and allows an expanding cone of complexity. However, this model does *not* preserve deterministic dynamics. It is an *annealed* approximation to a deterministic "grammar" or mapping of the infinite set of strings into itself. The lack of determinism is easy to see. Consider two rows of the input matrix in which strings $S1, \ldots, S5$ are present, but in the second of these rows, string $S6$ is also present. By determinism, in the second row, all the machines, input bundles and transformations which might occur in the first row are also present, hence must occur in the corresponding next state of the world. But under the annealed model above, filling the response matrix rows in all possible ways, such determinism is not guaranteed. Instead this model is an annealed approximation to a deterministic dynamics, whose statistical features may prove useful to analyze, as has proved the case with Boolean networks.[3]

The concepts of Jets, Fire Balls, Eggs, Traveling and Fixed Eggs, Filigreed Fogs and Pea Soups, etc., are all clear in either the deterministic or annealed picture as trajectories from a maintained source set (Jets, Mushrooms, etc.) or "free" dynamics (Eggs, Fire Balls, Filigreed Fogs, Pea Soups, etc.).

Since these systems are just infinite Boolean networks explored from an initial state invoking relatively short and few symbol strings, the concepts of *dynamical attractors, ordered, chaotic, and complex behaviors,* carry over directly. But, in addition, in comparison to a *fixed state space*, we have here the idea of attractors in *composition space*, namely the *sets of strings* which comprise the system as well as the dynamical behaviors evidenced among the set of strings, whether eggs, mushrooms or filligreed fogs.

One can begin to guess at the relation between dynamics and composition space. The Boolean idealization shows the set of *all possible transformations* from the current set of strings into the next set of strings. By contrast, in other dynamics not all machines and input bundle pairs at each moment, hence only a subset of all transitions will occur. In particular, as remarked above, one might want to model the presence of an "inhibitor" string which, when present, unites with a machine and *reliably blocks* its action, just as repressor molecules bind to cis acting DNA sites and block transcription. Note that this kind of dynamics does depart from the choice to allow all possible transformations induced by machines on inputs to occur. Were that most general choice made, then in the presence of the inhibitor string *both* the machine without the string and the machine with the string would be present. The former would carry out its prefigured transformation, the latter would not. In the modified dynamics, the presence of the inhibitor string "uses up" all copies of the machine and prevents the transformation which would otherwise occur. The importance is this: Once one allows inhibition of transitions in this way, then the dynamics which occurs can be chaotic, ordered, or complex.

Consider the case in which, on the infinite graph, the graph growth process creates a leaky egg which emits a narrow infinite jet. Will the infinite jet actually occur? The *dynamics itself can control the subset of the composition set explored.* For example, the dynamics of the system might cut off all transformations at the base of the infinite jet, so no strings will actually be formed which flow up and create the jet. Clearly, it will be easier to control this process if the dynamics of the system were in the ordered regime, rather than chaotic. In the ordered regime, all string processes from the egg to the jet entrance might be inhibited. Under chaotic dynamics in the egg, firing of strings at the base of the jet would be hard to prevent. Thus, control of which subset of the composition set actually occurs is clearly more readily done if there are ordered dynamics than if there are chaotic dynamics. But conversely, achievement of ordered dynamics in Boolean networks requires control over the number of inputs per variable, and over the biases in the Boolean functions. Both controls will be easier to maintain in a *finite egg, than in an infinite mushroom, filigreed fogy, or pea soup*. In these cases, the elaboration of feedback connections to each string is roughly unbounded. Thus, these systems are more likely to exhibit chaotic dynamics, and thus to explore fuller reaches of their possible composition set, than will finite eggs with orderly dynamics.

Obviously, finiteness in real physical chemical systems is also controlled by thermodynamics, in economies by the costs of production, aggregate demand, and budget constraints. However, in worlds of ideas, myths, scientific creations, cultural transformations, etc., no such bound may occur. Thus, it is of interest to see how such algorithmic string systems can control their own exploration of their possible composition set by dynamic control over their actual processes.

The generalization to the case with multiple chemostats or "boxes" is obvious. It is equivalent to a set of "linked" Boolean nets, that is, those which share some "external" variables.

RANDOM GRAMMARS

While infinite Boolean systems should prove useful, random grammars may be even more readily studied. Grammars range from simple regular languages to context insensitive and context sensitive to recursively enumerable. These most powerful grammars are known to be as powerful as universal Turing machines. Any grammar can be specified by a list of pairs of symbol strings with the interpretation that each instance of the "right" member of the pair in some "input string" is to be substituted by the corresponding "left" member of the pair. Thus, were the sequences (110011) and (0011) such a pair, then starting with a given input string, any instance of (0011) would be replaced by (110011). Effectively carrying out such a transformation on an initial string requires a precedence order among the pairs of symbol sequences in the grammar, and a means to limit the "depth" to which such substitutions are allowed to occur. For example, replacement of (0011) with (110011) creates a new (0011) sequence. Shall it be operated upon again by the rule? If so, recursion will generate an infinite string by repeated substitutions at that site. If not, the depth has been limited. Limiting depth limits the length of the transformed string with respect to the input string.

Recursively enumerable grammars, which can be defined by a finite list of pairs of symbol strings where the partner on the left can be shorter or longer than the partner on the right, are as powerful as universal Turing machines. Tuning the number of pairs of symbol strings, the lengths of those symbol strings, and their symbol sequence complexity tunes the power and character of the grammar. A further "amplification" parameter specifies by how much and whether always, or on average, substituted symbol sequences are longer than or shorter than the sequence substituted. Additional grammar rules allow strings to be cleaved or to be ligated. In short, a few simple parameters can be used to specify a grammar space. Using them, random grammars within each set of values of the parameters can be chosen and the resulting string dynamics studied.

A simplest approach is this: Use a random set of pairs of strings as the random grammar. Begin with a set of strings, and operate on each of these according to the grammar. Here, however, strings do not act on one another. Stuart Cowan has suggested the same approach.

A more useful approach suggested by Albert Wong, and closely related to Fontana's work, as well as our own origin of life model, is to define grammars of substitutions, gluing and cutting operations, but require that strings contain "enzymatic sites" such that the strings themselves are carriers of the grammatical operators. Thus, if the grammar specifies that string "ab" is replaced by string "cd-dcde," then an "enzyme string" with an "ab" enzymatic site would search target strings for a matching "ab" site, and if found, substitute "cddcde" in the target string at that site. Or, the enzymatic string might cut or glue strings as sites. Clearly, this can be implemented in binary strings, with matching as complements or identities. Any such grammar-chemistry must also make definite choices about the precedence order in which grammar rules are applied, and in which "depth" of recursive substitution or other actions at one site are pursued.

More complex "machine" and input bundle sets can also be built up by generalizing on the idea of enzymatic sites. Real proteins often cooperate by forming multimeric enzymes carrying out the same or even a succession of biochemical transformations. Here the constituent monomer proteins recognize one another and self-assemble within the cell to form the ordered protein aggregate which is the cooperative complex enzymatic machine. Similarly, we might extend the grammar rules to specify how ordered collections of strings self assemble and act as machines or input bundles to yield unique output sets of strings.

The use of grammars is likely to be very important in analyzing the emergence of functional adaptive systems. The Boolean idealization allows the set of all possible next strings to be followed. But it does not readily allow for growth in the numbers of copies of each string, for inhibitory interactions, hence competition, between strings, and so forth. In contrast, just such features emerge readily in models where strings interact with one another via grammatical rules. I return to this below in considering the implications of these ideas for mutualism, community structure and economics.

A major question is the relation between grammars and Boolean world transition matrices. That is, given the Boolean picture and a mapping of input to output matrices, what equivalent set of grammars does each such deterministic mapping correspond to? Are there some mappings which are not statable as grammars of M strings acting on I strings to yield O strings? In short, while all grammars will yield Boolean mappings of the set of strings into itself it is not yet clear to me that each such possible Boolean mapping is derivable from one or more *grammar*.

The relation between grammar complexity and the kinds of objects, jets, eggs, fogs, etc., which arise is a central object for analysis. Some points already seem plausible. A simple grammar may be more likely to give rise to finite eggs. A complex grammar may be more likely to give rise only to infinite mushrooms or filigreed fogs. The reason is intuitively clear. The first finite autocatalytic sets found were the hexamer single stranded RNA and its two trimer substrates, as noted above. The point-point complementarity due to base pairing allows this system to make an exact complement, then itself, in a closed cycle *which need not expand out into sequence space*. It is possible for this autocatalytic set to remain a two-cycle, and finite. Once overlapping "sticky" ends and ligation are allowed, this system can give rise at least to infinite filigreed fogs. Now consider a very complex grammar: the "fixed probability" rule for autocatalytic sets. Here, each string has a fixed probability of catalyzing any reaction. The grammar is complex in the sense that, after catalytic interactions are assigned, the set of "sites" in the enzyme which can be taken to act on sites in substrates is indefinitely complex. In due course in supracritical systems under the probability of catalysis "P" rule, the formation of all strings will be catalyzed, hence this system creates a pea soup. It seems highly likely that the more complex the grammar, the less easy it will be to limit string generation to finite eggs.

THE GROWTH AND ASYMPTOTIC FORM OF MUTUAL INFORMATION AS STRINGS ACT ON STRINGS

The action of strings on strings to produce strings according to a grammar should build up constraints in symbol sequences in the strings which are produced over time. Such constraints should show up in a measure of relations between symbols called mutual information. The mutual information between pairs of symbols S apart is defined as:

$$\sum P_{ab}(S) \log_2 \frac{P_{ab}(S)}{P_a P_b}$$

where P_a or P_b is the frequency of value "a" $= 1$ or "b" $= 0$, in the set of symbol sequences, and $P_{ab}(S)$ is the frequency of symbol value a at position 1 and symbol value b at position 2 at distance S from position 1. $P_{ab}(S)$ is averaged over all pairs of positions S apart in the set of symbol sequences under considerations.

In natural language texts, mutual information, $M(S)$ typically decreases as an inverse power law as S increases.[18] Thus, nearby symbols tend to be more strongly correlated than distant symbols.

Consider now a system of 1000 binary strings, each chosen at random among strings length 100. Because the set is chosen at random, the mutual information between sites at any distance, S, will be 0. Let the strings act upon one another in a chemostat such that 1000 strings are always maintained in the system. As these mutual interactions occur, the action of strings on one another creates correlations, hence mutual information. Preliminary studies with David Penkower in my laboratory at the University of Pennsylvania indicates that, in fact, in these systems mutual information begins very close to 0 and builds up as interactions take place to an asymptotic form which depends upon the grammar. Typically, simpler grammars appear to lead to higher mutual information. Typically, mutual information builds towards an asymptotic form which is high for small values of S and decreases as S increases.

These preliminary results suggest that the time course in which mutual information is built up to the asymptotic form as a function of numbers of string interactions, and that form itself, *give information about the complexity of the grammar.*

Interestingly, one can envision experiments in which random single stranded RNA molecules, perhaps length 100, are allowed to interact with one another. If these mediate ligation, cleavage, and transesterification reactions, as do hexamers and ribozymes, then over time the sequences in the system should build up mutual information as a function of S. This should be testable by using PCR amplification, cloning, and sequencing of the interacting RNA sequences over time. In turn, estimates of grammatical complexity are bulk estimates of enzymatic site complexity as RNA sequences act catalytically on one another. Similar efforts may prove useful for mixtures of initially random polypeptides or other potentially catalytic polymers. While the actual length distribution of RNA polymers in such a system will be sensitive to the thermodynamic factors such as hydration of the environment, it may well be the case that measures of mutual information among nucleotides as a function of distance apart is sensitive only to the grammatical complexity with which RNA sequences act upon one another.

STOCHASTIC GENERALIZATION

The model is fully deterministic. Expand it to include random bit mutations in strings to yield stochastic versions of the same basic model. Note that the grammar rules, applied to strings without reference to the use of other strings as tools, is the analogue of spontaneous reactions occurring without an enzyme, in the autocatalytic polymer set model. Hence, these are the natural form of spontaneous mutations in these systems.

APPLICATIONS TO BIOLOGICAL, NEURAL, AND ECONOMIC SYSTEMS

Random grammars and the resulting systems of interacting strings will hopefully become useful models of functionally integrated, and functionally interacting biological, neural, psychological, technological or cultural systems. The central image is that a string represents a polymer, a good or service, an element in a conceptual system, or role in a cultural system. Polymers acting on polymers produce polymers. Goods acting on goods produce goods. Ideas acting on ideas produce ideas. The aim is to develop a new class of models in which the underlying grammar implicitly yields the ways strings act on strings to produce strings, interpret such production as functional couplings, and study the emergent behaviors of string systems in these various contexts. I consider first some implications for biological models.

Part of my own interest in models of autocatalytic polymer systems, beyond the serious hope that they bear on the origin of life on earth and presumably elsewhere in the cosmos, lies in the fact that such systems afford a crystalline founding example of *functional wholeness*. Given the underlying model of chemical interactions, once an autocatalytic set of catalytic polymers emerges, it is a coherent whole by virtue of achieving *catalytic closure*. Given the underlying model chemistry and catalytic closure, the *functional role* played by each polymer or monomer in the continued existence and proliferation of the autocatalytic set is clear. Note that we here feel impelled, almost required, to begin to use functional language. This reflects the fact that such a self-reproducing system allows a natural definition of the "purpose" of any polymer part subservient to the overarching "purpose," abetted by natural selection, to persist and prevail. In this unconscious sense, an autocatalytic set becomes a locus of agency.

Model autocatalytic sets are natural testbeds to study the emergence of collaborative or competitive interactions. We need merely specify how such systems may export or import strings to one another, and we will find in consequence how they come to cope with such exchanges. As remarked above, such interactions are literally what it means for such systems to come to "know" one another. By studying these properties across grammars, it should be possible to understand how grammar structure, as well as the structure of interacting autocatalytic sets, governs the coupled coevolutionary structures which emerge. The ways model autocatalytic sets build internal models of one another may well mimic the ways *E. coli* and IBM know their worlds. In turn, this may well yield insight into the onset of mutualisms, symbiosis, and competition in the biological realm.

POTENTIAL NEURAL AND PSYCHOLOGICAL IMPLICATIONS

Artificial Intelligence has harbored a long debate between those who favored models of the mind based on sequential inference as exhibited by sequential computer programs, and parallel processing neural networks. The former are widely used in expert systems, in analyses of linguistic and inferential webs, and so forth. Parallel processing neural networks have re-emerged more recently as models of content addressable memories. Here a dynamical attractor is thought of as a memory or as the paradigm of a class. All initial states flowing to that attractor achieve the desired memory or class. Hence, such systems generalize from attractor to basin. Learning consists in sculpting attractor basins and attractors to store desired patterns of neural activity.[9]

Random grammars and the consequent models of strings acting algorithmically on strings to form jets, eggs, mushrooms, or fogs, seem to be a new and useful marriage of the two classes of AI models. Like sequential rule-based models where one action or classification triggers downstream cascades of actions, one string or a set of strings creates downstream cascades of strings. Like parallel-processing networks, here many strings can act on one another in parallel to create jets, mushrooms, eggs, or fogs. Unlike the more familiar AI models models, where the *couplings* among the elementary processes must be *defined by external criteria*, in grammar string models the coupling among those processes is *defined internally* by the grammatical rules which determine how strings generate one another. There is an important sense in which the "meaning" of one elementary process with respect to others is given by local production transformations and the global structure, jet, egg, mushroom and its natural dynamics. In this sense random grammars are quite like Holland's Classifier Systems, in which rules cast as binary strings trigger the firing of other rules, attain a fitness or "strength" dependent upon payoff in a mock economy, and coevolve with other rules by mutation, recombination and selection. In Holland's case, the couplings among the rules are governed by match criteria by which the action part of one rule acts on the message condition part of another rule. The match requirements are under the control of the rules themselves. In contrast, in random but fixed grammars, the couplings among strings is governed by the grammar. The two classes of models can be further melded by allowing the set of strings in the system to control which string pairs themselves count as grammar rules. In short, an extension of the basic framework would *require symbol strings to control the very grammatical rules by which they transform one another.* Jan Hutson made a similar suggestion during discussions at the Santa Fe Institute. The ways in which instantiation of grammatical rules by the dynamics of string transformations modify the kinds of string sets attainable appears to be an interesting question.

Another feature of grammar models is that the set of processes is open and potentially infinite, unlike more familiar parallel-processing models. Such systems may remain perpetually changing, always out of equilibrium, always adapting, rather than falling to simple dynamical attractors.

It is not entirely implausible that such grammar-string models may prove useful in thinking about the "schemas" by which personality elements are constructed. Consider, for example, the stunning phenomenon of multiple personalities. Typically each "self" has only faint or no awareness of the alternative personalities. The situation is rather like a gestalt shift when regarding a Necker cube. When seen in one way, one literally *cannot simultaneously* perceive the cube in the second way. The two are mutually exclusive perceptual organizations of the visual world. It seems of interest to consider an "egg" able to interact with an external world as a kind of "self" which knows and organizes its world in some self consistent way. But the same system may harbor more than one "egg," each mutually exclusive of other eggs, each living in its own self consistent world.

Another feature of these models seems useful and may relate "holism" in science to stability of ego-structures and "centrality" in the web of string processes. First, consider the thesis of holism in science. Suppose I hold the earth to be flat and you hold it to be round. We perform a critical experiment at the seashore watching a ship sail out to sea. I predict it will dwindle to a point. You predict the hull will lapse from sight before the superstructure. Your prediction is confirmed. "The world is round, admit it!" you claim in jubilation. "No," I respond, "light rays fall in a gravitational field, so of course the hull disappears first." The point, first stressed by Quine, is that any hypothesis confronts the world intertwined in a whole mesh of other hypotheses, laws, and statements of initial conditions. Given disconfirming evidence, consistency requires that *some* statement(s) of the premises be abandoned. But we are free to choose which premise we shall abandon and which we shall save. I can "save" my hypothesis that the earth is flat at the price of a very bizzare and convoluted physics. We cannot avoid Quine's point. Typically we choose to save those hypotheses that are the most central to our conceptual web, and give up peripheral hypotheses or claims about initial conditions. But that very choice renders those central claims very hard to refute, indeed, almost true by definition. Now the interesting point to add is that the hypotheses we choose to save are those which is a graph theoretic sense, are central to the conceptual web. Letting a string process create an egg, mushroom or other object connected via string exchange to an outside world, stand as our model of a conceptual framework, that egg entity will have more central and less central elements. If an "egg" is a "self" knowing its world, preservation of self becomes preservation of the central elements in the egg while a peripheral "metabolism" fluctuates into and out of existence. Indeed, one wonders if the concept of resistance in psychotherapy, a phenomenon familiar in practice if hard to quantify, can, in part, be made sense of in terms of preservation of core elements of the egg. One can consistently continue to maintain that the world is flat despite apparently enormous evidence to the contrary.

MODELS OF CULTURAL COHERENCE AND TRANSFORMATION

What did China's leaders know in the summer of 1989? What occurs when an isolated culture comes into contact with a world culture? What constitutes the integration and coherence of a culture and how do new ideas, myths, or production techniques, transform the culture? Just as it is a vast jump from grammar-string models to models of personality structure, so too is it hubris to leap to cultural models. Yet the phenomena feel the same. New strings are injected into an egg. It transforms to something different and coherent, even perhaps stable if unperturbed, another egg, another closed coherent culture. Conversely, modern society is open, explosive, changing, indefinitely expanding in ideas, goods, services, myths. Have we now become culturally supracritical? Can we construct models in which cultures can be stable Eggs, then transform into a different kind of object, a Fog? It seems worth serious consideration.

APPLICATION TO MODELS OF TECHNOLOGICAL EVOLUTION OF ECONOMIC WEBS

Grammar models may prove useful in developing a new class of theories about technological coevolution. Surprisingly, although technological evolution is thought by many to be a major, perhaps the preeminent factor driving modern global economic growth, economists lack a coherent theory of the phenomenon. The problem is that the issue is not merely economic, it is technological. In a sense which requires unpacking, the goods and services in an economy themselves offer new opportunities to invent yet further goods and services. In turn, new goods and services drive older goods and services out of the economy. Thus, the system transforms. For example, the invention of the automobile leds to requirements for a host of other goods and services, ranging from paved roads, traffic lights, and traffic police and courts, to oil refineries, gasoline stations, motels, automobile repair facilities, parts manufacturers, and emission control devices. And the advent of the automobile led to elimination of the horse for most transport. With the horse went stables, public watering troughs, smithies, the pony express, and a host of other goods and services. In grammar-string theories of economies with goods and services modeled as strings, injection of a new string engenders avalanches in which many new strings enter and drives a set of old strings from the structure.

This example states the problem faced by the economist. In order to understand the current "web" structure of the goods and services of an economy, and how that very structure governs its own possibilities of transformation by the invitation to invent new goods which intercalate into the web, transform it, and eliminate other goods, one needs a *theory* for which goods and services "fit" together technologically.

Economists call such "fitting" complementarity. Thus, the nut and bolt are complements, hammer and nail are complements, and so forth. Complements are sets of goods or services which are used jointly to produce a given other good, service, or consumer product. Substitutes are sets of goods which might substitute for one another in a given production technology or consumption good. Screws can substitute for nails, KCl can substitute for ordinary salt at dinner.

It is just such a theory of the ways an economic-technological web governs its own transformation which appears beyond current reach. And it is just such a role that random grammar models may play: If a good or service is modeled as a symbol string, then each grammar and its consequent string transformation rules amounts to a *model* of the technological couplings among goods and services. If broad regimes exist in grammar space which yield similar economic consequences, and one regime maps onto the real world, we may attain a model of the technological couplings among actual goods and services and the consequences for economic growth.

A concrete way to build grammar-string models of economic growth is the following: First, specify a grammar by which strings act on one another to produce strings. The set of strings which are a machine, M, that jointly act on a string or set of strings to produce an output set, are complements. All parts of M are needed to make the product. Alternative strings, or sets of strings which, as input to M, yield the same output set are substitutes. Weaker senses of complements and substitutes arise if output sets which are overlapping but not identical are considered. The machines and their input and output relations are the production technologies of the model economy. The transformations specify the numbers of each type of string required as input or machine part to make a specified number of each kind of output string. Thus, the grammar implies an input-output matrix. To this a formal economic model can add constraints on exogenous inputs to the economy, such as raw material mined from the ground. These might be supplied by a founder set of strings maintained at a constant "concentration." To carry economic analysis further, the utility of each string must be specified. A simple, if arbitrary choice is just the length of each string. Finally, a budget constraint must be specified. This can be the total utility of goods and services now in the economy plus the utility of exogenous input from the founder set. Given these constraints, the "equilibrium" for the current economy specified in terms of the linked set of goods and services is the *ratio* of production of all the goods and services which maximizes the total utility of all the goods and services in the economy subject to the budget constraint. That ratio can also be thought of as the price of the goods relative to one another, taking any single good as the unit.

The *growth* of the economy over time in terms of the introduction of new goods and services can be studied as follows: Start at the current equilibrium, with the current set of goods and services. Use the grammar rules to construct all possible new goods and services derivable by allowing the current goods and services to "act" on one another in all possible ways. This generates all possible new goods which are technologically "next to" those in the current economy. Alternatively, some random or non-random subset of these might be chosen as potential new goods. The "next economy" is constructed containing the potential new goods and services plus all the current goods and services. These new and old goods specify, via the grammar, a new input-output matrix for the economy. The equilibrium of the new economy, as derived from its modified input-output matrix, is then assessed. At that equilibrium, some of the new potential goods may "make a profit," hence be produced at a positive rate. Others may make a loss, hence not be produced at a finite rate. Similarly, some old goods will still make a profit, others will now make a loss. The new economy is comprised only of those old and new goods which jointly make a profit. Hence, cascades of new goods enter the economy, cascades of old goods are driven from it.

ECONOMIC MODELS OF THIS TYPE WOULD SEEM OF INTEREST IN A NUMBER OF REGARDS

First, they actually model the growth of economies due to the growth in niches afforded by goods to create new goods.

Second, phase transitions occur and may model economic take off. Economies which have too few goods may not be supracritical, hence may never take off. But if the initial economy is more complex, or if several economies come into contact and exchange goods and services, the coupled system may jump from one in which each separate economy makes a small finite jet to a supracritical mushroom which explodes into the space of potential goods and services. Hence, this is a model for economic take off, and even, perhaps, a model for the Industrial Revolution.

Third, these model economies can be expected to exhibit enormous *historical contingency* coupled with *law-like behavior*. If, at each stage in the growth of the kinds of goods and services in the economy, a random subset of potential new goods are "tried" and some are accepted, that sampling process will strongly bias the future directions of growth in the economy. Hence, the goods which actually emerge and integrate into the system will become frozen accidents guiding the future evolution of the system in ways similar to biological evolution. Yet the statistics of the process, the size distribution of new sectors of the economy, the numbers of new goods entering, and old goods leaving, the changes in richness of interconnection within the web, may all be stable given membership in a regime or class of grammars.

Fourth, the decidability problems in filigreed fogs and other objects imply that it may be *logically* impossible to deduce that a given good is not ultimately producible from the current technologies. This implies that markets must be *incomplete*.

Fifth, the same failure of decidability may imply that economic agents must, logically, be boundedly rational. Both these points cut at the core of neoclassical economics, hence may invite its extension.

SUMMARY

We are interested in how complex systems come to exist and adapt. Current theories are based on dynamical systems in which the elementary variables of the system may be molecules, genes, organisms, neurons, cognitive elements, or cultural roles. Yet in these theories there is no underlying microscopic theory or account of the actual *functional couplings* among the variables. Rather these couplings must be postulated or defined based on some external criteria which might range from observation to *ad hoc*.

In this seminar, I introduce a new class of models based on random samples from "grammar space." This approach is a generalization of studies of autocatalytic peptide and RNA systems and of Fontana's recent extension of that work to AlChemy, an algorithmic chemistry. In my further generalization, I consider the denumerably ordered power set of symbol strings whose lengths may be arbitrary. Ordered sets of strings act as machines on ordered sets of strings as input bundles to yield unique output bundles of strings. Limitation on the number and length of output strings as a function of the number and lengths of input strings assures that the system, if started with a finite number of finite length strings, remains denumerably infinite at most. Each grammar specifies a unique mapping of this power set into itself.

In this string world new kinds of dynamical objects, here dubbed jets, lightning balls, mushrooms, fixed or traveling eggs, filligreed fogs, and pea soups, arise and inhabit string space. In addition to the compositional character of such sets, we may consider their dynamical behavior. An "egg" a collective identity operator in string space, may recreate its strings in a steady state, limit cycle, or on a chaotic attractor.

Eggs and autocatalytic mushrooms and other sets seem natural models of integrated function and even of agency. In these models, strings of symbols can stand for molecules, neural activities, cognitive elements, goods or services, or cultural roles. Thus such objects may be useful models of prebiotic organisms among which mutualistic or competitive interactions may coevolve, models of parallel processing neural systems, ego-systems, economic systems, and cultural systems. In all cases, we attain models of functional integration and transformation.

The central intellectual agenda is based on the presumption that by analyzing a variety of grammars from regions of a parameterized grammar space, a few broad regimes each with quite characteristic consequences will emerge. Where we can map such generic behaviors onto molecular, organismic, neural, psychological, economic, or cultural data, we may have found the proper functional universality class to explain phenomena in these areas of chemistry, biology, and the social sciences. Perhaps *E. coli* and IBM know their worlds in much the same way.

ACKNOWLEDGMENTS

This work was performed almost exclusively at the Santa Fe Institute. I have bene-fited from discussion with many colleagues. Walter Fontana's paper was a significant stimulation to these ideas. Discussions with Stuart Cowan, Albert Wong, and Jan Hutson from the 1990 Summer School on Complexity at Santa Fe were useful. The economic modeling that derives from the random grammar model grows out of ex-tensive discussions on earlier string models of economic technological webs with the economic "web" group at the Santa Fe Institute, including Luca Anderlini, Ken Ar-row, Brian Arthur, Michele Boldrin, John Miller, José Sheinkman, Hal Varian and others. The most recent work, based on random grammars and economic models is being pursued with Paul Romer, who pointed out the connection between grammar models and input-output matrices. This work was partially funded by NIH, GM, and ONR.

REFERENCES

1. Bak, P., C Tang, and K Wiesenfeld. "Self-Organized Criticality." *Phys. Rev. A.* **38(1)** (1988): 364–374.
2. Cech, T. R. "The Generality of Self-Splicing RNA: Relationship to Nuclear RNA Splicing." *Cell* **44** (1986): 207–210.
3. Derrida, B., and Y. Pomeau. "Random Networks of Automata: A Simple Annealed Approximation." *Europhys. Lett.* **1(2)** (1986): 45–49.
4. Erdos P., and A. Renyi. " On the Random Graphs 1 vol 6." Debrecar, Hungary: Inst. Math. Univ. DeBreceniens, 1959.
5. Erdos, P., and A. Renyi. "On the Evolution of Random Graphs, Publ. No. 5." Math. Inst. Hungar Acad. Sci., 1960.
6. Farmer, J. D., S. A. Kauffman, and N. H. Packard. "Autocatalytic Replication of Polymers." *Physica* **22D** (1986): 50–67.
7. Fontana, W. "Algorithmic Chemistry." In *Artificial Life II*, edited by C. Langton, et al. In press
8. Holland, J. H. "Escaping Brittleness: The Possibilities of General Purpose Learning Algorithms Applied to Parallel Rule-Based Systems." In *Machine Learning II*, edited by R. S. Michalski, J. G. Carbonell, and T. M. Mitchell. Los Altos, CA: Morgan Kauffman, 1986.
9. Hopfield, J. J. "Neural Networks and Physical Systems with Emergent Collective Computational Abilities." *Proc. Natl. Acad. Sci. USA.* **79** (1982): 1554–2558.
10. Joyce, J. "Nonenzymatic Template-Directed Synthesis of Informational Macromolecules." In *Cold Spring Harbor Symposia on Quantitative Biology*, Vol LII, 41–52. Cold Spring Harbor Laboratory, 1987.
11. Kauffman, S. A. "Metabolic Stability and Epigenesis in Randomly Connected Nets." *J. Theoret. Biol.* **22** (1969): 437–467.
12. Kauffman, S. A. "The Large Scale Structure and Dynamics of Gene Control Circuits: An Ensemble Approach." *J. Theoret. Biol.* **44** (1974): 167–190.
13. Kauffman, S. A. "Emergent Properties in Random Complex Automata." *Physica* **10D** (1984): 145–156.
14. Kauffman, S. A. "Autocatalytic Sets of Proteins." *J. Theoret. Biol.* **119** (1986): 1–24.
15. Kauffman, S. A. "The Evolution of Economic Webs." In *The Economy as A Evolving Complex System*, edited by P. W. Anderson, K. J. Arrow, and D. Pines. Santa Fe Institute Sciences in the Sciences of Complexity, 125–146. Reading, MA: Addison Wesley, 1988.
16. Kauffman, S. A. *Origins of Order: Self Organization and Selection in Evolution.* Oxford: Oxford University Press, In press.
17. Langton, C. ed. In *Artificial Life II*. Redwood City, CA: Addison Wesley , In press.
18. Li W. "Measures of Mutual Information in Natural Languages." Santa Fe Institute Preprint Series, 1990.

19. Packard, N. H. "Adaptation to the Edge of Chaos." 1988.

Andreas S. Weigend,† Bernardo A. Huberman‡ and David E. Rumelhart*

†Physics Department, Stanford University, Stanford, CA 94305; ‡Dynamics of Computation Group, Xerox PARC, Palo Alto, CA 94304; *Psychology Department, Stanford University, Stanford, CA 94305

Forecasting Chaotic Computational Ecosystems[1]

We investigate connectionist networks for short-term prediction of time series from computational ecosystems. The networks are able to predict the fraction of agents choosing a given strategy for hundreds of steps forward in time. The eventual forecasting function after several million iterations exhibits a frequency spectrum very similar to the original data. Furthermore, we show how quantities describing the system, such as the Lyapunov exponent, can be estimated via the network solution.

[1]This paper is an abridged version of "Predicting the future: a connectionist approach" by Weigend et al., published in the *International Journal of Neural Systems* 1.[6] The algorithm as well as its application to predicting sunspot data are described in more detail in "Back-propagation, weight-elimination and time series prediction" by Weigend et al.,[7] published in the *Proceedings of the 1990 Connectionist Summer School.*

1. INTRODUCTION

In many instances, the desire to predict the future is the driving force behind the search for laws that explain the behavior of certain phenomena. Examples range from forecasting the weather and anticipating currency exchange rates to Newton's laws of motion.

The ability to forecast the behavior of a given system hinges on two types of knowledge. The first and most powerful one is knowledge of the laws underlying a given phenomenon. When this knowledge is expressed in the form of deterministic equations that can in principle be solved, one can predict the future outcome of an experiment once the initial conditions are completely specified.

A second, albeit less powerful, method for predicting the future relies on the discovery of strong empirical regularities in observations of the system. The motion of the planets, the small amplitude oscillations of a pendulum, or the rhythm of the seasons carry within them the potential for predicting their future behavior from knowledge of their cycles without resorting to knowledge of the underlying mechanism.

There are problems, however, with the latter approach. Periodicities are not always evident, and they are often masked by noise. Even worse, there are phenomena—although recurrent in a generic sense—that seem random, without apparent periodicities.

We analyze a time series of resource allocation in computational ecosystems: data generated by Monte Carlo simulations of such systems were used to make future predictions which were then compared with experiments. Competition for resources in the presence of imperfect information can lead to chaotic behavior, possibly superimposed on periodic behavior.

Since the hallmark of chaos is exponential amplification of uncertainty, accurate predictions arbitrarily far into the future cannot be expected. Nevertheless, connectionist networks of the type introduced by Lapedes and Farber[4] are able to predict the behavior of the system very well for hundreds of steps forward in time.

2. DETERMINISTIC CHAOS AND TIME SERIES PREDICTION

We begin our analysis with a review of some issues of deterministic chaos that are relevant to the prediction of time series. Deterministic chaos is characterized by an exponential divergence of nearby trajectories. For time series prediction, there are two related consequences. On the one hand, since the uncertainty of the prediction increases exponentially with time, chaos precludes any *long-term* predictability (beyond the distribution of the data). On the other hand, it allows for *short-term* predictability: a random-looking series might have been produced by a deterministic system and actually be predictable in the short run. A prediction algorithm for chaotic systems thus has to capture the short-term structure of the time series.

The short-range structure of chaotic behavior can be captured by expressing the present value x_t as a function of the previous d values of the time series itself,[1,2]

$$x_t = f(\text{past values}) = \begin{cases} I\!R^d \to I\!R \\ (x_{t-1}, x_{t-2}, ..., x_{t-d}) \mapsto x_t \, . \end{cases} \tag{1}$$

The vector $(x_{t-1}, x_{t-2}, ..., x_{t-d})$ lies in the d-dimensional time delay space or *lag space*.

This embedding in lag space is to be contrasted to a stronger approach that is possible for non-chaotic systems, where one usually builds a model from first principles, determines boundary conditions and initial values, and finds the solution x_t for all times t,

$$x_t = f(\text{time}) = \begin{cases} I\!R \to I\!R \\ t \mapsto x_t \, . \end{cases} \tag{2}$$

This amounts to a "shortcut in time." Returning to possibly chaotic deterministic systems, a predictive model is specified by three ingredients:

1. Choose an embedding for the time series $\{x_t\}$, i.e., choose a lag space.
2. Approximate past relationships $\{x_t(x_{t-1}, x_{t-2}, ..., x_{t-d})\}$ by a smooth surface. Different approaches in time series prediction differ in the choice of primitives (polynomials, splines, sigmoids, radial basis functions,...) and in the choice between one global fit in lag space vs. many local fits.

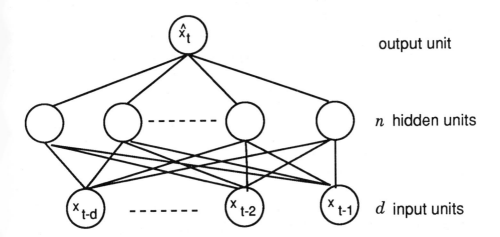

FIGURE 1   Architecture of a feed-forward network with one hidden layer. Units are shown as circles, connections as lines.

3. Choose a cost function that evaluates how well the points are approximated by the surface, such as the sum of the squared differences or more robust error measures.

Then, given the embedding, the primitives and the cost function, find the parameters for the surface that minimize the cost function.

Once the surface has been determined, the prediction for the value following a point in lag space is given by the value of the surface above that point. The problem of prediction, usually framed as *extrapolation* in time, is re-framed for time-invariant systems as *interpolation* in lag space. Following the approach by Lapedes and Farber,[4] we train connectionist networks on examples from the past to find such surfaces.

3 NETWORKS FOR TIME SERIES PREDICTION

3.1 ARCHITECTURE

The networks are feed-forward networks with one hidden layer, as shown in Figure 1. The abbreviation *d-n-*1 denotes the following network:

- The *d input units* are given the values $x_{t-1}, x_{t-2}, \ldots, x_{t-d}$.
- The *n nonlinear hidden units* are fully connected to the input units.
- The linear *output unit* is fully connected to the hidden units, producing the prediction $\hat{x}_t$ as the weighted sum of the activations of the hidden units.
- Output and hidden units have adjustable *biases*.
- The *weights* can be positive, negative, or zero.
- There are no direct connections from input to output that skip the hidden layer.

The nonlinearities are located in the *sigmoid activation function*

$$S_h = S(\xi_h) = \frac{1}{1 + e^{-a\xi_h}} = \frac{1}{2}\left(1 + \tanh\frac{a}{2}\xi_h\right), \tag{3}$$

where ξ_h denotes the input of the network (including bias b_h) into hidden unit h,

$$\xi_h = \sum_{i=1}^{d} w_{hi}x_i + b_h = \vec{w}_h \cdot \vec{x} + b_h. \tag{4}$$

x_i stands for x_{t-i}, the value of input i, and w_{hi} is the weight between input unit i and hidden unit h. The contribution $\vec{x} \cdot \vec{w}_h$ is the projection of the input vector $\vec{x} = (x_1, x_2, ..., x_d)$ on the weight vector $\vec{w}_h = (w_{h1}, w_{h2}, ..., w_{hd})$. The sigmoid performs a smooth mapping $(-\infty, +\infty) \rightarrow (0, 1)$. The slope of the sigmoid, a, can be absorbed into weights and biases without loss of generality and is set to one. The desired overall input-output behavior is obtained by superimposing such sigmoids globally.

3.2 TRAINING

We use the error back propagation algorithm of Rumelhart et al.[5] to train the network: the parameters are changed by gradient descent on the cost surface over the space of weights and biases. In the simplest case, assuming Gaussian noise for the misfit between data and model, the cost function is the total residual variance for a set of examples, $\mathcal{S}$,

$$\sum_{k \in \mathcal{S}} \left(\text{target}_k - \text{prediction}_k\right)^2 = \sum_{k \in \mathcal{S}} \left(x_k - \widehat{x}_k\right)^2 , \qquad (5)$$

where x_k ("target$_k$") is the true value of the time series at time k, and $\widehat{x}_k$ ("prediction$_k$") is the output of the network for time k. This fitting error and describes how well the points $\{x_k, \ k \in \mathcal{S}\}$ are approximated by the surface over the input space. In the simplest case, there are two data sets. The first set is used to train the network, the second set to establish the quality of the prediction.

4 COMPUTATIONAL ECOSYSTEMS

An important example of distributed computation is provided by computer networks in which nearly independent processes operate concurrently with no central controls, with incomplete, sometimes inconsistent information, and with a high degree of communication.

In a recent study of such computational ecosystems, Kephart et al.[3] showed that when computational agents compete for resources by making independent decisions based upon incomplete knowledge and delayed information, their overall behavior can become chaotic. This departure from optimality is generated by the appearance of a strange attractor in the asymptotic dynamics of the system.

In the hope of eliminating this undesirable collective behavior, one would like to endow the computational agents with the ability to predict the future dynamics of the system. They can then adapt and improve their predictions of the behavior of the rest of the system. However, such increased predictability and subsequent modification of their choices may in turn affect the dynamics of the system.

4.1 LEARNING THE TIME SERIES

We investigated the time series generated by the computational ecosystem of Kephart et al.[3] for the use of resources in such a computational ecosystem. It is reproduced in Figure 2. The vertical axis denotes the fraction of agents choosing a given strategy; the horizontal axis corresponds to time in units of the updates. The entire data set consists of 501 data points. It was split arbitrarily in a training set of the first $301 - d$ points of the time series and a prediction set, consisting of the remaining 200 points.

The input dimension d was varied from 6 to 40. The network did not learn the series for $d = 6$, but started to learn for $d = 10$. Improvement was still found by increasing d from 10 to 20. Not much was gained by a further increase to 40. The number of hidden units was varied from 6 to 100. Fourteen sigmoid hidden units were found to be sufficient.

To achieve predictions several steps into the future, the predicted output is fed back as input for the next prediction and all other input units are shifted back one unit. Hence, the inputs consist of *predicted* values as opposed to actual observations of the original time series. The ability to predict the behavior of the computational ecosystem by multi-step predictions is demonstrated in Figure 2.

The 20-14-1 network was trained without seeing any of the target values beyond time 300. It then predicted the values starting at time 301 by iteration, i.e., it used its own predictions as input, rather than the original data.

The bottom panel of Figure 2 shows the increase of the variance with prediction time. The straight line, superimposed on the logarithmic plot, is a simple exponential least squares fit ($\rho = 0.7$) for the prediction variance, σ^2, as a function of the number of iterations, I,

$$\sigma^2 \propto 10^{\,0.016\,I} \propto e^{\,0.036\,I} \propto 2^{\,0.051\,I}. \tag{6}$$

The exponential increase of the prediction error is typical for a chaotic system; the exponent (0.036) gives an upper estimate of the Lyapunov coefficient. This value corresponds to an average information loss of 0.051 bits per iteration.

4.2 EVENTUAL FORECASTING FUNCTION

Despite the exponential loss of predictive power in a chaotic system with prediction time, qualitative features of the long-term behavior can be investigated. The network was iterated several million times. The 300 subsequent values and their frequency spectrum are plotted along with the spectrum of the original data in Figure 3. Note that the spectra turn out to be very similar, although no attempt was made to draw the network's attention to the concept of frequency. The spectra consist of a relatively broad background with some peaks due to the time delays in the computational ecosystem.

In closing, we point out that the solution is stable. Although the linear output unit in principle allows for values outside the unit interval, the output of the trained network remains in the correct range. Even when the iteration is started with random inputs between 0 and 5, the output quickly settles to a behavior very similar to the original one.

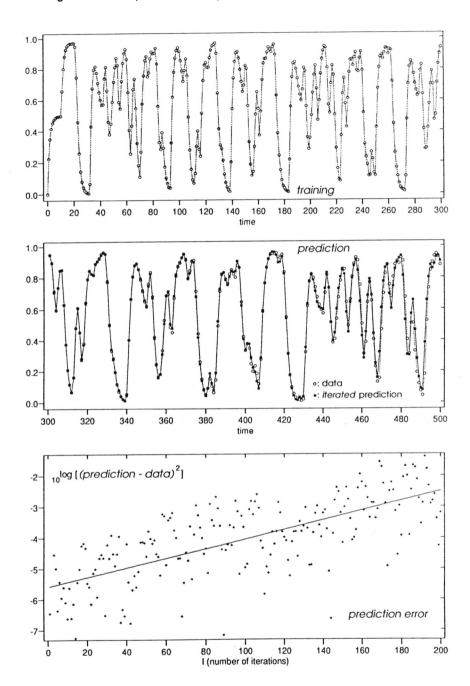

FIGURE 2 Fraction of agents choosing a given strategy vs. time. Top: Training data. Middle: *Iterated* prediction (squares) of a 20-14-1 network and data (circles). Bottom: Prediction error as function of the prediction time. The line corresponds to a least-squares fit.

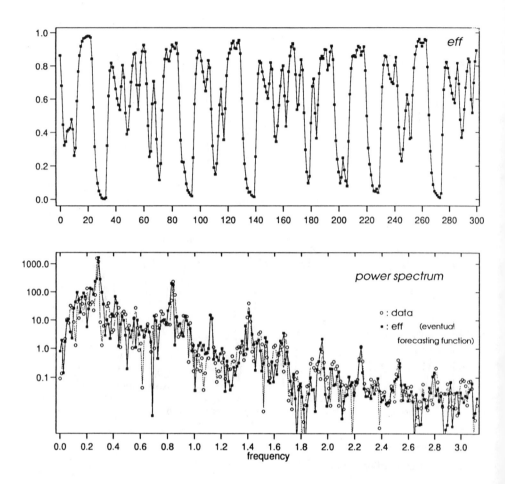

FIGURE 3 Top: eventual forecasting function of the network trained on the ecosystem series. Bottom: frequency spectra (in arbitrary units) of the original series (circles) and the eventual forecasting function (squares).

5 SUMMARY

We investigated connectionist networks for short-term prediction of a time series from a computational ecosystem. They showed excellent predictive ability. We found that the basin of attraction of the solutions was broad. The eventual forecasting function after several million iterations exhibited a frequency spectrum very similar to the original data, although the network was never given any indication of the concept of frequency. Furthermore, we showed how the Lyapunov exponent can be estimated via the network solution.

We are presently investigating training on additional tasks, different cost functions, fully recurrent networks, and architectures for non-stationary time series. The problem of overfitting, serious for short and noisy data, is addressed by Weigend et al.[6,7] Possible further applications of these methods are in economics and finance, protein sequencing, seismic data, nonlinear predictive coding, and music.

We thank Martin Casdagli and Neil Gershenfeld for the discussions during the 1990 Complex Systems Summer School in Santa Fe, New Mexico. This work was supported in part by a grant from the Office of Naval Research (N00014-87-K-0671).

REFERENCES

1. Eubank, Stephen, and J. Doyne Farmer. "An Introduction to Chaos and Randomness." In *1989 Lectures in Complex Systems, SFI Studies in the Sciences of Complexity*, edited by Erica Jen, Lect. Vol. II. Redwood City, CA: Addison-Wesley, 1990.
2. Farmer, J. Doyne, and John J. Sidorowich. "Exploiting Chaos to Predict the Future and Reduce Noise." *Evolution, Learning and Cognition*, edited by Y. C. Lee. Singapore: World Scientific, 1989.
3. Kephart, Jeffrey O., Tad Hogg, and Bernardo A. Huberman. "Dynamics of Computational Ecosystems." *Phys. Rev. A* **40** (1989): 404.
4. Lapedes, Alan S., and Robert M. Farber. "Nonlinear Signal Processing Using Neural Networks: Prediction and System Modelling." Technical Report LA-UR-87-2662, Los Alamos National Laboratory, 1987.
5. Rumelhart, David E., Geoffrey E. Hinton, and Ronald J. Williams. "Learning Internal Representations by Error Propagation." In *Parallel Distributed Processing*, edited by D. E. Rumelhart and J. L. McClelland, 319. Cambridge: MIT Press, 1986.
6. Weigend, Andreas S., Bernardo A. Huberman, and David E. Ruemlhart. "Predicting the Future: A Connectionist Approach." *Interna'l. J. Neural Systems* **1** (1990): 193.
7. Weigend, Andreas S., David E. Rumelhart, and Bernardo A. Huberman. "Back-Propagation, Weight-Elimination and Time Series Prediction." In *Proceedings of the 1990 Connectionist Models Summer School*, edited by D. S. Touretzky, J. L. Elman, T. J. Sejnowski, and G. E. Hinton, 105. Morgan Kaufmann, 1990.

Russel E. Caflisch
Department of Mathematics, University of California, Los Angeles, CA 90024-1555; e-mail
address: caflisch@math.ucla.edu

Singularities for Complex Hyperbolic Equations

Singularities occur in solutions of algebraic equations and partial differential equations and are important for many problems in mathematical physics. A general theory is described here for singularities in the solution of a hyperbolic equations in a complex space variable. The solution is defined on a Riemann surface and the singularities are branch points for this surface. The generic form for a singularity is found to be a square root. There are two types of collisions between singularities. At a transverse collision the singularity remains of square-root type generically. At a tangential collision the singularity is generically of cube-root type.

INTRODUCTION

This lecture will describe singularities in systems of partial differential equations (pde's) for which the dependent and independent variables are complex (as opposed to real). The singularities will be branch points on an appropriate Riemann surface.

Examples of fluid dynamic flows for which singularities arise are the Kelvin-Helmholtz problem for vortex sheets,[3,6,9,10,11,12,13] the Rayleigh-Taylor problem,[15,16] and unsteady Prandtl boundary layers.[14] The most interesting singularity problem for fluid dynamics is the possibility of singularity formation from smooth initial data for the three-dimensional Euler equations.[1,2]

In each of these problems, the singularity would be smoothed out if viscosity, or other smoothing mechanisms, were included. The effects of the singularity would remain, however, even if the singularities would not. These effects include roll-up of vortex sheets in the Kelvin-Helmholtz and Rayleigh-Taylor problems, and onset of separation in unsteady Prandtl boundary layers. Thus, a singularity is an indicator for the onset of complex behavior and serves as a simple localized phenomenon which may be easier to understand than a more realistic, complicated flow.

Singularities are most naturally described through analytic functions and complex variables. The singularity is a branch point at which the analytic function becomes multi-valued. The singularity can be "unfolded" by introducing a "uniformizing variable" in terms of which the function is single valued. The geometric description of this unfolding is the subject of catastrophe theory. We shall use both the analytic and geometric viewpoints in our study of singularities.

SINGULARITIES FOR THE LAPLACE EQUATION

The simplest example of singularities for a pde is in the initial value problem for the Laplace equation in space versus time

$$u_{tt} + u_{xx} = 0 \tag{1}$$

with

$$u(x,0) = u_0(x) \qquad u_t(x,0) = u_1(x) \tag{2}$$

prescribed initially. This problem with both u_0 and u_1 prescribed is well known to be ill-posed.[8] Part of the ill-posedness is that a singularity can form in finite time from initial data u_0, u_1 that is arbitrarily smooth.

Through complex extension of this equation, the singularity formation is easy to understand. Set $x = iy$ and denote $\tilde{u}(y,t) = u(iy,t)$. Then $\partial_x = -i\partial_y$, so that (1) becomes the wave equation

$$\tilde{u}_{tt} - \tilde{u}_{yy} = 0 \tag{3}$$

with initial data

$$\tilde{u}(y,0) = u_0(iy) \qquad \tilde{u}_t(y,0) = u_1(iy). \tag{4}$$

The solution of this equation can be written in the form

$$\tilde{u}(y,t) = \tilde{f}(y+t) + \tilde{g}(y-t) \tag{5}$$

in which $\tilde{f}$ and $\tilde{g}$ and simply related to u_0 and u_1. Denote

$$f(x) = \tilde{f}(-ix) \qquad g(x) = \tilde{g}(-ix) \tag{6}$$

Then

$$u(x,t) = f(x - it) + g(x + it). \tag{7}$$

Now we can easily understand the process of singularity formation for the Laplace Eq. (1). Suppose that f and g are analytic on the real line, i.e., arbitrarily smooth there, but that g has a singularity off in the complex plane at a point $z_0 = x_0 + iy_0$. Then at time $t = y_0$ the solution u given by Eq. (7) will have a singularity at the point x_0.

This can be stated in a way that is more global and geometric. The formula (7) says that the Laplace equation has complex characteristics, which are the lines $z = z_0 \pm it$. Singularities, as well as other features of the solution, will move along those lines. This also shows the advantage of describing singularities in terms of complex variables. For complex space variables there is no difference between elliptic and hyperbolic equations since they both have characteristics. In a later section this result will be generalized to nonlinear hyperbolic systems.

Finally, note that for Laplace's equation the singularities on the real line at later times are of the same type as the initial singularities in the complex plane. In order to understand the generic form of these singularities, we must find the generic form of singularities for analytic functions. This is discussed in the next section.

SINGULARITIES FOR ALGEBRAIC EQUATIONS

Square-root singularities are generic for algebraic equations. Suppose that a function $f(z)$ is defined through an algebraic equation, such as

$$f^n + \varepsilon f = z. \tag{8}$$

For $\varepsilon = 0$ this functions has an nth root; i.e., $f(z) = z^{1/n}$. For $\varepsilon \neq 0$ all of the singularities are of square-root type. This is seen by noting that a singularity is a point at which $\partial_z f = \infty$, i.e., $\partial_f z = 0$. If $\partial_{ff} z \neq 0$, then the singular point is of square-root type; while an nth root is characterized by the conditions

$$\partial_f z = \ldots = \partial_f^{n-1} z = 0 \qquad \partial_f^n z \neq 0. \tag{9}$$

Since $\partial_f z = n f^{n-1} + \varepsilon$ and $\partial_{ff} z = n(n-1) f^{n-2}$ in this example, the only multiple-order singularities occur for $\varepsilon = 0$.

In general the solution f of an algebraic equation will generically have square-root singularities (i.e., branch points), since zeroes of $\partial_f z$ and $\partial_{ff} z$ will only rarely coincide.

Next, consider a collision between two square-root branch points. Although a collision is a non-generic event, it might occur due to some symmetry requirements or other constraints. Collision of two roots of $\partial_f z = 0$ implies that $\partial_{ff} z = 0$, but generically $\partial_{fff} z \neq 0$. Therefore a generic collision of square-root branch points will form a cube-root branch point.

We next claim that the same is true for solutions of complex hyperbolic pde's.

SINGULARITIES FOR BURGERS' EQUATION

The simplest example of a nonlinear hyperbolic equation is Burgers' equation, which is used as a test problem for all nonlinear hyperbolic theories. Burgers' equation is

$$\partial_t f + f \partial_z f = 0 \qquad f(z, 0) = f_0(z). \tag{10}$$

The solution is given implicitly by the formula

$$f(z, t) = f_0(z_0) \tag{11}$$
$$z = z_0 + t f_0(z_0). \tag{12}$$

This says that f is constant and equal to its initial value $f_0(z_0)$ on the characteristic line of points z with initial point z_0 and slope $f_0(z_0)$.

Singularities are points at which

$$\infty = \partial_z f = (\partial_{z_0} f_0)(\partial_{z_0} z)^{-1}. \tag{13}$$

If the initial data f_0 is analytic, then this is possible only if

$$\partial_{z_0} z = 0 \tag{14}$$

which is exactly the condition that the characteristic lines $z(t; z_0)$ have an envelope as the parameter z_0 is varied. At such an envelope the generic behavior is again of square-root type, since the next derivative $\partial_{z_0}^2 z$ will almost always be nonzero.

From Eq. (12) the condition for a singularity is

$$1 + t f_0'(z_0) = 0. \tag{15}$$

If f_0 is real analytic and entire entire in z_0, then for small t the solutions of Eq. (15) must be near ∞. Moreover, since $\overline{f_0(z)} = f_0(\overline{z})$, then solutions of Eq. (15) must come inconjugate pairs. Thus, singularities for Burgers' equation start come from ∞ in the complex plane at $t = 0$. When a singularity appears on the real line, it must correspond to a collision of two singularities, one coming from above the real

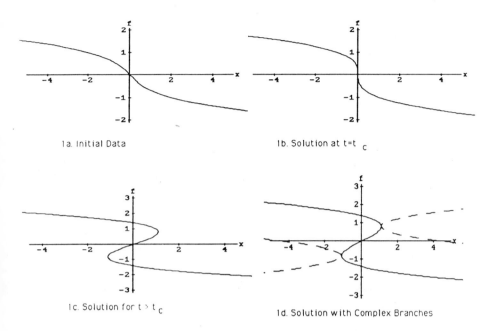

1a. Initial Data

1b. Solution at t=t$_c$

1c. Solution for t > t$_c$

1d. Solution with Complex Branches

FIGURE 1   Solution of Burgers' Equation.

line and the second, its complex conjugate, coming from below. As described in the previous section, at such a collision the generic form of the singularity is a cube root.

The nature of the singularities for Burgers' equation can be understood also from the implicit solution (11) and (12). Consider initial data f_0 that is decreasing as in Figure 1(a). Since points on the solution move at speed f_0, then points on the upper part of the initial curve will overtake points on the lower part. At some time $t = t_c$, the solution $f(x, t_c)$ will have a vertical tangent at some point x_c as in Figure 1(b). At later times $t > t_c$, the curve will turn over on itself as in Figure 1(c), so that $f(x, t)$ will be multi-valued. These turning points x_1 and x_2 are singular points as described above; so that they are actually branch points for the solution. There are branches of the solution that extend beyond these branch points, but on these branches the solution f is complex valued. This is indicated by the dashed lines in Figure 1(d).

The multi-valuedness of the solution and its branch points can also be understood from the description of the solution through characteristics. In Figure 2 starting at time t_c, the real characteristics are shown with their envelopes. Outside the characteristics, the solution f is single valued; while inside it has three real values. When f is analytically extended, it must have three values everywhere, but two of the values will be complex outside the envelopes. The real values of f come

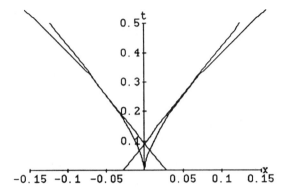

FIGURE 2   Characteristics
(straight lines) and envelopes
for Burgers' Equation.

from characteristics that start on the real line and stay real. The complex values of
f at time t come from characteristics that start off in the complex plane and pierce
through the real line at that time.

The effect of viscosity on complex singularities for Burgers' equation is analyzed
by Bessis and Fournier.[17]

SINGULARITIES FOR A TWO-BY-TWO SYSTEM WITH RIEMANN INVARIANTS

The results above for Burgers' equation can now be generalized to systems of pde's
for which there are exactly two characteristic speeds. First, this will be done for a
system of two equations in diagonal form, i.e., in Riemann invariant form. Later it
will be stated for more general systems.

Consider a system of two equations in Riemann invariant form

$$\partial_t f + \lambda(f, g)\partial_z f = 0 \qquad (16)$$

$$\partial_t g + \mu(f, g)\partial_z g = 0 \,. \qquad (17)$$

For this system singularities will again move along envelopes for the characteristics.
Thus there are two types of singularities, those for the λ-characteristics and those for
the μ-characteristics. The generic form of singularities is again of square-root type,
but now we are also interested in interactions, i.e., collisions, of the singularities.

There are two types of collisions: When two λ singularities (or two μ singularities) collide, they are traveling at the same speed λ (or μ), so that they meet
tangentially. We call this a *tangential collision*. On the other hand, when a λ singularity hits a μ singularity, they will meet transversely (since $\lambda \neq \mu$), and we call
it a *transverse collision*. This is indicated in Figure 3.

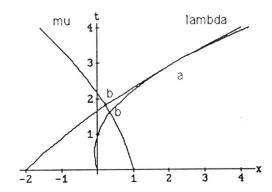

FIGURE 3 Characteristics for
2×2 System with Tangential (a)
and Transverse (b) collisions.

The generic behavior of f and g at these collisions can be analyzed using a generalization of the hodograph transformation[7] in which the independent variables x, t are replaced by new independent variables marking the characteristics. The result from this is that at a tangential collision the generic form of the singularity is a cube root, as for a collision of singularities in Burgers' equation. This is to be expected since for singularities on the same characteristic family, the local variation of the other characteristic family could be ignored.

On the other hand, at a transverse collision, the generic singularity form is a square root. This is rather surprising since the square-root form is not generic for singularities of algebraic equations, as described in section 3. The reason for the square-root behavior for the transverse collision is that the differential equation entails some constraints that force the singularity to be special at a transverse collision.

Although this result can be proved analytically, it is most easily seen geometrically by looking at the relevant catastrophe surface. For the collision of two square roots, the catastrophe surface is the nondegenerate elliptic. We can show that all other singularity types are either unrealizeable or unstable. By saying that other types of singularities are unstable, we mean that if some solution has such a singularity, then after a slight change in initial data or in the coefficients λ and μ the singularity will be changed to a stable type.

GENERAL SYSTEMS WITH TWO CHARACTERISTIC SPEEDS

Finally, we describe a generalization of these results for systems of equations for which there are exactly two characteristic speeds. Consider the system

$$\partial_t F + M(F)\partial_z F = 0 \tag{19}$$

in which F is a vector and $M(F)$ is a matrix. The characteristic speeds for Eq. (19) are the eigenvalues of the matrix M. The generic behavior of singularities for this system is stated in the following theorem:

THEOREM. (Singularities for Complex Hyperbolic PDE's).[5]

Suppose that the system (19) satisfies the following conditions:

1. F is a nearly constant.
2. M(F) is analytic.
3. M(F) has exactly two (or one) distinct eigenvalues λ and μ and a full set of eigenvectors.

   Then

4. The generic form of singularities for F is of square-root type.
5. Singularities for F move on envelopes of characteristics; in particular a singular point z_s moves at velocity either λ or μ.
6. The generic form of F at a collision of singularities is square-root type for a transverse collision and cube-root type for a tangential collision.

The precise meaning of the term "generic" is made clear by Caflisch, Hou, and Ercolani.[5] The proof of this theorem is based on the abstract Cauchy-Kowalewski theorem[4] and on a modification of the hodograph transformation, which causes the restriction to two speeds. The behavior of systems with more than two speeds is an open question. Also the analysis of this theorem is based on assumption of singularities in the initial data. Another outstanding problem is how singularities would form from entire initial data. For example, a singularity might form at points z where $\lambda = \mu$.

ACKNOWLEDGMENTS

AMS Classification 35L67, 14B05. Research supported in part by the Air Force Office of Scientific Research under grant number AFOSR 90-0003.

REFERENCES

1. Beale, J. T., T. Kato, and A. Majda. "Remarks on the Breakdown of Smooth Solutions for the 3-D Euler Equations." *Comm. Math. Phys.* **94** (1984): 61–66.
2. Brachet, M. E., D. Meiron, S. Orszag, B. Nickel, R. Morf, and U. Frisch. "Small-Scale Structure of the Taylor-Green Vortex." *J. Fluid Mech.* **130** (1983): 411–452.
3. Caflisch, R. E., and O. F. Orellana. "Singularity Formulation and Ill-Posedness for Vortex Sheets." *SIAM J. Math. Anal.* **20** (1988): 293–307.
4. Caflisch, R. E. "A Simplified Version of the Abstract Cauchy-Kowalewski Theorem with Weak Singularities." *Bull. AMS* **23** (1990): 495–500.
5. Caflisch, R. E., T. Y. Hou, and N. Ercolani. "Multi-Valued Solutions of Nonlinear Hyperbolic Systems." 1990, in preparation
6. Caflisch, R. E., T. Y. Hou, and N. Ercolani. "Singularity Formation on Vortex Sheets." 1990, in preparation.
7. Courant, R., and K. O. Friedrichs. *Supersonic Flow and Shock Waves* Wiley.
8. Garabedian, P. R. *Partial Differential Equations.* Wiley, 1964.
9. Krasny, R. "On Singularity Formation in a Vortex Sheet and the Point Vortex Approximation." *J. Fluid Mech.* **167** (1986): 65–93.
10. Meiron, D. I., G. R. Baker, and S. A. Orszag. "Analytic Structure of Vortex Sheet Dynamics, Part 1., Kelvin-Helmholtz Instability." *J. Fluid Mech.* **114** (1982): 283–298.
11. Moore, D. W. "The Spontaneous Appearance of a Singularity in the Shape of an Evolving Vortex Sheet." *Proc Roy. Soc. London A* **365** (1979): 105–119.
12. Moore, D. W. "Numerical and Analytical Aspects of Helmholtz Instability." *Theoretical and Applied Mechanics, Proc. XVI ICTAM*, edited by Niordson and Olhoff, 629–633. North-Holland, 1984.
13. Shelley, M. J. "A Study of Singularity Formation in Vortex Sheet Motion by a Spectrally Accurate Vortex Method." *J. Fluid Mech.* 1989, to appear.
14. Van Dommelen, L. L., and S. F. Shen. "The Spontaneous Generation of the Singularity in a Separating Laminar Boundary Layer." *J. Comp. Phys.* **38** (1980): 125–140.
15. Pugh, David. "Development of Vortex Sheets in Boussinesq Flows—Formation of Singularities."Ph.D. Thesis, Imperial College, 1989.
16. Siegel, M. "An Analytical and Numerical Study of Singularity Formation in the Rayleigh-Taylor Problem." Ph.D. Thesis, New York University, 1989.
17. Bessis, D., and J. D. Fournier. "Pole Condensation and the Riemann Surface Associated with a Shock in Burgers' Equation." *J. Physique Lett.* **45** (1984): L833–L841.

III Student Contribution

Kevin Atteson
GRASP Lab, University of Pennsylvania, 3401 Walnut Street, Suite 3006, Philadelphia, PA 19104

An Overview of the Minimum Description Length Principle

1. INTRODUCTION

In this report, we give an overview of the Minimum Description Length (MDL) principle for statistical inference proposed by Rissanen.[2,3] The MDL principle follows the general philosophy for inductive inference put forth by algorithmic complexity theory.[4] In algorithmic complexity theory, we model a string of data by a smallest description of that data, or, in other words, the smallest program which outputs the data. However, it can be shown that no algorithm can find a smallest program which outputs the data.

With the MDL principle, we restrict attention to a class of probabilistic models rather than all possible algorithms as in algorithmic complexity. Given a model class, we choose the specific model which allows the data to be described most concisely. In order to describe the data in a decodable way using a model from the model class, it is necessary to encode information so that the specific model chosen can be determined by the decoder.

Let the sequence of data be denoted by $x^n = (x_1, \ldots, x_n)$.[1] The class of models is parametrized by a k-vector of parameters $\theta^k = (\theta_1, \ldots, \theta_k)$ from some set Θ^k.

[1] x^0 is the empty sequence of data.

The models are given by the distributions $f_{\theta^k}(x^n)$.[2] By Shannon's first coding theorem, a sequence, x^n, with probability $f_{\theta^k}(x^n)$, can be encoded in length:

$$-\log_2(f_{\theta^k}(x^n)) = -\sum_{i=0}^{n-1}\log_2 f_{\theta^k}(x_{i+1}|x^i)$$

as n goes to infinity. In order to minimize the description length, we choose the value of θ^k which minimizes the above. Since each model has the same number of parameters, we assume that each model requires the same number of bits to encode and so we can ignore it in the minimization. In this case, for a fixed number of parameters, this is equivalent to the maximum likelihood method of estimation.

Now consider an extended model class in which the number of parameters, k, can also vary. In this class of models, the number of parameters is a parameter and the distribution of the models is $f_{(k;\theta^k)}(x^n)$. Suppose also that the models are nested, that is, all distributions with k parameters can be represented as distributions with $k+1$ parameters:

$$\{f_{k;\theta^k)}(x^n):\theta^k\in\Theta^k\}\subset\{f_{(k+1;\theta^{k+1})}(x^n):\theta^{k+1}\in\Theta^{k+1}\}.$$

Many standard methods of statistical inference are inappropriate for nested models. We consider the maximum likelihood method of estimation. Since the models are nested, the likelihood of some model with $k+1$ parameters will be at least the likelihood of any model with k parameters. Thus, maximum likelihood typically chooses the largest number of parameters. In the extreme case, when one of the models assigns probability 1 to the observed data, maximum likelihood will always choose this model. However, MDL was specifically designed to handle such nested model classes in a uniform manner. With MDL, we must encode the parameters of the model as well as encoding the data within that model. In the case where a model gives probability 1 to the data, there must be a model in the class for each possible sequence of data since the model class is chosen before the data is observed. Thus, encoding the model is no better than encoding the data in the first place and so this model has no worth in the MDL setting.

2. THE PREDICTIVE COMPLEXITY

Now consider the use of MDL with nested models as described above. In order to encode enough information so that the model is decodable, we proceed in a predictive manner as follows. We encode each element of data x_{t+l} by choosing a

[2] In practice, the data space must be discretized.

model for it based on the preceding data x^t. Given a fixed number of parameters k, we choose the value of θ^k which minimizes the description length:

$$\hat{\theta}^k(t) = \arg\min_{\theta_k} \left(-\sum_{i=0}^{t} \log_2 f_{(k;\theta^k)}(x_{i+1}|x^i) \right).$$

We then choose the value of k which would have produced the minimum description length for the past data:

$$\hat{k}(t) = \arg\min_{k} \left(-\sum_{i=0}^{t} \log_2 f_{(k;\hat{\theta}^k(i))}(x_{i+1}|x^i) \right).$$

The model $(\hat{k}(t); \hat{\theta}^k(t))$ is chosen at time t. The total description length is:

$$-\sum_{i=0}^{t} \log_2 f_{(\hat{k}(i);\hat{\theta}^k(i))}(x_{i+1}|x^i).$$

The description length given above is called the predictive complexity and can be used to compare different model classes.

3. AR PROCESSES

We now consider MDL with the model class of Gaussian autoregressive (AR) processes. An AR process of order k is obtained from a white-noise process (independent and identically distributed) by filtering with a *stable* infinite-impulse response filter. An infinite-impulse response filter is a system which given an input sequence $\{e_t\}$, yields an output sequence $\{x_t\}$ such that:

$$x_t = \sum_{j=1}^{k} \theta_j x_{t-j} + e_j.$$

The filter is stable if for any bounded input sequence, the output sequence is also bounded. For an AR process, the input sequence, $\{e_t\}$, is a white-noise sequence. We consider Gaussian AR processes in which e_t is Gaussian with unit variance.

In order to choose an AR model of order k, $\hat{\theta}^k(t)$, for a sequence of data, x^n, we minimize:

$$\hat{\theta}^k(t) = \arg\min_{\theta^k} \left(\sum_{i=0}^{t} \left(x_{i+1} - \sum_{j=1}^{\min(k,i)} \theta_j x_{i+1-j} \right)^2 \right).$$

This is the least squares criterion. Care must be taken to ensure that the chosen model is stationary, that is, that the filter is stable. We then choose $\hat{k}(t)$ to minimize:

$$\hat{k}(t) = \arg\min_k \left(\sum_{i=0}^{t} \left(x_{i+1} - \sum_{j=1}^{\min(k,i)} \hat{\theta}_j(i) x_{i+1-j} \right)^2 \right).$$

This is the sum of the actual squared errors known as the accumulated prediction errors.[1]

4. CONCLUSION

The MDL principle is a method of statistical inference in which we choose the model, from some class of models, which allows the data to be described as concisely as possible. The MDL principle handles nested models in a uniform manner. The predictive complexity gives a method of coding the data within the model in order to choose models when iterative prediction is required. The predictive complexity also gives an objective criterion for comparison of model classes. The predictive complexity can be applied to Gaussian AR processes and yields an inference procedure which is equivalent to least squares for each fixed number of parameters.

REFERENCES

1. Cover, Thomas M., Peter Gacs, and Robert M. Gray. "Kolmogorov's Contributions to Information Theory and Algorithmic Complexity." *The Annals of Probability* **17(3)** (1989): 840-865.
2. Rissanen, Jorma. "Order Estimation by Accumulated Prediction Errors." In *Essays in Time Series Analysis*, edited by J. Gani and M. B. Priestley, 55–61. Sheffield, England: Applied Probability Trust, 1986.
3. Rissanen, Jorma. "Stochastic Complexity and Modeling." *The Annals of Statistics* **14(3)** (1986):1080–1100.
4. Shannon, C. E. "A Mathematical Theory of Communication." *Bell Systems Technical Journal* **27** (1948): 379–423.

Joseph L. Breeden†and Alfred Hübler‡
Center for Complex Systems Research–Beckman Institute, and the Physics Department, University of Illinois, 405 North Mathews Avenue, Urbana, IL 61801; email address: †breeden@complex.ccsr.uiuc.edu and ‡alfred@complex.ccsr.uiuc.edu

Equations of Motion from Data with Hidden Variables

In the area of nonlinear modeling, many competing techniques have been developed with various advantages and disadvantages. Reconstructions with delay coordinates,[9] state space partitioning for local fitting,[5,6,8] and neural nets[7] are among the commonly used methods. The most obvious difficulty in using any of these is the issue of interpreting the results physically. If the goal of modeling is forecasting alone, any approach is acceptable given that the predictions are accurate. However, the primary motivation for the technique developed here is to increase knowledge about the dynamics of the system being studied. Because of this, the reconstruction of equations of motion with hidden variables will utilize the *trajectory method* as developed by Cremers and Hübler[2] and Eisenhammer et al.,[4] similar to that of Crutchfield and McNamara.[3]

The trajectory method was developed for situations where all the variables required for an embedding of the dynamics can be measured. Unfortunately, this is almost never the case in real experiments. The experimentor frequently cannot measure one or more of the variables, i.e., some variables are *hidden*. Within the framework of the trajectory method, hidden variables and any preknowledge of the dynamics may be easily incorporated. This procedure works for models based upon coupled maps or sets of ordinary differential equations. For purposes of discussion, the technique is outlined for ODE's. The modification to maps is straightforward.

To explain the reconstruction of the dynamics of systems with hidden variables, we assume that only one variable, w, is hidden, i.e, $N_h = 1$; and our experimental data, $\mathbf{x}$, contains N_o observables, $\mathbf{x} \in R^{N_o}$. The dynamics throughout the state space are represented by a single, globally deterministic trial model $\dot{y}_i(t) = f_i(\mathbf{y}(t); \mathbf{c})$, $i = 1, \ldots, N$, $\mathbf{y} \in R^N$ where $N = N_o + N_h$ is assumed sufficient to embed the dynamics. The functions $\{f_i\}$ may be of any form, but are often taken to be a series expansion in $\mathbf{y}$. The goal is to find the best coefficients, $\mathbf{c}$, to reproduce the experimental data. In some situations the form of the functions $\{f_i\}$ is known, but the coefficients are unknown. This added information greatly simplifies the modeling process.

To optimize $\mathbf{c}$, trial values are used to predict $\mathbf{y}(t_{n+1})$ given $\mathbf{x}(t_n)$ from

$$y_i(t_{n+1}) = x_i(t_n) + \int_{t_n}^{t_{n+1}} f_i(\mathbf{x}(t'), w(t'); \mathbf{c}) \, dt', \quad i = 1, \ldots, N. \tag{1}$$

The value of the hidden variable $w(t_n)$ is thus required. We can solve for $w(t_n)$ as $f_i(\mathbf{x}(t_n), w(t_n); \mathbf{c}) - x_i(t_{n+1}) = 0$, $i = 1, \ldots, N_o$. Having one hidden variable, only one of these equations is needed to solve for $w(t_n)$, but since we do not expect our first guess for $\mathbf{c}$ to be correct, each of the N_o equations is solved, each possibly producing multiple solutions. To be accepted, these roots are required to be real and to satisfy any physical bounds upon their value. Aside from these constraints, there is no *a priori* method to determine which root is correct. Therefore, each is tried in turn with the best root chosen according to predictive accuracy.

Using one such $w(t_n)$, $w(t_{n+1})$ is determined from Eq. (1). The predicted values $\mathbf{y}(t_{n+2})$ are then calculated from $(\mathbf{x}(t_{n+1}), w(t_{n+1}))$ and compared to $\mathbf{x}(t_{n+2})$. The error in the model is thus

$$\chi_\nu^2 = \frac{1}{N_o(M-2) - N_c} \sum_{i=2}^{M} \sum_{j=1}^{N_o} \frac{1}{\sigma_{ij}^2} (y_j(t_i) - x_j(t_i))^2 \tag{2}$$

where N_c is the number of free coefficients, M is the number of data points, and σ_{ij} is the experimental error in the jth vector component of the ith measurement. The task of finding the optimal model parameters has now been reduced to a chi-squared minimization problem. Thus, the best parameters are determined by $\partial \chi_\nu / \partial c_i = 0, \forall i$. Therefore, the ability to determine these coefficients rests upon the strength of the algorithm employed to search through the space of parameters. Since this has been formulated as a standard χ_ν^2 problem, all of the normal statistical tests can be applied. Typically, $\chi_\nu \simeq 1$ implies that the modeling was successful; however, more sophisticated tests can be applied as well, e.g., F-test, etc. In the presence of noise, several local minima with roughly equal χ_ν may occur; the false minima typically corresponding to a piecewise fit of the data. These false minima can often be discriminated by checking that the corresponding model is stable over the time scale of the experimental measurements.

The above analysis can be extended simply to handle an arbitrary number of hidden variables. If we have at least as many observables as hidden variables,

$N_o \geq N_h$, then rather than solving one of the model equations for a single $w(t_n)$, we must simultaneously solve N_h of the model equations for $\mathbf{w}(t_n), \mathbf{w} \in N_h$. Once $\mathbf{w}(t_n)$ has been determined, the analysis proceeds exactly as before. When $N_h > N_o$, we cannot generate enough equations from the first N_o model equations using only $\mathbf{x}(t_n)$ and $\mathbf{x}(t_{n+1})$ to uniquely determine the N_h initial conditions of $\mathbf{w}(t_n)$. However, additional observations $\mathbf{x}(t_{n+2}), \mathbf{x}(t_{n+3}), \dots$ can be utilized to generate the needed equations. This succeeds because the functional relationship between $\mathbf{w}(t_n)$ and $\mathbf{w}(t_{n+1})$ is known from the model equations. In general, $N_h([N_h/N_o] + 1)$ equations and at least $[(N_c + N_h)/n_o] + 1$ observations are necessary where [ ] indicates the greatest integer.

Unfortunately, the hidden variable reconstruction method detailed above is not a panacea. For some models, it may not be possible to uniquely determine the coefficients and hidden variables simultaneously. The following is an example of a model which does not exhibit unique solutions when both ε and $w(n)$ are unknown.

$$
\begin{aligned}
y(n+1) &= 2y(n) + \varepsilon w(n) , \\
w(n+1) &= w(n) .
\end{aligned}
\tag{3}
$$

Fortunately, prior to the modeling process a given model can be checked to confirm that it has unique solutions, assuming that the model contains no extraneous terms. This is done by computing the Wronskian matrix for the equations required for an analytic solution with respect to the free coefficients and hidden variables. If the

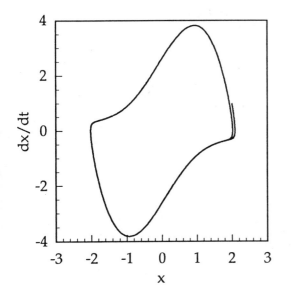

FIGURE 1 The experimental data is shown for the van der Pol oscillator with $\eta = 2.0$, i.e., Eq. 4 with $\alpha = 0.0$. This data was used to test the hidden variable reconstruction with $q(t)$ assumed hidden.

determinant of this matrix is not zero, the model will have unique solutions. This method is discussed in detail by Breeden and Hübler.[1]

To demonstrate the application of the hidden variable reconstruction, data was generated for the van der Pol equation, Eq. 4, with $\eta = 2.0$ and $\alpha = 0.0$.

$$\dot{p} = q \, ,$$
$$\dot{q} = \eta \left(1 - p^2\right) q - p + \alpha \, . \tag{4}$$

Clearly, α is an extraneous parameter. It is included only to demonstrate that spurious terms can be isolated during modeling. In Figure 1 the original system is shown. For the modeling, q was assumed to be a hidden variable. The value of $\dot{p}$ was computed from a second-order piecewise fit to $p(t)$. This was used to solve for the hidden variable and subsequently the unknown parameters. The χ_ν landscape in η and α is shown in Figure 2, revealing a single minimum corresponding to the correct parameter values. Note that $\chi_\nu \neq 1$, because of additional noise introduced during the calculation of the derivatives. These errors should be included in σ_{ij}, but were left out to illustrate the effect.

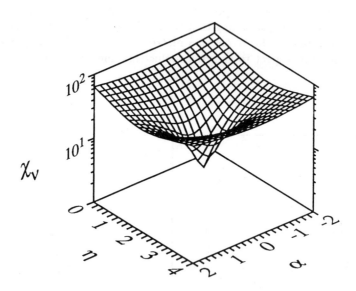

FIGURE 2   The χ_ν landscape for Eq. 4 is shown with η and α as unknown parameters. The minimum coincides with the proper value of η and also reveals that $\alpha \simeq 0.0$ indicating that it is an extraneous term in the model. $\chi_\nu \neq 1$ at the minimum because of errors arising from calculating $\dot{p}(t)$. Those errors should be incorporated into σ_{ij}, but were not so as to illustrate the effect.

The preceding analysis demonstrates that it is possible to reconstruct hidden variables in a dynamical system directly from the observables by simultaneously reconstructing the equations of motion for the system. The greatest advantage of this technique is that the resulting models can be interpreted physically so as to understand, as well as forecast, the dynamics. Moreover, this technique has been found to be applicable in the presence of noise, especially for models based upon ODE's. Since the resulting models are inherently deterministic, a successful model also indicates the embedding dimension of the experimental data.

REFERENCES

1. Breeden, J. L., and A. Hübler. "Reconstructing Equations of Motion Using Unobserved Variables." *Phys. Rev. A* **42** (1990): 5817–5826.
2. Cremers, J., and A. Hübler. "Construction of Differential Equations from Experimental Data." *Zeit. Naturforsch.* **42a** (1986): 797–802.
3. Crutchfield, J. P., and B. S. McNamara. "Equations of Motion from a Data Series" *J. Complex Sys.* **3** (1987): 417–452.
4. Eisenhammer, T., A. Hübler, N. H. Packard, and J. A. S. Kelso. "Modeling Experimental Time Series with Ordinary Differential Equations." Technical Report CCSR-89-7, Center for Complex Systems Research, 1989.
5. Farmer, J. D., and J. J. Sidorowich. "Exploiting Chaos to Predict the Future and Reduce Noise." In *Evolution, Learning, and Cognition*, edited by Y. C. Lee, 277. Singapore: World Scientific Press, 1988.
6. Meyer, T. P., F. C. Richards, and N. H. Packard. "Learning Algorithm for Modeling Complex Spatial Dynamics." *Phys. Rev. Lett.* **63** (1989): 1735–1738.
7. Müller, B., and J. Reinhart. *Neural Networks: An Introduction.* Berlin: Springer-Verlag, 1990.
8. Packard, N. H. "A Genetic Learning Algorithm for the Analysis of Complex Data." *Complex Systems* **4** (1990): 543.
9. Packard, N. H, J. P. Crutchfield, J. D. Farmer, and R. S. Shaw. "Geometry from a Time Series." *Phys. Rev. Lett.* **45** (1980): 712–716.

John C. Crepeau
Department of Mechanical Engineering, University of Utah, Salt Lake City, UT 84112

Spectral Entropy and Self-Organization

In many complex systems, self-organizing behavior is observed. These systems include neural networks, sand piles, chemical processes, and fluid flows. This type of behavior can be observed in everyday phenomena. A precise, universal definition of self-organization, however, remains elusive. Spatial patterns are created in neural networks; sand piles governed by local rules remain at some angle of repose; waves appear spontaneously when certain chemicals are mixed together; and two- and three-dimensional vortex structures are created from instabilities in laminar flows. Clearly, we observe interesting behavior. Can we somehow quantify it?

A natural order parameter is the entropy, proposed by Gibbs over one hundred years ago. In classical thermodynamics, the entropy is defined only for equilibrium states, and these states, when separated from external forcing parameters, approach an entropy maximum. In non-equilibrium processes, however, the entropy may fluctuate and have local minima. During these non-equilibrium situations, self-organization may occur.

In order to measure the level of self-organization in certain systems, one can use the spectral entropy[4] introduced by Powell and Percival. The spectral entropy is a convenient parameter for quantifying self-organization in time series or power-spectra data. It takes the usual form of the entropy,

$$S = -\sum_k P_k \ln(P_k).$$ (1)

However, the P_k terms are defined by

$$P_k = \frac{|f_k|^2}{\sum_{k'} |f_{k'}|^2},$$ (2)

where $|f_k|^2$ is the value of the power spectra at some wavenumber ($\propto$ frequency). The P_k term gives a measure of the energy distributed throughout the system.

Equation 2 is qualitatively similar to the classical thermodynamic probability distribution. In classical thermodynamics, the numerator of the Boltzmann probability is proportional to the energy at some state and the denominator is the partition function, a sum over all energy states of the system.

One of the advantages of the spectral entropy is its relative ease of computation. It is computed from the power spectra, which may come from a wide variety of experiments in many different fields. There need not be vast amounts of data points in order to compute the spectral entropy. If the power series contains only a few dominant spikes, the spectral entropy will be low; a broad-banded power spectra is characterized by a large spectral entropy.

In addition, the spectral entropy characteristics follow our intuitive notions of Gibbsian entropy: ordered structures have a low spectral entropy and a structure-less time series possesses a higher spectral entropy. For example, convection cells described by the Lorenz model for Rayleigh-Bénard convection[1] and vortex bursts in free shear layers[2] have a lower spectral entropy than their equilibrium or laminar states. Preliminary results also show that the spectral entropy decreases in the activity of neural networks as the synaptic order increases.[3]

The spectral entropy gives a quantitative measure of the power spectra. A power spectra where the energy is concentrated in a few wave numbers implies a higher information content and yields a lower spectral entropy. The spectral entropy may also be used to compare the localized order of a structure within a time series. If data suggests small regions of organization, one may compute the spectral entropy of the structure and compare its order with the surrounding flow.

A portion of this work was funded through the Thiokol Inc. University IR&D program under the direction of Professor W. K. Van Moorhem.

REFERENCES

1. Crepeau, J. C., and L. King Isaacson. "On the Spectral Entropy Behavior of Self-Organizing Processes." *J. Non-Equilib. Thermo.* **15** (1990): 115–126.
2. Crepeau, J. C., and L. K. Isaacson. "Spectral Entropy as a Measure of Self-Organization in Transition Flows." In *Proc. NATO Advanced Research Workshop on Self-Organization, Emerging Properties and Learning*, 12-14 March 1990, Austin, Texas, edited by A. Babloyantz. Plenum Press, to appear.
3. McGuire, P. Private communication, June 1990, Santa Fe, New Mexico.
4. Powell, G. C., and I. C. Percival. "A Spectral Entropy Method for Distinguishing Regular and Irregular Motion of Hamiltonian Systems." *J. Phys. A: Math. Gen.* **12** (1979): 2053–2071.

Daniel R. Greening
Computer Science Department, University of California, Los Angeles, CA 90024-1597;
email: dgreen@cs.ucla.edu

Asynchronous Parallel Simulated Annealing

The simulated annealing algorithm can approximately solve many difficult combinatorial optimization problems. To alleviate its long execution time, researchers have parallelized it. However, exploiting more than a few processors is difficult without fundamentally altering the algorithm's properties. One successful approach eliminates some interprocessor synchronization, increasing parallelism at the expense of calculation accuracy. A tradeoff appears: with no synchronization, parallel simulated annealing will rapidly converge to a poor solution; with complete synchronization, it will slowly converge to a good solution. Experimental and analytic research, reviewed here, sheds light on the success or failure of asynchronous, parallel, simulated annealing algorithms .

1. INTRODUCTION

Simulated annealing is a Monte Carlo technique often used to find nearly optimal solutions to NP-hard problems.[1] In particular, it is currently the most successful technique for solving VLSI circuit placement problems.[2] It has also been used to solve the traveling salesman problem, graph partitioning, various protein-configuration problems, etc.

Simulated annealing follows the general form shown in Figure 1. It considers several proposed solution states, s_i, in succession, each of which is an element of a state space S. Each state $s \in S$ has a cost, $\varepsilon(s)$; the algorithm seeks a minimum-cost state.

Simulated annealing starts with a randomly constructed state, s_0. Each state, $s \in S$, has a neighborhood defined by the move set, $\mu(s) \subset S$. At each step in the "annealing loop," it selects a state from the neighborhood, $s_i' \in \mu(s_i)$ with probability $\beta(s_i, s_i')$.

The cost difference $\delta_i = \varepsilon(s_i') - \varepsilon(s_i)$ determines the next state in the sequence. If $\delta_i \leq 0$, the next state, s_{i+1} will be s_i'. If $\delta_i > 0$, the next state is determined stochastically with a pseudo-random number generator: with probability $e^{-\delta_i/T_i}$ state $s_{i+1} \leftarrow S_i'$, and otherwise $s_{i+1} \leftarrow s_i$.

T_i, usually called the temperature, serves as a control parameter: a large temperature allows large cost-increasing moves; a small temperature only allows small cost increases. Simulated annealing programs initially set T to a value high enough that nearly all permutations are accepted, then slowly lower T until the cost remains stable. The sequence T_i is called the temperature schedule. It is generally considered that an optimum temperature schedule keeps the average observed cost close to its equilibrium value.

Under certain constraints, simulated annealing can be shown to converge to the Boltzmann distribution.[3] However, the use of inaccurate cost functions, both in parallel and sequential implementations, can lead to a non-Boltzmann distribution.

```
1.      T ← ∞;
2.      S ← s₀;
3.      while not done
4.          s' ← generate (s);
5.          ΔE ← ε(s') − ε(s);
6.          if (ΔE < 0)∨ (random () < e^−ΔE/T)
7.              s ← s';
8.          T ← reduce-temperature(T);
9.          end while;
```

FIGURE 1   Simulated Annealing Algorithm.

Because simulated annealing runs often require lengthy execution times, research efforts have focused on optimizing the temperature schedule, improving the sequence of permutations, parallelizing the algorithm, and using cheaper, approximate cost functions. Many parallel simulated annealing implementations naturally involve random cost functions—we call such algorithms *asynchronous*.[4,5]

This paper considers the effect of two kinds of cost-function errors on the simulated annealing algorithm: errors with fixed bounds and errors with normal distributions. Binomially distributed errors can occur in asynchronous algorithms, based on the interconnectedness of the state variables and the total number of processors: in the limit, these approach the normal (Gaussian) distribution.

Equilibrium properties can be bounded under such conditions. These results also hold for errors that accumulate in a sum. To our knowledge, the only related theoretical work analyzes instantaneous errors with fixed bounds, over a restricted class of move spaces.[6]

1.1 ASYNCHRONOUS COST FUNCTIONS

Asynchronous, parallel, simulated annealing programs typically operate as follows. The state variables are partitioned among several processors. A processor can only modify variables in its own partition. Each processor can also examine local read-only copies of variables from other partitions, but such variables are not guaranteed to be up-to-date. Periodically, the local copies are updated and the partitioning is changed. Figure 2 shows a typical example.

Because these algorithms calculate costs based on old variables, such calculations can be termed inaccurate or erroneous.[4,7] Allowing greater inaccuracies can often improve speed, but can also degrade results.[8]

Experiments have hinted that when errors are proportionally constrained to temperature, results improve. Invoking these observations, researchers have modified asynchronous algorithms to obtain better final states.[9,10] The analytic results presented here confirm a temperature dependence.

We will consider cost functions with fixed error bounds and with Gaussian distributions.

We can further classify errors into two general categories: instantaneous errors and accumulated errors. An instantaneous error is simply the difference between the true and inaccurate costs computed at a given time. An accumulated error is the sum of a stream of instantaneous errors.

Instantaneous errors cannot be easily observed. Doing so requires that we compute both the inaccurate and true costs of a move, annihilating any speed advantage conferred by the inaccurate cost function. Furthermore, though we can easily observe a sum of errors in *accepted* moves by computing the true and observed cost after executing a sequence of moves, we cannot account for errors in rejected moves.

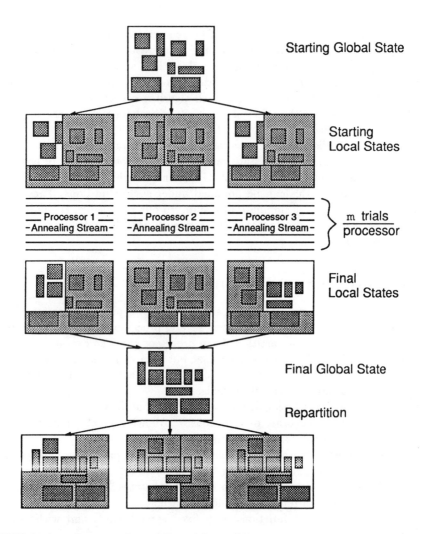

FIGURE 2   Asynchronous Stream-Based Annealing.

Thus, we cannot always rely on empirical observations to estimate accumulated error properties.

In this paper, we discuss a relationship between the equilibrium cost of an erroneous algorithm and that of an equivalent perfect algorithm, based on both bounded and Gaussian errors. These results are likely to lead to an appropriate schedule of error-control that promotes a good result, and maximizes the benefits of the inaccurate cost function.

2. ASYMPTOTIC CONVERGENCE

If we fix the temperature, the simulated annealing algorithm can be modeled as a homogeneous Markov chain. We call this model an *annealing chain*. To satisfy ergodicity and reversibility, it must satisfy a set of commonly accepted properties.[3] The equilibrium density under these conditions is the Boltzmann distribution, namely,

$$\rho(s,T) = \frac{e^{-\varepsilon(s)/T}}{\sum_{s' \in S} e^{-\varepsilon(s')/T}} \cdot \tag{1}$$

When errors appear in the cost function, as is the case with many parallel simulated annealing implementations, we refer to the true cost, $\varepsilon(s)$, of state s, and the erroneous cost $\varepsilon_\epsilon(s) = \varepsilon(s) + \varepsilon$, of state s. $\varepsilon_\epsilon(s)$ is typically a random variable.

The results shown in the following sections omit proofs for brevity. The interested reader can find proofs for fixed-bound results in Greening[11] and for Gaussian cost functions in Greening.[12]

2.1 BOUNDED ERRORS

If cost-function errors are bounded, then we have $\underline{\varepsilon} \le \epsilon \le \overline{\varepsilon}$. We can use those bounds to compare the equilibrium probability distributions of annealing with and without errors.

THEOREM 1. Let $\rho_\epsilon(s,T)$ and $\rho(s,T)$ be the equilibrium probabilities of state s at temperature T, with bounded errors and without errors, respectively. Then,

$$e^{(\underline{\varepsilon}-\overline{\varepsilon})/T}\rho(s,T) \le \rho_\epsilon(s,T) \le e^{(\overline{\varepsilon}-\underline{\varepsilon})/T}\rho(s,T). \tag{2}$$

Equilibrium behavior is often characterized by its "macroscopic properties." Any macroscopic property, $F(T)$, is the expected value of some function, $f(s)$, over the state space,

$$F(T) = \sum_{s \in S} f(s)\rho(s,T). \tag{3}$$

The equilibrium cost, usually denoted $E(T)$, is a macroscopic property over the function ε.

THEOREM 2. Let $F_\epsilon(T)$ and $F(T)$ be equivalent macroscopic properties, for function f, at temperature T, with cost functions ε_ϵ and ε, respectively. Then

$$e^{(\underline{\varepsilon}-\overline{\varepsilon})/T}F(T) \le F_\epsilon(T) \le e^{(\overline{\varepsilon}-\underline{\varepsilon})/T}F(T). \tag{4}$$

The most commonly measured macroscopic property, the average cost, is, of course, constrained by Theorem 2.

2.2 GAUSSIAN ERRORS

Cost errors can often be represented by assigning a Gaussian distribution to the cost of each state. We will show that when the variances of the state costs do not differ greatly, simulated annealing with inaccurate costs converges to a good solution.

LEMMA 1. If X is a Gaussian random variable with mean μ and standard deviation σ, then

$$E[e^{-X/T}] = e^{\sigma^2/2T^2 - \mu/T} \,. \tag{5}$$

THEOREM 3. Let $\varepsilon: S \to \mathbf{R}$ be a cost function, and let its stationary Boltzmann distribution be $\rho: S \to [0, 1]$. Consider a random cost function $\phi: S \to \mathbf{R}$, where each random variable $\phi(s)$ is an independent Gaussian distribution with mean $\varepsilon(s)$ and variance $\sigma^2(s)$. Let $\rho_\phi: S \to [0, 1]$ give its stationary Boltzmann distribution. Then $\rho_\phi(s)$ can be bounded by

$$e^{(\underline{\sigma}^2 - \overline{\sigma}^2)/2T^2} \rho(s) \le \rho_\phi(s) \le e^{(\overline{\sigma}^2 - \underline{\sigma}^2)2T^2} \rho(s) \,. \tag{6}$$

COROLLARY 2. If all variances are constant in Theorem 3, then $\rho_\phi(s) = \rho(s)$.

2.3 IDENTICALLY DISTRIBUTED ERRORS

We saw in Corollary 2 that errors with identical, independent Gaussian distributions over the state space have no effect on the Boltzmann distribution. This is generally true for any identical, independent probability measure.

THEOREM 4. For each $s \in S$, let X_s be a random variable with probability distribution function $p(x_s)$, and with all X_s identically and independently distributed. Consider the random cost function $\theta(s) = \varepsilon(s) + X_s$, and its associated Boltzmann probability measure $\rho_\theta: S \to [0, 1]$. If the resulting Markov chain is ergodic, then $\rho_\theta(s) = \rho(s)$.

2.4 ACCUMULATED ERRORS

When n erroneous moves are accepted in succession, the apparent total cost change can be accumulated in a sum,

$$\Delta_\epsilon = \sum_{i=0}^{n-1} \delta_\epsilon(s_i, s_{i+1}) \,. \tag{7}$$

We call this series a "move-sequence." Consider the move-sequence $s = \langle s_0, \ldots, s_n \rangle$. If we have calculated the true cost at the beginning and end of the sequence, $\varepsilon(s_0)$ and $\varepsilon(s_n)$, we can compute the true total cost change as

$$\Delta = \varepsilon(s_n) - \varepsilon(s_0). \tag{8}$$

The accumulated error for the sequence is

$$\epsilon_s = \Delta_\epsilon - \Delta = \sum_{i=0}^{n-1} \epsilon_i. \tag{9}$$

The instantaneous error results presented in the previous two sections can be extended to account for accumulated errors. Other researchers have argued for a distinction between instantaneous and accumulated errors.[6,13] When considering strictly bounded errors, this distinction is unnecessary. Gaussian errors, however, complicate matters.

Figure 3 shows an example sequence of four moves, running on two processors, and contrasts instantaneous and accumulated error. The second, third, and fourth moves exhibit non-zero instantaneous error, which is shown in the center column. The accumulated error for the entire sequence is shown in the rightmost column.

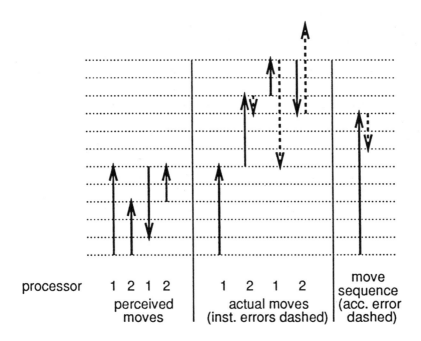

processor	1 2 1 2	1 2 1 2	move
perceived		actual moves	sequence
moves		(inst. errors dashed)	(acc. error dashed)

FIGURE 3 Instantaneous and Accumulated Errors.

Accumulated errors only reflect the instantaneous errors of *accepted* moves. No rejected moves are shown in Figure 3.

Researchers have measured accumulated errors in running parallel simulated annealing systems.[8,10,14,15] Some have conjectured a relationship between those measured errors and the equilibrium cost.

In order to establish a firm connection, we first extend the instantaneous error results to cover a move sequence.

THEOREM 5. The results of Sections 2.1 and 2.2 apply to accumulated errors.

A cautionary note: Gaussian distribution of instantaneous errors does not guarantee that the resulting accumulated error has a Gaussian distribution.

2.5 ESTIMATING BOUNDS

In the general case, the upper error bounds, $\bar{\epsilon}$ and $\bar{\epsilon}_s$, cannot be easily estimated at low temperatures. Some annealing algorithms have errors symmetric about the mean, in particular parallel implementations often have this property.[6] The upper bound can then be set to the lower bound estimate.

It is tempting to presume that the accumulated error is related to the instantaneous error by $\underline{\epsilon}_s = n\underline{\epsilon}$ and $\bar{\epsilon}_s = n\bar{\epsilon}$, where n is the number of steps in a move sequence. It is also tempting to assert that the instantaneous bounds will be tighter than the accumulated error bounds, as $\underline{\epsilon}_s \leq \underline{\epsilon}$ and $\bar{\epsilon}_s \geq \bar{\epsilon}$. Neither statement is true.

Our bounded-error analysis did not specify any properties of ϵ and ϵ_s, except the bounds. We used worst-case behavior to construct homogeneous Markov chains; other intermediate cases need not result in homogeneous Markov chains, but the results will still be constrained by the worst case.

Suppose, for example, that ϵ_i, the instantaneous error at step i, has behavior

$$\epsilon_i = \begin{cases} n & \text{if } i = n - 1; \\ 0 & \text{otherwise.} \end{cases} \tag{10}$$

In this case, $\underline{\epsilon} = 0 \neq \underline{\epsilon}_s = n$ and $\bar{\epsilon} = n = \bar{\epsilon}_s = n$. Oddly, the error bounds $\underline{\epsilon}_s$ and $\bar{\epsilon}_s$ narrow equilibrium properties tightly: $\rho_{\epsilon_s}(s) = \rho(s)$, and $F(T) = F_{\epsilon_s}(t)$. However, the bounds computed from the instantaneous errors do not:

$$e^{-n/T}\rho(s, T) \leq \rho_\epsilon(s, T) \leq e^{+n/T}\rho(s, T). \tag{11}$$

3. CONCLUSION

Calculation errors, resulting from parallel implementation or an approximate cost function, affect equilibrium properties in simulated annealing, and the quality of its results. Theorems derived in this manuscript constrain equilibrium annealing properties in the presence of bounded errors and Gaussian errors.

Algorithms can exhibit either instantaneous errors or accumulated errors. Instantaneous error is difficult to measure in parallel annealing programs. However, accumulated error, the sum of a sequence of instantaneous errors, can be measured with minimal overhead. Researchers have measured and reported accumulated error for various implementations. These measurements can motivate error estimates.

This paper analyzes the effects of both instantaneous and accumulated errors. One cannot infer much about instantaneous error behavior from accumulated error behavior. Under some circumstances, accumulated error bounds can generate tighter constraints than instantaneous error bounds can; an example of this counterintuitive finding was shown.

Researchers have speculated that accumulated errors influence macroscopic properties, and have tuned annealing programs based on those speculations. This paper has proved a direct connection between the errors that researchers have measured, and annealing properties, including average cost. The analysis reported here should help improve asynchronous, parallel, simulated annealing algorithms in the future.

ACKNOWLEDGMENTS

I am grateful to Mary Pugh for her thoughts on this work. Frederica Darema, Milŏs Ercegovac, and Ron Lussier provided much encouragement. IBM and the Santa Fe Institute provided a stimulating research environment. The author remains responsible for all mistakes.

REFERENCES

1. Kirkpatrick, Jr., S., C. D. Gelatt, and M. P. Vecchi. "Optimization by Simulated Annealing." *Science* **220(4598)** (1983): 671–680.
2. Sechen, Carl, Kai-Win Lee, Bill Swartz, Jimmy Lam, and Dahe Chen. "TimberWolfSC version 5.4: Row-Based Placement and Routing Package." Technical report, Yale University, New Haven, Connecticut, July 1989.
3. Otten, R. H. J. M., and L. P. P. P. van Ginneken. *The Annealing Algorithm.* Boston: Kluwer, 1989.
4. Greening, Daniel R. "Parallel Simulated Annealing Techniques." *Physica D: Non-linear Phenomena* **42** (1990): 293–306
5. Bertsekas, Dimitri P., and John N. Tsitsiklis. *Parallel and Distributed Computation: Numerical Methods.* Englewood Cliffs, NJ: Prentice Hall, 1989.
6. Durand, M.D., and Steve R. White. "Permissible Error in Parallel Simulated Annealing." Technical Report RC 15487, IBM T.J. Watson Research Center, Yorktown Heights, New York, 1990.
7. Durand, M. D. "Accuracy vs. Speed in Placement. " *IEEE Design and Test of Computers* **8(6)** (June 1989): 8–34.
8. Greening, Daniel R., and Frederica Darema. "Rectangular Spatial Decomposition Methods for Parallel Simulated Annealing." In *Proceedings of the International Conference on Supercomputing*, Crete, Greece: IEEE Computer Society Press, June 1989, 295-302.
9. Casotto, Andrea, Fabio Romeo, and Alberto Sangiovanni-Vincentelli. "A Parallel Simulated Annealing Algorithm for the Placement of Macro-Cells." *IEEE Transactions on Computer-Aided Design* **CAD-6(5)** (September 1987): 838–847.
10. Banerjee, Prithviraj, Mark Howard Jones, and Jeff S. Sargent. "Parallel Simulated Annealing Algorithms for Cell Placement on Hypercube Multiprocessors." *IEEE Transactions on Parallel and Distributed Systems* **1(1)** (January 1990): 91–106.
11. Greening, Daniel R. "Equilibrium Conditions of Asynchronous Parallel Simulated Annealing." In *MCNC International Workshop on Layout Synthesis.* Research Triangle Park, NC, 1990.
12. Greening, Daniel R. "Gaussian Cost Functions in Simulated Annealing." Technical report, IBM T.J. Watson Research Center, Yorktown Heights, NY, 1990.
13. Durand, M. D. "Cost Function Error in Asynchronous Parallel Simulated Annealing Algorithms." Technical Report CUCS-423-89, Columbia University Computer Science Department, New York, NY, June 1989.
14. Darema, Frederica, Scott Kirkpatrick, and Alan V. Norton. "Parallel Algorithms for Chip Placement by Simulated Annealing." *IBM Journal of Research and Development* **31(3)** (May 1987): 391–402.

15. Jayaraman, Rajeev, and Frederica Darema. "Error Tolerance in Parallel Simulated Annealing Techniques." In *Proceedings of the International Conference on Computer Design*, 545–548. IEEE Computer Society Press, 1988.

J. A. Hoffnagle,† L. G. Reyna,‡ and J. R. Sobĕhart*

†IBM Research Division, Almaden Research Center, Almaden, CA 10598; ‡IBM Research Division, T.J. Watson Research Center, Yorktown Heights, NY 10598; and *Instituto Balseiro, 8400 San Carlos de Bariloche, Argentina

Dynamics of a Trapped Ion Driven by Stochastic Optical Processes

The probability distributions for the ground and excited states of a two-level ion in an harmonic trap are studied. The incoming photons excite the ion which relaxes back to its ground state by either spontaneous or stimulated emission. We assume the arrival times of the photons is described by a Poisson process. Solutions of the model are obtained by numerical and analytical methods. The results obtained are similar to those of quantum mechanical computations for atoms in radiation fields, in the limit of heavy particles.

INTRODUCTION

The confinement of small clusters of ions in potential wells, or traps, has been the subject of intense study in recent years, especially since the demonstration by Neuhauser et al.[9] of optical detection of individual ions. The possibility of studying the behavior of small numbers of ions may help the understanding of the mechanisms involved in the phase transition between ordered, crystal-like states and chaotic, disorganized motion.

Ions trapped in potential wells can be cooled to very low temperatures by resonance radiation pressure, as reviewed by Stenholm.[11] The ions are illuminated by a laser beam having a frequency slightly less than that of the ions' resonance transition. Consequently, when an ion is moving against the laser beam, the frequency of the light in the ion's rest frame is Doppler-shifted towards resonance and the absorption of laser photons is increased. The subsequent spontaneous emission reduces on the average the ion momentum since the ion motion must compensate for the momentum change in the radiation field. Using this technique, Diedrich et al.[6] have cooled single ions to temperatures below a millidegree Kelvin.

The possibility of having ions at extremely low temperatures allows for very high precision spectroscopy, which may provide a basis for greatly improved time and frequency standards. To increase the signal-to-noise ratio involved in such measurements, it would be desirable to work with many ions; however, the Coulomb interaction between ions provides a route to heat ion clusters confined in oscillating trapping potential wells, producing a transition from a desirable condensed regime to a chaotic regime (see Blatt et al.[1]). Such transitions have been reported by Hoffnagle et al.[7] and Brewer et al.,[3] when studying the behavior of ions as function of the trap operating conditions. Similar observations were also reported by Blumel et al.[2]

The phase transitions exhibited by ion clusters are of interest in their own right as order-chaos transitions in an exceptionally simple system. It should be pointed out that ensembles of condensed ions behave differently than conventional solids, since the average distance separating the ions is about five orders of magnitude larger than the typical distance separating atoms in most solids. Furthermore, it is not the ion-electron interaction which binds the system but the confining potential well.

The actual transition has been analyzed as an order-chaos transition by Brewer et al.[3] and Hoffnagle et al.[7] Once the system is perturbed from equilibrium, the ions begin to experience very brief, "hard" collisions, which may lead to chaos. The initial perturbation that displaces the system from equilibrium could be due to either the discrete nature of the momentum transfer between the cooling laser and the ions, or other disturbances, such as collisions with residual gas molecules. The purpose of this contribution is to study the noise level associated with the cooling process.

The current model describes the dynamics of dilute ion ensembles in terms of a coupled system of stochastic differential equations, correct to all orders in the momentum transfer between the ion and the radiation field. The present calculations, in the heavy ion limit, reduce to previous results derived by Stenholm[11] by considering the Fokker-Planck description, and also by Javanainen and Stenholm[8] by using a quantum treatment.

DYNAMICS OF SINGLE ION IN A RADIATION FIELD

We consider the simplified case of the semiclassical description of a two-level ion trapped in a static, one-dimensional harmonic potential. The transition from the ground state to the excited state is due to either *absorption* of a photon from the laser beam; at this time there is a discrete transfer of momentum to the ion. The transition from the excited level to the ground state, however, takes place due to *stimulated emission* or *spontaneous emission*, and again the discrete momentum transfer is taken into account.

Absorption is assumed to take place at times t_r, distributed according to a Poisson process with velocity-dependent rate given by the Lorentzian

$$r(v) = \frac{\chi^2}{4} \frac{\gamma}{(\Delta + qv)^2 + \gamma^2/4} , \qquad (1)$$

where q is the light wave vector and $v = dx/dt$ is the ion velocity. The detuning is given by $\Delta = \nu - \Omega$, where Ω is the laser frequency and ν is the electronic transition frequency with line γ. The term $\Delta + qv$ in Eq. (1) accounts for the Doppler shift of the laser frequency Ω produced by the ion motion. The interaction parameter between the ion and the radiation field is given by $\chi = \mu E/\hbar$, where μ is the coupling dipole moment and E is the applied field strength. At times t_r, the ion velocity changes by $k = \hbar q/m$, where $\hbar q$ is the photon momentum and m is the ion mass.

Stimulated emission is assumed to take place at times $t_{r'}$, also governed by a Poisson process with rate $r(v)$. At these times the ion velocity changes by $-k$. Spontaneous emission takes place at times t_γ, again a Poisson process, but now with rate γ. The change in ion velocity in this case is given by either $k' = k$ or $k' = -k$, with the equal probabilities.

The dynamics of a two-level ion trapped in an harmonic potential well and driven by a noise, which strength and frequency depend upon the particular ion internal state, is described by the following equation

$$\frac{d^2x}{dt^2} + \omega^2 x = k \sum_{t_r} \delta(t - t_r), \qquad (2a)$$

when the ion is in the ground state, and by

$$\frac{d^2x}{dt^2} + \omega^2 x = -k \sum_{t_{r'}} \delta(t - t_{r'}) + k' \sum_{t_\gamma} \delta(t - t_\gamma), \qquad (2b)$$

when the ion is in the excited state. Here ω is the angular frequency of the ion in the trap.

In the absence of the absorption/emission processes, the ion follows a deterministic trajectory in phase space. The presence of small changes in the velocity is expected to produce fluctuations in the ion motion leading to an uncertainty in its localization. A probabilistic treatment is, therefore, needed to determine the general features of ions interacting stochastically with radiation fields.

The evolution equations for the probability distribution of the ion being in the ground state will be denoted $P_g(x, v, t)$, and for the excited state $P_e(x, v, t)$. The equations of motion for these probabilities are given by

$$
\begin{aligned}
P_g(x, v, t + \Delta t) = \; & P_g(x - v\Delta t, v + \omega^2 x \Delta t, t)(1 - r\Delta t) \\
& + P_e(x - v\Delta t, v + \omega^2 x \Delta t + k, t) r\Delta t \\
& + \frac{1}{2} \sum_{k' \in \{-k, k\}} P_e(x - v\Delta t, v + \omega^2 x \Delta t + k', t) \gamma \Delta t
\end{aligned}
\tag{3a}
$$

and

$$
\begin{aligned}
P_e(x, v, t + \Delta t) = \; & P_e(x - v\Delta t, v + \omega^2 x \Delta t, t)(1 - (r + \gamma)\Delta t) \\
& + P_g(x - v\Delta t, v + \omega^2 x \Delta t - k, t) r\Delta t.
\end{aligned}
\tag{3b}
$$

Using

$$
P(x - v\Delta t, v + \omega^2 x \Delta t, t) = P(x, v, t) + \Delta t \left(-\frac{\partial}{\partial x}(vP) + \frac{\partial}{\partial v}(\omega^2 x P) \right) + O(\Delta t^2)
\tag{4}
$$

results in the equation for the ground-state distribution

$$
\begin{aligned}
\frac{\partial P_g}{\partial t} + \frac{\partial}{\partial x}(vP_g) - \frac{\partial}{\partial v}(\omega^2 x P_g) = \; & -r(v)\, P_g + r(v + k)\, P_e(x, v + k, t) \\
& + \frac{\gamma}{2}\left(P_e(x, v + k, t) + P_e(x, v - k, t) \right)
\end{aligned}
\tag{5a}
$$

and the equation for the excited-state distribution

$$
\frac{\partial P_e}{\partial t} + \frac{\partial}{\partial x}(vP_e) - \frac{\partial}{\partial v}(\omega^2 x P_e) = -(r(v) + \gamma)\, P_e + r(v - k)\, P_g(x, v - k, t).
\tag{5b}
$$

Eqs. (5a) and (5b) form a nonlocal system of equations coupled in velocity coordinate.

The relevant moments of the random process corresponding to the transition from the ground state to the excited state are

$$
\overline{X - x} = v\Delta t + O(\Delta t^2), \quad \overline{V - v} = (-\omega^2 x + kr(v))\, \Delta t + O(\Delta t^2),
$$

$$
\overline{(V - v)^2} = r(v)k^2 \Delta t + O(\Delta t^2)
\tag{6a}
$$

and

$$
\overline{(V - v)^n} = r(v)k^n \Delta t + O(\Delta t^2),
\tag{6b}
$$

for the higher moments. The moments for the transition from the excited state to ground state are

$$\overline{X - x} = v\Delta t + O(\Delta t^2), \quad \overline{V - v} = -\left(\omega^2 x + kr(v)\right)\Delta t + O(\Delta t^2),$$

$$\overline{(V - v)^2} = (r(v) + \gamma)k^2 \Delta t + O(\Delta t^2) \tag{6c}$$

and

$$\overline{(V - v)^n} = \left(r(v)(-k)^n + \frac{\gamma}{2}(k^n + (-k)^n)\right)\Delta t + O(\Delta t^2). \tag{6d}$$

Expanding Eqs. (5a) and (5b) for small noise amplitude k, results in the following system of equations

$$\frac{\partial P_g}{\partial t} + \frac{\partial}{\partial x}(vP_g) - \frac{\partial}{\partial v}(\omega^2 x P_g) = -rP_g + \sum_{n=0}^{\infty} \frac{k^n}{n!} \frac{\partial^n}{\partial v^n}(rP_e) + \frac{\gamma}{2}\sum_{n=0}^{\infty} \frac{k^{2n}}{(2n)!} \frac{\partial^{2n} P_e}{\partial v^{2n}} \tag{7a}$$

and

$$\frac{\partial P_e}{\partial t} + \frac{\partial}{\partial x}(vP_e) - \frac{\partial}{\partial v}(\omega^2 x P_e) = -(r + \gamma)P_e + \sum_{n=0}^{\infty} \frac{(-k)^n}{n!} \frac{\partial^n}{\partial v^n}(rP_g). \tag{7b}$$

The successive terms in the above expansions (7a) and (7b) can be identified with the moments of the probability distributions derived in Eqs. (6a)–(6d). The system of Eqs. (7a) and (7b) can be described as a *generalized Fokker-Planck expansion*. The system takes into account the transfer of momentum between the driving noise and the ion to all orders.[11]

Solutions to Eqs. (5a) and (5b) are characterized by the presence of two different time scales. The time scale $1/\gamma$, associated with the absorption-emission processes scales with the decay time of the excited level[8,11] and the time scale of the ion cooling induced by the radiation field. The second time scale is slower due to the large number of spontaneous emissions necessary to significantly reduce the ion energy.

A rough estimate of the cooling time scale is given by the time needed to reduce the initial energy $E_0 = mv_0^2/2$ to the value $m\Delta^2/2q^2$, obtained from the resonant velocity $-\Delta/q$. Assuming that the initial ion velocity v_0 is larger than Δ/k, and taking into account that the absorption-emission process is only possible when the ion velocity is close to the resonant velocity, the energy change per absorption-emission cycle can be approximated by Blatt.[1]

$$\Delta E = \sum_{k'} \frac{m}{2}(v + k - k')^2 - \frac{m}{2}v^2 \approx \frac{\hbar\Delta\gamma}{r(\Delta/k) + \gamma}. \tag{8}$$

The time spent by the ion in the velocity interval near the resonant velocity $|\Delta + qv| \leq \gamma/(2q)$, is approximately given by $\delta t = \gamma/\omega v_0 q$. The number of absorptions, and subsequent spontaneous emissions, during this time interval is given by $N \approx r(\Delta/q)\delta t \approx \chi^2/\omega q v_0$. Thus, the time scale for the cooling process can be estimated by

$$\gamma t_c \approx \frac{mq v_0{}^3}{\hbar \gamma \Delta}(1 + \frac{\gamma^2}{\chi^2}).$$ (9)

Note that for the choice of parameters: $\chi = \gamma$, $\delta = \gamma$, $E_0 = 10^3 \hbar \gamma$ and $q\sqrt{2\hbar/\gamma m} = 0.1$, we obtain $\gamma t_c = 4.5 \ 10^3$, a time scale much longer than the lifetime of the excited state.

NUMERICAL RESULTS

In this section we present numerical solutions to the system of equations (5a)–(5b). The equations were discretized on stretched meshes, using second-order centered approximations. The time discretization chosen was a second-order backward difference method. The resulting matrix problems were solved using sparse matrix techniques.

In the numerical simulations, the following parameters were used

$$\gamma = 2.51 \ 10^7 \ 1/s, \quad \Delta = -\gamma, \quad \chi = \gamma, \quad \omega = 0.1 \ \gamma,$$

$$E_0 = 10^3 \gamma, \quad \text{and} \quad \sqrt{\frac{2\hbar}{m\gamma q}} = 0.11.$$

Figures 1 and 2 show contour plots of the probability distributions in phase space. The horizontal axis represents the position coordinate and the vertical axis the velocity coordinate. The vertical scale is taken about 100 times $\sqrt{2\hbar\gamma/m}$. The figures show the early formation of a dip in the ground-state probability distribution for velocities close to the resonant velocity. The time evolution shows that the saturation of the relative population of both states is completed after a time proportional to γ^{-1}. The subsequent cooling of the ion modifies the initial saturation value but in a time scale comparable to the laser cooling process.

Figure 3 shows the dependence of both populations in the center of the trap as function of the velocity evaluated at different times, stressing particular features of the cooling process. During the initial stage of the laser cooling, after thousands of spontaneous emission, the energy decays on a time scale in agreement with Eq. (9).

The frequency of photon absorptions is largely decreased after the ion velocity becomes smaller than the resonant value. The time scale to completely reduce the ion energy to its final value, where the interaction between the radiation field and the ion becomes stationary, can be found in Javanainen and Stenholm.[8]

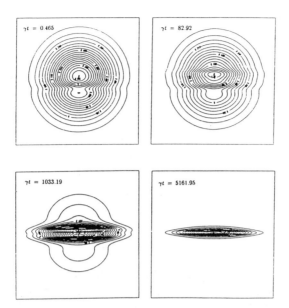

FIGURE 1   Contour plots of the ground-state distribution.

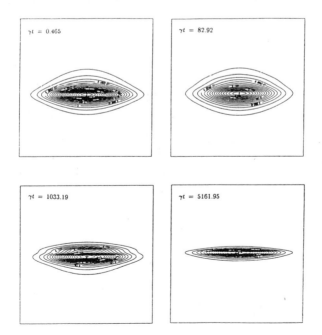

FIGURE 2   Contour plots of the excited-state distribution.

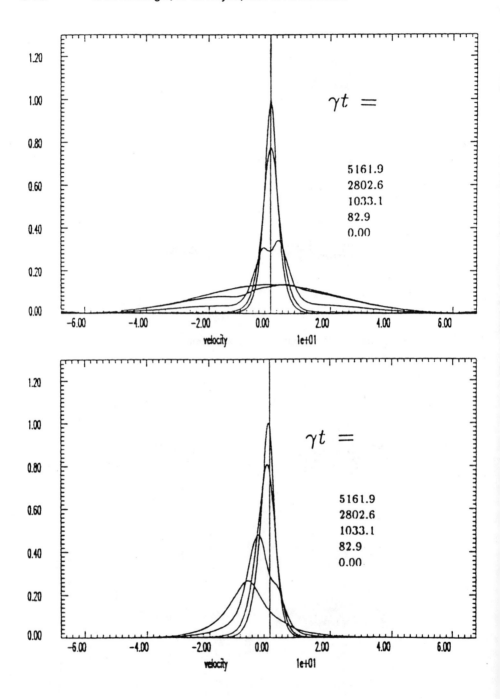

FIGURE 3 Ground-state and excited-state distributions as function of the ion velocity.

During the latter stage of the laser cooling, the probability distributions are characterized by a spreading in position and velocity coordinates resulting from the fluctuations produced by the driving noise. However, in the limiting case of negligible small noise amplitude, both distributions must be singularly localized over the ion orbit in phase space, which is described by the coordinates $R = \sqrt{\omega^2 x^2 + v^2}$ and $\theta = \arctan(v/\omega x)$.

ASYMPTOTIC ANALYSIS

An approximate solution to the *generalized Fokker-Planck* system can be found in the limit of small driving noise amplitude. We use the following singular approximation

$$P_g = e^{\phi/k}\,(f_o + kf_1 + ...) \quad \text{and} \quad P_e = e^{\phi/k}\,(g_o + kg_1 + ...) \tag{10}$$

into Eqs. (5a–b). Summing all terms of the same order in the expansion parameter k results in the steady-state equations

$$\omega\frac{\partial\phi}{\partial\theta} = 0 , \tag{11a}$$

$$-\omega\frac{\partial f_o}{\partial\theta} = -r(R\sin\theta)f_o + e^{\sin\theta\,\partial\phi/\partial R}r(R\sin\theta)g_o + \gamma\,\cosh(\sin\theta\frac{\partial\phi}{\partial R})g_o. \tag{11b}$$

and

$$-\omega\frac{\partial g_o}{\partial\theta} = -(r(R\sin\theta) + \gamma)g_o + e^{-\sin\theta\,\partial\phi/\partial R}r(v)f_o. \tag{11c}$$

Actually the singular approximation of Eq. (10) can also be used to describe the early stages of the cooling process.

The ion velocities occurring during the end of the laser cooling become small because of the dissipative nature of the process. This allows for the following expansion of the induced rate $r(v) \approx r_o - \beta v + ...$ Thus, the equation for the final probability distributions can be found

$$\frac{\partial}{\partial\theta}\begin{pmatrix} f_o \\ g_o \end{pmatrix} = \left[\begin{pmatrix} -r_o & r_o + \gamma \\ r_o & -(r_o + \gamma) \end{pmatrix} + \sin\theta \begin{pmatrix} -\beta R & \beta R + r_o\frac{d\phi}{dR} \\ \beta R - r_o\frac{d\phi}{dR} & -\beta R \end{pmatrix} \right.$$
$$\left. + \sin^2\theta \begin{pmatrix} 0 & \frac{r_o+\gamma}{2}(\frac{d\phi}{dR})^2 + \beta R\frac{d\phi}{dR} \\ -\beta R\frac{d\phi}{dR} + \frac{r_o}{2}(\frac{d\phi}{dR})^2 & 0 \end{pmatrix} + ... \right]\begin{pmatrix} f_o \\ g_o \end{pmatrix} \tag{12}$$

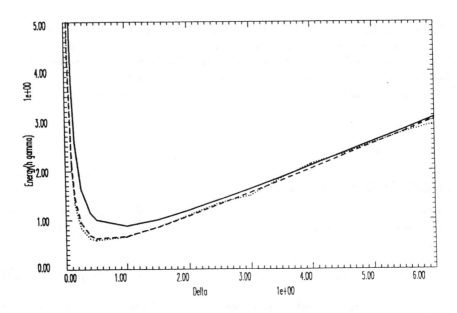

FIGURE 4 Final ion energy as function of Δ: Monte Carlo simulation . . ., analytical approximation $- - -$ and heavy particle limit ——.

Since solutions to Eq. (12) must be periodic in the angular variable θ, the function $\phi(R)$ is determined by

$$\phi(R) = \frac{\gamma\beta}{r_o}\frac{(2r_o + \gamma)}{(2r_o + \gamma)^2 + \gamma^2}(\frac{v^2}{2} + \frac{\omega^2 x^2}{2}), \tag{13}$$

which implies a final energy

$$E_f = \hbar\left(\frac{\Delta^2 + \gamma^2/4 + \chi^2/2}{4\Delta}\right)\left(1 + \left(\frac{\Delta^2 + \gamma^2/4}{\Delta^2 + \gamma^2/4 + \chi^2/2}\right)^2\right) \sim \hbar\gamma. \tag{14}$$

This final energy is independent of the noise amplitude provided $mk^2/2 \ll \hbar\gamma$. From Eq. (14), it can be seen that the ion energy is basically limited by the spontaneous emission line width, which is the energy uncertainty in the emitted radiation. When the line width of the excited level is very narrow, the energy associated with the transferred momentum, $k^2/2m$, becomes important and it sets up the energy limit of the final probability distribution spreading.

Eq. (14) is closely related to the result derived by Javanainen and Stenholm[8] using the quantum treatment in the heavy particle limit by Stenholm.[10]

The numerical solution of the steady state of the both levels and the results of a Monte Carlo (MC) simulation of the equation of motion (2) confirm the analytical estimation (14) derived for the ion energy. Figure 4 shows the final energy obtained in the MC simulation as well as the current analytical curve and the result derived from the quantum treatment in the heavy particle limit,[8] corrected to represent a discrete momentum transfer. It can be seen that our estimates and the MC simulation agree within a few percents in a wide range of values, but the agreement with the quantum estimate is limited to moderate detuning.

DISCUSSION

In the current model, the dynamics of an ion interacting stochastically with an electromagnetic radiation field has been analyzed. The absorption and emission of radiation is simulated with the exchange of photons of momentum $\hbar q$ between the field and the ion. The absorption and emission processes are assumed to occur randomly distributed in time according to Poisson processes, ignoring coherence or time correlation between the emitted photons.

As a result of the stochastic nature of the laser cooling, the ion does not only experience an average light-induced pressure force but also a random component associated with the spontaneous emission, which produces a spreading of the ion velocity.

The dependence of the absorption-emission rate on the ion velocity leads to a fast reduction of the population of ions in the lowest state close to the resonant velocity $v = \Delta/q$. This process saturates in times of the order of γ^{-1}, when both level populations reach a quasi-equilibrium state with the applied radiation field. The subsequent cooling of the ion to the energy scale $\hbar\gamma$ occurs after a large number of absorptions and emissions have taken place.

The results derived in the preceding sections indicate that the present semi-classical model accounts for the general features of the laser cooling of trapped ions when it is compared with the calculations obtained from the quantum treatment in the heavy particle limit.

ACKNOWLEDGMENTS

The work by J.R. Sobĕhart was partially supported by IBM-Argentina.

REFERENCES

1. Blatt, R., G. Lafyatis, W. D. Phillips, S. Stenholm, and D. J. Wineland. "Cooling in Traps." *Physica Scripta T* **22** (1988): 216–223.
2. Blümel, R., C. Kappler, W. Quint, and H. Walther. "Chaos and Order of Laser-Cooled Ions in a Paul Trap." *Phys. Rev. A* **40** (1989): 808–823.
3. Brewer, R. G., J. Hoffnagle, R. G. DeVoe, L. G. Reyna, and W. Henshaw. "Collision-Induced Two-Ion Chaos." *Nature* **344** (1990): 305–309.
4. DeVoe, R.G., J. Hoffnagle, and R. G. Brewer. "Role of Laser Damping in Trapped Ion Crystals." *Phys. Rev. A* **39** (1989): 4362–4365.
5. Diedrich, F., E. Peik, J. M. Chen, W. Quint, and H. Walter. "Observations of a Phase Transition of Stored Laser-Cooled Ions." *Phys. Rev. Lett.* **26** (1987): 2931–2934.
6. Diedrich, F., J. C. Bergquist, W. M. Itano, and D. J. Wineland. "Laser Cooling to the Zero-Point Energy of Motion." *Phys. Rev. Lett.* **62** (1989): 403–406.
7. Hoffnagle, J., R. G. DeVoe, L. G., Reyna, and R. G. Brewer. "Order-Chaos Transition of Two Trapped Ions." *Phys. Rev. Lett.* **61** (1988): 255–258.
8. Javanainen, J., and S. Stenholm. "Laser Cooling of Trapped Particles I." *Appl. Phys.* **21** (1980): 283–291.
9. Neuhauser, W., M. Hohenstatt, and P. E. Toschek. "Visual Observation and Optical Cooling of Electrodynamically Contained Ions." *Appl. Phys.* **17** (1978): 123–129.
10. Stenholm, S. "Distribution of Photons and Atomic Momentum in Resonance Fluorescence." *Phys. Rev. A* **27** (1983): 2513–2522.
11. Stenholm, S. "The Semiclassical Theory of Laser Cooling." *Rev. Mod. Phys.* **58** (1986): 699–739.

Jan E. Hutson
Department of Mathematics, University of Illinois, Urbana-Champaign, IL 61801

Undecidability in an Adaptive System

Adaptive systems can be modeled by a process of objects acting on objects to produce new objects. A model of this sort can have a rich structure; in fact, a study of the theory of such models, called Turing gases, shows that undecidable questions exist. This paper gives a general introduction to undecidability, Turing machines, and Turing gases, and then gives a formal proof of the undecidability of the membership problem for Turing gases.

INTRODUCTION

Models of adaptive behavior are used in many areas such as biology, chemistry, economics, and computer science. One possible model, the Turing gas, involves a set of objects that act upon one another to produce new objects. These objects may be thought to represent molecules in chemistry, goods and services in economics, or any appropriate objects in the system under study.

When this model is set up, one question we would like to answer is, "If I start with a certain set of objects, can the object '__' ever be created?" This question is analogous to the questions, "Can this particular molecule be created? Can this particular item become part of the economy?"

We show below that in general, this question is undecidable. This means that there is no procedure that will always answer such a question.

THE TURING GAS

The model we want to investigate is that of objects acting on objects. To make this formal, an object will be a finite string of symbols, where the symbols are taken from a fixed finite alphabet. Then these strings are thought of as "floating around in a box." Sometimes a string collides with another string, at which point a new string is produced, depending on those involved in the collision. The collision is considered to be asymmetric; one string acts on the other string. There are recursive (i.e., computable) rules for deciding what the product will be when a certain string acts on another particular string. Some of these rules may say "produce nothing for this type of collision"; that is, the collision will be elastic and produce no new string.

Following Fontana,[6] a system like that described above is called a *Turing gas*. For this, the *membership problem* is the problem of deciding, given a set of starting strings and an additional string in question, whether the string in question can ever appear if the Turing gas starts with the given starting set. If a certain string can appear, it is said to belong to the set of possible worlds derived from that starting set.

Additional structure on a Turing gas will depend on the particular application. One might allow strings to collide at random or with differing probabilities, although we always allow any string to collide with any other even if the probability is small. Also, the fate of the two original strings involved in a collision depends on the application. In the model considered here, the colliding strings do not disappear, but remain floating in the box. This is an alternative to having an infinite supply of the original strings, which would make the practical aspects of some applications difficult.

Of course, the membership problem for any of the versions above is the same, because it only questions the possible appearance of strings, and not probabilities or numbers.

UNDECIDABILITY AND THE TURING MACHINE

Ever since the work of Gödel, Church, Turing, and others in the first half of the century, many different problems have been found to be undecidable. Informally, undecidability simply means that there are always questions that cannot be answered in a finite amount of time. Of course, in reality this means these questions cannot be answered at all.

It turns out that the Turing gas membership problem is undecidable. To show this, we must first give some general background on Turing machines. There are many equivalent versions of Turing machines, one of which is presented here.

A *Turing machine* has a tape that is infinitely long in both directions, and a movable tape head which can read from and write to the tape. Along the tape are cells in which are written symbols from some fixed *tape alphabet*. This alphabet consists of a finite number of distinct symbols, including a special *blank* symbol, sometimes denoted B. At any given point, the tape head is on a certain cell, and must decide whether to move one cell to the right or one cell to the left. At the same time it must rewrite a symbol (possibly the same one) onto the cell it is leaving. The tape head's decision depends on two things, the symbol it is currently reading and the current *state* of the Turing machine. The state can be thought of as a symbol kept inside the tape head. This symbol belongs to the *set of states*, an alphabet distinct from the tape alphabet.

There are certain *transition rules* governing the action of the tape head. For example, a rule might say, "If reading symbol 2 while in state Q, then write symbol 6, move LEFT, and change to state P." So in addition to writing and moving, the tape head possibly changes state at each move. On the other hand, for some combinations of symbol and state, there might be NO transition rule given. If the tape head encounters a symbol and state with no rule, it simply stops, and the Turing machine is said to *halt*.

To run the Turing machine, the tape starts with a string of non-blank symbols in the middle, surrounded by blanks on the rest of the tape. The tape head starts on the leftmost non-blank symbol, then by following the rules, the Turing machine does one of three things: (1) eventually halts from lack of a needed rule; (2) runs forever; or (3) hits a special state called an *accept state*, in which case it halts automatically, and is said to *accept*. This special state belongs to the set of states just like the others, so transition rules might send the Turing machine to this state during the operation.

Now, a problem is undecidable if there is no procedure which when given the input (the question) will always give the answer in a finite amount of time. An example of a known undecidable problem is that of whether a Turing machine will accept. That is, there are Turing machines for which acceptance is impossible to predict. If a string is put on the tape, and one wants to know whether the Turing machine will ever accept, there is actually no way to make an accurate prediction. It is true that someone can start running the Turing machine, but after a while it may still be running so one still does not know if it will ever accept. Even after it runs for ten or twenty or a hundred years, there is no guarantee that it won't run for another thousand or million or more years and *then* suddenly accept.

THE MEMBERSHIP PROBLEM

We now show rigorously the undecidability of the membership problem.

THEOREM. The membership problem for Turing gases is undecidable.

PROOF This will be done by encoding the Turing machine in a Turing gas. So Turing gases are just as powerful as Turing machines; if someone wanted to run a Turing machine, they could just set up the appropriate Turing gas and let that "run" instead.

To begin, we must be able to give the tape contents, head position, and state of a Turing machine at any given time in a simple form. To do this, write down the tape contents excluding leading and trailing blanks, then just to the left of the current symbol being read insert a symbol indicating the current state. Of course, symbols used for states must be different from the tape symbols. Also, in some cases the tape head will be among either the leading or trailing blanks, in which case, enough blanks should be listed to reach the head position.

Such an encoding is called a Turing machine *configuration*. For example, if a Turing machine has a tape containing "...BB1351BB...," tape head on the second 1, and is in state Q, the configuration would be 135Q1. Some of the strings in our Turing gas will be of this form. Call them *configurations*.

Other strings floating around will encode the transition rules. Each rule can be encoded with a string of five symbols. The rule, "If reading symbol 2 while in state Q, then write 6, move LEFT, and change to state P," has string 2Q6LP. Call such a string a *rule*.

One last type of string for our Turing gas is the string consisting of a single symbol, $, and is called the *accept string*.

Now the Turing gas must simulate a Turing machine, and this is done as follows: At the beginning, a string for each transition rule is floating around, as well as a string for the initial Turing machine configuration. Then strings start colliding. The vast majority of collisions are elastic. But when a configuration collides with the rule that applies to it (in the Turing machine sense), the product is a string that gives the next Turing machine configuration, i.e., the tape contents, head position, and state that such a Turing machine would enter after following that transition rule from the previous configuration.

This process continues so that an updated configuration is formed each time the previous one collides with the applicable rule. Eventually, perhaps, a configuration containing an accepting state will be produced. In this case, any time that it collides with any other string, the product is the accept string, $.

One may be concerned about all the former configurations still floating around, possibly colliding yet again with their applicable rules. However, this simply causes more copies of each configuration to be formed, but does not affect the simulation of the Turing machine.

So there is an equivalence between the Turing gas and the Turing machine. The accept string belongs to the set of possible worlds of the Turing gas if and only if the Turing machine accepts.

Now, since there are Turing machines for which acceptance cannot be predicted, there are Turing gases such as that above, for which the membership of a string in the set of possible worlds cannot be predicted.

Q.E.D.

The proof above actually gives a stronger result. Notice that the collisions described in the the proof could have been considered to be symmetric, with the same result. So a Turing gas with only symmetric collisions is just as powerful, and still has undecidable membership problem.

As in most proofs of undecidability, the construction seems contrived, probably because it *is*. However, the demonstration of a single, albeit contrived, example indicates the existence of many more undecidable processes lurking, often in the problems of most interest to us.

While we have shown that the general problem under investigation is undecidable, there are always particular instances that happen to be decidable. A Turing gas could be constructed with some simple rules, such as, "Every collision between any two strings produces the string 7Q75." Then the solution to the membership problem is easy: If the string in question is the same as a start string or is 7Q75, then the answer is "YES." Otherwise, the answer is "NO." This is a particular decidable case, but for the general Turing gas, the problem is still undecidable.

CONCLUSIONS

Under rigorous treatment, the membership problem for Turing gases is seen to be undecidable. One interpretation of this result is that the study of Turing gases is genuinely interesting, and does not collapse to some simple theory.

At the same time, the undecidability of the membership problem does not imply undecidability of every problem. One can still study entire classes of problems on Turing gases and perhaps find general solutions to major problems of interest. In fact, one area not addressed here is that of probabilities and frequencies of strings and collisions, which is an area of great interest to people studying systems like these. Some questions in these areas are decidable, while probably many are formally undecidable.

REFERENCES

1. Bagley, R. J., J. D. Farmer, S. A. Kauffman, N. H. Packard, A. S. Perelson, and I. M. Stadnyk. "Modeling Adaptive Biological Systems." *BioSystems* **23** (1989): 113–138.
2. Davis, M., ed. *The Undecidable.* New York: Raven Press, 1965.
3. Church, A. "An Unsolvable Problem of Elementary Number Theory." *Amer. J. Math.* **58** (1936): 345–363.
4. Farmer, J. D., S. A. Kauffman, and N. H. Packard. "Autocatalytic Replication of Polymers." *Physica D* **22** (1986): 50–67.
5. Gödel, K. "Über formal unentscheidbare Sätze der Principia Mathematica und verwandter System I." *Monatshefte für Mathematik und Physik.* **38** (1931): 173–198. English translation by Elliott Mendelson in *The Undecidable.*[2]
6. Fontana, W. "Algorithmic Chemistry: A Model for Functional Self-Organization." Preprint.
7. Hopcroft, J. E., and J. D. Ullman. *Introduction to Automata Theory, Languages, and Computation.* Reading, MA: Addison-Wesley, 1979.
8. Kauffman, S. A. "Cellular Homeostasis, Epigenesis and Replication in Randomly Aggregated Macromolecular Systems." *J. Cybernetics* **1** (1971): 71–96.
9. Kauffman, S. A. *Origins of Order.* Oxford: Oxford University Press, 1990.
10. Turing, A. M. "On Computable Numbers with an Application to the Entscheidungsproblem." *Proc. London Math. Soc.* ser. 2 **42** (1936): 230–265. See Correction in *ibid.* **43** (1937): 544–546.

Jérôme Losson
Department of Physics and Center For Non-Linear Dynamics, McGill University, McIntyre Building, Room 1123a, 3655 Drummond, H3G-IY6, Montréal P. Q, CANADA

Bifurcations and Chaos in a Paradigm Equation for Delayed Mixed Feedback

1. INTRODUCTION

The instantaneous transmission of information between two systems is physically impossible. The presence of delays poses a fundamental constraint on any theory describing physical interactions. If the time scale of the delays is comparable to the time scale of the processes under consideration, then it becomes essential to include the delay explicitly in the models. The investigators have to deal with delay differential equations (DDE's). The fields of investigation in which such situations occur are as varied as neurophysiology and laser physics, studies of dynamical diseases and nuclear engineering.[1,2,4]

We are concerned here with one such equation. It arose in the context of biomathematical modeling, and was first used as a paradigm for mixed delayed feedback systems.[5] When the feedback is mixed, the response of the system is maximal for some intermediate value of the variable, extremal values being considered damaging. These control mechanisms arise most often in the biological sciences, since they regulate and maintain living organisms.

1990 Lectures in Complex Systems, SFI Studies in the Sciences of Complexity, Lect. Vol. III, Eds. L. Nadel and D. Stein, Addison-Wesley, 1991   **527**

To obtain the equation, the rate of change of a quantity x is equated to the difference between a production rate P and a destruction rate D:

$$\frac{dx}{dt} = P - D. \tag{1.1}$$

If the production rate depends on the delayed variable $x(t-\tau)$ while the destruction rate is proportional to $x(t)$, we have

$$\frac{dx}{dt} = -\alpha x(t) + P(x(t-\tau)) \tag{1.2}$$

We consider the feedback function P to be piecewise constant:

$$P(y) = \begin{cases} c & \text{if } y \in [\theta_1, \theta_2] \\ 0 & \text{if } y \notin [\theta_1, \theta_2] \end{cases} \tag{1.3}$$

This "box" shape is the idealization of a smooth hump, characteristic of mixed feedback control functions. In general if P is a smooth function, then Eq. (1.2) is not solvable analytically. The piecewise constant P makes analytic integration possible. The solutions are piecewise exponentials increasing or decreasing depending on whether the delayed variable is inside or outside the interval $[\theta_1, \theta_2]$.

$$x(t) = \begin{cases} x(t_0) \exp[-\alpha(t - t_0)] & \text{if } x(s - \tau) \notin [\theta_1, \theta_2] \forall s \in [t_0, t] \\ \gamma - (\gamma - x(t_0)) \exp[-\alpha(t - t_0)] & \text{if } x(s - \tau) \in [\theta_1, \theta_2] \forall s \in [t_0, t] \end{cases} \tag{1.4}$$

where $\gamma = c/\alpha$ the asymptote for the rising exponential.

2. THE ANALOG COMPUTER

Analog computation might appear an archaic way to study the dynamics of a DDE. In fact analytic and numerical integration schemes for Eq. (1.2) already exist.[3,5] The design of a task specific electronic analog computer allows us to study *in real time* a physical system built so that it will be described accurately by a DDE. It facilitates rapid scans of parameter space, and can serve as a guide for the analytic investigation of such problems as the influence of noise or non-constant IF's on solution behavior.

A. CIRCUIT DESIGN

The circuit is a closed loop oscillator. It amplifies the present signal, delays it and transforms it simultaneously according to the desired feedback function, sums the resulting signals, integrates the sum, and then equates the result of the integration with the initial signal.

The amplification, summing, and integration are performed by standard operational amplifiers. The delay is provided by an analog delay line. This component is a sampling device known as a "Bucket Brigade Device" or BBD. Comparators and logic switches provide the desired feedback function.

The initial preparation of the circuit (equivalent to the initial function for the equation) is given by the turn-on state of the BBD. Control of this state is of crucial importance if we are to study the sensitivity of our system to initial conditions. While some results in the literature indicate the presence of multistability in DDE's,[3,4] no work has been directed at understanding this phenomenon *per se*. By interfacing the analog computer with a digital computer we can program the desired IF into the delay line, close the loop with a switching circuit, and observe the onset of oscillation.

B. CIRCUIT PERFORMANCE

With constant IF's, the signals obtained using the analog computer display quantitative agreement with the analytic solutions[5] when the limit cycles are simple (i.e., when the solutions possess less than four extrema per delay).

When the limit cycles are more complicated, the strict quantitative agreement with the theory is kept up to a shift in parameter space. This is due to the fact that dynamically relevant parameters cannot be measured accurately on the circuit when the signals vary rapidly (i.e., when the solutions possess more than four extrema per delay).

Because it has a finite cut-off frequency the circuit cannot reproduce the most erratic solutions of Eq. (1.2). This is not a serious problem. We can use this novel tool to further our understanding of delayed dynamics by focusing our attention on regions of parameter space in which the solutions oscillate with relatively low frequencies.

As stated above, the turn-on state of the BBD is the initial function. The oscillator can be started with constant initial functions. The bifurcation diagram obtained as θ_1 is increased slowly towards θ_2 (keeping all other parameters in the equation constant) can then be studied. After increasing θ_1 we lower it and find *hysteresis* in the bifurcation diagram. That is, different solutions of Eq. (1.2) coexist at the same parameter values depending on whether θ_1 is being raised or lowered. This indicates the possible existence of multistability in the system. Previously multistability had been observed in DDE's but only by the variation of constant IF's.[4] In our case the IF's are non-constant since they are solutions of Eq. (1.2).

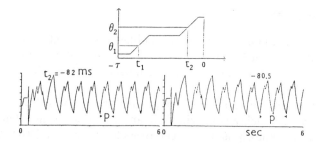

FIGURE 1 A change in the single parameter t_2 of the initial function (top) leads to a doubling of the period (P) of the electronically generated solutions (left vs. right) with all other parameters held constant. The change in t_2 was only 1.8%.

The next question that we address is whether or not multistability can be observed using simple non-constant IF's described by a small set of parameters. The reason is the following: If multistability is observed in the physical system, then it is of interest to pursue our analytic investigation of this phenomenon. We test the circuit with simple IF's hoping that the fewer the control parameters, the simpler the analysis will be.

We illustrate the possible existence of multistability with IF's which cross both θ_1 and θ_2 only once. The crossing times of θ_1 and θ_2 on the initial interval are labeled t_1 and t_2. Figure 3 shows two solutions corresponding to the same parameter values for different initial conditions. This result is not a *proof* of the existence of multistability, but a strong indication of its presence in the system.

This result is interesting because period-doubling bifurcations are usually produced by changes in the parameters, not by changes in the initial conditions. To confirm the existence of these bifurcations and to understand their exact nature further analytical work is required. The analog computer is not a precise tool of investigation, rather it has given us a flavor for the complexity of the dynamics displayed by a relatively simple DDE. Its use has highlighted the sensitivity of these systems on their initial preparation, and will guide the preliminary analysis of the problem.

3. THEORY

A. REDUCTION TO ONE-DIMENSIONAL MAPS

We are going to describe here some techniques first presented by Mackey and an der Heiden.[5] They consist of reducing a class of DDE's to discrete one-dimensional maps and studying these maps to obtain an insight into the dynamics of the original delay differential equation.

The idea is to construct a transformation T mapping a time interval onto a point, and then to follow the evolution of this point under the action of a diffeomorphism. The sequence obtained by iterating the map completely describes the solution of the DDE. To understand the procedure it is useful to realize that the information contained in a time interval of length τ is redundant. This is due to the piecewise constant nature of P. To describe completely the solution on such a time interval, one only needs to know the following:

1. When the solution crossed the thresholds θ_1 and θ_2 within the last delay.
2. Which region of phase-space the solution entered at these crossing points. (The region above, inside, or below the interval $[\theta_1, \theta_2]$).

Restricting ourselves to regions of parameter space in which the solutions cross a given threshold at most once within a delay, only one variable is necessary to describe a solution on a time interval of length τ. We can illustrate the procedure with Eq. (1.2) along with the following normalizations: $\tau = 1$, and $\theta_1 = 1$. This simplifies the analysis without loss of generality. Let us define a set D of initial functions ϕ, on which the transformation T will act:

DEFINITION 1. $\phi \in D$ iff there exists a $w \in [0, 1]$ such that $\forall t \in [-1, -1 + w)$, $\phi(t) > 1$, $\forall t \in (-1 + w, 0)$, $\phi(t) < 1$ and $\phi(0) = 1$. (see Figure 2(a)).

> N. B. The solution $x(t)$ is completely determined on $[0,1]$ by w. The solution is labeled $: x_w(t)$.
>
> We note here without proof that there exists a unique w_l such that $\forall w < w_l$, the solution will tend to the lower asymptote without ever crossing θ_1, and a unique w_2 such that if $w > w_2$, the solution will tend to the upper asymptote and cross θ_2 before ever crossing θ_1. The proof can be found in Mackey and an der Heiden.[5]

From Figure 2(a) it is clear that the solution on the interval $I_2 = [w' - 1, w']$ belongs to the set D. If we now go back to the original interval I_1, $(t \in [1, 0])$ the time spent above the first threshold is w. In I_2, the time spent above the same threshold is now $w_3 - w' + 1$. Formally,

$$T(I_1) = w$$
$$T(I_2) = w_3 - w' + 1 \tag{2.1}$$

and this information is enough to describe the evolution of the solution $x_w(t)$ for all t.

The one-dimensional application F of the unit interval onto itself,

$$F(w) = w_3 - w' + 1 \tag{2.2}$$

is a discrete-time system which describes the evolution of $x_w(t)$ for all $t > w''$.

Note that if $F(w) > 1$, the solution escapes to θ_2. In the simplest case (illustrated in Figure 2(a)) it spends more than a delay above θ_2 and is then "reinjected" towards $\theta_1 = 1$. Because the solution remains above θ_1 longer than one delay, it *loses memory* of its behavior in the neighborhood of θ_1. As a result, on I_3 the time spent above θ_1 is independent of the way the solution reached θ_2. In other words, with this choice of parameters, the transformation F is constant on the interval $(w_2, 1]$. Figure 2(b) shows an example of a transformation F derived with a set of parameters for which the solution looks like the one displayed in Figure 2(a). A detailed derivation of this map can be found in Mackey and an der Heiden.[5]

The dynamics of the map are directly related to the behavior of the original DDE. For example, a fixed point of the application corresponds to a periodic solution of the equation, and the linear stability analysis around the fixed point tells us about the stability of the corresponding limit cycle. The presence of chaotic orbits in the map would imply the existence of chaotic solutions of the continuous time system. Similarly, the basins of attraction of the map characterize some of the basins of attraction in the infinite-dimensional phase space of the delay differential equation.

In regions of parameter space where the solutions cross a given threshold more than once within a delay, the reduction to a discrete time system is in principle possible. However, the dimension of the map increases with the number of variables necessary to describe the solution on an interval of length τ; in practice, this approach is not efficient for the more complex solutions of Eq. (1.2). Indeed, in some regions of parameter space, the complexity of the dynamics makes a deterministic investigation not only arduous but perhaps meaningless.

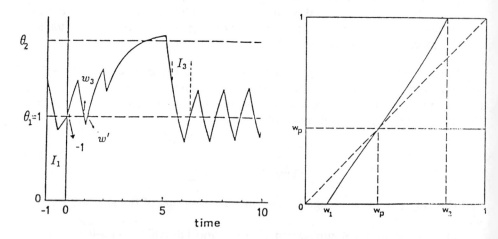

FIGURE 2   (a) is a spiral type solution of Eq. (1.2) with $\theta_1 = 1$, $\theta_2 = 1.9$, $\alpha = 1$, $\gamma = 2$, $\tau = 1$ and $w = 0.55$. The solution is slowly repelled upwards from the unstable limit cycle around θ_1. After a short excursion around θ_2, it decays back to θ_1 and the cycle repeats. This kind of behavior is described completely by the map shown in (b). (Figure courtesy of A. Longtin.[3])

3. CONCLUSIONS

The purpose of this exposé was to describe some of the techniques that have been used to study the wide range of behaviors displayed by an idealized, delayed, mixed feedback control system.

Most often, delayed feedback systems are part of complex biological regulation processes and the use of the oscillator has given us the opportunity to work with a physical system accurately described by a mathematical model. The electronic analog simulation indicated strongly the presence of multistability in a physical delayed feedback system, and will be used as a guide to direct further analysis of these mechanisms.

The reduction to a finite-dimensional map has proven to be a useful tool to classify some of the solutions of the equation even if it is limited in practice to the study of restricted regions of parameter space. The literature on finite-dimensional maps is extensive and we hope that further analysis will increase our understanding of delayed dynamics.

REFERENCES

1. Guevara, M. R., L. Glass, M. C. Mackey, and A. Shrier. "Chaos in Neurobiology." *IEEE Trans.* **13** (1983): 5.
2. Ikeda, K. "Multiple-Valued Stationary State and Its Instability of the Transmitted Light by a Ring Cavity System." *Optics Comm.* **30** (1979): 257.
3. Longtin, A. "Nonlinear Oscillations, Noise and Chaos in Neural Delayed Feedback." Ph.D. dissertation, Department of Physics, McGill University, 1990.
4. Mackey, M. C., and U. an der Heiden. "Dynamical Diseases and Bifurcations: Understanding Functional Disorders in Physiological Systems." *Funkt. Biol. Med.* **1** (1982): 156.
5. Mackey, M. C., and U. an der Heiden. "The Dynamics of Production and Destruction: Analytic Insight into Complex Behavior." *J. Math. Biol.* **16** (1982): 75.

Nicholas Provatas
Department of Physics, and Center for Nonlinear Dynamics, McGill University, 3600 University Street, Montreal, Quebec, Canada H3A-2T8

Asymptotic Periodicity in One-Dimensional Maps

Density evolution in one-dimensional maps is examined, associating the macroscopic state of these dynamical systems with a phase-space density. A type of density evolution known as asymptotic periodicity is studied. For asymptotically periodic systems, density evolution becomes periodic in time, as do also some macroscopic properties calculated from them. The properties of asymptotic periodicity are examined and then used to calculate the time correlation function and the conditional entropy of maps displaying asymptotic periodicity. We illustrate asymptotic periodicity in the hat map and the quadratic map at the parameters where these maps generate *banded chaos* or *quasi-periodicity*.

INTRODUCTION

Following the numerical (computer-generated) demonstration of the existence of highly irregular dynamical behavior in relatively simple nonlinear systems, the study of so-called "chaotic" systems has captured the attention of literally thousands of scientists in the physical, mathematical, biological, and social sciences. Since many studies have some or all of their basis in numerical simulations, it is natural that many investigators have focused attention on the irregular behavior displayed by the trajectories of these nonlinear systems.

However, the irregular and apparently unpredictable nature of trajectory evolution in many nonlinear dynamical systems can be greatly simplified if one looks at their behavior in terms of density evolution.[7] This alternative viewpoint has particular appeal when applying the concepts of nonlinear dynamics to many problems in statistical physics,[9] and offers an immediate connection with the mathematical discipline of ergodic theory which developed from the early work of Boltzman and Gibbs.

This paper examines the recently discovered[8] property of asymptotic periodicity in the density evolution of one-dimensional maps. The concepts of asymptotic periodicity are illustrated through two example systems. The first map we study is the hat map

$$S_a(x) = \begin{cases} ax & 0 \leq x \leq \frac{1}{2} \\ a(1-x) & \frac{1}{2} < x \leq 1, \end{cases} \tag{1}$$

where $0 < a \leq 2$. For $2^{1/2^{n+1}} < a \leq 2^{1/2^n}$ ($n = 0, 1, \cdots$), it has been shown[12] that Eq. (1) generates banded chaos of period 2^n. The second map is the quadratic map

$$S(x) = rx(1-x) \qquad 0 < r \leq 4, \tag{2}$$

which has been extensively studied because of its wealth of dynamical properties.[1] Maps like Eq. (2) often arise as approximations to Poincaré sections of low-dimensional strange attractors of dissipative systems. Examples include Couette flow, Bérnard instability,[4] and the truncated approximation of the Navier Stokes equations.[2] For the quadratic map there exists a set of parameters, denoted here as $r = r_n$, where banded chaos of period 2^n emerges.[3]

Qualitatively the phenomenon of banded chaos in Eq. (1) and Eq. (2) is characterized by the emergence of a phase-space attractor, on $[0, 1]$, comprising 2^n bands. A trajectory of the time series of Eq. (1), Eq. (2) will periodically visit each band. The motion within each band, however, is chaotic, possessing a positive Lyapunov exponent. More precisely, denoting the ith band by J_i, the trajectory of $S^{2^n} : J_i \rightarrow J_i$ is chaotic. Lorenz called this phenomenon *noisy periodicity*. The construction of the bands J_i and the maps $S^{2^n} : J_i \rightarrow J_i$ are given by Yoshida et al.[12] and Grossmann and Thomae,[3] for the hat map and the quadratic map respectively.

TRAJECTORY VERSUS DENSITY EVOLUTION

Knowing the phase-space attractor of a chaotic dynamical system, but not the actual solution of the trajectory through it, is reminiscent to the situation encountered when dealing with the N-body problem. To make the study of the evolution of a chaotic dynamical system more tractable, we may argue, as Gibbs did when dealing with the N-body problem, that a macroscopic state of a system is not in general described by a single point in phase space, but rather a collection, or *ensemble* of points. At a given time t, this ensemble is distributed according to some density f_t. The macroscopic state of the system is associated with f_t, and hence the evolution of a system, in this formalism, is therefore given by the evolution (or flow) of densities f_t. In this approach, exact values are replaced by ensemble averages or expectations weighted by the phase-space density, f_t.

For systems of statistical mechanics that are in the thermodynamic limit (number of particles $N \to \infty$), it is assumed that the evolution of densities attain the density of the canonical ensemble Z. For low-dimensional systems such as Eq. (1) and Eq. (2) however, the evolution of densities can display several types of non-equilibrium behavior.

THE EVOLUTION OF DENSITIES AND MARKOV OPERATORS

The evolution of densities under the action of a dynamical system, S, is described by a Markov operator which we denote by P. Formally any linear operator P^t : $L^1 \to L^1$ that satisfies

$$P^t f > 0 \qquad \text{and} \qquad \int_X P^t f(x) dx = \int_X f(x) dx$$

is called a *Markov operator*,[7] where X denotes the phase space on which S operates. Throughout this paper we deal with a subset of L^1 functions which are everywhere nonnegative and normalized to one. This is the set of densities and is denoted by D. It is clear that, when a Markov operator acts on a density, it yields another density. Beginning with an ensemble of phase-space points representing some macroscopic state of a system, and distributed according to a density f_0, one unit of time (iteration) later the new density state of the system, f_1, is given by $f_1 = P f_0$. For deterministic one-dimensional dynamical systems P is also known as the *Frobenius-Perron* operator, and is given by

$$P f(x) = \frac{d}{dx} \int_{S^{-1}(0,x)} f(y) dy. \tag{3}$$

Markov operators may also possess a *stationary density* f^*. This density satisfies $P f^* = f^*$ and may be associated with a state of thermodynamic equilibrium of a dynamical system.

The evolution of densities under Eq. (3) characterizes P as well as the dynamical map[7] S. Three general behaviors may be displayed by the sequence $\{P^t f_0\}$, where f_0 represents the density of initial preparation of the system. These are *ergodicity*, *mixing*, and *exactness*. In all three cases the system possesses an invariant density

f^*. However where the three behaviors differ is in the way the sequence $\{P^t f_0\}$ converges to f^*.

Of the three cases above, exactness implies the strongest form of convergence of $\{P^t f_0\}$. Mathematically a system is said to be exact if and only if

$$\lim_{t\to\infty} |P^t f_0 - f^*| = 0$$

as $t \to \infty$, for all initial densities f_0. Exactness may be considered as the analogue of an approach to equilibrium from all initial preparations of a system.

Mixing implies a weak form of convergence of $\{P^t f_0\}$. In particular for any function $\mathcal{F}$ a mixing system satisfies

$$\lim_{t\to\infty} \langle P^t f_0, \mathcal{F} \rangle = \langle f^*, \mathcal{F} \rangle,$$

for all initial densities f_0. Mixing systems spread densities throughout the accessible phase space, as determined by the support of f^*.

Ergodicity implies the weakest form of convergence of $\{P^t f_0\}$. For ergodic systems

$$\lim_{t\to\infty} \frac{1}{t} \sum_{n=1}^{t-1} \langle P^n f_0, \mathcal{G} \rangle = \langle f^*, \mathcal{G} \rangle$$

for all $f_0 \in D$ and any function $\mathcal{G}$. Exactness implies mixing which in turn implies ergodicity. However, ergodicity alone does not constrain the sequence $\{P^t f_0\}$ to become asymptotically equal to f^*.

ASYMPTOTIC PERIODICITY

Asymptotic periodicity is a type of density evolution that may be displayed by one-dimensional maps. Without loss of generality the phase space of these maps will be taken as $[0, 1]$. For asymptotically periodic systems the sequence $\{P^t f_0\}$ satisfies the Komornik Spectral Decomposition (see Komornik and Lasota,[6] Theorem 1). This theorem states that for any initial density f_0,

$$Pf_0(x) = \sum_{i=1}^{r} \lambda_i(f_0)g_i(x) + Qf_0(x), \tag{4}$$

where the functions g_i form a sequence of r densities satisfying $g_i g_j = 0$ if $i \neq j$, $i, j = 1, \cdots, r$. This condition implies that the supports of the g_i densities, denoted supp$\{g_i\}$, are disjoint. Also the g_i satisfy $Pg_i = g_{\alpha(i)}$, where $\alpha(i)$ is a permutation on the numbers $\{1, 2, \cdots, r\}$. For ergodic systems $\alpha(i)$ must be a cyclic permutation.[7]

The scaling coefficients $\lambda_i(f_0)$ are linear functionals of the initial density f_0 given by

$$\lambda_i(f_0) = \int_0^1 Z_i(x)f_0(x)dx,$$

where $\{Z_i(x)\}$ is a sequence of L^∞ functions. The symbol Q is called the transient operator, and satisfies $\|Q^t f_0\| \to 0$ as $t \to \infty$. The tth iterate, $P^t f_0$, may be written as

$$P^t f_0(x) = \sum_{i=1}^r \lambda_i(f_0)g_{\alpha^t(i)}(x) + Q^t f_0(x). \tag{5}$$

Allowing the transient operator to decay and noting that the permutation $\alpha(i)$ is invertible, we may write

$$P^t f_0(x) = \sum_{i=1}^r \lambda_{\alpha^{-t}(i)}(f_0)g_i(x). \tag{6}$$

From Eq. (6), it is easy to verify that the (necessarily unique) invariant density of ergodic asymptotically periodic systems is given by

$$f^*(x) = \frac{1}{r}\sum_{i=1}^r g_i(x). \tag{7}$$

Equation (6) describes a density evolution that is periodic in time. At a given time, $P^t f_0$ may be visualized as a linear combination in the basis states g_i, each scaled by a probabilistic weighting factor $\lambda_{\alpha^{-t}(i)}(f_0)$. Since we are dealing with densities, the $\lambda_i(f_0)$ sum to 1. Each coefficient $\lambda_i(f_0)$ gives a probabilistic measure of an asymptotically periodic system described by the map S, being in a basis state g_i. Scaling of more than one basis state implies that the system has a probability of being in more than one basis state. When only one term is present in the sum of Eq. (6), the system will be found in one of the g_i states at all times.

Expectation values of a measurable quantity, O, at a time t are given by weighting over the density $P^t f_0$. It is clear from Eq. (6) that $\langle O \rangle$ will generally be periodic in time. The time dependence of the oscillation is found from Eq. (6) to be

$$\langle O \rangle(t) = \lambda_{\alpha^{-t}(i)}(f_0)\langle O(x)\rangle_i, \tag{8}$$

where

$$\langle O(x)\rangle_i = \int_{\text{supp}\{g_i\}} O(x)g_i(x)dx.$$

ASYMPTOTIC PERIODICITY IN THE HAT AND QUADRATIC MAPS

In this section we illustrate numerically the properties of asymptotic periodicity using the hat map Eq. (1) and the quadratic map Eq. (2). From Eq. (3) it is easy to show that the Frobenius-Perron operators of these two maps are respectively

$$Pf(x) = \frac{1}{a} \left[f\left(\frac{x}{a}\right) + f\left(1 - \frac{x}{a}\right) \right] \tag{9}$$

and

$$Pf(x) = \frac{1}{\sqrt{1 - \frac{4x}{r}}} \left[f\left(\frac{1}{2} + \frac{1}{2}\sqrt{1 - \frac{4x}{r}}\right) + f\left(\frac{1}{2} - \frac{1}{2}\sqrt{1 - \frac{4x}{r}}\right) \right]. \tag{10}$$

It can be shown[10] that, at the parameters where these maps display period 2^n banded chaos, the evolution of $\{P^t f_0\}$ becomes asymptotically periodic with period 2^n. Thus, asymptotically, the members of the sequence $\{P^t f_0\}$ will be linear combination in 2^n g_i densities. The scaling coefficient associated with each g_i in this decomposition is just $\lambda_{\alpha^{-t}(i)}(f_0)$, where $\alpha^t(i) = (i + t) \mod 2^n$.

When the hat map parameter a is in the range $2^{\frac{1}{4}} < a \le 2^{\frac{1}{2}}$, the sequence $\{P^t f_0\}$ displays period-two banded chaos. In this parameter range the scaling coefficients $\lambda_i(f_0)$ and the densities g_i of the spectral decomposition, Eq. (4) can be determined analytically. Figure 1 shows the sensitivity of the evolution of $\{P^t f_0\}$ to f_0, for three choices of initial density f_0, when $a = \sqrt{2}$. Figure 1(c) involves a transient, since f_0 is not already in a linear combination of g_1 and g_2.

A similar asymptotic decomposition exists for the quadratic map at the parameter values $r = r_n$, where banded chaos is displayed. Figure 2 illustrates one cycle in the evolution of $\{P^t f_0\}$ for the quadratic map when $r = r_1$. Once again we see that different choices of f_0 (shown in the inset) lead to different spectral decompositions of $\{P^t f_0\}$, in the basis states g_i, and hence different statistical descriptions of the subsequent physical evolution of the system characterized by this map.

THE CONDITIONAL ENTROPY FOR ASYMPTOTICALLY PERIODIC SYSTEMS

Assuming the existence of a density f describing a thermodynamic state of a system at a time t, Gibbs introduced the concept of the *index of probability*, given by $-\log f(x)$. Weighting the index of probability by the density f, he introduced what is now known as the *Boltzmann-Gibbs entropy*, given by

$$H(f) = -\int_X f(x) \log f(x) dx .$$

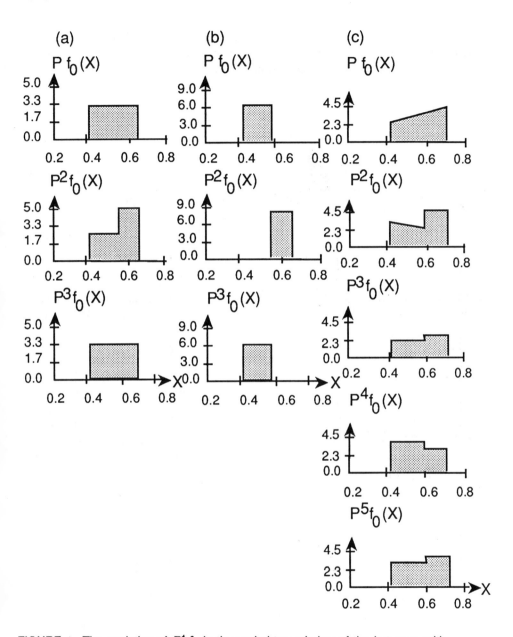

FIGURE 1 The evolution of $P^t f_0$ in the period-two window of the hat map, with $a = \sqrt{2}$. In (a), f_0 is uniform over $J_1 \cup J_2$. Since the g_i are uniform over J_i, $i = 0, 1$, $P^t f_0$ sets into immediate oscillations without transients. In (b), f_0 is uniform over the subspace J_2. Again $P^t f_0$ sets into immediate oscillations through the states g_1 and g_2. In (c) $f_0(x) = (4(5 + \sqrt{2})/7)x$, restricted to $J_1 \cup J_2$. Now $P^t f_0$ evolves through two transient densities before settling into a periodic oscillation.

It can be shown[5,11] that the Boltzmann-Gibbs entropy is the only entropy definition satisfying the property of being an extensive quantity, which a mathematical analog of the thermodynamic entropy should have.

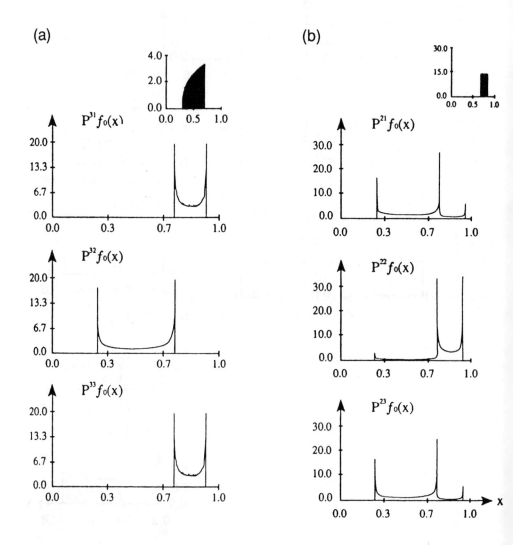

FIGURE 2 A numerical illustration of one periodic cycle of the asymptotic sequence $\{P^t f_0\}$ for the parameter $r = r_1 = 3.678573508$. A transient of 20 densities has been discarded, and the iterates $P^{21} f_0$, $P^{22} f_0$, and $P^{23} f_0$ are shown. Since $P^{21} f_0 = P^{23} f_0$, the sequence $\{P^t f_0\}$ asymptotically repeats with period two. In (a), the initial density f_0, shown in the inset, has the form $f_0(x) = 4.52(x - 0.3)^{1/3}$, while in (b) $f_0(x)$ is uniform on $[0.7, 0.8]$.

The Boltzmann-Gibbs entropy can be generalized by introducing the *conditional entropy*. If f and g are two densities such that supp $f \subset$ supp g, then the conditional entropy of the density f with respect to the density g is defined as

$$H_c(f|g) = - \int_X f(x) \log \left[\frac{f(x)}{g(x)} \right] dx. \tag{11}$$

As $H_c(f|g) = 0$ when $f = g$, this implies that the conditional entropy is a measure of how close the functional form of f is to g. Moreover, using an identity known as the Gibbs inequality, it can be shown[7] that the conditional entropy satisfies

$$H_c(f|g) \leq 0. \tag{12}$$

When dealing with asymptotically periodic systems, the limiting conditional entropy takes on a particularly transparent form, clearly expressing that $\lim_{t\to\infty} H_c \times (P^t f_0|f^*)$ is dependent on the initial preparation of the system through f_0. To see this, use the invariant density Eq. (7) along with the asymptotic decomposition Eq. (6), and the orthogonality of the g_i to obtain

$$\lim_{t\to\infty} H_c(P^t f_0|f^*) = \sum_{i=1}^{r} \int_X \lambda_{\alpha^{-t}(i)}(f_0) g_i(x) \log(r\lambda_{\alpha^{-t}(i)}(f_0)) dx$$

$$= \sum_{i=1}^{r} \int_X (\lambda_{\alpha^{-t}(i)}(f_0) g_i(x)) \left[\log(\lambda_{\alpha^{-t}(i)}(f_0)) + \log(r) \right] dx. \tag{13}$$

Also since the permutation $\alpha(i)$ is invertible, we have

$$\sum_{i=1}^{r} \lambda_{\alpha^{-t}(i)}(f_0) = \sum_{i=1}^{r} \lambda_i(f_0). \tag{14}$$

Thus, defining

$$H_c^{\infty}(P^t f_0|f^*) \equiv \lim_{t\to\infty} H_c(P^t f_0|f^*), \tag{15}$$

we may re-express the limiting conditional entropy as

$$H_c^{\infty}(P^t f_0|f^*) = -\log(r) - \sum_{i=1}^{r} \lambda_i(f_0) \log \lambda_i(f_0). \tag{16}$$

Noting that the $0 \leq \lambda_i(f_0) \leq 1$ for all i, we obtain

$$-\log(r) \leq H_c^{\infty}(P^t f_0|f^*) \leq 0. \tag{17}$$

When an initial density f_0 is localized over one of supp $\{g_i\}$, then $\{P^t f_0\}$ will asymptotically cycle through the sequence $\{g_i\}$. In this case there is only one component to the spectral decomposition Eq. (5) at any time t. According to Eq. (16) this situation is one of lowest conditional entropy and $H_c^\infty(P^t f_0|f_0) = -\log(r)$. Physically this implies that the initial ensemble, described by f_0, will evolve through supp $\{g_i\}$ in the most localized manner possible. At any time t, only a "pure state" g_i is needed to describe the statistical properties of the system. In general any f_0 whose support runs over the boundary of some g_i will cause $P^t f_0$ to decompose into a linear combination of several densities g_i. At a time t the members of an ensemble are now less localized, and more information is required to determine their distribution through the phase space. As a result, the conditional entropy of

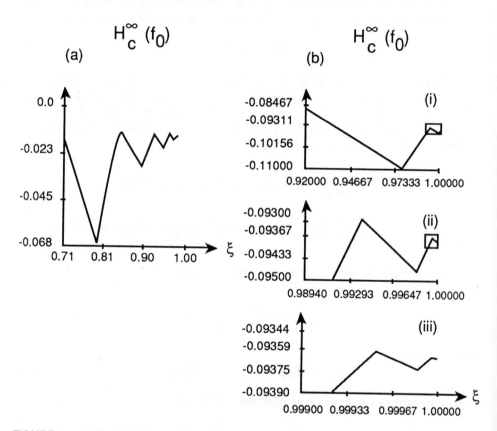

FIGURE 3   (a) The limiting conditional entropy, $H_c^\infty(f_0)$, versus the spreading parameter ξ for the hat map at $a = \sqrt{2}$. ξ is equal to the width of the support of an initial density f_0 which is uniform over $[L_b, L_b + \xi]$. The local maxima in the figure are all equal. (b) shows a graph of the limiting conditional entropy $H_c^\infty(f_0)$ versus ξ for the quadratic map at $r = r_1$. The parameter ξ plays the same role as in (a). Variations in $H_c^\infty(f_0)$ occur over smaller ξ scale for the quadratic map. $b(ii)$ is a blow-up of the inset box in frame $b(i)$. $b(iii)$ is a blow-up of the inset box in $b(ii)$.

a linear combination of several densities g_i has a higher conditional entropy than a single pure state g_i.

Equation (16) shows that the unique, limiting conditional entropy of an asymptotically periodic system settles down to a value uniquely determined by the density of the initial preparation of the system, while the iterates $P^t f_0$ remain asymptotically periodic. This implies that all density states within the cycle to which $\{P^t f_0\}$ converges are of the same entropy with respect to the stationary density Eq. (7).

The dependence of conditional entropy on f_0 is illustrated in Figure 3(a) for the hat map at $a = \sqrt{2}$ and in Figure 3(b) for the quadratic map at $r = r_1$. The figure shows a plot of limiting conditional entropy versus a parameter ξ where ξ is such that

$$f_0(x) = \frac{1}{\xi - L_b} \qquad L_b \le x \le \xi \tag{18}$$

and L_b specifies a lower bound chosen to coincide with the minimum of supp $\{g_2\}$, where g_2 is the leftmost density of the components of Figure 1 or Figure 2.

REFERENCES

1. Feigenbaum, M. "Universal Behavior in Non-Linear Systems." *Los Alamos Science* **1** (1988): 4.
2. Franceschini, V., and C. Tebaldi. "Sequences of Infinite Bifurcations and Turbulence in a Five Mode Truncation of the Navier Stokes Equations." *J. Stat. Phys.* **21** (1979): 707.
3. Grossmann, S., and S. Thomae. "Correlations and Spectra of Periodic Chaos Generated by the Logistic Parabola." *J. Stat. Phys.* **26** (1981): 485.
4. Hellman, R. H. G. "Self-Generated Chaotic Behavior in Non-Linear Mechanics." In *Fundamental Problems in Statistical Mechanics*, edited by E. G. D. Cohen, vol. 5, 165, 1980.
5. Khinchin, A. I. "Asymptotic Decomposition of Markov Operators." In *Mathematical Foundations of Statistical Mechanics*. New York: Dover, 1949.
6. Komornik, J., and A. Lasota. *Bull. Polish. Acad. Sci. (Math.)* **35** (1987): 321.
7. Lasota, A., and M. Mackey. *Probabilistic Properties of Deterministic Systems.* Cambridge: Cambridge University Press, 1985.
8. Lasota, A., and M. Mackey. "Noise and Statistical Periodicity." *Physica D* **22** (1985): 143.
9. Mackey, M. C. "The Dynamic Origin of Increasing Entropy." *Rev. Mod. Phys.* **61** (1989): 981.
10. Provatas, N. "Inherent and Noise Induced Asymptotic Periodicity." M.Sc. thesis, McGill University, 1990.
11. Skagerstam, B. S. "On the Mathematical Definition of Entropy." *Z. Naturforsch.* **29a** (1974): 1239.
12. Yoshida, T., H. Mori, and H. Shigematsu. "Analytic Study of Chaos of the Tent Map: Bond Structures, Power Spectra, and Critical Behaviors." *J. Stat. Phys.* **31** (1982): 278.

Albert J. Wong
Physics Department, Princeton University, Princeton, NJ 08544

Theme and Variations: Spin Glasses, Neural Networks, and Prebiotic Evolution

Prebiotic organic soups and neural networks appear to have strongly analogous traits. Both may be mathematically modeled by spin glasses. Certain pattern incorporation techniques of neural network models (the Hopfield model and the temporal-associative memory model) provide mathematical constructions that may be fruitfully applied to prebiotic model systems.

1. INTRODUCTION

Life on Earth has been wrought from what was a once lifeless planet. We, as living beings, stand as testimony to the emergence of order out of what seems to be an initial chaos. How can a disordered system self-organize itself into an ordered one?

Our minds, too, can forge order out of disorder. Anyone who has been puzzled and then suddenly recognizes a clear pattern, must know at least intuitively that we, as thinking beings, can create lucid, ordered understandings from what before were disordered, muddled thoughts. This mental process of insight, of suddenly seeing disordered information become ordered, seems analogous to the chemical processes whereby a disordered genetic soup evolves into an information-laden ordered structure. In this brief discussion, I elaborate upon certain analogies between the prebiotic evolution of ordered polymers from a disordered soup and neural network pattern recognition.

One distinction between these two systems of prebiotic soups and neural networks however is critical. A neural network is fully occupied at a given moment by a single information string, i.e., the instantaneous collective firing state of the network. On the other hand, a polymer ensemble at any instant contains a large number of distinct information strings—each polymer represents one information string. As a result of the prebiotic soup's multitude of strings, a rich feedback structure may arise.

2. SPIN GLASSES AND GENETIC POLYMERS

Assume that there are only two types of genetic bases (called monomers A and B), instead of the standard set of four "real world" RNA or DNA bases (adenine, uracil or thymine, cytosine, guanine). A polymer can then be represented by a directional sequence of A's and B's (e.g., A-B-B-A-A-...). A "species" of polymer is defined here as a set of polymers of nearly identical sequence patterns.

Evolutionary systems should be characterized by a high selectivity among diverse *possible* species to a small number of *realized* species. If the total number of evolutionarily accessible species is small, then our evolutionary universe would be boringly simple, uniform, and without subtle variations. Our model must allow for a diverse set of possible species. However, if among this diverse set of possible species, every single species were also evolutionarily realized, then the system would present an unselected bounty of species and would be boringly diverse.[1]

Spin glasses are one example of a physical system that evolve among diverse possible states into a small number of realized ones. As such, a spin-glass model seems particularly appropriate for modeling both the selectivity and diversity we desire in prebiotic evolution. Spin glasses are systems in which a non-magnetic host substance is "polluted" with quantized magnetic spin moments $S_i = \pm 1$, $i = 1, \ldots, n$, of a magnetic substance. These spins are localized in space at sites $i = 1, \ldots, n$. We may describe the overall state of the spin-glass system with a vector $\mathbf{S} \in \aleph^n \subset \Re^n$, where $\aleph^n$ is the n-dimensional Hamming space and where the elements of $\mathbf{S}$ are $S_1, S_2, \ldots, S_n$. The interactions J_{ij} between the spin moments S_i, S_j are random with mean zero. The J_{ij} coupling constants compete to cause a spin S_i to be aligned parallel (positive J_{ij}) or anti-parallel (negative J_{ij}) to each of the other S_j spins. Because of the large number of competing J_{ij} interactions, all of the interactions are rarely simultaneously "satisfied" for any state $\mathbf{S}$. Consequently, spin-glass systems are called *frustrated* systems—there is no single ground state for which all interactions are satisfied. As a consequence of their high degree of frustration, spin glasses have no unique ground state. However, starting from a random initial state $\mathbf{S}$, the spin-glass system will freeze below a temperature T_f into a particular state $\mathbf{S}'$, which is highly metastable (i.e., a local energy minima), but only one of a number of possible final configurations $\{\mathbf{S}^\alpha\}$. The free-energy surface in the state space $\aleph^n$ consists of numerous hills and valleys with many possible ground states. The Hamiltonian that defines the energy landscape is

$$H = \sum_{i,j=1}^{n} J_{ij} S_i S_j = -\mathbf{S}^t \cdot \mathbf{J} \cdot \mathbf{S}, \tag{1}$$

where S_i, S_j are the spins at sites i, j respectively, J_{ij} is the interaction between spins S_i and S_j, $\mathbf{S} \in \aleph^n$ is the state of the system (a column vector with elements $S_1, S_2, \ldots, S_n$), $\mathbf{J}$ is a linear transformation from $\Re^n \to \Re^n$ represented by a $n \times n$ matrix with elements J_{ij}, and $\mathbf{S}^t$ is the transpose of $\mathbf{S}$. Neural networks may be modeled by a similar Hamiltonian.[2]

The energy landscape as defined in Eq. (1) describes a system which has both diversity (many possible outcomes $\sim 2^n$) and selection (only a few realized outcomes). A natural analogy may be made between spin-glass systems and polymer ensembles by mapping each polymer onto a distinct Ising spin-glass chain ($A \to +1$, $B \to -1$). A polymer α of length n may now be represented as a vector $\mathbf{S}^\alpha \in \aleph^n$. Many complex chemical interactions will determine the robustness of a polymer to the stresses in a given environment.[1] We may summarize the effects of all environmental factors on a polymer chain $\mathbf{S}^\alpha$ of length n by a "death function" $D_n(\mathbf{S}^\alpha)$. In analogy with the spin-glass Hamiltonian, we have as previously proposed (see Anderson,[1] Stein and Anderson,[6] and Rokhsar et al.[4]).

$$D_n(\mathbf{S}^\alpha) = \sum_{i,j=1}^{n} J_{ij} S_i^\alpha S_j^\alpha = (\mathbf{S}^\alpha)^t \cdot \mathbf{J} \cdot \mathbf{S}^\alpha,$$

where J_{ij} describes abstractly the interaction between monomers at sites i and j. As before, J_{ij} may take on both positive and negative values. The death function may take on values between $-\infty, \infty$ as $n \to \infty$. This value corresponds to the negative of the Hamiltonian energy for spin-glass systems. Polymers with large values of D_n are interpreted as being more robust and stable, i.e., more likely to survive.

We can transform the robustness of polymer α, (D_n^α) into a probability (d_n^α) per unit cycle that the polymer sequence $\mathbf{S}^\alpha$ of length n will live. We wish to choose a smooth monotonic function that transforms the real line (the range of D_n^α) to the unit interval (the desired range of d_n^α). As long as the mapping is smooth and monotonic the "universality class" characteristics of the model should remain similar. (For an example see Rokhsar et al.[4]).

The exact physical interpretation of J_{ij} is not particularly important; many environmental factors might influence the J_{ij} interconnections,[1] and it is impossible to enumerate fully the complex and varied manner in which the environment may influence the robustness of a given polymer. What is important is that the

J_{ij} interconnections represent a highly complex environment that gives rise to a robustness landscape containing many local maxima, i.e., many metastable states, in the space of all polymers. We shall interpret the J_{ij} values as an abstract holistic representation of the environment's action on the set of polymers. In a broad sense, J constitutes the polymer ensemble's environment.

3. METHOD OF COMPUTER SIMULATION

An established procedure for modeling the behavior of genetic polymer ensembles[4] begins the simulation with a random initial "soup base" of monomers, dimers, and trimers. One polymer is chosen successively among the members of the polymer ensemble to serve as a template. One of the remaining polymers in the soup is compared to this polymer to see if it is sufficiently complementary to the template to bind and form a double-stranded complex. If, to each A or B on the template polymer, there corresponds respectively a B or A on the complement polymer, the complement fits the template exactly. The complement polymer binds to the template polymer with a probability dependent upon this degree of complementarity. Mismatches are allowed but are probabilistically disfavored. A probabilistic allowance is also made for complements to "hang off" of the end of the template, thus permitting the elongation of polymer strands. (For details see Rokhsar et al.[4])

Once one complement has bound to a template, the remaining open sites of the template polymer then may bind to other complements. After the template is fully matched (or a predetermined time has elapsed), a new template is chosen. The process is repeated until there are no more possible templates remaining.

Each of the template-complement complexes are next checked for inter-complement polymerization. If two complements are bonded at neighboring sites on a template, there is a probability determined by input probabilities p(A=A), p(A=B), p(B=B), that a covalent bond can form between two neighboring complements, hence forming a single long complement from two smaller ones. After all the template-complement double strands have been checked for inter-complement polymerization, the system is denatured (i.e., the double-stranded complexes are separated into two single strands.). These single strands are then exposed to the environment for a period of time during which each strand has a probability of "dying" as determined by the death function. The strands which "die" are discarded from the evolutionary soup. A new set of raw materials (monomers, dimers, and trimers) are added to the set of surviving polymers to constitute the new soup base. The cycle is then repeated up to arbitrarily large times.

In order for a polymer to reproduce itself, a string S first must make a complementary copy −S to be used as a template for S polymer production. Since, as was observed,[4] S and −S counter-balance to maintain approximately the same population levels even under substantial replication error probabilities (10–15%), the model may be simplified without loss of essential character, if the need for an intermediate complement for polymer replication is removed. We do so. The restoring force of complementary kinetics is gone, as the polymers S and $-$S no longer depend upon each other for replication. S and $-$S may now be viewed as two wholly independent polymers.

When the J_{ij} values are taken randomly, the genetic polymer ensemble demonstrated certain characteristics of real evolutionary systems, namely, selection among diverse possible "species" and competition between selected "species." More explicitly, after $\sim$100 generations, the polymer ensemble became dominated by a few types of long polymer strands with highly common sequence patterns. When the environment $\mathbf{J}$ was gradually and linearly modified to a new environment $\mathbf{J}'$, the "species" evolved and adapted to the varying environment.[4]

4. NEURAL NETWORKS AND GENETIC POLYMERS: HOPFIELD MODEL

In the spin-glass model of neural networks, a firing (or non-firing) neuron corresponds to a $+1$ (or -1) spin value. Each J_{ij} value represents the neural connection strength from neuron j to neuron i; positive (negative) J_{ij} values are excitatory (inhibitory) connections. A network evolves asynchronously in time towards a local energy minimum of the network's Hamiltonian. The Hamiltonian is defined as in Eq. (1) for spin glasses.

If a set of states $\{E^\alpha\}$ is chosen such that all of its members are mutually orthogonal to one another, each member of this set of states may be incorporated as a local energy minimum of the Hamiltonian by defining the J_{ij} interconnections as

$$ J_{ij} \equiv \frac{1}{n} \sum_\alpha E_i^\alpha E_j^\alpha . $$

More complicated prescriptions for the J_{ij} remove the requirement for mutual orthogonality of the desired stable states (for example, Personnaz et al.[3]). Using these prescriptions, we may control the locations of the energy minima for the spin-glass Hamiltonian. For neural networks, this ability to control the local energy minima translates into an ability to control what states a neural network will evolve towards, i.e., what pattern an initial state might be "recognized" as being. For prebiotic soups, control over energy minima allows the creation of energetically favorable niches into which a polymer soup will tend to evolve. That is, through the choice of J_{ij} values and niches $\{E^\alpha\}$, we may predetermine what polymer species will arise. In computer simulations that implement this Hopfield prescription, approximately half of the polymers evolved into the selected niches after about 200 generations.[7]

5. NEURAL NETWORKS AND GENETIC POLYMERS: TEMPORAL-ASSOCIATIVE MODEL

A temporal-associative memory neural network has been proposed which cycles from one "stable" state to the next.[5] This model utilizes a time-dependent Hamiltonian in which the instantaneous state of the network itself acts on the Hamiltonian. The J_{ij} interconnections of the network are continuously being updated and a local energy minimum at one instant may no longer be one at a later time. This model of neural network only allows transiently stable states because any state slowly modifies the J_{ij} interconnections to make other states (including eventually some of its own nearest neighbors) more energetically favorable.

Because only one information string exists at any given instant in a neural network system, neural networks may only sequentially cycle in time among particular stable states. However, as many information strings are present in an organic soup at any given time, a polymer ensemble may cycle in a resonant, concerted manner and we may establish a stable system of species-species interdependences. Every polymer may act upon the energy landscape simultaneously to stabilize or destabilize other polymers. Formally, we define the polymer ensemble's feedback through the recursion

$$\mathbf{J}^{\text{new}} = \mathbf{J}^{\text{old}} + \frac{\epsilon}{n} \sum_{\{S\}} \Delta \mathbf{J_S} \tag{2a}$$

where

$$\Delta \mathbf{J_S} \equiv (\mathbf{MS})(\mathbf{MS})^t = \mathbf{M}(\mathbf{SS}^t)\mathbf{M}^t, \tag{2b}$$

where ϵ is a constant that measure the overall magnitude of polymer's effects on the environment, where the sum is made over all the members of the ensemble, and where $\mathbf{M}$ is a linear transformation from $\Re^n \to \Re^n$. (In the matrix multiplication of Eq. (2b), the polymer vectors are truncated or extended as necessary.) The environment is updated after each cycle from $\mathbf{J}^{\text{old}} \to \mathbf{J}^{\text{new}}$.

If we define $\mathbf{M}$ as

$$\mathbf{M} \equiv \frac{1}{n} \sum_{\alpha=1}^{p} \mathbf{E}^{\alpha+1} \mathbf{E}^{\alpha t}, \tag{3}$$

where $\{\mathbf{E}^\alpha\}$, $\alpha = 1, \dots, p$ are a set of *ordered* mutually orthogonal niches and we define $\mathbf{E}^{p+1} \equiv \mathbf{E}^1$, then a concerted cycle of species dependences may arise. If $\mathbf{S}$ is in niche γ, then $\mathbf{S}$ acts to stabilize polymers in niche $\gamma + 1$. Species γ stabilizes species $\gamma + 1$, for $\gamma = 1, \dots, p$. Since we have designated $\mathbf{E}^{p+1} = \mathbf{E}^1$, we have a closed loop,

$$\mathbf{E}^1 \to \mathbf{E}^2 \to \mathbf{E}^3 \to \dots \to \mathbf{E}^p \to \mathbf{E}^1 \to \dots.$$

The chain may be akin to a food chain or species. γ might produce an enzyme useful for the operation of species $\gamma + 1$. The exact nature of the interaction is unimportant.

A single loop chain of four orthogonal species was simulated. Beginning with a set of random environmental coupling constants J_{ij}, the environment was recursively modified in accordance with the prescription from Eq. (2). We defined the molding operator **M** from Eq. (3). As similar to the case in which the J_{ij} are initially configured to provide energetically stable niches, this feedback model also caused the polymers to evolve from a disordered soup into an ordered set of species. Here the species were cyclicly dependent upon and interrelated with other species members. These species constituted about 50% of the organic soup after 100 generations.[7] The mathematical constructs used in neural networks have provided a manner of describing evolutionary niches and interspecies dependences in prebiotic evolution.

ACKNOWLEDGMENTS

The author would like to thank P. W. Anderson for helpful discussions.

REFERENCES

1. Anderson, P. W. "Suggested Model for Prebiotic Evolution: The Use of Chaos." *Proceedings of the National Academy of Sciences* **80** (1983): 3386.
2. Hopfield, J. J. "Neural Networks and Physical Systems with Emergent Collective Computational Abilities." *Proceedings of the National Academy of Sciences* **79** (1982): 2554.
3. Personnaz, L., I. Guyon, and G. Dreyfus. "Information Storage and Retrieval in Spin-Glass-Like Neural Networks." *J. Phys. Lett.* **46** (1985): L-359.
4. Rokhsar, D. S., P. W. Anderson, and D. L. Stein. "Self-Organization in Prebiological Systems: Simulations of a Model for the Origin of Genetic Information." *J. Molecular Evol.* **23** (1986): 119.
5. Sompolinsky, H., and I. Kanter. "Temporal Association in Asymmetric Neural Networks." *Phys. Rev. Lett.* **57** (1986): 2861.
6. Stein, D. L., and P. W . Anderson. "A Model for the Origin of Biological Catalysis." *Proceedings of the National Academy of Sciences* **81** (1984): 1751.
7. Wong, A. J. "Development of a Spin-Glass Model of Prebiotic Evolution." *J. Theor. Biol.* **146** (1990): 523.

Index

The Addison-Wesley **Advanced Book Program** and the SANTA FE INSTITUTE would like to offer you the opportunity to learn about our new "Studies In the Sciences of Complexity" titles and workshops in advance. To be placed on our mailing list and receive pre-publication notices and special offers, just **fill out this card completely** and return to us.

Title, Author, and Code # of this book: **Date purchased:**

_____

Name _____

Title _____

School/Company _____

Department _____

Street Address _____

City _____ State _____ Zip _____

Telephone (   )

Where did you buy this book?

- ☐ Bookstore
- ☐ Mail Order
- ☐ School (Required for Class)
- ☐ Campus Bookstore (individual Study)
- ☐ Toll Free # to Publisher
- ☐ Professional Meeting
- ☐ Publisher's Representative

☐ Other _____

Please define your primary professional involvement:

- ☐ Academic: Professor
- ☐ Academic: Student
- ☐ Academic: Researcher
- ☐ Industry: Administrator
- ☐ Industry: Researcher
- ☐ Industry: Technician
- ☐ Government: Administrator
- ☐ Government: Researcher
- ☐ Government: Technician

Check your areas of interest.

200 ☑ **SFI**

201 ☐ Agriculture	209 ☐ Communication Sciences	217 ☐ Information Sciences
202 ☐ Anthropology	210 ☐ Dentistry	218 ☐ Mathematics
203 ☐ Artificial Intelligence	211 ☐ Economics	219 ☐ Medical Sciences
204 ☐ Astronomy	212 ☐ Education	220 ☐ Pharmaceutical Sciences
205 ☐ Atmospheric Sciences	213 ☐ Engineering	221 ☐ Physics
206 ☐ Biological Sciences	214 ☐ Geology/Geography	222 ☐ Political Sciences
207 ☐ Chemistry	215 ☐ History/Philosophy Science	223 ☐ Psychology
208 ☐ Computer Sciences	216 ☐ Industrial Science	224 ☐ Social Sciences
226 ☐ OTHER _____	(please specify)	225 ☐ Statistics

Of which professional scientific associations are you an active member?

_____ _____ _____ _____ _____

_____ _____ _____ _____ _____

Would you like to be sent information about the SANTA FE INSTITUTE and its workshops?

☐ Yes ☐ No

fold and staple

BUSINESS REPLY MAIL
FIRST CLASS PERMIT NO. 828   REDWOOD CITY, CA 94065

Postage will be paid by Addressee:

ADDISON-WESLEY
PUBLISHING COMPANY, INC.®

Advanced Book Program
350 Bridge Parkway
Redwood City, CA 94065-1522